Archaeoastronomy in East Asia

Archaeoastronomy in East Asia

Historical Observational Records of Comets and Meteor Showers From China, Japan, and Korea

《东亚考古天文学》第二卷

David W. Pankenier, Zhentao Xu, and Yaotiao Jiang

CAMBRIA PRESS

Amherst, New York

Printed in the United States of America

Requests for permission should be directed to:
permissions@cambriapress.com, or mailed to:
Cambria Press
20 Northpointe Parkway, Suite 188
Amherst, NY 14228

Library of Congress Cataloging-in-Publication Data

Pankenier, David W. (David William)
Archaeoastronomy in East Asia : Historical observational records of comets and meteor showers from China, Japan, and Korea / by David W. Pankenier, Xu Zhentao, Jiang Yaotiao.
p. cm.
Includes bibliographical references.
ISBN 978-1-60497-587-1 (alk. paper)
1. Archaeoastronomy—East Asia. I. Xu, Zhentao. II. Jiang, Yaotiao. III. Title.

GN799.A8P36 2008
523.6—dc22

2008047979

Before the work on this latest catalog was finished, our coauthor Zhentao Xu passed away suddenly in the summer of 2004. Without the tireless efforts he and Yaotiao Jiang devoted over several years to searching a great many diverse sources and painstakingly copying out the original records, this work could never have been completed. It is to Zhentao Xu's memory that this volume is fondly dedicated.

Table of Contents

List of Figures and Table

Illustration

FOREWORD

It has long been known that the ancient Chinese were rather thorough astronomers. During an age when the European countries were providing little valuable information on events occurring in the heavens above, the Chinese were painstakingly monitoring the sky. Virtually everything they saw that was out of the ordinary was logged. But what really set the Chinese apart from their European counterparts was not just the vast number of accounts of comets, meteor showers, novae, and eclipses, but the details they documented for each event, such as dates and locations in the sky. The latter are of considerable significance.

European astronomers long ago adopted the constellations of the Greeks and, as they explored the world, these constellations became universal. But the Chinese remained mostly isolated and their astronomers adopted a different set of constellations. So, where the universally adopted constellations are 88 in number, the Chinese actually divided the sky up into 283 constellations. The Chinese constellations were also adopted by the Japanese and Koreans.

The ultimate benefit of the Chinese sky divisions was that when something new was spotted, the constellations they were said to be in were relatively small areas of the sky. Given the difficulty of accurately measuring the position of moving objects in early times, this was the next best way of giving a position. Astronomers of the 19th century were the first to fully appreciate these Chinese astronomical accounts, when investigations into the long-term motion of periodic comet 1P/Halley revealed several Chinese comets that were in the right area of the sky on the right date. The oldest account is from 240 BC! Other periodic comets to be successfully linked to Chinese comets include 109P/Swift-Tuttle and 153P/Ikeya-Zhang, with the former having first been seen in 69 BC!

My initial encounter with the valuable East Asian accounts came while researching my first books on comets and meteor showers during the mid-1970s. My initial haphazard methods first brought me to H. A. Newton's early research into past appearances of the Leonid meteor shower. Newton's papers in the 1863 and 1864 issues of the *American Journal of Science* revealed several observations made by the Chinese as early as the 10th century. Not long afterwards, I came across J. Williams' 1871 book titled *Observations of Comets From B. C. 611 to A. D. 1640 From the Chinese Annals.* Both of these works seemed good attempts to systematically represent what the Chinese had observed, but they each fell short of being truly definitive works by using few sources, which were mostly second-hand accounts.

My next encounter with East Asian comet observations was an article that Ho Peng Yoke wrote in 1962 for the periodical *Vistas in Astronomy.* This catalog was more complete than Williams' and typically used historical texts that were contemporary to the observations. He also took care in correcting apparent inaccuracies in some of the dates. Although Ho Peng Yoke also included accounts from Japanese and Korean historical texts, most of the Japanese accounts were not from contemporary sources. Nevertheless, the catalog became a reliable source that many researchers consulted when investigating ancient and medieval comets.

I first heard about the manuscript for *Archaeoastronomy in East Asia* a couple of years ago and encouraged its publication. Now seeing the complete manuscript, I believe its authors have created the most complete English catalog of ancient East Asian comets and meteor showers that has ever been published. This makes it a valuable resource for future research, and I have no doubt that with its help future comets will be identified with ancient East Asian comets.

Gary W. Kronk
Author of *Cometography*

PREFACE

East Asian Archaeoastronomy: Historical Records of Astronomical Observations of China, Japan and Korea, Volume 1, was published in the fall of 2000 and was well received. As one reviewer remarked,

> Our knowledge of the complexity of the development of astronomy in Asian cultures seems to grow each year as more texts are translated and researchers cross linguistic and social barriers to share their work. This volume will provide an invaluable resource for scholars working primarily in English who wish to open the door to such understanding further.
>
> —Steven Renshaw, 2002, *Isis*, *2*, pp. 296–297

Since Comet Shoemaker-Levy 9's spectacular bombardment of Jupiter in July 1994, astronomers have devoted increasing attention to the risks of objects entering the inner solar system at irregular intervals—so-called near-earth objects (NEOs) and fast-moving objects (FMOs). These include comets, meteors, and asteroids, mainly debris left over from the formation of the solar system. As the historical record of catastrophic impacts on the Earth and their implications become increasingly clear, the work of identifying, cataloging, and predicting the future orbits of such objects has assumed greater urgency. Many such bodies have an orbital history; for example, scientists know that certain important meteor streams like the Leonids began as short-term comets that disintegrated in much the same way as Shoemaker-Levy 9. Here, the historical records of comets and meteor showers from China, Japan, and Korea have the potential to make a contribution to historical research on the evolution of earth-crossing orbits. Since publication of the previous volume in 2000, important statistical analyses of certain records have been published (e.g., Ahn, 2000, 2002, 2003; Yang et al., 2005) but without complete translations. As a result, such research remains the exclusive province of those with expertise in the relevant languages. With this catalog, we, the authors, hope to ameliorate that situation. In addition to astronomers, scholars have communicated with us to tell us how the East Asian astronomical records have materially aided their research in fields as diverse as mythology, medieval iconography, ancient chronology, and Hawaiian oral history. This response encouraged us to redouble our efforts to complete the task of making the records readily accessible to cultural astronomers, as well as for use in historical and comparative research.

Applying the same rigorous selection criteria and style of presentation as in our previous catalog, we have assembled and translated here some 1,500 individual observations of comets and meteor showers from China, Japan, and Korea spanning nearly three millennia. With the publication of this volume, most of the historical records of East Asian astronomical observations are now accessible in English. It is our hope that the observations painstakingly recorded by those ancient astronomers will contribute to the progress of scholarship and be an inspiration to all.

Acknowledgments

I would like to gratefully acknowledge Usui Tadashi's timely assistance with the transcription of old Japanese titles, personal names, and with problems of interpretation, and also that of Hann Seewoo for similar help with Korean place names. Finally, I would also like to acknowledge with thanks the helpful suggestions of an anonymous reviewer, as well as those of Peter Jenniskens.

David W. Pankenier

ARCHAEOASTRONOMY IN EAST ASIA

Part 1

Comets

1.1. Ancient East Asian Observations of Comets

In "An Investigation of Cometary Records in the Yin-Shang Period" and elsewhere (Xu & Jiang, 1993), the authors discussed possible records of cometary apparitions found as early as the oracle-bone inscriptions from the Shang Dynasty (1554–1046 BC), which suggest that observers may already have taken note of different kinds of comets in the late 2nd millennium BC. From that time on, the historical accounts of comets gradually become increasingly systematic, with by far the largest number being recorded in the standard dynastic histories. Others were occasionally noted down in the writings of private individuals and local gazetteers. For example, representative early accounts appear in the "Basic Annals of the First Emperor of Qin" in the *Shiji* (*Grand Scribe's Records*), China's first comprehensive history compiled by Sima Qian at the end of the 2nd century BC, "in the 7th year of the First Emperor of Qin; a broom star initially emerged in the east, was observed in the north, then in the 5th month appeared in the west" (see figure 1.1); and again, "[in the 9th year]; a broom star appeared in the north, from the Dipper southward for eighty days." Both accounts provide detailed and informative descriptions of the

FIGURE 1.1. An account from *Shiji*, "Basic Annals of First Emperor of Qin," reporting the appearance of Comet Halley in 240 BC.

七年彗星先出東方見北方五月見西方正義彗音似歲反見並音
行綠反孝經內記云彗出北斗兵大起彗在三台臣害
君彗在大微君害臣彗在天象諸侯作亂所指其處大
惡彗在日旁子欲殺父將軍驁死以攻龍孤慶都集解徐廣曰慶一作應正義括
地志云定州恆陽縣西南四十里有白龍水又有挾龍
山又定州唐縣東北五十四里有孤山蓋都山也帝王
紀云望堯母慶都所居張晏云堯山在北堯母慶都山
在南相去五十里北登堯山南望慶都山也注水經云
望都故城東有山不連陵名之曰孤孤都聳
相近故望都山孤山及望都故城三處相近還兵攻汲
彗星復見西方十六日夏太后死索隱莊襄王所生母正義子楚母也

circumstances of the comets' appearance, disappearance, and actual movements.

Even greater attention was paid to the appearance of comets in the Han Dynasty (206 BC–AD 220), mainly because they were considered inauspicious. In addition to written records, detailed illustrations of different kinds of comets have been preserved that also display an artistic sensibility. Besides the famous cometary atlas on the 2nd century BC silk manuscript from Mawangdui, among other images inscribed on stone and wood are also to be found lively renderings of comets. Recently, in a work titled "*Tianyuan yuli xiangyi fu*," authored by the Ming Emperor Renzong (Zhu Gaochi; r. 1425–1434), the authors discovered a color illustration of a so-called heavenly lance (*tianqiang*). According to a gloss by the early astronomer Gan De (4th c. BC), "a lance star is basically a broom star, with a pointed tip 2 to 3 *zhang* long," so that the heavenly lance in the illustration is no doubt intended to represent a comet. This book was written in the first year of the Hongxi reign period (AD 1425), perhaps making this the world's earliest color depiction of a comet (Plate 1).

Ancient Korean and Japanese books, besides containing a great number of cometary records, also preserve numerous invaluable illustrations of cometary observations. Figure 1.2 is an ancient Korean sketch of a cometary observation in which a comet has just appeared between the two lunar mansions Yi [LM 27] and Zhen [LM 28]. The tail of the comet has split into several branches and looks just like the sweep of a broom suspended in the night sky.

In recent times, numerous, even more detailed, hand-drawn illustrations have been preserved in Japan. As shown in figure 1.3, from the second 10-day week of the 1st month of the 2nd year of the Kampō reign period (1742) through the 7th day of the 2nd month, the movements of a comet near the North Pole are realistically rendered with a writing brush.

FIGURE 1.2. Korean sketch of the comet of 28 October 1664.

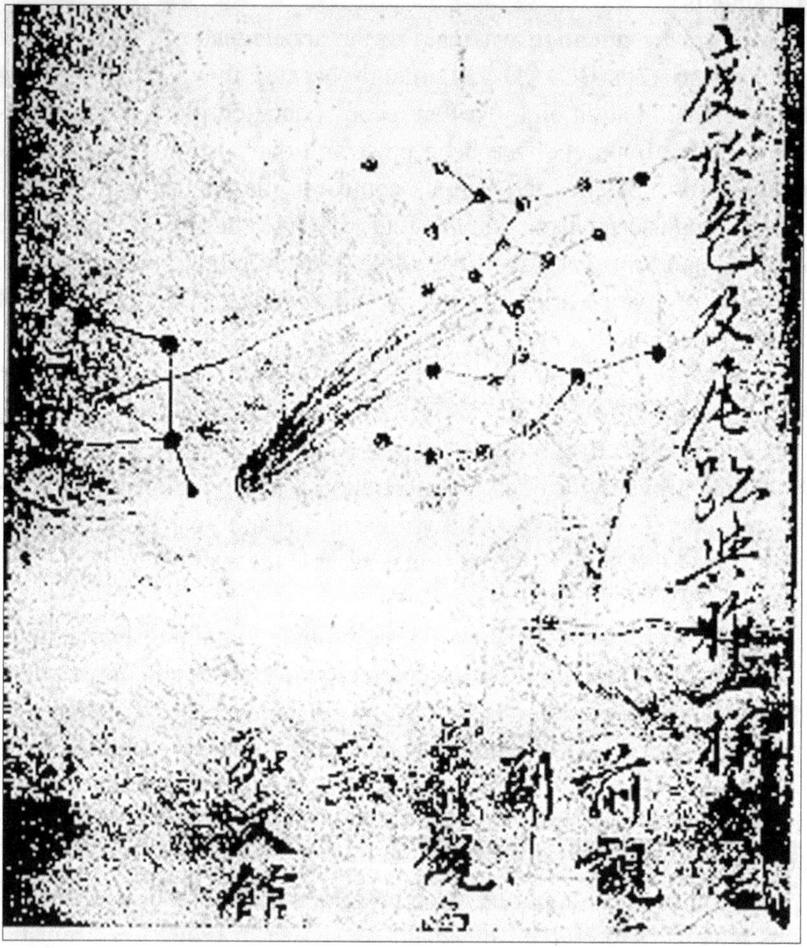

Source. After Rufus (1936).

FIGURE 1.3. Japanese chart of the comet of 1742.

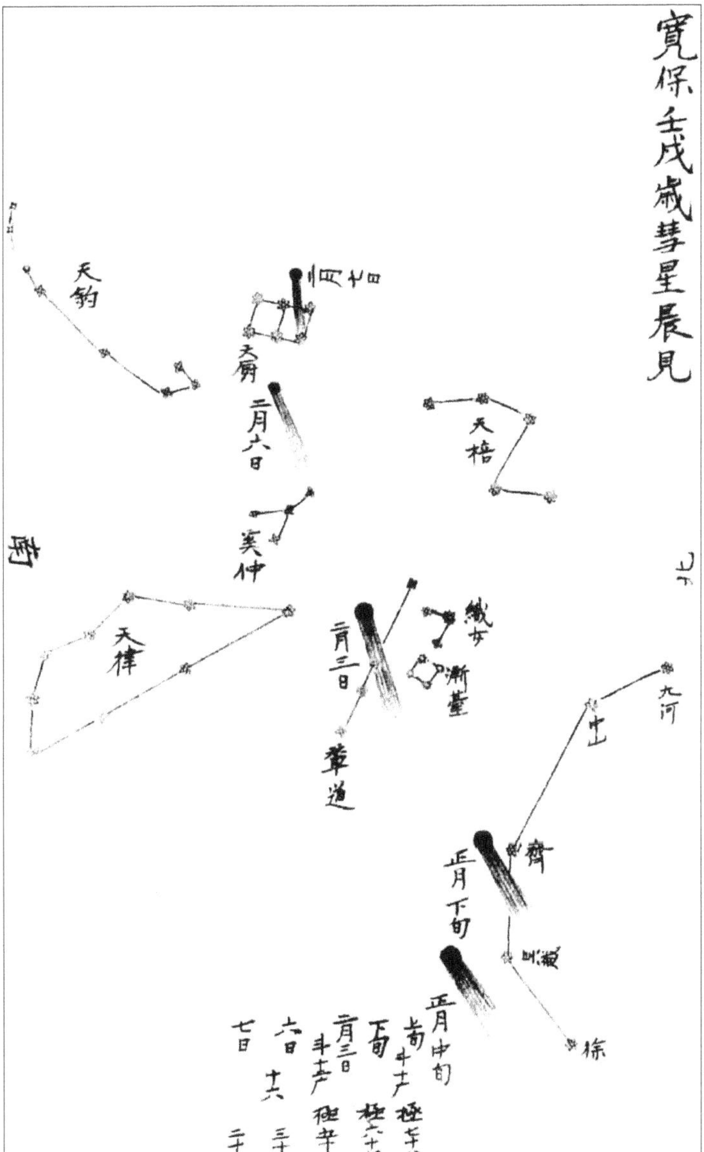

1.2. Notes on the Translations

In this volume, we use *pinyin* transliteration to transcribe Chinese; the Hepburn system to transcribe Japanese reign periods, book titles, and proper names; and the Revised Romanization of Korean (2002) to transcribe Korean book titles and proper names. Original Chinese technical terms such as the sexagesimal day-dates, names of asterisms, and the like are left in *pinyin* transcription. An expanded finding list of star names giving the western equivalents of Chinese stars and asterisms is in appendix A.

The technical terms for *comet* most commonly encountered in the records are *kexing*, "guest star" (anomalous astral body, either comet or nova); *huixing*, "broom star"; *xingbo*, "fuzzy star" (i.e., a tailless comet); and *changxing*, "long star," which are uniformly translated this way. In the past, some scholars have been perplexed by the compound *xingbo*, which appears to defy Chinese grammatical conventions by having *xing* "star/celestial body" modify *bo* "be fuzzy/bristle." However, *bo* has a verbal sense here, meaning "to become fuzzy or bushy." This is entirely consistent with cometary records where it is generally used to describe the appearance of tail-less comets or the changed aspect of a comet that has grown a tail. This usage appears in some of the earliest records and probably derives from a conception in which stars were thought to be capable of spontaneously becoming fuzzy, growing a tail, and moving about. Rendering *xing bo* as "bushy star" by analogy with *hui xing* ("broom" plus "star"), which is quite properly translated "broom star," is misleading in that it obscures the possibility that *bo* may imply a change of appearance. Thus, when one reads in a record of the comet of 9 January 595,

> 14th year of the Kaihuang reign period of Emperor Wen of the Sui Dynasty, 11th month, day *guiwei* [20]; there was a broom star that became fuzzy (有彗星孛) in Xu [LM 11] and Wei [LM 12], and then reached Kui [LM 15] and Lou [LM 16]. (*Sui shu: Tianwen zhi*, ch. 21, 612)

One suspects that it is the comet's reappearance without a tail that is being reported in shorthand. Indeed, comparison with European reports of the behavior of this comet confirms this interpretation (Kronk, 1999).

Comets are frequently described as having a *wei*, "tail"; *mang*, "rays"; *guangmang*, "bright rays"; *mangjiao*, "ray-like horns"; and *yan*, "flames," which change in length, brightness, or direction. The term *mang* literally refers to the "beard" or awn of a head of grain, a more descriptive word than the conventional expedient "rays" adopted here. Meteor showers are commonly recorded as *liuxing*, "falling/streaming stars," and this term is uniformly translated as "meteors." *Liu*, "flow/stream," used to refer to their motion, is generally translated as either "fell" or "streamed." Of course, "fell" implies a downward trajectory, but this is not necessarily the implication of the records. English equivalents of Chinese technical terms—azimuths, linear measures (i.e., *du*, *cun*, *chi*, *zhang*); double hours of the day and night; night watches; interactions between celestial bodies—as well as geographical coordinates of historical capitals of China, Japan, and Korea are provided in appendices B and C. In the records from all three traditions, the appearance of a comet or meteor may be likened to a cup, bowl, bottle, melon, peach, pear, papaya, hen's egg, and so on. There are no reliable indications of angular size for these common similes.

Detailed discussion of the historical sources for astronomical observations from China, Japan, and Korea and the terminology conventionally employed in Chinese observational records may be found in the references (Ho, 1966; Xu, Pankenier & Jiang, 2000; Stephenson & Green, 2002, Yang et al., 2005). Detailed star charts showing the locations of Chinese asterisms and their Western equivalents may be found in Ho (1966) and Stephenson (1994) and for the earliest period in Sun and Kistemaker (1997). The charts from Ho (1966) are reproduced in appendix D.

Throughout the period during which the Japanese and Korean observations were made, the calendars of both were essentially identical to the calendars in use in China at the time. From the Former Han Dynasty (206 BC–AD 23) on, year counts of Chinese emperors were periodically restarted with the initiation of so-called reign periods, as in 104 BC when commencement of the Taichu "Grand Inception" reign period

ILLUSTRATION 1. A "celestial lance" comet from *Tianyuan yuli xiangyi fu* (1425).

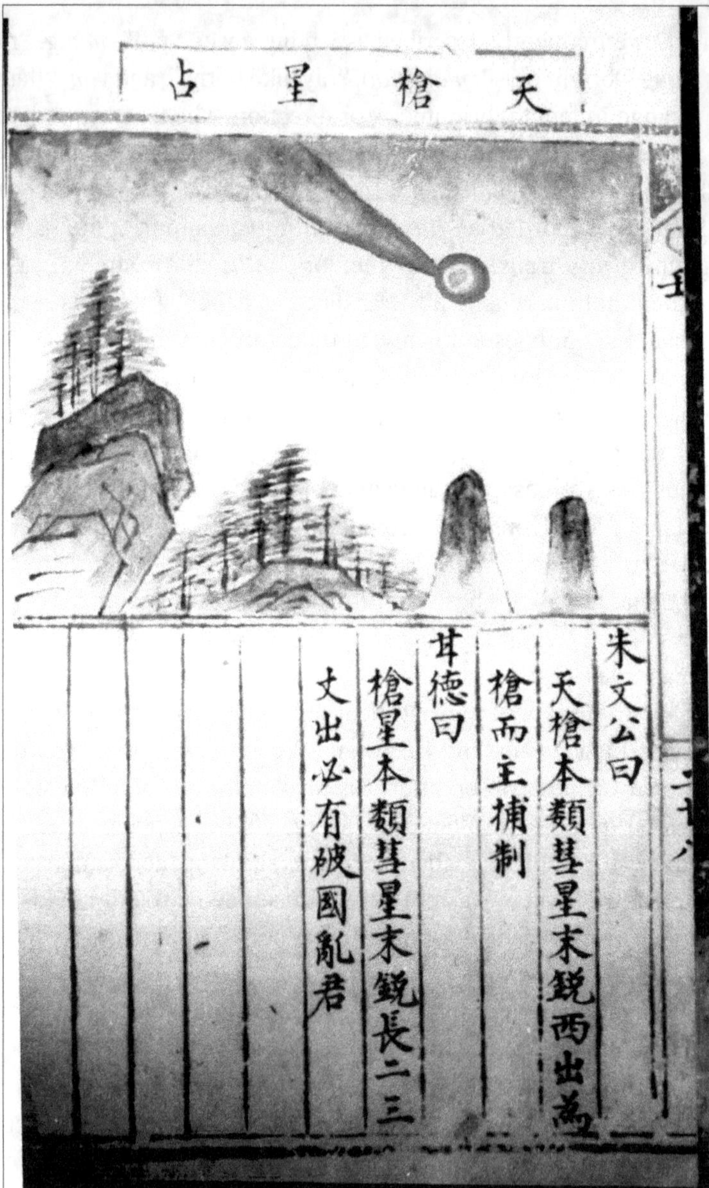

commemorated the beginning of a new cosmological epoch and calendrical system. Occasionally, reign periods commemorate astrological omens as, for example, in 134 BC when the Yuanguang "Primal Brilliance" reign period was inaugurated to mark the appearance of a spectacular comet the preceding year and also in AD 1056 when the Jiayou "Auspicious Aid" reign period was initiated to mark the appearance—and especially the disappearance nearly 2 years later—of the supernova of 1054, progenitor of the Crab Nebula. The Japanese later followed this practice of restarting year counts with the beginning of each reign period; though the Koreans did not, so that in the Korean case, year counts refer to the year of accession.

In contrast to other compilations of this kind, the authors provide the original classical Chinese text for every observational record translated. Readers who are so inclined may consult the original language. We strove to be as faithful as possible to the plain meaning of the original sources, and we welcome comments on the translations. We have endeavored to be as thorough as possible in culling the observational reports from a wide array of primary sources, including standard histories, local gazetteers, and private writings. To our knowledge, no comparable catalog of meteor shower observations exists in translation. In the case of the cometary apparitions, although many have been published in the past, there are gaps in the record and a rigorous distinction has not always been made between comets and supernovae, the latter also often denoted by the term *guest star*. For convenience, we provide cross-references to the translations in Ho (1962), Ho and Ang (1970), and Ahn (2004). In every case, we have checked the primary source, retranslated the record, and made note of errors or omissions we detected.

1.3. Modern Applications of Cometary Records

During the past few decades, an accumulation of evidence has made it abundantly clear that impacts by NEOs have played a major role in shaping the history of the Earth. The dinosaurs, as well as nearly three quarters of all species in existence 65 million years ago, appear to have been wiped out by the impact in the Yucatan of an object with an estimated diameter up to 10 kilometers. North American megafauna may have met a similar fate at the end of the last Ice Age. The 1908 Tunguska event in Siberia, recently declassified United States Air Force data on objects entering Earth's atmosphere, the discovery of increasing numbers of impact craters on the Earth, and the smashing into Jupiter of nearly two dozen fragments of Comet Shoemaker-Levy 9 in 1994 all demonstrate that collisions with NEOs pose a real and present danger. As a result, the study of historical records has assumed greater importance as a way of advancing research, for example, into the dynamic evolution of the orbits of short-period comets (Hasegawa, 1995).

Ho Peng-yoke (Ho, 1962; Ho & Ang, 1970) and Chen Zungui (Chen, 1984) made important contributions to the work of compiling historical records of cometary observations. The *Comprehensive Catalog of Ancient Chinese Records of Astronomical Phenomena* (*Zhongguo gudai tianxiang jilu zong ji*), compiled by the Beijing Observatory in 1988, expanded the scope of the historical sources surveyed to include local gazetteers from the entire country; however, this ambitious compilation is not without error.

To facilitate future research on the orbital evolution of NEOs, it was necessary to establish new selection criteria in order to collate all the potentially useful cometary records. The selection criteria established in compiling the present catalog of comet sightings are

1. First and foremost, the record must contain a description of the comet's location that can serve to determine its path.
2. Records prior to the Han Dynasty (206 BC–AD 220), although vague, are included for their historical significance.
3. Records of "guest stars," "anomalous stars," and so forth are included on a case-by-case basis if the description mentions a tail or movement.

4. Reports in which observers mention sighting only the comet's tail but no head, often using "vapor" (*qi*) to describe its appearance, are included on a case-by-case basis.

The original records of Chinese, Japanese, and Korean cometary observations that conform to these selection criteria were then checked in various ways and their dates converted to the Western calendar. Observations of Comet Halley from return –29 through –8, originally translated in chapter 5 of Xu et al. (2000), are reprinted here for convenience.

The range of terminating dates for the records varies. For example, the latest date of a comet in this catalog is in 1784. The *Draft History of the Qing Dynasty* has no comet reports after 1770. Ho and Ang (1970) continued their listing through the end of the Qing Dynasty in 1911. For this recent period, the existing records from China, Japan, and Korea are fairly routine and unremarkable, and since Western parallels exist, they have not been reproduced here.

1.4. East Asian Cometary Records

Nota bene (NB): For reference, records not meeting our selection criteria are included in italics, along with the number assigned by Ho (1962) and Ho and Ang (1970). Only primary sources are included among the original texts in appendix E. Secondary sources used by Ho and Kanda (1935, 1936) are not included among the Japanese texts, as they are not originals.

BC 1059 [China]
When King Wu [of Zhou] conquered King Zhou [of Shang], a broom star appeared with its tail pointing toward the Kingdom of Yin.
[Huainanzi] ch. 15

BC 1034 spring [China]
19th year of King Zhao, in spring; a star appeared in Ziwei.
[Zhushu jinian] xia

BC 974 Feb–Apr (Ho 3)

19th year of King Zhao of Western Zhou, spring; a star became fuzzy in Ziwei.

[Zhushu jinian]

BC 632 (Ho 4)

When [Duke] Wen of Jin was about to battle King Cheng of Chu at Chengpu, a broom star appeared in [the astral space of] Chu, with Chu holding the handle. [Duke Wen] asked Jiu Fan about it. Jiu Fan said, "When there is a broom star during combat, the one on the contrary [i.e., sweep] end gains the victory."

[Lunheng jiaoshi] ch. 5

NB: Ordinarily, a broom star in one's astral space is favorable, however, not in this case, according to Jiu Fan.

BC 613 [China] (Ho 5)

(a) 14th year of Duke Wen of Lu; a broom star entered Beidou.
[Shiji: shi-er guo nian biao] ch. 14

(b) 14th year of Duke Wen of Lu, 7th month (Jun 7–Jul 6); a broom star appeared and entered Beidou.
[Han shu: wu xing zhi] ch. 27

BC 525 [China] (Ho 7)

17th year of Duke Zhao of Lu; a broom star appeared in Chen.

[Shiji: shi-er guo nian biao] ch. 14

17th year of Duke Zhao of Lu, in the winter; a star appeared fuzzy in Dachen.

[Han shu: wu xing zhi] ch. 27

BC 516 [China] (Ho 8)

(a) 32nd year of Duke Jing of Qi; a broom star appeared.
[Shiji: shi-er guo nian biao] ch. 14

(b) 32nd year of Duke Jing of Qi; a broom star appeared in the northeast, corresponding to the astrological division of the state of Qi.
[Shiji] ch. 32

BC 500 [China] (Ho 9)

1st year of Duke Hui of Qin; a broom star appeared.

[Shiji: shi-er guo nian biao] ch. 14

BC 482 Oct–Dec [China] (Ho 10)

13th year of Duke Ai of Lu of the Springs and Autumns period, winter, 11th month; a star became fuzzy in the east.

[Han shu: wu xing zhi] ch. 27

[Chunqiu: Ai Gong 13]

BC 481 Oct–Dec [China] (Ho 11)

14th year of Duke Ai of Lu of the Springs and Autumns period, winter; a star became fuzzy.

[Han shu: wu xing zhi] ch. 3, ch. 27

BC 469 [China] (Ho 12)

7th year of Duke Li of Qin of the Warring States; a broom star was seen.

[Shiji: liu guo nian biao] ch. 15

NB: Ho mistakenly gave 470 BC.

BC 467 [China] (Ho 13)

10th year of Duke Li of Qin of the Warring States; a broom star appeared.

[Shiji: liu guo nian biao] ch. 15

NB: Regarded as Comet Halley, return –32.

BC 433 [China] (Ho 14)

8th year of King Kao of Zhou; a broom star appeared.

[Wenxian tongkao] ch. 286

BC 361 [China] (Ho 15)

(a) 10th year of King Hui of Wei of the Warring States; a broom star appeared.

[Shiji: Wei shi jia] ch. 44

(b) *1st year of Duke Xiao of Qin of the Warring States; a broom star appeared in the west.*
[Shiji: liu guo nian biao] ch. 15

BC 352 [China]
10th year of Duke Xia of Qin of the Warring States; when the town of Pilao in the state of Zhao was occupied, a broom star appeared.
[Shiji] ch. 44

BC 305 [China] (Ho 16)
2nd year of King Zhaoxiang of Qin of the Warring States; a broom star appeared.
[Shiji: liu guo nian biao] ch. 15
[Shiji: Qin benji] ch. 5

BC 303 [China] (Ho 17)
4th year of King Zhao of Qin of the Warring States; a broom star appeared.
[Shiji: liu guo nian biao] ch. 15
[Shiji: Qin benji] ch. 5

BC 296 [China] (Ho 18)
11th year of King Zhaoxiang of Qin of the Warring States; a broom star appeared.
[Shiji: Qin benji] ch. 5
[Shiji: liu guo nian biao] ch. 15

BC 240 [China] (Ho 19)

(a) 7th year of Emperor Qin Shihuang of the Warring States; a broom star first appeared in the east; it was then seen in the north; in the 5th month (May 26–Apr 24), it was seen in the west…the broom star was again seen in the west for 16 days.
[*Shiji: Qin Shihuang benji*] ch. 6

(b) 7th year of Emperor Qin Shihuang; a broom star was seen in the north and in the west.

[*Shiji: liu guo nian biao*] ch. 15

NB: This is Comet Halley, return –29.

BC 238 [China] (Ho 20)

(a) 9th year of the First Emperor of Qin of the Warring States; a broom star appeared, at times, stretching across the sky; in the 4th month…the broom star appeared in the west and again in the north, [extending] southward from Dou for 80 days.
[*Shiji: Qin Shihuang benji*] ch. 6

(b) 9th year of the First Emperor of Qin of the Warring States period; a broom star appeared stretching across the sky. Lao Ai fomented rebellion; his retainers were exiled to Shu. The broom star reappeared.
[*Shiji: liu guo nian biao*] ch. 15

NB: Ho (1962) mistranslated *huo*, "sometimes," as "possibly" and also misunderstood the direction of movement vis-à-vis the Dipper in (a).

BC 234 Feb 19–Mar 20 [China] (Ho 21)

13th year of the First Emperor of Qin of the Warring States, 1st month; a broom star appeared in the east.
[Shiji: Qin Shihuang benji] ch. 6

BC 214 [China] (Ho 22)

33rd year of the First Emperor of Qin of the Warring States period; a bright star emerged in the west. [Shiji jijie] (commentary) quoting Xu Guang says, "Huangfu Mi (AD 214–AD 282) says, 'a broom star appeared.'"
[Shiji: Qin Shihuang benji] ch. 6

BC 204 Aug 14–Sep 12 [China] (Ho 23)

(a) 3rd year Emperor Gaozu of the Han Dynasty, 7th month; a star became fuzzy at Dajiao.
[*Han shu :Gao di ji*] ch. 1

(b) 3rd year of Emperor Gaozu of the Han Dynasty, 7th month; a star became fuzzy at Dajiao for over 10 days before it set.
[*Han shu :wu xing zhi*] ch. 27

BC 172 [China] (Ho 24)

8th year of the Qianyuan reign period of Emperor Wen of the Han Dynasty; a "long star" emerged in the east.
[Han shu: Wen di ji] ch. 4

BC 162 Feb 6 [China] (Ho 25)

2nd year of the Houyuan reign period of Emperor Wen of the Han Dynasty, 1st month, day renyin [39]; in the evening a "heavenly tree" (i.e., comet) appeared in the southwest.
[Han shu: tianwen zhi] ch. 26
NB: This is considered by some to be a record of Comet Halley's return –28. The term tianchan, "heavenly tree," is idiosyncratic.

BC 157 Oct 2–31 [China] (Ho 26)

(a) 7th year of the Houyuan reign period of Emperor Wen of the Han Dynasty, 9th month; a star became fuzzy in the west.
[*Han shu: Jing di ji*] ch. 5

(b) 7th year of the Houyuan reign period of Emperor Wen of the Han Dynasty, 9th month; a star became fuzzy in the west. Its root pointed toward WEI [LM 6] and JI [LM 7], and its tip pointed toward XU [LM 11] and WEI [LM 12]. It was over 1 *zhang* long and reached the Milky Way; after 16 days, it was no longer seen.
[*Han shu: wuxing zhi*] ch. 27

BC 155 Sep 11–Oct 10 [China] (Ho 27)

2nd year of the Qianyuan reign period of Emperor Jing of the Han Dynasty, 8th month; a broom star emerged in the northeast.
[*Shiji: Xiao Jing ben ji*] ch. 11

BC 155 Nov 20–BC 154 Nov 8 [China] (Ho 28)

(a) 2nd year of the Qianyuan reign period of Emperor Jing of the Han Dynasty, winter, 12th month (BC 154 Jan 18–Feb 16); a star became fuzzy in the southwest.
[*Han shu: Jing di ji*] ch. 5

(b) 2nd year of the Qianyuan reign period of Emperor Jing of the Han Dynasty (BC 155 Nov 20–BC 154 Nov 8); in this year, a broom star emerged in the southwest.
[*Han shu: tianwen zhi*] ch. 26

NB: Ho (1962) lacked both records.

BC 154 Feb 27 [China] (Ho 29)

3rd year of the Qianyuan reign period of Emperor Jing of the Han Dynasty, 1st month, day *yisi* [42]; a "long star" emerged in the west.
[*Shiji: Xiao Jing ben ji*] ch. 11
NB: Ho (1962) gave an incomplete date.

BC 148 Apr 29–May 27 [China] (Ho 30)

2nd year of the Zhongyuan reign period of Emperor Jing of the Han Dynasty, 4th month; a star became fuzzy in the northwest.
[*Han shu: Jing di ji*] ch. 5

BC 147 Apr 18–May 17 [China] (Ho 31)

3rd year of the Zhongyuan reign period of Emperor Jing of the Han Dynasty, 3rd month; a broom star emerged in the northwest.
[*Shiji: Jing di ji*] ch. 11

BC 147 May 13 [China] (Ho 31)

3rd year of the Zhongyuan reign period of Emperor Jing of the Han Dynasty, 3rd month, day *dingyou* [34]; a broom star appeared at night in the northwest. It was white, 1 *zhang* long, and located in [lunar mansion] Zuixi [LM 20]. By dawn when it left (i.e., was obscured in daylight), it had decreased in size; after 15 days, it was no longer seen.
[*Han shu: tianwen zhi*] ch. 26

BC 147 Aug 6 [China] (Ho 32)

3rd year of the Zhongyuan reign period of Emperor Jing of the Han Dynasty, 6th month, day *renxu* [59]; a [white] fluffy star [*pengxing*] as big as a 2 peck measure appeared in the southwest, south of Fang [LM 4], at a distance of about 2 *zhang*. On day *guihai* [60], it was

northeast of XIN [LM 5] and about 1 *zhang* long. On day *jiazi* [1], it was north of WEI [LM 6] and about 6 *zhang* long. On day *ding-mao* [4], it was north of JI [LM 7], near the Milky Way, and slightly smaller; by dawn when it left (i.e., was obscured in daylight), it was as big as a peach; by its departure on day *renshen* [9], [it had been visible] for 10 days in all.
[*Han shu: tianwen zhi*] ch. 26
[*Xi Han huiyao*] ch. 28

BC 147 Oct 12–Nov 10 [China] (Ho 33)
3rd year of the Zhongyuan reign period of Emperor Jing of the Han Dynasty, 9th month; a star became fuzzy in the northwest.
[*Han shu: Jing di ji*] ch. 5

BC 138 Apr 9–May 8 [China] (Ho 34)
3rd year of the Jianyuan reign period of Emperor Wu of the Han Dynasty, 3rd month; a star became fuzzy in ZHU [LM 24] and ZHANG [26] that passed through TAIWEI, interfered with ZIGONG, and reached the Milky Way.
[*Han shu: tianwen zhi*] ch. 26

BC 138 May 9–Jun 6 [China] (Ho 35)
3rd year of the Jianyuan reign period of Emperor Wu of the Han Dynasty, 4th month; a star became fuzzy in TIANJI that reached ZHINÜ.
[*Han shu: tianwen zhi*] ch. 26

BC 138 Aug 5–Sep 3 [China] (Ho 36)
3rd year of the Jianyuan reign period of Emperor Wu of the Han Dynasty, 7th month; a star became fuzzy in the northwest.
[*Han shu: tianwen zhi*] ch. 26

BC 137 Sep 22–Oct 20 [China] (Ho 37)
4th year of the Jianyuan reign period of Emperor Wu of the Han Dynasty, 9th month; a star became fuzzy in the northeast.
[*Han shu: Wu di ji*] ch. 6

BC 135 Jul 3–Aug 1 [China] (Ho 38)

6th year of the Jianyuan reign period of Emperor Wu of the Han Dynasty, 6th month; a star became fuzzy in the north.
[*Han shu: wu xing zhi*] ch. 27

BC 135 Aug 31–Sep 29 [China] (Ho 39)

(a) 6th year of the Jianyuan reign period of Emperor Wu of the Han Dynasty, 8th month; a star became fuzzy in the east that stretched across the sky.
[*Han shu: Wu di ji*] ch. 6

(b) 6th year of the Jianyuan reign period of Emperor Wu of the Han Dynasty, 8th month; a long star emerged in the east, so long that it stretched across the sky; after 30 days, it departed.
[*Han shu: wu xing zhi*] ch. 27

NB: Ho (1962) lacked (b).

BC 120 Feb–Apr [China] (Ho 42)

3rd year of the Yuanshou reign period of Emperor Wu of the Han Dynasty, spring; a star became fuzzy in the east.
[*Han shu: Wu di ji*] ch. 6

BC 119 Feb–Apr [China] (Ho 43)

4th year of the Yuanshou reign period of Emperor Wu of the Han Dynasty, spring; a star became fuzzy in the northeast.
[*Han shu: Wu di ji*] ch. 6

BC 119 May 8–Jun 6 [China] (Ho 43)

4th year of the Yuanshou reign period of Emperor Wu of the Han Dynasty, 4th month; a long star emerged again in the northwest.
[*Han shu: wu xing zhi*] ch. 27

BC 110 May 29–Jun 26 [China] (Ho 44)

1st year of the Yuanfeng reign period of Emperor Wu of the Han Dynasty, 5th month; a star became fuzzy in DONGJING [LM 22], and then, it became fuzzy again in SANTAI.
[*Han shu: wu xing zhi*] ch. 27

BC 110 Aug–Oct [China] (Ho 44)

(a) 1st year of the Yuanfeng reign period of Emperor Wu of the Han Dynasty, autumn; a star became fuzzy in DONGJING [LM 22]; more than 10 days later, the star became fuzzy in SANTAI.
[*Shiji: feng shan shu*] ch. 28

(b) 1st year of the Yuanfeng reign period of Emperor Wu of the Han Dynasty, autumn; a star became fuzzy in DONGJING [LM 22] and then again became fuzzy in SANTAI.
[*Han shu: Wu di ji*] ch. 6

(c) 1st year of the Yuanfeng reign period of Emperor Wu of the Han Dynasty, autumn; a star became fuzzy in DONGJING [LM 22]; more than 10 days later, the star became fuzzy in SANTAI.
[*Han shu: jiao si zhi*] ch. 25

BC 108–107 [China] (Ho 45)

In the middle of the Yuanfeng reign period of Emperor Wu of Han, a star became fuzzy in HESHU.
[Han shu: tianwen zhi] ch. 26

BC 102 [China] (Ho 46)

In the middle of the Taichu reign period of Emperor Wu of Han, a star became fuzzy in ZHAOYAO.
[Han shu: tianwen zhi] ch. 26

BC 87 Aug 10–Sep 8 [China] (Ho 47)

2nd year of the Houyuan reign period of Emperor Zhao of the Han Dynasty, 7th month (Aug 10–Sep 8); a star appeared fuzzy in the east.
[*Han shu*] ch. 7
NB: This is Comet Halley, return –27.

BC 84 Mar 12–Apr 10 [China] (Ho 48)

3rd year of the Shiyuan reign period of Emperor Zhao of the Han Dynasty, spring, 2nd month; a star became fuzzy in the northwest.
[*Han shu: Zhao di ji*] ch. 7
NB: Ho (1962) did not cite this source.

BC 69 Jan 27–Feb 24 [China] (Ho 52)

(a) 1st year of the Dijie reign period of Emperor Xuan of the Han Dynasty, 1st month; a star became fuzzy in the west.
[*Han shu: Xuan di ji*] ch. 8

(b) 1st year of the Dijie reign period of Emperor Xuan of the Han Dynasty, 1st month; a star became fuzzy in the west, about 2 *zhang* from Venus.
[*Han shu: wu xing zhi*] ch. 27

BC 69 Jul 23 [China] (Ho 53)

1st year of the Dijie reign period of Emperor Xuan of the Han Dynasty, 6th month, day *wuxu* [35], first watch of the night; a white guest star about 2 *chi* long again stayed between ZUOJIAO and YOUJIAO [LM 1], pointing southeast.
[*Han shu: tianwen zhi*] ch. 26

BC 69 Aug 20 [China] (Ho 54)

1st year of the Dijie reign period of Emperor Xuan of the Han Dynasty, 6th month, day *bingyin* [3]; a guest star reappeared northeast of GUANSUO, moving southward, until the 7th month, on the night of day *guiyou* [10], when it entered TIANSHI, its rays of flame pointing southeast and its color was white.
[*Han shu: tianwen zhi*] ch. 26

BC 61 Jul 24–Aug 21 [China] (Ho 55)

1st year of the Shenjue reign period of Emperor Xuan of the Han Dynasty, 6th month; a star became fuzzy in the east.
[*Han shu: Xuan di ji*] ch. 8

BC 49 Apr 14–May 12 [China] (Ho 56)

(a) 1st year of the Huanglong reign period of Emperor Xuan of the Han Dynasty, 3rd month; a star became fuzzy in WANGLIANG and GEDAO then entered ZIGONG.
[*Han shu: Xuan di ji*] ch. 8

(b) 1st year of the Huanglong reign period of Emperor Xuan of the Han Dynasty, 3rd month; a guest star over 1 *zhang* long stayed about 9 *chi* northeast of WANGLIANG, pointing west; it emerged within GEDAO and reached as far as ZIGONG.
[*Han shu: tianwen zhi*] ch. 26

BC 49 Apr 14–May 12 [Korea]
9th year of King Hyeokgeose of Silla, spring, 3rd month; there was a fuzzy star in WANGLIANG.
[*Samguk sagi*] ch. 1 (Ho 56)
[*Jeungbo munheon bigo*] ch. 6

BC 47 Jun 20–Jul 18 [China] (Ho 58)
2nd year of the Chuyuan reign period of Emperor Yuan of the Han Dynasty, 5th month; a guest star appeared in the space of MAO [LM 18] and stayed about 5 *chi* east of JUANSHE; its color was greenish-white, its flames 3 *cun* long.
[*Han shu: tianwen zhi*] ch. 26

BC 44 Mar 25 [Korea] (Ho 59)
14th year of King Hyeokgeose of Silla, spring, 2nd month, day *jiyou* [46]; there was a fuzzy star in HEGU.
[*Jeungbo munheon bigo*] ch. 6

BC 44 May 18–Jun 16 [China] (Ho 59)
5th year of the Chuyuan reign period of Emperor Yuan of the Han Dynasty, 4th month; a reddish-yellow broom star about 8 *chi* long emerged in the northwest. After a few days, it was over 1 *zhang* long, pointing northeast, within the space of SHEN [LM 21].
[*Han shu: tianwen zhi*] ch. 26

BC 44 May 18–Jun 16 [Korea]
14th year of King Hyeokgeose of Silla, summer, 4th month; there was a fuzzy star at the fifth star of SHEN [LM 21].

[*Samguk sagi*] ch. 1 (Ho 59)
[*Jeungbo munheon bigo*] ch. 6

BC 32 Feb 6–Mar 7 [China] (Ho 60)

(a) 1st year of the Jianshi reign period of Emperor Cheng of the Han Dynasty, 1st month; a star became fuzzy in YINGSHI [LM 13].
[*Han shu: Cheng di ji*] ch. 10

(b) 1st year of the Jianshi reign period of Emperor Cheng of the Han Dynasty, 1st month; a star became fuzzy in YINGSHI [LM 13]. It was greenish-white, 6 or 7 *zhang* long, and over 1 *chi* wide.
[*Han shu: wu xing zhi*] ch. 27

BC 12 Aug 26–Oct 20 [China] (Ho 61)

1st year of the Yuanyan reign period of Emperor Cheng of the Han Dynasty, 7th month, day *xinwei* [8] (Aug 26); a star became fuzzy in DONGJING [LM 22]; it was treading on WU ZHUHOU; it appeared to the north of HEXU and passed through XUANYUAN and TAIWEI; later, it traveled more than 6 *du* daily. In the morning, it appeared in the east. On the evening of the 13th day (Sep 7), it was seen in the west; it trespassed against CIFEI, CHANGQIU, [BEI-]DOU, and Saturn; its "swarming flames" again penetrated ZIGONG, with DAHUO right behind. It reached TIANHE (the Milky Way), sweeping the region of HOU and FEI; then it moved south, crossing and trespassing against DAJIAO and SHETI. When it reached TIANSHI, it moved slowly at a regular pace. Its flames entered TIANSHI. After a further 10 days, it went toward the west. On the 56th day (Oct 20), it went out of sight together with CANGLONG.
[*Han shu*] ch. 27
NB: This is Comet Halley, return –26.

BC 5 Jan 10–Feb 7 [China] (Ho 62)

1st year of the Jianping reign period of Emperor Ai of the Han Dynasty, 12th month; a white vapor emerged in the southwest, reaching from the ground up to the sky. It emerged beneath SHEN [LM 21]

and penetrated TIANCE, as wide as a bolt of cloth and over 10 *zhang* long. It lasted more than 10 days before departing.
[*Han shu: tianwen zhi*] ch. 26

BC 5 Mar 9–Apr 6 [China] (Ho 63)
2nd year of the Jianping reign period of Emperor Ai of the Han Dynasty, 2nd month; a broom star emerged in QIANNIU [LM 9] for over 70 days.
[*Han shu: tianwen zhi*] ch. 26

BC 4 Apr 24 [China] (Ho 64)
3rd year of the Jianping reign period of Emperor Ai of the Han Dynasty, 3rd month, day *jiyou* [46]; a star became fuzzy in HEGU.
[*Han shu: Ai di ji*] ch. 11

BC 4 [Korea] (Ho 64)
54th year of King Hyeokgeose of Silla, 2nd month, day jiyou [46] (Feb 23); a fuzzy star was seen at HEGU.
[*Samguk sagi*] ch. 1

AD 22 Nov 13–Dec 12 [China] (Ho 65)
(a) 3rd year of the Dihuang reign period of Wang Mang of the Xin Dynasty, 11th month; a star became fuzzy in ZHANG [LM 26].
[*Hou Han shu: Guangwu di ji*] ch. 1
(b) 3rd year of the Dihuang reign period of Wang Mang of the Xin Dynasty, 11th month; a star became fuzzy in ZHANG [LM 26] then traveled southeastward for 5 days before disappearing.
[*Hou Han shu: tianwen zhi*] ch. 20
(c) 3rd year of the Dihuang reign period of Wang Mang of the Xin Dynasty, 11th month; a star became fuzzy in ZHANG [LM 26] then traveled southeastward for 5 days before disappearing.
[*Han shu: Wang Mang zhuan*] ch. 99
NB: Ho (1962) did not cite (a) and (c).

AD 39 Mar 13 [China] (Ho 68)
(a) 15th year of the Jianwu reign period of Emperor Guangwu of the Han Dynasty, 1st month, day *dingwei* [44]; a star became

fuzzy in MAO [LM 18], and a star became fuzzy in YINGSHI [LM 13].
[*Hou Han shu: Guangwu di ji*] ch. 1

(b) 15th year of the Jianwu reign period of Emperor Guangwu of the Han Dynasty, 1st month, day *dingwei* [44]; a broom star appeared in MAO [LM 18], traveled slowly northwestward, and entered YINGSHI [LM 13], trespassing on LIGONG. On day *yiwei* [32], in the 3rd month, it reached DONGBI [LM 14] where it was extinguished, having been visible for 49 days.
[*Hou Han shu: tianwen zhi*] ch. 20

NB: *Dong Han hui yao*, ch. 16, has "2nd month, day *yiwei* [32]."

AD 46 Dec 17–AD 47 Jan 15 [Korea] (Ho 69)
3rd year of King Minjung of Goguryeo, winter, 11th month; there was a fuzzy star in the south; after 20 days, it was extinguished.
[*Jeungbo munheon bigo*] ch. 6
[*Samguk sagi*] ch. 14

AD 54 Feb 27–Mar 27 [Korea] (Ho 70)
31st year of King Yuri of Silla, spring, 2nd month; there was a fuzzy star in ZIGONG.
[*Samguk sagi*] ch. 1
[*Jeungbo munheon bigo*] ch. 6

AD 54 Apr 29 [China] (Ho 70)
30th year of the Jianwu reign period of Emperor Guangwu of the Han Dynasty, intercalary month, day *guichou* [50]; a star became fuzzy in ZIGONG.
[*Hou Han shu: Guangwu di ji*] ch. 1

AD 54 [China] (Ho 70)
30th year of the Jianwu reign period of Emperor Guangwu of the Han Dynasty, intercalary month, day *jiawu* [31]; "Mercury" was at 20 *du* in DONGJING [LM 22] when it produced a white vapor pointing southeast with flames 5 *chi* long. It became a comet that traveled northeast

as far as the western wall of ZIGONG where it stopped. On day *jiazi* [1] in the 5th month, it was not seen; in all, it was visible for 31 days.
[*Hou Han shu: tianwen zhi*] ch. 20
NB: In this year, the 3rd intercalary month did not contain day *jiawu*.

AD 55 Dec 6–AD 56 Apr 6 [China] (Ho 71)
31st year of the Jianwu reign period of Emperor Guangwu of the Han Dynasty, 10th month, day *jihai* [36]; there was a guest star in the space of QIXING [LM 25], with flames about 2 *chi* long. It traveled southwestward until the next year, the 22nd day of the 2nd month (Apr 6), when it was extinguished about 6 *chi* northeast of YUGUI [LM 23]; in all, it was visible for 113 days.
[*Hou Han shu: tianwen zhi*] ch. 20
NB: Ho (1962) misdated the record to Dec 12–Mar 27.

AD 59 Jun 30–Jul 28 [Korea]
3rd year of King Talhae of Silla, summer, 6th month; a star became fuzzy in TIANCHUAN.
[*Samguk sagi*] ch. 1 (Ho 72)
[*Jeungbo munheon bigo*] ch. 6

AD 60 Aug 9 [China] (Ho 73)
(a) 3rd year of the Yongping reign period of Emperor Ming of the Han Dynasty, 6th month, day *dingmao* [4]; a star became fuzzy north of TIANCHUAN.
[*Hou Han shu: Ming di ji*] ch. 2
(b) 3rd year of the Yongping reign period of Emperor Ming of the Han Dynasty, 6th month, day *dingmao* [4]; a broom star about 2 *chi* long emerged north of TIANCHUAN, traveling slowly northward to the south of KANG [LM 2]; it was visible for 35 days then departed.
[*Hou Han shu: tianwen zhi*] ch. 21
NB: *Fuhou gujin zhu* said, "the broom star was about 3 *chi* long and visible for 35 days, whereupon it departed." *Dong Han huiyao* had "135 days and then departed."

AD 61 Sep 27 [China] (Ho 74)

4th year of the Yongping reign period of Emperor Ming of the Han Dynasty, 8th month, day *xinyou* [58]; a guest star emerged northwest of GENGHE pointing toward GUANSUO; after 70 days, it departed.
[*Hou Han shu: tianwen zhi*] ch. 21

AD 65 Jul 29 [China] (Ho 76)

8th year of the Yongping reign period of Emperor Ming of the Han Dynasty, 6th month, day *renwu* [19]; a long star emerged in LIU [LM 24] and ZHANG [LM 26] and extended some 37 *du*; it trespassed on XUANYUAN, pierced TIANCHUAN, and approached TAIWEI, its vapor reaching SHANGJIE. In all, it was visible 56 days and then departed.
[*Hou Han shu: tianwen zhi*] ch. 21
NB: *Ling*, "mound," should be *ling*, "approach."

AD 66 Feb 20 [China] (Ho 78)

(a) 9th year of the Yongping reign period of Emperor Ming of the Han Dynasty, 1st month, day *wushen* [45] (Feb 20); a guest star 8 *chi* long appeared at QIANNIU [LM 9]. It passed JIANXING and reached the south of FANG [LM 4], where it was extinguished. It was visible for 50 days.
[*Hou Han shu*] ch. 11

(b) 9th year of the Yongping reign period of Emperor Ming of the Han Dynasty, 1st month, day *wushen* [45] (Feb 20); a guest star passed [NAN-]DOU [LM 8], JIANXING, JI [LM 7], and FANG [LM 4]. It (also) passed JIAO [LM 1] and KANG [LM 2] and reached YI [LM 27]. Its rays pointed to the east.
[*Hou Han shu* note quoting *Gu jin zhu*] *zhi* 11

NB: This is Comet Halley, return –25.

AD 71 Mar 6 [China] (Ho 80)

14th year of the Yongping reign period of Emperor Ming of the Han Dynasty, 1st month, day *wuzi* [25]; a guest star emerged in MAO [LM 18], was visible for 60 days, and then slowly extinguished at the right corner of XUANYUAN.
[*Hou Han shu: tianwen zhi*] ch. 21

AD 75 Jul 14 [China] (Ho 81)

(a) 18th year of the Yongping reign period of Emperor Ming of the Han Dynasty, 6th month, day *jiwei* [56]; a star became fuzzy in TAIWEI.
[*Hou Han shu: Ming di ji*] ch. 2

(b) 18th year of the Yongping reign period of Emperor Ming of the Han Dynasty, 6th month, day *jiwei* [56]; a broom star 3 *chi* long emerged in ZHANG [LM 26], turned at LANGJIANG, and entered TAIWEI to the south; both [asterisms] belong to the space of ZHANG [LM 26].
[*Hou Han shu: tianwen zhi*] ch. 21

AD 76 Oct 7 [China] (Ho 82)

(a) 1st year of the Jianchu reign period of Emperor Zhang of the Han Dynasty, 8th month, day *gengyin* [27]; a star became fuzzy in TIANSHI.
[*Hou Han shu: Zhang di ji*] ch. 3

(b) 1st year of the Jianchu reign period of Emperor Zhang of the Han Dynasty, 8th month, day *gengyin* [27]; a broom star about 2 *chi* long emerged in TIANSHI then slowly moved and entered 3 *du* within QIANNIU [LM 9]. It lasted 40 days and then slowly extinguished.
[*Hou Han shu: tianwen zhi*] ch. 21

AD 78 Jan 18 [China]

(a) 2nd year of the Jianchu reign period of Emperor Zhang of the Han Dynasty, 12th month, day *wuyin* [15]; a star became fuzzy in ZIGONG.
[*Hou Han shu: Zhang di ji*] ch. 3

(b) 2nd year of the Jianchu reign period of Emperor Zhang of the Han Dynasty, 12th month, day *wuyin* [15]; a broom star 8 or 9 *chi* long emerged at 3 *du* in LOU [LM 16] then slowly entered the middle of ZIGONG. It was visible for 106 days and then extinguished.

[*Hou Han shu: tianwen zhi*] ch. 21

NB: *Dong Han huiyao* made it the 1st year.

AD 79 Mar 22–Apr 20 [Korea]

23rd year of King Talhae of Silla, spring, 2nd month; a broom star was seen in the east then was seen again in the north, and after 20 days, it was extinguished.

[*Samguk sagi*] ch. 1 (Ho 84)

[*Jeungbo munheon bigo*] ch. 6

AD 85 May 20 [China] (Ho 85)

2nd year of the Yuanhe reign period of Emperor Zhang of the Han Dynasty, 4th month, day *dingsi* [54]; a guest star 3 *chi* long emerged at dawn in the east at 8 *du* in WEI [LM 17], passed GEDAO, and entered ZIGONG. It lasted 40 days and slowly extinguished.

[*Hou Han shu: tianwen zhi*] ch. 21

NB: *Dong Han hui yao* had "1st year." Based on what follows in the text, the original's "1st year, 4th month" should be "2nd year, 4th month." Ho (1962) gave the date as May 25.

AD 101 Jan 12 [China] (Ho 87)

12th year of the Yongyuan reign period of Emperor He of the Han Dynasty, 11th month, day *guiyou* [10]; during the night, there was a grayish vapor 3 *zhang* long that arose in TIANYUAN, pointing to JUNSHI in the northeast; it was visible for 10 days.

[*Hou Han shu: tianwen zhi*] ch. 21

AD 104 May 30 [China] (Ho 89)

16th year of the Yongyuan reign period of Emperor He of the Han Dynasty, 4th month, day *dingwei* [44]; a white vapor like silk floss developed in ZIGONG. On day *wuwu* [55], a guest star emerged in ZIGONG and traveled westward as far as MAO [LM 18]; on day *renshen* [9] in the 5th month, it was extinguished.

[*Hou Han shu: tianwen zhi*] ch. 21

AD 110 Jan 9–Feb 6 [China] (Ho 91)

3rd year of the Yongchu reign period of Emperor An of the Han Dynasty, 12th month; a grayish broom star 6 to 7 *chi* long arose south of TIANYUAN and pointed northeast.
[*Hou Han shu: tianwen zhi*] ch. 21

AD 110 Jan 27 [China] (Ho 91)

3rd year of the Yongchu reign period of Emperor An of the Han Dynasty, 12th month, day *yihai* [12]; a star became fuzzy in TIANYUAN.
[*Hou Han shu: An di ji*] ch. 5

AD 110 Jul 15 [China] (Ho 92)

4th year of the Yongchu reign period of Emperor An of the Han Dynasty, 6th month, day *jiazi* [1]; a grayish guest star big as a plum, with rayed vapors 2 *chi* long, pointed toward SHANGJIE asterism in the southwest.
[*Hou Han shu: tianwen zhi*] ch. 21
NB: *Dong Han huiyao* had "6th month, day *bingzi* [13]." Ho (1962) dated the record July 27.

AD 117 Jan 9 [China] (Ho 93)

3rd year of the Yuanchu reign period of Emperor An of the Han Dynasty, 11th month, day *jiawu* [31]; a guest star appeared in the west. On day *jihai* [36], it was in XU [LM 11] and WEI [LM 12] reaching WEI [LM 17] and MAO [LM 18] to the south.
[*Hou Han shu: tianwen zhi*] ch. 21
NB: Ho (1962) misidentified the year as AD 116 and so proposed to emend "11th month" to read "12th month."

AD 126 Mar 23 [China] (Ho 95)

1st year of the Yongjian reign period of Emperor An of the Han Dynasty, 2nd month, day *jiawu* [31]; [*Gujin zhu*] says, "1st year of the Yongjian reign period, 2nd month, day *jiawu* [31]; a guest star entered TAIWEI..." [*Li shi jia shu*] says, "At the time, in the sky there was an

anomalous vapor and Li He submitted a letter remonstrating"...then on the 13th of the month a guest star with vapor resembling a broom star or fuzzy star passed by TIANSHI, GENGHE, ZHAOYAO, QIANG, and BANG. On the 16th day, it entered ZIGONG and approached BEICHEN. On the 17th day, it further passed by WENCHANG and TAILING, reaching a point between TIANCHUAN and JISHUI where it slowly faded from sight.
[*Hou Han shu: tianwen zhi*] ch. 21

AD 128 Sep 12–Oct 11 [Korea]
17th year of King Jima of Silla, autumn, 8th month; a long star spanned the entire sky.
[*Samguk sagi*] ch. 1 (Ho 96)
[*Jeungbo munheon bigo*] ch. 6

AD 132 Jan 29 [China] (Ho 97–98)
(a) 6th year of the Yongjian reign period of Emperor Shun of the Han Dynasty, 12th month, day *renshen* [21]; a guest star emerged in QIANNIU [LM 9].
[*Hou Han shu: Shun di ji*] ch. 6
(b) 6th year of the Yongjian reign period of Emperor Shun of the Han Dynasty, 12th month, day *renshen* [21]; a grayish guest star with rayed vapors over 2 *chi* long and pointing southwest was at 6 *du* in QIANNIU [LM 9].
[*Hou Han shu: tianwen zhi*] ch. 21
NB: Ho (1962) misdated the record to AD 131.

AD 132 [China] (Ho 98)
1st year of the Yangjia reign period of Emperor Shun of the Han Dynasty, 12th month, day *wuzi* [25]; a guest star emerged in TIANYUAN.
[*Hou Han shu: Shun di ji*] ch. 6
NB: The 12th month lacked day *wuzi* [25]; one suspects this is an error for intercalary 10th month.

AD 133 Feb 8 [China] (Ho 99)

1st year of the Yangjia reign period of Emperor Shun of the Han Dynasty, intercalary month, day *wuzi* [25]; a guest star 2 *chi* wide and 5 *zhang* long arose southwest of TIANYUAN.

[*Hou Han shu: tianwen zhi*] ch. 21

AD 133 Feb 9 [China] (Ho 99)

1st year of the Yangjia reign period of Emperor Shun of the Han Dynasty, intercalary month, 17th day *jichou* [26]; at night, there was a white vapor from TIANYUAN that hurried toward ZUOZU [?] and entered YUJING. After several days, it was extinguished.

[*Hou Han shu: lie zhuan*] ch. 30

NB: Possibly the same as the preceding.

AD 141 Mar 27–Apr 23 [China] (Ho 100)

(a) 6th year of the Yonghe reign period of Emperor Shun of the Han Dynasty, 2nd month, day *dingsi* [54] [Mar 27]; a broom star was seen in the east. It was 6 or 7 *chi* long, and its color was greenish white. It pointed southwestward toward YINGSHI [LM 13] and the FENMU star. On day *dingchou* [14] (Apr 16), the broom star was 1 *du* within KUI [LM 15] and was 6 *chi* in length. On day *guiwei* [20] (Apr 22), it was seen at dusk. It passed to the northwest of MAO [LM 18] and BI [LM 19]. On day *jiashen* [21] (Apr 23), it was seen at DONGJING [LM 22]. Gradually it passed YUGUI [LM 23], LIU [LM 24], QIXING [LM 25], and ZHANG [LM 26]. Its rays extended to SANTAI. When it reached the middle of XUANYUAN, it was extinguished.

[*Hou Han shu*] *zhi*. 11

(b) Emperor Shun, 6th year of the Yonghe reign period, 2nd month, day *dingsi* [54] (Mar 27); a star appeared fuzzy in YINGSHI [LM 13].

[*Hou Han shu*] ch. 6

NB: This is Comet Halley, return –24.

AD 149 Sep 20–Oct 19 [Korea]

16th year of King Ilseong of Silla, autumn, 8th month; there was a fuzzy star in TIANSHI.
[*Samguk sagi*] ch. 1 (Ho 101)
[*Jeungbo munheon bigo*] ch. 6

AD 149 Oct 19 [China] (Ho 101)

(a) 3rd year of the Jianhe reign period of Emperor Huan of the Han Dynasty, 8th month, day *yichou* [2]; a star became fuzzy in TIANSHI.
[*Hou Han shu: Huan di ji*] ch. 7

(b) 3rd year of the Jianhe reign period of Emperor Huan of the Han Dynasty, 8th month, day *yichou* [2]; a yellowish-white broom star with rays 5 *chi* long pointing southeast appeared in the middle of TIANSHI; on day *wuchen* [5] of the 9th month, it was not seen.
[*Hou Han shu: tianwen zhi*] ch. 22

AD 153 Nov 4–Dec 3 [Korea] (Ho 102)

20th year of King Ilseong of Silla, winter, 10th month; a broom star was seen in the east and then again in the northeast.
[*Jeungbo munheon bigo*] ch. 6
[*Samguk sagi*] ch. 1

AD 161 Jun 14 [China] (Ho 105)

(a) 4th year of the Yanxi reign period of Emperor Huan of the Han Dynasty, 5th month, day *xinyou* [58]; a star became fuzzy in XIN [LM 5].
[*Hou Han shu: Huan di ji*] ch. 7

(b) 4th year of the Yanxi reign period of Emperor Huan of the Han Dynasty, 5th month, day *xinyou* [58]; a guest star was in YINGSHI [LM 13]. It moved slowly forward and produced rays about 5 *chi* long. When it reached 1 *du* in XIN [LM 5], it turned into a broom star.
[*Hou Han shu: tianwen zhi*] ch. 22

AD 178 Aug 31–Sep 29 [China] (Ho 106)

(a) 1st year of the Guanghe reign period of Emperor Ling of the Han Dynasty, 8th month; a star became fuzzy in TIANSHI.
[*Hou Han shu: Ling di ji*] ch. 8

(b) 1st year of the Guanghe reign period of Emperor Ling of the Han Dynasty, 8th month; a red broom star several *chi* long emerged north of KANG [LM 2] and entered within TIANSHI then slightly lengthened to 5 or 6 *zhang*. It traversed more than 10 lunar mansions in over 80 days and then disappeared in TIANYUAN.
[*Hou Han shu: tianwen zhi*] ch. 22

AD 180 Oct–Dec [China] (Ho 107)

(a) 3rd year of the Guanghe reign period of Emperor Ling of the Han Dynasty, winter; a star became fuzzy in LANG and HU.
[*Hou Han shu: Ling di ji*] ch. 8

(b) 3rd year of the Guanghe reign period of Emperor Ling of the Han Dynasty, winter; a broom star emerged in LANG and HU, traveled eastward as far as ZHANG [LM 26], and then departed.
[*Hou Han shu: tianwen zhi*] ch. 22

NB: *Hou Han shu: Ling di ji* made it "3rd year, winter, intercalary month." On checking, there was no intercalary month that year.

AD 182 Aug 17–Sep 15 [China] (Ho 108)

(a) 5th year of the Guanghe reign period of Emperor Ling of the Han Dynasty, 7th month; a star became fuzzy in TAIWEI.
[*Hou Han shu: Ling di ji*] ch. 8

(b) 5th year of the Guanghe reign period of Emperor Ling of the Han Dynasty, 7th month; a broom star emerged below SANTAI, traveled eastward and entered TAIWEI, and reached TAIZI and XINCHEN, lasting over 20 days before disappearing.
[*Hou Han shu: tianwen zhi*] ch. 22

AD 182 Aug 17–Sep 15 [Korea]

4th year of King Gogukcheon of Goguryeo, autumn, 7th month; there was a fuzzy star in TAIWEI.

[*Jeungbo munheon bigo*] ch. 6
[*Samguk sagi*] ch. 16 (Ho 108)

AD 186 Oct 31–Nov 28 [Korea]
21st year of King Chogo of Baekje, winter, 10th month; there was a fuzzy star in the northwest; after 20 days, it was extinguished.
[*Jeungbo munheon bigo*] ch. 6
[*Samguk sagi*] ch. 23 (Ho 110)

AD 188 Mar 16–Apr 13 [China] (Ho 111)
(a) 5th year of the Zhongping reign period of Emperor Ling of the Han Dynasty, 2nd month; a star became fuzzy in ZIGONG.
[*Hou Han shu: Ling di ji*] ch. 8
(b) 5th year of the Zhongping reign period of Emperor Ling of the Han Dynasty, 2nd month; a broom star emerged in KUI [LM 15] then retrograded and entered ZIGONG. Later it reemerged three times and disappeared only after more than 60 days.
[*Hou Han shu: tianwen zhi*] ch. 22

AD 188 Jul 28 [China] (Ho 112)
5th year of the Zhongping reign period of Emperor Ling of the Han Dynasty, 6th month, day *dingmao* [4]; a guest star as large as a 3 *sheng* ("quart") bowl emerged at GUANSUO. It traveled southwestward and entered TIANSHI then disappeared when it reached WEI [LM 6].
[*Hou Han shu: tianwen zhi*] ch. 22

AD 191 Oct 6–Nov 4 [China] (Ho 113)
2nd year of the Chuping reign period of Emperor Xian of the Han Dynasty, 9th month; Chiyou's Banner appeared south of JIAO [LM 1] and KANG [LM 2]. It was white and over 10 *zhang* long.
[*Hou Han shu: tianwen zhi*] ch. 22

AD 191 Oct 6–Nov 4 [Korea] (Ho 113)
8th year of King Beolhyu of Silla, 9th month; Chiyou's banner was seen at JIAO [LM 1] and KANG [LM 2].
[*Samguk sagi*] ch. 2, 23

AD 193 Dec 3 [China] (Ho 114)

(a) 4th year of the Chuping reign period of Emperor Xian of the Han Dynasty, 10th month, day *xinchou* [38]; a star became fuzzy in TIANSHI.
[*Hou Han shu: Xian di ji*] ch. 9

(b) 4th year of the Chuping reign period of Emperor Xian of the Han Dynasty, 10th month; a fuzzy star emerged between the two JIAO, [i.e., DAJIAO and JIAO (LM 1)] traveled northeastward, and entered within TIANSHI where it was extinguished.
[*Hou Han shu: tianwen zhi*] ch. 22

AD 200 Nov 6 [China] (Ho 115)

5th year of the Jian'an reign period of Emperor Xian of the Han Dynasty, 10th month, day *xinhai* [48]; a star became fuzzy in DALIANG.
[*Hou Han shu: Xian di ji*] ch. 9
[*Hou Han shu: tianwen zhi*] ch. 22

AD 204 Nov 10–Dec 9 [China] (Ho 116)

9th year of the Jian'an reign period of Emperor Xian of the Han Dynasty, 10th month; a star became fuzzy in DONGJING [LM 22].
[*Hou Han shu: Xian di ji*] ch. 9

AD 204 Nov 10–Dec 9 [Korea]

39th year of King Chogo of Baekje, winter, 10th month; there was a fuzzy star in DONGJING [LM 22].
[*Jeungbo munheon bigo*] ch. 6
[*Samguk sagi*] ch. 23 (Ho 116)

AD 204 Dec 10–AD 205 Jan 7 [China] (Ho 117)

9th year of the Jian'an reign period of Emperor Xian of the Han Dynasty, 11th month; a star became fuzzy in DONGJING [LM 22] and YUGUI [LM 23] then entered XUANYUAN and TAIWEI.
[*Hou Han shu: tianwen zhi*] ch. 22

AD 206 Jan 27–Feb 25 [China] (Ho 118)

(a) 11th year of the Jian'an reign period of Emperor Xian of the Han Dynasty, 1st month; a star became fuzzy in BEIDOU.
[*Hou Han shu: Xian di ji*] ch. 9

(b) 11th year of the Jian'an reign period of Emperor Xian of the Han Dynasty, 1st month; a star became fuzzy in BEIDOU, its head inside the bowl [of the dipper] and its tail penetrating ZIGONG as far as BEICHEN.
[*Hou Han shu: tianwen zhi*] ch. 22

AD 207 Nov 10 [China] (Ho 119)

12th year of the Jian'an reign period of Emperor Xian of the Han Dynasty, 10th month, day *xinmao* [28]; a star became fuzzy in CHUNWEI [LM 27–28].
[*Hou Han shu: Xian di ji*] ch. 9
[*Hou Han shu: tianwen zhi*] ch. 22

AD 213 Jan 10–Feb 7 [China] (Ho 120)

17th year of the Jian'an reign period of Emperor Xian of the Han Dynasty, 12th month; a star became fuzzy in WUZHUHOU.
[*Hou Han shu: Xian di ji*] ch. 9
[*Hou Han shu: tianwen zhi*] ch. 22

AD 217 Oct–Dec [China]

22nd year of the Jian'an reign period of Emperor Xian of the Han Dynasty, winter; a star became fuzzy in the northeast.
[*Hou Han shu: Xian di ji*] ch. 9
NB: Ho (1962) provided only the Korean record of this comet.

AD 217 Nov 17–Dec 15 [Korea]

21st year of King Sansang of Goguryeo, winter, 10th month; there was a fuzzy star in the northeast.
[*Jeungbo munheon bigo*] ch. 6
[*Samguk sagi*] ch. 16 (Ho 121)

AD 218 Apr 13–May 12 [China] (Ho 122)

23rd year Jian'an reign period of Emperor Xian of the Han Dynasty, 3rd month; a star appeared fuzzy in the east in the morning for more than 20 days. In the evening, it appeared in the west. It trespassed against and passed WUCHE, DONGJING [LM 22), WU ZHUHOU, WENCHANG, XUANYUAN, HOUFEI, and TAIWEI. Its sharp rays pointed toward DIZUO.

[*Hou Han shu*] *zhi* 12

NB: This is Comet Halley, return –23.

AD 225 Dec 9 [China] (Ho 124)

6th year of the Huangchu reign period of Emperor Wen of the Wei Dynasty, 10th month, day *yiwei* [32]; a star became fuzzy in SHAOWEI then passed through XUANYUAN.

[*Song shu: tianwen zhi*] ch. 23

AD 232 Dec 3 [China] (Ho 125)

6th year of the Taihe reign period of Emperor Ming of the Wei Dynasty, 11th month, day *bingyin* [3]; a star became fuzzy in YI [LM 27], close to the SHANGJIANG star of TAIWEI.

[*San guo zhi: Wei shu*] ch. 3

[*Song shu: tianwen zhi*] ch. 23

AD 236 [China] (Ho 126)

4th year of the Qinglong reign period of Emperor Ming of the Wei Dynasty, 11th month, day *jihai* [36]; a broom star appeared, trespassing on the asterisms HUANZHE and TIANJI.

[*San guo zhi: Wei shu: Ming di ji*] ch. 3

[*Song shu: tianwen zhi*] ch. 23

NB: There was no day *jihai* [36] in the 11th month.

AD 236 Nov 30 [China] (Ho 126)

(a) 4th year of the Qinglong reign period of Emperor Ming of the Wei Dynasty, 10th month, day *jiashen* [21]; a star became

fuzzy in DACHEN; on day *yiyou* [22], it became fuzzy again in the east.
[*San guo zhi: Wei shu: Ming di ji*] ch. 3

(b) 4th year of the Qinglong reign period of Emperor Ming of the Wei Dynasty, 10th month, day *jiashen* [21]; a star 3 *chi* long became fuzzy in DACHEN; on day *yiyou* [22], it became fuzzy again in the east.
[*Song shu: tianwen zhi*] ch. 23
[*Jin shu: tianwen zhi*] ch. 13

AD 238 Jul 29–Aug 27 [China] (Ho 127)
2nd year of the Jingchu reign period of Emperor Ming of the Wei Dynasty, 8th month; a broom star 3 *chi* long appeared in ZHANG [LM 26] that "retrograded" toward the west; after 41 days, it was extinguished.
[*Song shu: tianwen zhi*] ch. 23

AD 238 Aug 21 [China] (Ho 127)
2nd year of the Jingchu reign period of Emperor Ming of the Wei Dynasty, 8th month, day *guichou* [50]; a broom star appeared in lunar mansion ZHANG [LM 26].
[*San guo zhi: Wei shu: Ming di ji*] ch. 3

AD 238 Sep 30 [China] (Ho 128)
2nd year of the Jingchu reign period of Emperor Ming of the Wei Dynasty, 10th month, day *guisi* [30]; a guest star appeared in WEI [LM 12] that "retrograded" to the north of LIGONG and south of TENGSHE. On day *jiachen* [41], it trespassed on the ZONG star. On day *jiyou* [46], it was extinguished.
[*Song shu: tianwen zhi*] ch. 23

AD 240 Nov 10 [China] (Ho 129)
1st year of the Zhengshi reign period of King Qi of Wei, 10th month, day *yiyou* [22]; a broom star 3 *zhang* long appeared in the west in WEI

[LM 6], brushed past QIANNIU [LM 9], and trespassed on Venus. In the 11th month, day *jiazi* [1], it advanced and trespassed on YULIN. [*Song shu: tianwen zhi*] ch. 23

AD 245 Sep 18 [China] (Ho 130)

6th year of the Zhengshi reign period of King Qi of Wei, 8th month, day *wuwu* [55]; a white broom star 2 *chi* long appeared in QIXING [LM 25]. It advanced as far as ZHANG [LM 26] for 23 days in total then was extinguished.
[*Song shu: tianwen zhi*] ch. 23

AD 246 Jan 16 [China] (Ho 131)

7th year of the Zhengshi reign period of King Qi of Wei, 11th month, day *guihai* [60]; (a broom star) 1 *chi* long appeared again in ZHEN [LM 28] for a total of 156 days, then it was extinguished.
[*Song shu: tianwen zhi*] ch. 23
NB: There was no day *guihai* [60] in the 11th month. Consequently, Ho made it Jan 16, 247.

AD 248 Apr 11–May 9 [China] (Ho 132)

9th year of the Zhengshi reign period of King Qi of Wei, 3rd month; there was another greenish-white [broom star] 6 *chi* long in MAO [LM 18], with rays pointing toward the southwest.
[*Song shu: tianwen zhi*] ch. 23

AD 248 Aug 7–Sep 4 [China] (Ho 133)

9th year of the Zhengshi reign period of King Qi of Wei, 7th month; (a broom star) 2 *chi* long appeared again in YI [LM 27] and advanced as far as ZHEN [LM 28]. It lasted for 42 days and then was extinguished.
[*Song shu: tianwen zhi*] ch. 23

AD 251 [China] (Ho 134)

3rd year of the Jiaping reign year of King Qi of Wei, 11th month, day *guihai* [60]; a star became fuzzy in YINGSHI [LM 13] then traveled westward for 90 days and then was extinguished.

[*Jin shu: tianwen zhi*] ch. 13
[*Song shu: tianwen zhi*] ch. 23
NB: *Song shu* has *guiwei* [20], but there was no day *guiwei* in the 11th month.

AD 252 Mar 24 [China] (Ho 135)
4th year of the Jiaping reign year of King Qi of Wei, 2nd month, day *dingyou* [34]; a white broom star 5 to 6 *zhang* long appeared in the west in WEI [LM 17], its rays pointing south and penetrating SHEN [LM 21]; it lasted 20 days and then was extinguished.
[*Song shu: tianwen zhi*] ch. 23

AD 253 Dec 8–AD 254 Jan 5 [China] (Ho 136)
5th year of the Jiaping reign year of King Qi of Wei, 11th month; a broom star 5 *zhang* long appeared again in ZHEN [LM 28], west of ZUOZHIFA of TAIWEI, pointing southeast for 190 days, and then it was extinguished.
[*Song shu: tianwen zhi*] ch. 23

AD 254 Nov 27–Dec 26 [China] (Ho 137)
(a) 1st year of the Wufeng reign period of King Kuaiji of Wu, 11th month; there was a bushy star in DOU [LM 8] and NIU [LM 9].
[*San guo zhi: Wu shu*] ch. 48
(b) 1st year of the Zhengyuan reign period of Duke Gaoguixiang of Wei, 11th month; a white vapor appeared to emerge from the side of NANDOU [LM 8]. It was several *zhang* broad and stretched across the sky.
[*Jin shu: tianwen zhi*] ch. 13
[*Song shu: tianwen zhi*] ch. 23

AD 255 Jan 25–Feb 23 [China] (Ho 138)
(a) 2nd year of the Zhengyuan reign period of Duke Gaoguixiang of Wei, 1st month; a broom star several tens of *zhang* long arose in the space (allocated to) Wu and Chu [LM 27–LM 9].
[*San guo zhi: Wei shu: Muqiujian zhuan*] ch. 28

(b) 2nd year of the Zhengyuan reign period of Duke Gaoguixiang of Wei, 1st month; a broom star appeared in the space (allocated to) Wu and Chu [LM 27–LM 9] and stretched across the sky in the northwest. [*Jin shu: tianwen zhi*] ch. 13

AD 257 Dec 23–AD 258 Jan 21 [China] (Ho 139)
2nd year of the Ganlu reign period of Duke Gaoguixiang of Wei, 11th month; a white broom star appeared in JIAO [LM 1].
[*Song shu: tianwen zhi*] ch. 23

AD 259 Nov 23 [China] (Ho 140)
4th year of the Ganlu reign period of Duke Gaoguixiang of Wei, 10th month, day *dingychou* [14]; a guest star appeared in the middle of TAIWEI that turned and traveled southeastward, passing through lunar mansion ZHEN [LM 28], in all, for 7 days, and then it was extinguished.
[*Song shu: tianwen zhi*] ch. 23
NB: *Tong zhi* made it "4th year" and "70 days and then it was extinguished."

AD 260 Jul 25–Aug 23 [Korea]
14th year of King Cheomhae of Silla, autumn, 7th month; there was a fuzzy star in the east; after 25 days, it was extinguished.
[*Jeungbo munheon bigo*] ch. 6
[*Samguk sagi*] ch. 2 (Ho 141)

AD 262 Dec 2 [China] (Ho 142)
3rd year of the Jingyuan reign period of King Chenliu of Wei, 11th month, day *renyin* [39]; a white broom star 5 *cun* long appeared in KANG [LM 2], which turned and traveled northward. It lasted 45 days and then was extinguished.
[*Song shu: tianwen zhi*] ch. 23
NB: *San guo hui yao* made it "5 *zhang* long."

AD 265 Jun 1–Jun 30 [China] (Ho 143)
2nd year of the Xianxi reign period of King Chenliu of Wei, 5th month; a white broom star over 1 *zhang* long and pointing southeast

appeared in WANGLIANG, lasting for 12 days, and then it was extinguished.
[*Song shu: tianwen zhi*] ch. 23

AD 268 Feb 18 [China] (Ho 144)

(a) 4th year of the Taishi reign period of Emperor Wu of Jin, 1st month, day *bingxu* [23]; a star became fuzzy in ZHEN [LM 28].
[*Jin shu: Wu di ji*] ch. 3

(b) 4th year of the Taishi reign period of Emperor Wu of Jin, 1st month, day *bingxu* [23]; a greenish-white broom star appeared in ZHEN [LM 28]. It traveled northwestward then turned and traveled eastward.
[*Song shu: tianwen zhi*] ch. 23
[*Jin shu: tianwen zhi*] ch. 13

AD 269 Oct 13–Nov 10 [China] (Ho 145)

5th year of the Taishi reign period of Emperor Wu of Jin, 9th month; a star became fuzzy in ZIGONG.
[*Song shu: tianwen zhi*] ch. 23
[*Jin shu: Wu di ji*] ch. 3
[*Jin shu: tianwen zhi*] ch. 13

AD 269 Oct 13–Nov 10 [Korea]

36th year of King Goi of Baekje, autumn, 9th month; there was a fuzzy star in ZIGONG.
[*Jeungbo munheon bigo*] ch. 6
[*Samguk sagi*] ch. 24 (Ho 145)

AD 275 Jan 14–Feb 12 [China] (Ho 146)

10th year of the Taishi reign period of Emperor Wu of Jin, 12th month; a star became fuzzy in ZHEN [LM 28].
[*Song shu: tianwen zhi*] ch. 23
[*Jin shu: Wu di ji*] ch. 3
[*Jin shu: tianwen zhi*] ch. 13

AD 276 Jun 29–Jul 27 [China] (Ho 147)

(a) 2nd year of the Xianning reign period of Emperor Wu of Jin, 6th month; a star became fuzzy in DI [LM 3]…7th month, a star became fuzzy in DAJIAO…8th month, a star became fuzzy in TAIWEI and reached YI [LM 27], BEIDOU, and SANTAI.
[*Song shu: tianwen zhi*] ch. 23

(b) 2nd year of the Xianning reign period of Emperor Wu of Jin, 6th month, day *jiaxu* [11]; a star became fuzzy in DI [LM 3]…7th month, a star became fuzzy in DAJIAO…8th month, a star became fuzzy in TAIWEI that reached YI [LM 27], BEIDOU, and SANTAI.
[*Jin shu: tianwen zhi*] ch. 13

(c) 2nd year of the Xianning reign period of Emperor Wu of Jin, 6th month, day *jiaxu* [11]; a star became fuzzy in DI [LM 3]…7th month, a star became fuzzy at DAJIAO…8th month, a star became fuzzy in TAIWEI. In the 9th month, it became fuzzy again in YI [LM 27].
[*Jin shu: Wu di ji*] ch. 3

NB: There was no day *jiaxu* [11] in the 6th month.

AD 277 Feb 20–Jun 17 [China] (Ho 148)

(a) 3rd year of the Xianning reign period of Emperor Wu of Jin, 1st month (Feb 20–Mar 21); a star became fuzzy in the west; in the 3rd month, a star became fuzzy in WEI [LM 17]…in the 4th month, a star became fuzzy in NÜYU…in the 5th month, fuzzy again in the east. In the 7th month, a star became fuzzy in ZIGONG.
[*Jin shu: tianwen zhi*] ch. 13

(b) 3rd year of the Xianning reign period of Emperor Wu of Jin, 4th month (May 20–Jun 17); a star became fuzzy in NÜYU… 5th month, fuzzy again in the east. In the 7th month, a star became fuzzy in ZIGONG.
[*Song shu: tianwen zhi*] ch. 23

AD 278 May 9–Jun 7 [China] (Ho 149)

4th year of the Xianning reign period of Emperor Wu of Jin, 4th month; Chiyou's Banner appeared in DONGJING [LM 22].
[*Jin shu: tianwen zhi*] ch. 13

[*Song shu: tianwen zhi*] ch. 23
NB: The *Song shu* record lacks "in DONGJING."

AD 279 Mar 30–Apr 27 [China] (Ho 150)

(a) 5th year of the Xianning reign period of Emperor Wu of Jin, 3rd month; a star became fuzzy in LIU [LM 24].
[*Song shu: tianwen zhi*] ch. 23

(b) 5th year of the Xianning reign period of Emperor Wu of Jin, 3rd month; a star became fuzzy in LIU [LM 24]. In the 4th month, it became fuzzy again in NÜYU.
[*Jin shu: Wu di ji*] ch. 3
[*Jin shu: tianwen zhi*] ch. 13

AD 279 Jul 26–Aug 24 [China] (Ho 150)

5th year of the Xianning reign period of Emperor Wu of Jin, 7th month; a star became fuzzy in ZIGONG.
[*Jin shu: Wu di ji*] ch. 3
[*Jin shu: tianwen zhi*] ch. 13

AD 281 Sep 1–29 [China] (Ho 151)

2nd year of the Taikang reign period of Emperor Wu of Jin, 8th month; a star became fuzzy in ZHANG [LM 26].
[*Song shu: tianwen zhi*] ch. 23
[*Jin shu: Wu di ji*] ch. 3
[*Jin shu: tianwen zhi*] ch. 13

AD 281 Nov 28–Dec 27 [China] (Ho 152)

2nd year of the Taikang reign period of Emperor Wu of Jin, 11th month; a star became fuzzy in XUANYUAN.
[*Song shu: tianwen zhi*] ch. 23
[*Jin shu: Wu di ji*] ch. 3
[*Jin shu: tianwen zhi*] ch. 13

AD 283 Apr 22 [China] (Ho 153)

4th year of the Taikang reign period of Emperor Wu of Jin, 3rd month, day *wushen* [45]; a star became fuzzy in the southwest.

[*Song shu: tianwen zhi*] ch. 23
[*Jin shu: tianwen zhi*] ch. 13

AD 287 Oct 24–Nov 22 [China] (Ho 154)

8th year of the Taikang reign period of Emperor Wu of Jin, 9th month; a star several tens of *zhang* long became fuzzy in NANDOU [LM 8] for over 10 days and then was extinguished.
[*Song shu: tianwen zhi*] ch. 23
[*Jin shu: tianwen zhi*] ch. 13

AD 295 May 1–30 [China] (Ho 156)

(a) 5th year of the Yuankang reign period of Emperor Hui of Jin, 4th month; a star appeared fuzzy in KUI [LM 15]. It reached XUANYUAN and TAIWEI. It passed SANTAI and DALING.
[*Jinshu*] ch. 13 and [*Songshu*] ch. 24

(b) 5th year of the Yuankang reign period of Emperor Hui of Jin, 4th month; a broom star was seen in the west. It emerged fuzzy at KUI [LM 15] and reached XUANYUAN.
[*Jin shu*] ch. 4

NB: This is Comet Halley, return –22.

AD 300 Dec 12–AD 301 Jan 25 [Korea] (Ho 159)

1st year of King Micheon of Goguryeo, 12th month; a fuzzy star was seen in the east.
[Samguk sagi] ch. 5

AD 300 Dec 28–AD 301 Jan 25 [China] (Ho 159)

(a) 1st year of the Yongkang reign period of Emperor Hui of Jin, 12th month; a broom star emerged west of NIU [LM 9] pointing toward TIANSHI.
[*Jin shu: tianwen zhi*] ch. 13
[*Song shu: tianwen zhi*] ch. 24

(b) 1st year of the Yongkang reign period of Emperor Hui of Jin, 12th month; a broom star appeared in the east.
[*Jin shu: Hui di ji*] ch. 4

AD 301 May 24–Jun 22 [China] (Ho 160)
2nd year of the Yongkang reign period of Emperor Hui of Jin, 4th month; a broom star appeared in the astral space of the state of Qi [LM 10–11].
[*Jin shu: tianwen zhi*] ch. 13
[*Song shu: tianwen zhi*] ch. 24

AD 302 May 14–Jun 11 [China] (Ho 161)
1st year of the Tai'an reign period of Emperor Hui of Jin, 4th month; a broom star appeared in daylight.
[*Jin shu: Hui di ji*] ch. 4
[*Jin shu: tianwen zhi*] ch. 13

AD 302 May 14–Jun 11 [Korea]
5th year of King Bunseo of Baekje, summer, 4th month; a broom star was seen in daylight.
[*Jeungbo munheon bigo*] ch. 6
[*Samguk sagi*] ch. 24 (Ho 161)

AD 303 Apr 3–May 2 [China] (Ho 162)
2nd year of the Taian reign period of Emperor Hui of Jin, 3rd month; a broom star appeared in the east pointing toward SANTAI.
[*Song shu: tianwen zhi*] ch. 24
[*Jin shu: tianwen zhi*] ch. 13

AD 305 Sep 5–Oct 4 [China] (Ho 164)
2nd year of the Yongxing reign period of Emperor Hui of Jin, 8th month; a star became fuzzy in MAO [LM 18] and BI [LM 19].
[*Song shu: tianwen zhi*] ch. 24
[*Jin shu: tianwen zhi*] ch. 13

AD 305 Nov 21 [China] (Ho 165)
2nd year of the Yongxing reign period of Emperor Hui of Jin, 10th month, day *dingchou* [14]; a star became fuzzy in BEIDOU.
[*Song shu: tianwen zhi*] ch. 24
[*Jin shu: Hui di ji*] ch. 4
[*Jin shu: tianwen zhi*] ch. 13

AD 315 Sep–Oct [Korea] (Ho 166)

16th year of King Micheon of Goguryeo, 8th month, autumn; a fuzzy star was seen in the northeast.

[Samguk sagi] ch. 17

AD 329 Aug 11–Sep 9 [China] (Ho 167)

(a) 4th year of the Xianhe reign period of Emperor Cheng of the Jin Dynasty, 7th month; a star became fuzzy in the northwest for 23 days and then was extinguished.
[*Song shu: tianwen zhi*] ch. 24

(b) 4th year of the Xianhe reign period of Emperor Cheng of the Jin Dynasty, 7th month; a star became fuzzy in the northwest.
[*Jin shu: Cheng di ji*] ch. 7

(c) 4th year of the Xianhe reign period of Emperor Cheng of the Jin Dynasty, 7th month; a star became fuzzy in the northwest, trespassed on DOU. It lasted 23 days and then was extinguished.
[*Jin shu: tianwen zhi*] ch. 13

AD 336 Feb 16 [China] (Ho 168)

(a) 2nd year of the Xiankang reign period of Emperor Cheng of the Jin Dynasty, 1st month, day *xinsi* [18]; in the evening, a broom star appeared in the west in KUI [LM 15].
[*Song shu: tianwen zhi*] ch. 24
[*Jin shu: tianwen zhi*] ch. 13

(b) 2nd year of the Xiankang reign period of Emperor Cheng of the Jin Dynasty, 1st month, day *xinsi* [18]; a broom star appeared in KUI [LM 15].
[*Jin shu: Cheng di ji*] ch. 7

AD 336 Feb 16 [Korea] (Ho 168)

33rd year of King Biryu of Baekje, spring, 1st month, day *xinsi* [18]; a broom star was seen in KUI [LM 15].

[*Jeungbo munheon bigo*] ch. 6 [China]

AD 340 Mar 25 [China] (Ho 169)

6th year of the Xiankang reign period of Emperor Cheng of the Jin Dynasty, 2nd month, day *gengchen* [17]; a star became fuzzy in TAIWEI.

[*Jin shu: Cheng di ji*] ch. 7

[*Jin shu: tianwen zhi*] ch. 13

AD 343 Dec 8 [China] (Ho 170)

(a) 1st year of the Jianyuan reign period of Emperor Kang of the Jin Dynasty, 11th month, 6th day; a broom star 7 *chi* long with a white tail appeared in KANG [LM 2].

[*Song shu: tianwen zhi*] ch. 24

(b) 1st year of the Jianyuan reign period of Emperor Kang of the Jin Dynasty, 11th month, 6th day; a white broom star 7 *chi* long appeared in KANG [LM 2].

[*Jin shu: tianwen zhi*] ch. 13

AD 349 Dec 2–AD 350 Mar 30 [China] (Ho 171)

(a) 5th year of the Yonghe reign period of Emperor Mu of the Jin Dynasty, 11th month, day *yimao* [52]; a white broom star 1 *zhang* long appeared in KANG [LM 2], its rays pointing west. In the 6th year, 1st month, day *dingchou* [14], the broom star appeared again in KANG [LM 2].

[*Jin shu: tianwen zhi*] ch. 13

(b) 5th year of the Yonghe reign period of Emperor Mu of the Jin Dynasty, 11th month, day *yimao* [52]; a white broom star 1 *zhang* long appeared in KANG [LM 2], its rays pointing west.

[*Song shu: tianwen zhi*] ch. 24

(c) 6th year of the Yonghe reign period of Emperor Mu of the Jin Dynasty, intercalary month, day *dingchou* [14] (Mar 30); a broom star appeared in KANG [LM 2].

[*Jin shu: Mu di ji*] ch. 8

NB: (1) In the 11th month, there was no day *yimao* [52]; day *jimao* [16] in the 11th month was Dec 2. (2) There was an intercalary 2nd month in the 6th year.

AD 358 Jun 26 [China] (Ho 172)

(a) 2nd year of the Shengping reign period of Emperor Mu of the Jin Dynasty, 5th month, day *dinghai* [24]; a broom star emerged in TIANCHUAN within the space of WEI [LM 17].
[*Song shu: tianwen zhi*] ch. 24

(b) 2nd year of the Shengping reign period of Emperor Mu of the Jin Dynasty, 5th month, day *dinghai* [24]; a broom star emerged in TIANCHUAN in WEI [LM 17].
[*Jin shu: tianwen zhi*] ch. 13

AD 363 Aug 26–Sep 23 [China] (Ho 173)

(a) 1st year of the Xingning reign period of Emperor Ai of the Jin Dynasty, 8th month; a star became fuzzy in DAJIAO and KANG [LM 2] then entered TIANSHI.
[*Song shu: tianwen zhi*] ch. 24

(b) 1st year of the Xingning reign period of Emperor Ai of the Jin Dynasty, 8th month; a star became fuzzy in JIAO [LM 1] and KANG [LM 2] then entered TIANSHI.
[*Jin shu: Ai di ji*] ch. 8
[*Jin shu: tianwen zhi*] ch. 13

AD 374 Mar 4–Nov 19 [China] (Ho 175)

(a) 2nd year of the Ningkang reign period of Emperor Xiaowu of Jin, 1st month (should be 2nd month), day *dingsi* [54] (Mar 4); star appeared fuzzy in NÜ [LM 10] and XU [LM 11]. It passed DI [LM 3], KANG [LM 2], JIAO [LM 1], ZHEN [LM 28], YI [LM 27], and ZHANG [LM 26]. In the 3rd month, day *bingxu* [23] (Apr 2), a broom star was seen at DI [LM 3]. In the 9th month, day *dingchou* [14] (Nov 19), a star appeared fuzzy in TIANSHI.
[*Jin shu*] ch. 13

(b) 2nd year of the Ningkang reign period of Emperor Xiaowu of Jin, 2nd month, day *dingsi* [54] (Mar 4); a star appeared fuzzy in NÜ [LM 10] and XU [LM 11]...In the 3rd month, day *bingxu* [23] (Apr 2), a broom star was seen in DI [LM 3]...In the 9th

month, day *dingchou* [14] (Nov 19), a star appeared fuzzy in TIANSHI.
[*Jin shu*] ch. 9

(c) 2nd year of the Ningkang reign period of Emperor Xiaowu of the Jin Dynasty, 9th month (Nov 19), day *dingchou* [14]; a star became fuzzy in TIANSHI.
[*Song shu: tianwen zhi*] ch. 25
[*Jin shu: Xiaowu di ji*] ch. 9
[*Jin shu: tianwen zhi*] ch. 13

NB: This is Comet Halley, return –21. Ho noted that (c) may be a postperihelion observation of Comet Halley that year.

AD 390 Jul 28–Aug 26 [Korea] (Ho 178)
6th year of King Jinsa of Baekje, autumn, 7th month; there was a fuzzy star in BEIHE.
[*Jeungbo munheon bigo*] ch. 6

AD 390 Aug 7 [China] (Ho 178)
15th year of the Taiyuan reign period of Emperor Xiaowu of the Jin Dynasty, 7th month, day *dingsi* [54]; a star became fuzzy in BEIHE.
[*Jin shu: Xiaowu di ji*] ch. 9

AD 390 Aug 22 [China] (Ho 178)

(a) 15th year of the Taiyuan reign period of Emperor Xiaowu of the Jin Dynasty, 7th month, day *renshen* [9]; a star over 10 *zhang* long became fuzzy at the BEIHE garrison, passed TAIWEI, SANTAI, WENCHANG, and entered BEIDOU. In the 8th month, on day *wuxu* [35], it entered ZIWEI and then was extinguished.
[*Song shu: tianwen zhi*] ch. 25

(b) 15th year of the Taiyuan reign period of Emperor Xiaowu of the Jin Dynasty, 7th month, day *renshen* [9]; a white star over 10 *zhang* long became fuzzy at BEIHE garrison, passed TAIWEI, SANTAI, WENCHANG, and entered BEIDOU. In the 8th month, on day *wuxu* [35], it entered ZIGONG and then was extinguished.
[*Jin shu: tianwen zhi*] ch. 13

AD 395 Mar 8–Apr 5 [Korea]

4th year of King Asin of Baekje, spring, 2nd month; there was a fuzzy star in the northwest; after 20 days, it was extinguished.
[*Jeungbo munheon bigo*] ch. 6
[*Samguk sagi*] ch. 25 (Ho 180)

AD 395 Sep 30–Oct 29 [China] (Ho 181)

20th year of the Taiyuan reign period of Emperor Xiaowu of the Jin Dynasty, 9th month; a fuzzy star-like silk floss traveled southeastward, passed Nü [LM 10] and XU [LM 11], and reached KUXING.
[*Song shu: tianwen zhi*] ch. 25
[*Jin shu: tianwen zhi*] ch. 13
NB: *Jin shu: tianwen zhi* had "arrived at KUXING."

AD 396 Jul 22–Aug 19 [China] (Ho 182)

1st year of the Huangshi reign period of Emperor Taizu of the Northern Wei Dynasty, 6th month; a star became broom-like in MAOTOU [i.e., MAO (LM18)].
[*Wei shu: tianxiang zhi*] ch. 105

AD 400 Mar 12–Apr 9 [Korea]

9th year of King Asin of Baekje, spring, 2nd month; there was a fuzzy star in KUI [LM 15] and LOU [LM 16].
[*Jeungbo munheon bigo*] ch. 6
[*Samguk sagi*] ch. 25 (Ho 183)

AD 400 Mar 19 [China] (Ho 183)

(a) 4th year of the Long'an reign period of Emperor An of the Jin Dynasty, 2nd month, day *jichou* [26]; a star 3 *zhang* long became fuzzy in KUI [LM 15]. It ascended as far as GEDAO and the western enclosure of ZIGONG, entered the bowl of DOU, and reached SANTAI, TAIWEI, DIZUO, and DUANMEN.
[*Song shu: tianwen zhi*] ch. 25

(b) 4th year of the Long'an reign period of Emperor An of the Jin Dynasty, 2nd month, day *jichou* [26]; a star became fuzzy in KUI [LM 15] and LOU [LM 16], advancing as far as ZIWEI.
[*Jin shu: An di ji*] ch. 10

(c) 4th year of the Long'an reign period of Emperor An of the Jin Dynasty, 2nd month, day *jichou* [26]; a star 3 *zhang* long became fuzzy in KUI [LM 15]. It ascended as far as GEDAO and the western enclosure of ZIGONG, entered the bowl of BEIDOU, and reached SANTAI. In the 3rd month, it passed by TAIWEI, DIZUO, and DUANMEN.
[*Jin shu: tianwen zhi*] ch. 13

AD 400 Apr 10–May 9 [China] (Ho 183)
3rd year of the Tianxing reign period of Emperor Taizu of the Northern Wei Dynasty, 3rd month; a star became fuzzy in KUI [LM 15] then passed by GEDAO, reaching the western enclosure of ZIWEI. It entered the bowl of BEIDOU, trespassed on TAIYANGSHOU, rounded XIATAI, overran NANGONG, stepped on DIZUO, and then emerged via DUANMEN.
[*Wei shu: tianxiang zhi*] ch. 105

AD 400 Apr 10–May 9 [China] (Ho 183)
4th year of the Long'an reign period of Emperor An of the Jin Dynasty, 3rd month; a broom star appeared in TAIWEI.
[*Jin shu: An di ji*] ch. 10

AD 400 Sep 5–Oct 3 [Korea] (Ho 184)
45th year of King Naemul of Silla, autumn, 8th month; a fuzzy star was seen in the east.
[Samguk sagi] ch. 3

AD 401 Jan 2 [China] (Ho 185)

(a) 4th year of the Long'an reign period of Emperor An of the Jin Dynasty, 12th month, day *wuyin* [15]; a star became fuzzy in GUANSUO, TIANSHI, and TIANJIN.

[*Song shu: tianwen zhi*] ch. 25
[*Jin shu: tianwen zhi*] ch. 13

(b) 4th year of the Long'an reign period of Emperor An of the Jin Dynasty, 12th month, day *wuyin* [15]; a star became fuzzy in TIANSHI.
[*Jin shu: An di ji*] ch. 10

AD 402 Nov 11–Dec 10 [China] (Ho 186)

1st year of the Yuanxing reign period of Emperor An of the Jin Dynasty, 10th month; there was a white guest star like silk floss to the west of TAIWEI; by the 12th month, it entered TAIWEI.
[*Song shu: tianwen zhi*] ch. 25
[*Jin shu: tianwen zhi*] ch. 13

AD 415 Jun 21–24 [China] (Ho 188)

(a) 2nd year of the Shenrui reign period of Emperor Taizong of the Northern Wei Dynasty, 4th month, day *xinsi* [18] (Jun 21); a star became fuzzy in TIANSHI. In the 5th month, on day *jiashen* [21] (Jun 24), the broom star emerged at TIANSHI, swept DIZUO, and was located to the north of FANG [LM 4] and XIN [LM 5].
[*Wei shu: tianxiang zhi*] ch. 105

(b) 11th year of the Yixi reign period of Emperor An of the Jin Dynasty, 5th month, day *jiashen* [21] (Jun 24); a broom star emerged from TIANSHI, swept DIZUO, and was located by FANG [LM 4] and XIN [LM 5].
[*Song shu: tianwen zhi*] ch. 25

(c) 11th year of the Yixi reign period of Emperor An of the Jin Dynasty, 5th month, day *jiashen* [21] (Jun 24); a broom star emerged twice from TIANSHI, swept DIZUO, and was located north of FANG [LM 4] and XIN [LM 5].
[*Jin shu: tianwen zhi*] ch. 13

AD 415 Jun 24 [Korea] (Ho 188)

11th year of King Jeonji of Baekje, 5th month, day jiashen [21]; a broom star was seen.
[Samguk sagi] ch. 25

AD 415 Aug 8 [China] (Ho 187)

2nd year of the Shenrui reign period of Emperor Taizong of the Northern Wei Dynasty, 6th month, day *jisi* [6]; a star became fuzzy south of Mao [LM 18].

[*Wei shu: tianxiang zhi*] ch. 105

NB: Ho misdated this record to the 1st year of Shenrui, AD 414.

AD 418 Jun 24–Jul 6 [China] (Ho 191)

(a) 14th year of the Yixi reign period of Emperor An of the Jin Dynasty, 5th month, day *gengzi* [37] (Jun 24); a star became fuzzy in the bowl of Beidou. In the 7th month, on day *guihai* [60] (Sep 15), a broom star emerged west of Taiwei, its handle rising up from below the Shangjiang star. Its rays gradually grew to over 10 *zhang* in length. It proceeded to sweep Beidou, Ziwei, and Zhongtai.

[*Jin shu: tianwen zhi*] ch. 13

(b) 14th year of the Yixi reign period of Emperor An of the Jin Dynasty, 5th month, day *renzi* [49] (Jul 6); a star became fuzzy in the bowl of Beidou...In the 7th month...on day *guihai* [60] (Sep 15), a broom star emerged west of Taiwei, its handle rising up from below the Shangjiang star. Its rays gradually grew to over 10 *zhang* in length. It proceeded to sweep Beidou, Ziwei, and Zhongtai.

[*Song shu: tianwen zhi*] ch. 25

AD 418 Oct 16–Nov 13 [China]

3rd year of the Taichang reign period of Emperor Taizong of the Northern Wei Dynasty, 9th month; a long broom star became fuzzy in Beidou and overran Ziwei. On day *xinyou* [58] (Nov 22), it entered Nangong, in all (appearing) for over 80 days.

[*Wei shu: tianxiang zhi*] ch. 105

AD 419 Jan 12–Feb 10 [China] (Ho 192)

3rd year of the Taichang reign period of Emperor Taizong of the Northern Wei Dynasty, 12th month; a broom star emerged from Tian-jin, entered Taiwei, bisected Beidou, disrupted Zigong, trespassed on

TIANBANG, in all for over 80 days. When it reached the Milky Way, it was extinguished.
[*Wei shu: tianxiang zhi*] ch. 105
[*Wei shu: Cui Hao zhuan*] ch. 35
NB: Ho misdated this record to Dec 416–Jan 417.

AD 419 Feb 17 [China] (Ho 192)
1st year of the Yuanxi reign period of Emperor Gong of the Jin Dynasty, 1st month, day *wuxu* [35]; a star became fuzzy by the western fence of TAIWEI.
[*Jin shu: Gong di ji*] ch. 10
[*Jin shu: tianwen zhi*] ch. 13

AD 419 Feb 17 [Korea]
15th year of King Jeonji of Baekje, spring, 1st month, day *wuxu* [35]; there was a fuzzy star in TAIWEI.
[*Jeungbo munheon bigo*] ch. 6
[*Samguk sagi*] ch. 25 (Ho 192)

AD 422 Mar 16–26 [China] (Ho 195)
(a) 7th year of the Taichang reign period of Emperor Taizong of the Northern Wei Dynasty, 2nd month, day *xinsi* [18] (Mar 16); a star became fuzzy in XU [LM 11] and WEI [12], pointing toward HEJIN.
[*Wei shu: tianxiang zhi*] ch. 105
(b) 3rd year of the Yongchu reign period of Emperor Wu of Song of the Southern Dynasties, 2nd month, day *bingxu* [23] (Mar 21); a star became fuzzy in XU [LM 11] and WEI [12].
[*Nan shi: Song benji*] ch. 1
(c) 3rd year of the Yongchu reign period of Emperor Wu of Song of the Southern Dynasties, 2nd month, day *xinmao* [28] (Mar 26); a star became fuzzy in XU [LM 11] and WEI [12], pointed toward HEJIN, and swept HEGU.
[*Song shu: tianwen zhi*] ch. 26

AD 422 Dec 14 [China]

7th year of the Taichang reign period of Emperor Taizong of the Northern Wei Dynasty, 11th month, day *jiayin* [51]; a broom star emerged in Shi [LM 13], swept Beidou, and reached as far as the opening in the wall [of Ziwei].
[*Wei shu: tianxiang zhi*] ch. 105

AD 422 Dec 18 [China] (Ho 196)

(a) 3rd year of the Yongchu reign period of Emperor Wu of Song of the Southern Dynasties, 11th month, day *wuwu* [55]; a star became fuzzy in Yingshi [LM 13].
[*Song shu: Shao di ji*] ch. 4
[*Nan shi: Song benji*] ch. 1

(b) 3rd year of the Yongchu reign period of Emperor Wu of Song of the Southern Dynasties, 11th month, day *wuwu* [55]; a star became fuzzy in Shi [LM 13] and Bi [LM 14].
[*Song shu: tianwen zhi*] ch. 26

AD 423 Jan 28–Feb 25 [China]

8th year of the Taichang reign period of Emperor Taizong of the Northern Wei Dynasty, 1st month; a broom star 3 *zhang* long emerged south of Kui [LM 15] and swept He in the southeast.
[*Wei shu: tianxiang zhi*] ch. 105

AD 423 Feb 13 [China] (Ho 197)

(a) 1st year of the Jingping reign period of Emperor Shao of Song of the Southern Dynasties, 1st month, day *yimao* [52]; a star became fuzzy in Dongbi [LM 14].
[*Song shu: Shao di ji*] ch. 4
[*Nan shi: Song benji*] ch. 1

(b) 1st year of the Jingping reign period of Emperor Shao of Song of the Southern Dynasties, 1st month, day *yimao* [52]; a white star over 2 *zhang* long became fuzzy south of Dongbi [LM 14], brushed Tianyuan, and then was extinguished after 20 days.
[*Song shu: tianwen zhi*] ch. 26

AD 423 Dec 13–14 [China] (Ho 198)

(a) 1st year of the Jingping reign period of Emperor Shao of Song of the Southern Dynasties, 10th month, day *wuwu* [55] (Dec 13); a star became fuzzy north of DI [LM 3] with a tail 4 *zhang* long pointing to the northwest. It penetrated SHETI and headed toward DAJIAO traveling eastward. Each day it grew longer by 6 to 7 *chi*; after more than 10 days, it was extinguished.
[*Song shu: tianwen zhi*] ch. 26

(b) 1st year of the Jingping reign period of Emperor Shao of Song of the Southern Dynasties, 10th month, day *jiwei* [56] (Dec 14); a star became fuzzy in DI [LM 3], its tail pointing toward WEI [LM 6]; it penetrated SHETI heading toward DAJIAO. By the 11th month, it was in WEI [LM 12], and in the 12th month, it swept TIANCANG, after which it was extinguished.
[*Song shu: Shao di ji*] ch. 4

NB: Ho's date for this record contains a misprint; the date should be AD 423, not 473.

AD 423 Dec 19–AD 424 Jan 16 [China] (Ho 198)

8th year of the Taichang reign period of Emperor Taizong of the Northern Wei Dynasty, 11th month; a broom star became fuzzy at TUSIKONG.
[*Wei shu: tianxiang zhi*] ch. 105

AD 435 Jun 12–Jul 11 [China]

1st year of the Taiyan reign period of Emperor Taiwu of the Northern Wei Dynasty, 5th month; a broom star emerged in XUANYUAN.
[*Wei shu: tianxiang zhi*] ch. 105

AD 436 Jun 21 [China] (Ho 199)

2nd year of the Taiyan reign period of Emperor Taiwu of the Northern Wei Dynasty, 5th month, day *renshen* [9]; a star became fuzzy in FANG [LM 4].
[*Wei shu: tianxiang zhi*] ch. 105

AD 442 Oct 20–Nov 18 [China] (Ho 201)

(a) 19th year of the Yuanjia reign period of Emperor Wen of Song of the Southern Dynasties, 9th month, day *bingchen* [53] (Nov 1); there was a guest star in BEIDOU which then became a broom star. It entered WENCHANG, penetrated WUCHE and swept BI [LM 19], brushed TIANJIE, and passed TIANYUAN before being extinguished in the last month of winter.
[*Nan shi: Song benji*] ch. 2

(b) 3rd year of the Taiping zhenjun reign period of Emperor Taiwu of Northern Wei, 9th month, day *yichou* [2] (Nov 10); a star became fuzzy in TIANLAO that entered WENCHANG and WUCHE, passed between MAO [LM 18] and BI [LM 19], and reached TIANYUAN; for over 100 days it set in the west together with those lunar mansions.
[*Wei shu: tianxiang zhi*] ch. 105

(c) 19th year of the Yuanjia reign period of Emperor Wen of Song of the Southern Dynasties, 9th month (Oct 20–Nov 18); a guest star appeared in BEIDOU that gradually became a broom star. It reached the end of TIANYUAN where it was extinguished.
[*Song shu: tianwen zhi*] ch. 26

AD 449 Jun 7–Jul 5 [China] (Ho 202)

10th year of the Taiping zhenjun reign period of Emperor Taiwu of Northern Wei, 5th month; a broom star emerged north of MAO [LM 18].
[*Wei shu: tianxiang zhi*] ch. 105

AD 449 Nov 1–Nov 30 [China] (Ho 203)

(a) 26th year of the Yuanjia reign period of Emperor Wen of Song of the Southern Dynasties, day *guimao* [40] (Nov 11); a broom star appeared in TAIWEI.
[*Nan shi: Song benji*] ch. 2

(b) 26th year of the Yuanjia reign period of Emperor Wen of Song of the Southern Dynasties, 10th month (Nov 1–Nov 30); a broom star entered TAIWEI.
[*Song shu: tianwen zhi*] ch. 26

(c) 10th year of the Taiping zhenjun reign period of Emperor Taiwu of Northern Wei, 10th month, day *xinsi* [18]; a broom star appeared in TAIWEI.
[*Wei shu: tianxiang zhi*] ch. 105
NB: There was no day *xinsi* (18) in the 10th month.

AD 451 Jun 10–Jul 13 [China] (Ho 204)
(a) 28th year of the Yuanjia reign period of Emperor Wendi of Song, 4th month, day *jimao* [16] (Jun 10); a broom star was seen in MAO [LM 18]...In the 5th month, day *renzi* [49] (Jul 13), the broom star was seen within TAIWEI facing DIZUO.
[*Nan shi*] ch. 2
(b) 1st year of the Zhengping reign period of Emperor Taiwu of Wei, 5th month (Jun 15–Jul 14); a broom star was seen at QUANSHE, it entered TAIWEI. In the 6th month, day *xinyou* [58] (Jul 22), the broom star approached DIZUO. In the 7th month, day *yiyou* [22] (Aug 15), it trespassed against SHANGXIANG, swept PINGXING, and emerged (from TAIWEI) at DUANMEN. It was extinguished at YI [LM 27] and ZHEN [LM 28].
[*Wei shu*] ch. 105
(c) Emperor Wen, 28th year of the Yuanjia reign period, 5th month (Jun 15–Jul 14); a broom star was seen at QUANSHE, it entered TAIWEI, approached DIZUO, it trespassed against SHANGXIANG, swept PINGXING, and emerged at DUANMEN. It was extinguished at YI [LM 27] and ZHEN [LM 28].
[*Song shu*] ch. 26
NB: This is Comet Halley, return –20.

AD 453 Feb 24–Mar 25 [China] (Ho 205)
2nd year of the Xing'an reign period of Emperor Wencheng Gaozong of the Northern Wei Dynasty, 2nd month; a star became fuzzy in the west.
[*Wei shu: tianxiang zhi*] ch. 105

AD 454 Feb 14–AD 455 Feb 2 [Korea]
28th year of King Biyu of Baekje; there was a fuzzy star more than 2 *zhang* long in the northwest.

[*Jeungbo munheon bigo*] ch. 6
[*Samguk sagi*] ch. 25 (Ho 206)

AD 460 Oct 31–Nov 28 [China] (Ho 207)

1st year of the Heping reign period of Emperor Wencheng of the Northern Wei Dynasty, 10th month; a long star over 1 *zhang* long emerged together with TIANCANG.
[*Wei shu: tianxiang zhi*] ch. 105

AD 461 Apr 20 [China] (Ho 208)

2nd year of the Heping reign period of Emperor Wencheng of the Northern Wei Dynasty, 5th month, day *xinsi* [18]; a red long star as long as a bolt of cloth appeared at TIANJIN. It was extinguished and then reappeared, its size in the hundreds.
[*Wei shu: tianxiang zhi*] ch. 105

AD 464 Dec 15–AD 465 Jan 12 [China] (Ho 209)

5th year of the Heping reign period of Emperor Wencheng of the Northern Wei Dynasty, 11th month; a pure white long star emerged in ZHINÜ; it was the very image of a broom star.
[*Wei shu: tianxiang zhi*] ch. 105

AD 498 Nov 29–Dec 28 [China] (Ho 212)

22nd year of the Taihe reign period of Emperor Xiaowen of the Northern Wei Dynasty, 11th month; a broom star arose in XUANYUAN and passed south of GUI [LM 23] as far as the Milky Way.
[*Wei shu: tianxiang zhi*] ch. 105

AD 501 Feb 13–Apr 14 [China] (Ho 214)

1st year of the Zhongxing reign period of Emperor He of the Southern Qi Dynasty, 3rd month, day yisi [42]; on this night, a broom star stretched across the sky.
[Nan shi: Qi benji] ch. 5

AD 507 Aug 15 [China] (Ho 215)

4th year of the Zhengshi reign period of Emperor Xuanwu of the Northern Wei Dynasty, 7th month, day *jimao* [16]; a star became fuzzy in the northeast.

[*Wei shu: tianxiang zhi*] ch. 105

AD 520 Oct 1 [China] (Ho 216)

1st year of the Putong reign period of Emperor Wu of the Liang Dynasty, 9th month, day *yihai* [12]; a star appeared at dawn in the east, flaming brightly like fire.

[*Liang shu: Wu di ji*] ch. 3

[*Nan shi: Liang benji*] ch. 7

[*Sui shu: tianwen zhi*] ch. 21

AD 520 Oct 7 [China] (Ho 216)

1st year of the Zhengguang reign period of Emperor Xiaoming of the Northern Wei Dynasty, 9th month, day *xinsi* [18]; there was a broom star with bright flames like fire that emerged in the east.

[*Wei shu: tianxiang zhi*] ch. 105

NB: Possibly the same comet as the preceding.

AD 530 Aug 29–Sep 27 [China] (Ho 217)

3rd year of the Yong'an reign period of Emperor Xiaozhuang of Northern Wei, 7th month, day *jiawu* [31] (Aug 29); a broom star was seen in the morning in the northeast. It was 1 *zhang* to the east of ZHONGTAI. It was 6 *chi* long, and its color was pure white. It moved northeast but pointed toward the southwest. On day *dingyou* [34] (Sep 1), it was 1 *chi* to the northwest of the upper star of XIATAI then went out of sight in the morning. On day *gengzi* [37] (Sep 4), in the evening, it was seen in the northwest. It was 1 *chi* long and pointed to the southeast. It gradually moved and entered DI [LM 3]. In the 8th month, day *jiwei* [56] (Sep 23), it was barely seen. On day *guihai* [60] (Sep 27), it was extinguished.

[*Wei shu*] ch. 105

NB: This is Comet Halley, return –19.

AD 533 Mar 1 [China] (Ho 218)

5th year of the Zhongtong reign period of Emperor Wu of the Liang Dynasty, 1st month, day *jiyou* [46]; a long star appeared.
[*Liang shu: Wu di ji*] ch. 3
[*Sui shu: tianwen zhi*] ch. 21

AD 535 Feb 18–AD 536 Feb 7 [China] (Ho 219)

2nd year of the Tianping reign period of Emperor Xiaojing of the Eastern Wei Dynasty; a star became fuzzy in TAIWEI; it passed by XIATAI and reached as far as SHI [LM 13] and BI [LM 14] where it was extinguished.
[*Wei shu: tianxiang zhi*] ch. 105

AD 539 Nov 17 [China] (Ho 221)

(a) 1st year of the Xinghe reign period of Emperor Xiaojing of the Eastern Wei Dynasty, 10th month, day *xinchou* [38]; a broom star over 1 *zhang* long emerged from NANDOU [LM 8]. By the 11th month, day *bingxu* [23], it was 3 *chi* from Venus, over 1 *zhang* long, and pointing southeast. Not until the 2nd month, day *yimao* [52], when it reached LOU [LM 16], was it extinguished.
[*Wei shu: tianxiang zhi*] ch. 105

(b) 5th year of the Datong reign period of Emperor Wu of the Liang Dynasty, 10th month, day *xinchou* [38]; a broom star over 1 *chi* long emerged from NANDOU [LM 8], pointing southeast, then gradually grew to over 1 *zhang* in length. By the 11th month, day *yimao* [52], it had reached LOU [LM 16] where it was extinguished.
[*Sui shu: tianwen zhi*] ch. 21

AD 560 Oct 9 [China] (Ho 223)

(a) 1st year of the Tianjia reign period of Emperor Wen of the Chen Dynasty, 9th month, day *guichou* [50]; a broom star appeared.
[*Chen shu: Shi zu ji*] ch. 3
[*Nan shi: Chen benji*] ch. 9

(b) 1st year of the Tianjia reign period of Emperor Wen of the Chen Dynasty, 9th month, day *guichou* [50]; a broom star 4 *chi* long appeared, its rays pointing southwest.
[*Sui shu: tianwen zhi*] ch. 21

AD 565 Jul 22–24 [China] (Ho 225–226)

(a) 5th year of the Baoding reign period of Emperor Wu of the Northern Zhou Dynasty, 6th month, day *gengshen* [57] (Jul 22); a broom star emerged in SANTAI and entered WENCHANG. It trespassed on SHANGJIANG and, after that, passed the western wall of ZIGONG, entered WEI [LM 12], gradually lengthening to over 1 *zhang* and pointing toward SHI [LM 13] and BI [LM 14]. After over 100 days, it shortened somewhat to 2 *chi* 5 *cun* in length and was extinguished in XU [LM 11] and WEI [LM 12].
[*Zhou shu: Wu di ji*] ch. 5
[*Sui shu: tianwen zhi*] ch. 21

(b) 5th year of the Baoding reign period of Emperor Wu of the Northern Zhou Dynasty, 6th month, day *gengshen* [57] (Jul 22); a broom star emerged in SANTAI and entered WENCHANG. It trespassed on SHANGJIANG and then passed ZIGONG and entered WEI [LM 12], gradually lengthening to over 1 *zhang*. After over 100 days, it was extinguished.
[*Bei shi: Zhou ben ji*] ch. 10

(c) 6th year of the Tianjia reign period of Emperor Wen of the Chen Dynasty, 6th month, day *xinyou* [58] (Jul 23); a broom star appeared north of SHANGTAI.
[*Nan shi: Chen ben ji*] ch. 9

(d) 1st year of the Tiantong reign period of Emperor Houzhu of the Northern Qi Dynasty, 6th month, day *renxu* [59] (Jul 24); a broom star as large as a hand emerged northeast of WENCHANG, afterward gradually lengthening until it reached over 1 *zhang* in length; then after 100 days, it was extinguished.
[*Bei Qi shu: di ji*] ch. 8
[*Bei shi: Qi ben ji*] ch. 8

(e) 1st year of the Tiantong reign period of Emperor Houzhu of the Northern Qi Dynasty, 6th month, day *renxu* [59] (Jul 24); a broom star several *cun* long appeared at WENCHANG. It entered WENCHANG, trespassed on SHANGJIANG, and afterward went through the western wall of ZIGONG and entered WEI [LM 12], gradually lengthening to over 1 *zhang* and pointed toward SHI [LM 13] and BI [LM 14]. After over 100 days, it was extinguished in XU [LM 11] and WEI [LM 12].
[*Sui shu: tianwen zhi*] ch. 21

AD 568 Jul 20 [China] (Ho 227)

(a) 3rd year of the Tianhe reign period of Emperor Wu of the Northern Zhou Dynasty, 6th month, day *jiaxu* [11]; a star became fuzzy in DONGJING [LM 22] that traveled northward for a month, reaching YUGUI [LM 23] whereupon it was extinguished.
[*Zhou shu: Wu di ji*] ch. 5

(b) 3rd year of the Tianhe reign period of Emperor Wu of the Northern Zhou Dynasty, 6th month, day *jiaxu* [11]; a star became fuzzy in DONGJING [LM 22].
[*Bei shi: Zhou ben ji*] ch. 10

(c) 3rd year of the Tianhe reign period of Emperor Wu of the Northern Zhou Dynasty, 6th month, day *jiaxu* [11]; a fuzzy star 1 *zhang* long appeared in DONGJING [LM 22]. On top, it was white, and below it was red and pointed; it gradually traveled eastward until the 7th month, day *guimao* [40], when it was extinguished about 8 *cun* north of GUI [LM 23].
[*Sui shu: tianwen zhi*] ch. 21

AD 568 Aug 9–Sep 6 [China] (Ho 228)

(a) 3rd year of the Tianhe reign period of Emperor Wu of the Northern Zhou Dynasty, 7th month, day *jiwei* [56] (Sep 3); a guest star appeared in FANG [LM 4]. It gradually traveled eastward and entered TIANSHI, trespassed on YINGSHI [LM 13], and reached KUI [LM 15]; then after over 40 days, it was extinguished.
[*Zhou shu: Wu di ji*] ch. 5

(b) 3rd year of the Tianhe reign period of Emperor Wu of the Northern Zhou Dynasty, 7th month, day *jiwei* [56] (Sep 3); a guest star as white as silk floss and as big as a peck measure appeared in FANG [LM 4] and XIN [LM 5], growing gradually larger as it traveled eastward. By the 8th month, it entered TIANSHI and was about as long as a bolt of cloth. It again traveled eastward where it trespassed on HEGU and YOUJIANG. On day *guiwei* [20], it trespassed on HUGUA, entered SHI [LM 13], and trespassed on LIGONG. In the 9th month, on day *renyin* [39], it entered KUI [LM 15] and shrank. On day *renxu* [59], it arrived at a point about 1 *chi* north of LOU [LM 16] where it was extinguished after 69 days in all.
[*Sui shu: tianwen zhi*] ch. 21

(c) 4th year of the Tiantong reign period of Emperor Houzhu of the Northern Qi, 7th month (Aug 8–Sep 6); a fuzzy star as white as silk floss and as big as a peck measure appeared in FANG [LM 4] and XIN [LM 5], travelling eastward. By the 8th month, it entered TIANSHI and gradually lengthened to about 4 *zhang*, then trespassed on HUGUA, passed through XU [LM 11] and WEI [LM 12], entered SHI [LM 13], and trespassed on LIGONG. In the 9th month, it entered KUI [LM 15], reached LOU [LM 16], and was extinguished.
[*Sui shu: tianwen zhi*] ch. 21

AD 574 Apr 4 [China] (Ho 229)

3rd year of the Jiande reign period of Emperor Wu of the Northern Zhou Dynasty, 2nd month, day *wuwu* [55]; a guest star as big as a peach and greenish-white in color emerged about 3 *chi* southeast of WUCHE. It gradually traveled eastward and lengthened slowly to about 2 *chi*. By the 4th month, day *renchen* [29], it entered WENCHANG. By day *dingwei* [44], it entered the bowl of BEIDOU, afterward emerging from the bowl and gradually growing smaller; in all, it was visible for 93 days.
[*Sui shu: tianwen zhi*] ch. 21

AD 574 May 31–Jun 2 [China] (Ho 230)

(a) 3rd year of the Jiande reign period of Emperor Wu of the Northern Zhou Dynasty, 4th month, day *yimao* [52] (May 31); a reddish-white star as large as a fist became fuzzy outside the wall of ZIGONG. It pointed toward WUDIZUO and gradually traveled south-eastward, lengthening slowly to 1 *zhang* 5 *chi*. By the 5th month, day *jiazi* [1], it arrived north of SHANGTAI and was extinguished.
[*Sui shu: tianwen zhi*] ch. 21

(b) 3rd year of the Jiande reign period of Emperor Wu of the Northern Zhou Dynasty, 4th month, day *dingsi* [54] (Jun 2); a star became fuzzy in the northeast at the wall of ZIGONG. It was 7 *chi* long.
[*Zhou shu: Wu di ji*] ch. 5

(c) 3rd year of the Jiande reign period of Emperor Wu of the Northern Zhou Dynasty, 4th month, day *dingsi* [54] (Jun 2); a star became fuzzy in DONGJING [LM 22].
[*Bei shi: Zhou ben ji*] ch. 10

AD 575 Apr 27 [China] (Ho 231)

7th year of the Taijian reign period of Emperor Xuan of the Chen Dynasty, 4th month, day *bingxu* [23]; a star became fuzzy at DAJIAO.
[*Chen shu: Xuan di ji*] ch. 5
[*Nan shi: Chen ben ji*] ch. 10
[*Sui shu: tianwen zhi*] ch. 21

AD 579 Nov 5–Dec 3 [Korea] (Ho 232)

26th year of King Wideok of Baekje, winter, 10th month; a long star spanned the entire sky; after 20 days, it was extinguished.
[*Jeungbo munheon bigo*] ch. 6
[*Samguk sagi*] ch. 27

AD 588 Nov 22 [China] (Ho 235)

8th year of the Kaihuang reign period of Emperor Wen of the Sui Dynasty, 10th month, day *jiazi* [1]; a star became fuzzy in QIANNIU [LM 9].

[*Bei shi: Sui ben ji*] ch. 11
[*Sui shu: Gaozu ji*] ch. 2

AD 595 Jan 9 [China] (Ho 236)

(a) 14th year of the Kaihuang reign period of Emperor Wen of the Sui Dynasty, 11th month, day *guiwei* [20]; a star became fuzzy in JIAO [LM 1] and KANG [LM 2].
[*Bei shi: Sui ben ji*] ch. 11
[*Sui shu: Gaozu ji*] ch. 2

(b) 14th year of the Kaihuang reign period of Emperor Wen of the Sui Dynasty, 11th month, day *guiwei* [20]; there was a broom star that became fuzzy in XU [LM 11] and WEI [LM 12] and then reached KUI [LM 15] and LOU [LM 16].
[*Sui shu: tianwen zhi*] ch. 21

AD 595 Jan 9 [Korea]

41st year of King Wideok of Baekje, winter, 11th month, day *guiwei* [20]; there was a fuzzy star in JIAO [LM 1] and KANG [LM 2].
[*Jeungbo munheon bigo*] ch. 6
[*Samguk sagi*] ch. 27 (Ho 236)

AD 608 Jan 23 [China] (Ho 237)

(a) 3rd year of the Daye reign period of Emperor Yang of Sui, 3rd month, day *xinhai* [48] (Apr 4); a long star was seen in the west. It stretched across the sky. It trespassed against and passed KUI [LM 15], LOU [LM 16], JIAO [LM 1], and KANG [LM 2], then it disappeared. In the 9th month, day *xinwei* [8] (Oct 21), it returned and was seen in the south. It again extended across the sky from JIAO [LM 1] and KANG [LM 2]. It swept DIZUO of TAIWEI. It trespassed against all of the other lunar mansions except SHEN [LM 21] and DONGJING [LM 22]. After the new year (AD 608 Jan 23), it was extinguished.
[*Sui shu*] ch. 21

(b) 3rd year of the Daye reign period of Emperor Yang of Sui, spring, 1st month, day *bingzi* [13] (Feb 28); a long star extended across the sky. It appeared at DONGBI [LM 14]. After 20 days, it ceased… In the 2nd month, day *jichou* [26] (Mar 13), a broom star was seen at DONGJING [LM 22] and WENCHANG. It passed DALING, WUCHE, and BEIHE. It entered TAIWEI and swept DIZUO. It ceased after more than 100 days.
[*Bei shi*] ch. 12

NB: This is Comet Halley, return –18.

AD 607 Jun 25 [China] (Ho 238)

3rd year of the Daye reign period of Emperor Yang of the Sui Dynasty, 5th month, day *guiyou* [10]; a star became fuzzy in WENCHANG and SHANGJIANG.
[*Bei shi: Sui ben ji*] ch. 12

AD 608 Oct 22 [China] (Ho 239)

4th year of the Daye reign period of Emperor Yang of the Sui Dynasty, 9th month, day *wuyin* [15]; a broom star emerged at WUCHE, swept WENCHANG, and reached FANG [LM 4] where it was extinguished.
[*Bei shi: Sui ben ji*] ch. 12
[*Sui shu: Yang di ji*] ch. 3

AD 615 Jul 2–30 [China] (Ho 240)

11th year of the Daye reign period of Emperor Yang of the Sui Dynasty, 6th month; there was a black and pointed fuzzy star 5 to 6 *cun* long southeast of WENCHANG. During the night, it wavered then traveled northwestward, after several days reaching WENCHANG at a distance of 4 to 5 *cun* from [ZI]GONG, which it did not enter but moved on and was extinguished.
[*Sui shu: tianwen zhi*] ch. 21

AD 617 Jul 9–Aug 6 [China] (Ho 241)

13th year of the Daye reign period of Emperor Yang of the Sui Dynasty, 6th month; a star became fuzzy at WUDIZUO of TAIWEI. It was

yellowish-red and about 3 to 4 *chi* long; after several days, it was extinguished.
[*Sui shu: tianwen zhi*] ch. 21

AD 617 Oct 6–Nov 3 [China] (Ho 242)
13th year of the Daye reign period of Emperor Yang of the Sui Dynasty, 9th month; a broom star appeared in YINGSHI [LM 13].
[*Sui shu: Yang di ji*] ch. 4
[*Bei shi: Sui ben ji*] ch. 12

AD 626 Mar 26 [China] (Ho 243)
(a) 9th year of the Wude reign period of Emperor Gaozu of the Tang Dynasty, 2nd month, 23rd day, at night; a star became fuzzy between WEI [LM 17] and MAO [LM 18], in all for 28 days, then it became fuzzy again in JUANSHE.
[*Jiu Tang shu: tianwen zhi*] ch. 36
NB: *Jiao kan ji* said *fan*, "in all," should be *zhi*, "until," making the translation "through the 28th day."
(b) 9th year of the Wude reign period of Emperor Gaozu of the Tang Dynasty, 2nd month, day *renwu* [19]; a star became fuzzy in WEI [LM 17] and MAO [LM 18]. On day *dinghai* [24], it became fuzzy in JUANSHE.
[*Xin Tang shu: Gaozu ji*] ch. 1
(c) 9th year of the Wude reign period of Emperor Gaozu of the Tang Dynasty, 2nd month, day *renwu* [19]; a star became fuzzy between WEI [LM 17] and MAO [LM 18]. On day *dinghai* [24], it became fuzzy in JUANSHE.
[*Xin Tang shu: tianwen zhi*] ch. 32

AD 634 Aug 29–Sep 27 [Japan] (Ho 244)
6th year of Emperor Jomei; a long star was seen in the south; people at the time said it was a broom star.
[*Nihon shoki*]; [*Ichidai yōki*]

AD 634 Sep 20–22 [China] (Ho 244)

(a) 8th year of the Zhen'guan reign period of Emperor Taizong of the Tang Dynasty, 8th month, 23rd day (Sep 20); a star became fuzzy in XU [LM 11] and WEI [LM 12] then passed through XUANXIAO [LM 10–12]. It lasted 11 days in all and then was extinguished.
[*Jiu Tang shu: tianwen zhi*] ch. 36

(b) 8th year of the Zhen'guan reign period of Emperor Taizong of the Tang Dynasty, 8th month, day *jiazi* [1] (Sep 22); a star became fuzzy in XU [LM 11] and WEI [LM 12] then passed through DI [LM 3]; by the first 10-day week of the 11th month, it was extinguished.
[*Jiu Tang shu: Taizong ji*] ch. 3

(c) 8th year of the Zhen'guan reign period of Emperor Taizong of the Tang Dynasty, 8th month, day *jiazi* [1] (Sep 22); a star became fuzzy in XU [LM 11] and WEI [LM 12].
[*Xin Tang shu: Taizong ji*] ch. 2

(d) 8th year of the Zhen'guan reign period of Emperor Taizong of the Tang Dynasty, 8th month, day *jiazi* [1] (Sep 22); a star became fuzzy in XU [LM 11] and WEI [LM 12] and passed through XUANXIAO [LM 10–12]; on day *yihai* [12], it no longer appeared.
[*Xin Tang shu: tianwen zhi*] ch. 32

AD 635 Jan 24–Feb 22 [Japan] (Ho 244)

7th year of Emperor Jomei, 1st month; the broom star reappeared in the east.
[*Nihon shoki*] ch. 23; [*Ichidai yōki*] *kō*
[*Nihon kiryaku*] *zen*
NB: *Nihon kiryaku mae* and *Ichidai yōki* both made it the 3rd month.

AD 639 Jan 29–Feb 26 [Korea] (Ho 245)

41st year of King Mu of Baekje, 1st month; a fuzzy star was seen in the northwest.
[Samguk sagi] ch. 27

AD 639 Mar 5 [Japan] (Ho 245)

(a) 11th year of Emperor Jomei, 1st month, day *jisi* [6]; a long star was seen in the northwest, and at the time Master Min said, "It is a broom star."
[*Nihon shoki*]

(b) 11th year of Emperor Jomei, 1st month; a long star was seen in the northwest.
[*Fusō ryakuki*]

AD 639 Apr 30 [China] (Ho 245)

(a) 13th year of the Zhen'guan reign period of Emperor Taizong of the Tang Dynasty, 3rd month, day *yichou* [2]; a star became fuzzy in BI [LM 19] and MAO [LM 18].
[*Jiu Tang shu: Taizong ji*] ch. 3
[*Xin Tang shu: Taizong ji*] ch. 2
[*Xin Tang shu: tianwen zhi*] ch. 32

(b) 13th year of the Zhen'guan reign period of Emperor Taizong of the Tang Dynasty, 3rd month, 22nd day, at night; a star became fuzzy in BI [LM 19] and MAO [LM 18].
[*Jiu Tang shu: tianwen zhi*] ch. 36

AD 641 Aug 1 [China] (Ho 246)

(a) 15th year of the Zhen'guan reign period of Emperor Taizong of the Tang Dynasty, 6th month, day *jiyou* [46]; a star became fuzzy in TAIWEI, then trespassed against LANGWEI…In the autumn, 7th month, on day *jiaxu* [11], the fuzzy star was extinguished.
[*Jiu Tang shu: Taizong ji*] ch. 3

(b) 15th year of the Zhen'guan reign period of Emperor Taizong of the Tang Dynasty, 6th month, day *jiyou* [46]; a star became fuzzy in TAIWEI.
[*Xin Tang shu: Taizong ji*] ch. 2

(c) 15th year of the Zhen'guan reign period of Emperor Taizong of the Tang Dynasty, 6th month, day *jiyou* [46]; a star became fuzzy

in TAIWEI that trespassed against LANGWEI; in the 7th month, on day *jiaxu* [11], it was not seen.
[*Xin Tang shu: tianwen zhi*] ch. 32

(d) 15th year of the Zhen'guan reign period of Emperor Taizong of the Tang Dynasty, 6th month, 19th day; a star became fuzzy in TAIWEI that trespassed against LANGWEI; in 7th month, on day *jiaxu* [11], it was extinguished.
[*Jiu Tang shu: tianwen zhi*] ch. 36

AD 647 Sep 5–Oct 3 [Korea] (Ho 248)
1st year of Queen Jindoek of Silla, 8th month; a broom star was seen in the south; many stars drifted northward.
[Samguk sagi] ch. 5

AD 663 Sep 29 [China] (Ho 250)

(a) 3rd year of the Longshuo reign period of Emperor Gaozong of the Tang Dynasty, 8th month, day *guimao* [40]; a broom star appeared at ZUOSHETI.
[*Jiu Tang shu: Gaozong ji*] ch. 4
[*Xin Tang shu: Gaozong ji*] ch. 3

(b) 3rd year of the Longshuo reign period of Emperor Gaozong of the Tang Dynasty, 8th month, day *guimao* [40]; there was a broom star over 2 *chi* long at ZUOSHETI; on day *yisi* [42], it did not appear.
[*Xin Tang shu: tianwen zhi*] ch. 32

AD 667 May 24 [China] (Ho 251)
2nd year of the Qianfeng reign period of Emperor Gaozong of the Tang Dynasty, 4th month, day *bingchen* [53]; there was a broom star in the northeast, between WUCHE, BI [LM 19], and MAO [LM 18]; by day *yihai* [12], it did not appear.
[*Xin Tang shu: tianwen zhi*] ch. 32
NB: There was no day *yihai* [12] in the 4th month of the 2nd year.

AD 668 May 17–Jun 14 [China] (Ho 252)

(a) 1st year of the Zongzhang reign period of Emperor Gaozong of the Tang Dynasty, 4th month, day *bingchen* [53] (May 18); a broom star appeared between BI [LM 19] and MAO [LM 18]… on day *yihai* [12], the broom star was extinguished.
[*Jiu Tang shu: Gaozong ji*] ch. 5

(b) 1st year of the Zongzhang reign period of Emperor Gaozong of the Tang Dynasty, 4th month, day *bingchen* [53] (May 18); a broom star emerged at WUCHE.
[*Xin Tang shu: Gaozong ji*] ch. 3

(c) 1st year of the Zongzhang reign period of Emperor Gaozong of the Tang Dynasty, 4th month (May 17–Jun 14); a broom star appeared at WUCHE…the star subsequently became fuzzy and its bright rays smaller…the star became fuzzy in the northeast…on the 22nd day, the star was extinguished.
[*Jiu Tang shu: tianwen zhi*] ch. 36

AD 668 May 17–Jun 14 [Korea]

(a) 8th year of King Munmu of Silla, 4th month; a broom star guarded TIANCHUAN.
[*Samguk sagi*] ch. 6 (Ho 252)
[*Jeungbo munheon bigo*] ch. 6

(b) 27th year of King Bojang of Goguryeo, summer, 4th month; a broom star was seen between BI [LM 19] and MAO [LM 18].
[*Jeungbo munheon bigo*] ch. 6

AD 672 Sep 27–Oct 25 [Korea]

12th year of King Munmu of Silla, autumn, 9th month; a broom star emerged seven times in the north.
[*Samguk sagi*] ch. 7 (Ho 253)
[*Jeungbo munheon bigo*] ch. 6

AD 675 Oct 24–Nov 22 [China] (Ho 254)

(a) 2nd year of the Shangyuan reign period of Emperor Gaozong of the Tang Dynasty, 10th month, day *renwu* [19] (Nov 4); a star

became fuzzy south of JIAO [LM 1] and KANG [LM 2]. It was 5 *chi* long.
[*Jiu Tang shu: Gaozong ji*] ch. 5

(b) 2nd year of the Shangyuan reign period of Emperor Gaozong of the Tang Dynasty, 10th month, day *renwu* [19] (Nov 4); a broom star emerged at JIAO [LM 1] and KANG [LM 2].
[*Xin Tang shu: Gaozong ji*] ch. 3

(c) 2nd year of the Shangyuan reign period of Emperor Gaozong of the Tang Dynasty, 10th month (Oct 24–Nov 22); a broom star 5 *chi* long appeared south of JIAO [LM 1] and KANG [LM 2].
[*Jiu Tang shu: tianwen zhi*] ch. 36

AD 676 Aug 15–Sep 12 [Japan]

(a) 4th year of Emperor Temmu, 7th month; in this month, a star 7 to 8 *chi* long was seen in the east; by the 9th month, it stretched across the sky.
[*Nihon shoki*]
[*Dai Nihon shi*] (Ho 255)

(b) 4th year of Emperor Temmu, 7th month; a star 8 *chi* long was seen in the east.
[*Fusō ryakuki*]

AD 676 Aug 15–Sep 12 [Korea]

16th year of King Munmu of Silla, autumn, 7th month; a broom star over 6 to 7 *bu* (paces) long emerged between BEIHE and JISHUI.
[*Jeungbo munheon bigo*] ch. 6
[*Samguk sagi*] ch. 7 (Ho 255)

AD 676 Sep 4 [China] (Ho 255)

(a) 3rd year of the Shangyuan reign period of Emperor Gaozong of the Tang Dynasty, 7th month, 21st day; a broom star over 3 *chi* long appeared in DONGJING [LM 22], pointing toward NANHE and JIXIN. It gradually traveled northeastward, its bright rays grew more sinuous and 3 *zhang* in length; [they] swept ZHONGTAI

and pointed at WENCHANG. It lasted 58 days before being extinguished.
[*Jiu Tang shu: tianwen zhi*] ch. 36

(b) 3rd year of the Shangyuan reign period of Emperor Gaozong of the Tang Dynasty, 7th month, day *dinghai* [24]; there was a broom star over 3 *chi* long in DONGJING [LM 22], pointing toward BEIHE. It traveled northeastward, its bright rays increased in luxuriance, and 3 *zhang* in length; it swept ZHONGTAI and pointed at WENCHANG. By the 9th month, day *yiyou* [22], it was no longer seen.
[*Xin Tang shu: tianwen zhi*] ch. 32

AD 681 Oct 17 [China] (Ho 256)

(a) 2nd year of the Yonglong reign period of Emperor Gaozong of the Tang Dynasty (1st year of the Kaiyao reign period), 9th month, day *bingshen* [33] (1st day); a broom star 5 *chi* long appeared in TIANSHI.
[*Jiu Tang shu: Gaozong ji*] ch. 5

(b) 2nd year of the Yonglong reign period of Emperor Gaozong of the Tang Dynasty (1st year of the Kaiyao reign period), 9th month, day *bingshen* [33] (1st day), at night; a broom star 5 *chi* long appeared in the west in TIANSHI. It gradually grew smaller and traveled eastward, emerging from TIANSHI and reaching the YOUQI of HEGU; after 17 days, it was extinguished.
[*Jiu Tang shu: tianwen zhi*] ch. 36

(c) 2nd year of the Yonglong reign period of Emperor Gaozong of the Tang Dynasty (1st year of the Kaiyao reign period), 9th month, day *bingshen* [33] (1st day); a broom star emerged in TIANSHI.
[*Xin Tang shu: Gaozong ji*] ch. 3

(d) 2nd year of the Yonglong reign period of Emperor Gaozong of the Tang Dynasty (1st year of the Kaiyao reign period), 9th month, day *bingshen* [33] (1st day); there was a broom star 5 *chi* long within TIANSHI that gradually grew smaller and

traveled eastward as far as HEGU; by day *guichou* [50], it was extinguished.
[*Xin Tang shu: tianwen zhi*] ch. 32

AD 681 Nov 2 [Japan] (Ho 256)
9th year of Emperor Temmu, 9th month, 16th day; a broom star was seen.
[Dai Nihon shi]; Kanda

AD 683 Apr 20 [China] (Ho 257)
(a) 2nd year of the Yongchun reign period of Emperor Gaozong of the Tang Dynasty (1st year of the Hongdao reign period), 3rd month, day *bingwu* [43] (18th day); a broom star appeared north of WUCHE; after 25 days, it was extinguished.
[*Jiu Tang shu: Gaozong ji*] ch. 5
(b) 2nd year of the Yongchun reign period of Emperor Gaozong of the Tang Dynasty (1st year of the Hongdao reign period), 3rd month, day *bingwu* [43] (18th day); a broom star appeared north of WUCHE; after 25 days in all, it was extinguished.
[*Jiu Tang shu: tianwen zhi*] ch. 36
(c) 2nd year of the Yongchun reign period of Emperor Gaozong of the Tang Dynasty (1st year of the Hongdao reign period), 3rd month, day *bingwu* [43] (18th day); there was a broom star north of WUCHE; by the 4th month, day *xinwei* [8], it was no longer seen.
[*Xin Tang shu: tianwen zhi*] ch. 32
(d) 2nd year of the Yongchun reign period of Emperor Gaozong of the Tang Dynasty (1st year of the Hongdao reign period), 3rd month, day *bingwu* [43] (18th day); a broom star emerged at WUCHE.
[*Xin Tang shu: Gaozong ji*] ch. 3

AD 683 Oct 25–Nov 23 [Korea]
3rd year of King Sinmun of Silla, 10th month; a broom star emerged in WUCHE.

[*Samguk sagi*] ch. 8 (Ho 257)
[*Jeungbo munheon bigo*] ch. 6

AD 684 Sep 6–Oct 9 [China] (Ho 258)

(a) 1st year of the Wenming reign period of Empress Wu-ze-tian of Tang, 7th month, 22nd day (*xinwei* [8]) (Sep 6); in the evening, there was a broom star in the west. It was more than 1 *zhang* long. In the 8th month, on day *jiachen* [41] (Oct 9), it was not seen…in the 1st year of the Guangzhai reign period, 9th month, day *dingchou* [14] (Nov 11)…a star like a half moon was seen in the west.
[*Xin Tang shu*] ch. 32

(b) 1st year of the Wenming reign period of Empress Wu-ze-tian of Tang, 7th month, 22nd day (*xinwei* [8]) (Sep 6); there was a broom star in the northwest. It was more than 1 *zhang* long. It lasted for a total of more than 42 days, then it disappeared.
[*Tang huiyao*] ch. 43

NB: This and the following are Comet Halley, return –17.

AD 684 Sep 7; AD 684 Dec 12–AD 685 Jan 10 [Japan]

(a) 12th year of Emperor Temmu, 11th month; in this month, there was a fuzzy star in the center of the sky that accompanied MAO [LM 18] and moved with it; by the end of the month, it was gone.
[*Nihon shoki*] 29

(b) 12th year of Emperor Temmu, 7th month, 23rd day, *renshen* [9] (Sep 7); a broom star over 10 *zhang* long emerged in the west.
[*Ichidai yōki*]

(c) 12th year of Emperor Temmu, 7th month, 23rd day (Sep 7); a broom star over 10 *zhang* long was seen. During the 11th month (Dec 12–Jan 10), there was a fuzzy star in the zenith that kept company with MAO [LM 18] and moved with it; by the end of the month, it was gone.
[*Dai Nihon shi*] (Ho 258)

AD 699 Mar 7–Apr 5 [Korea]

8th year of King Hyoso of Silla, spring, 2nd month; a white vapor spanned the entire sky, and there was a fuzzy star in the east.

[*Jeungbo munheon bigo*] ch. 6
[*Samguk sagi*] ch. 8 (Ho 259)

AD 701 Mar 14–Apr 12 [Korea]
10th year of King Hyoso of Silla, spring, 2nd month; a broom star entered the Moon.
[*Samguk sagi*] ch. 8 (Ho 260)
[*Jeungbo munheon bigo*] ch. 6

AD 707 Nov 16 [China] (Ho 261)
(a) 1st year of the Jinglong reign period of Emperor Zhongzong of the Tang Dynasty, 10th month, day *renwu* [19] (18th day); a broom star appeared at azimuth *you* (W); after more than a month, it was extinguished.
[*Jiu Tang shu: Zhongzong ji*] ch. 7
(b) 1st year of the Jinglong reign period of Emperor Zhongzong of the Tang Dynasty, 10th month, day *renwu* [19] (18th day); a broom star appeared in the west; after 43 days in all, it was extinguished.
[*Jiu Tang shu: tianwen zhi*] ch. 36
(c) 1st year of the Jinglong reign period of Emperor Zhongzong of the Tang Dynasty, 10th month, day *renwu* [19] (18th day); a broom star emerged in the west.
[*Xin Tang shu: Zhongzong ji*] ch. 4
(d) 1st year of the Jinglong reign period of Emperor Zhongzong of the Tang Dynasty, 10th month, day *renwu* [19] (18th day); there was a broom star in the west; by the 11th month, day *jiayin* [51], it was not seen.
[*Xin Tang shu: tianwen zhi*] ch. 32

AD 708 Jul 28 [China] (Ho 262)
(a) 2nd year of the Jinglong reign period of Emperor Zhongzong of the Tang Dynasty, 7th month, 7th day (*dingyou* [34]); a star became fuzzy between Wei [LM 17] and Mao [LM 18].
[*Jiu Tang shu: tianwen zhi*] ch. 36

(b) 2nd year of the Jinglong reign period of Emperor Zhongzong of the Tang Dynasty, 7th month, 7th day (*dingyou* [34]); a star became fuzzy between WEI [LM 17] and MAO [LM 18].
[*Xin Tang shu: Zhongzong ji*] ch. 4

(c) 2nd year of the Jinglong reign period of Emperor Zhongzong of the Tang Dynasty, 7th month, 7th day (*dingyou* [34]); a star became fuzzy between WEI [LM 17] and MAO [LM 18].
[*Xin Tang shu: tianwen zhi*] ch. 32

AD 709 Sep 16 [China] (Ho 263)

(a) 3rd year of the Jinglong reign period of Emperor Zhongzong of the Tang Dynasty, 8th month, day *renchen* [29] (8th day); a star became fuzzy in ZIGONG.
[*Jiu Tang shu: Zhongzong ji*] ch. 7
[*Jiu Tang shu: tianwen zhi*] ch. 36
[*Xin Tang shu: Zhongzong ji*] ch. 4
[*Xin Tang shu: tianwen zhi*] ch. 32

(b) 3rd year of the Jinglong reign period of Emperor Zhongzong of the Tang Dynasty, 8th month, day *renchen* [29] (8th day); a star became fuzzy at the wall of ZIWEI.
[*Tang huiyao*] ch. 43

AD 712 Jul 9–Aug 6 [China] (Ho 264)

(a) 1st year of the Taiji reign period of Emperor Ruizong of the Tang Dynasty, 7th month, 4th day (Aug 10); a broom star entered TAIWEI.
[*Jiu Tang shu: tianwen zhi*] ch. 36

(b) 1st year of the Xiantian reign period of Emperor Ruizong of the Tang Dynasty, 7th month, day *xinwei* [9] (Aug 10); there was a broom star that entered TAIWEI.
[*Xin Tang shu: Ruizong ji*] ch. 5

(c) 1st year of the Taiji reign period of Emperor Ruizong of the Tang Dynasty, 6th month (Jul 9–Aug 10); there was a broom star that entered TAIWEI from XUANYUAN; it continued as far as DAJIAO, where it was extinguished.
[*Xin Tang shu: tianwen zhi*] ch. 32

AD 718 Dec 8 [Japan] (Ho 265)

2nd year of the Yōrō reign period, 11th month, day *renyin* [39]; a broom star guarded the Moon.
[*Shoku Nihongi*]

AD 725 Feb 11 [Japan] (Ho 267)

2nd year of the Jinki reign period, 1st month, 24th day; a fuzzy star was seen.
[Dai Nihon shi]

AD 730 Jun 30 [China] (Ho 268)

(a) 18th year of the Kaiyuan reign period of Emperor Xuanzong of the Tang Dynasty, 6th month, day *jiazi* [1] (11th day); a broom star appeared at WUCHE.
[*Jiu Tang shu: Xuanzong ji*] ch. 8

(b) 18th year of the Kaiyuan reign period of Emperor Xuanzong of the Tang Dynasty, 6th month, day *jiazi* [1] (11th day); a broom star appeared at WUCHE.
[*Jiu Tang shu: tianwen zhi*] ch. 36

(c) 18th year of the Kaiyuan reign period of Emperor Xuanzong of the Tang Dynasty, 6th month, day *jiazi* [1] (11th day); a broom star emerged at WUCHE.
[*Xin Tang shu: Xuanzong ji*] ch. 5

(d) 18th year of the Kaiyuan reign period of Emperor Xuanzong of the Tang Dynasty, 6th month, day *jiazi* [1] (11th day); there was a broom star at WUCHE.
[*Xin Tang shu: tianwen zhi*] ch. 32

AD 730 Jul 9–19 [China] (Ho 268)

(a) 18th year of the Kaiyuan reign period of Emperor Xuanzong of the Tang Dynasty, 6th month, day *guiyou* [10] (Jul 9); a star became fuzzy in BI [LM 19] and MAO [LM 18].
[*Jiu Tang shu: Xuanzong ji*] ch. 8
[*Xin Tang shu: Xuanzong ji*] ch. 5

(b) 18th year of the Kaiyuan reign period of Emperor Xuanzong of the Tang Dynasty, 6th month, day *guiyou* [10] (Jul 9); a broom star became fuzzy in BI [LM 19] and MAO [LM 18].
[*Xin Tang shu: tianwen zhi*] ch. 32

(c) 18th year of the Kaiyuan reign period of Emperor Xuanzong of the Tang Dynasty, 6th month, 30th day (Jul 19); a star became fuzzy in BI [LM 19] and MAO [LM 18].
[*Jiu Tang shu: tianwen zhi*] ch. 36

AD 738 Apr 1 [China] (Ho 269)

(a) 26th year of the Kaiyuan reign period of Emperor Xuanzong of the Tang Dynasty, 3rd month, day *bingzi* [13] (8th day); a star became fuzzy within the enclosure of ZIWEI then passed by the bowl of DOU. After more than 10 days, it was no longer seen due to overcast skies.
[*Jiu Tang shu: Xuanzong ji*] ch. 9
[*Jiu Tang shu: tianwen zhi*] ch. 36

(b) 26th year of the Kaiyuan reign period of Emperor Xuanzong of the Tang Dynasty, 3rd month, day *bingzi* [13] (8th day); a star became fuzzy in ZIWEI.
[*Xin Tang shu: Xuanzong ji*] ch. 5

(c) 26th year of the Kaiyuan reign period of Emperor Xuanzong of the Tang Dynasty, 3rd month, day *bingzi* [13] (8th day); a star became fuzzy in the ZIWEI enclosure then passed by the bowl of BEIDOU. After more than one 10-day week, it was no longer seen due to overcast skies.
[*Xin Tang shu: tianwen zhi*] ch. 32

AD 745 Jan 8 [Japan] (Ho 271)

16th year of the Tempyō reign period, 12th month, 2nd day; a fuzzy star was seen.
[Dai Nihon shi]

AD 759 Apr 2–May 1 [Korea]

18th year of King Gyeongdoek of Silla, spring, 3rd month; a broom star was seen and was not extinguished until autumn.

[*Jeungbo munheon bigo*] ch. 6
[*Samguk sagi*] ch. 9 (Ho 272)

AD 760 May 16 [China] (Ho 273)

(a) 3rd year of the Qianyuan reign period of Emperor Suzong of Tang (i.e., 1st year of the Shangyuan reign period), 4th month, day *dingsi* [54] (27th day, May 16); a broom star was seen in the east. It was between Lou [LM 16] and Wei [LM 17]. Its color was white, and it was 4 *chi* long. It moved rapidly toward the northeast. It passed Mao [LM 18], Bi [LM 19], Zui [LM 20], Shen [LM 21], [Dong-]jing [LM 22], [Yu-]*gui* [LM 23], Liu [LM 24], and Xuanyuan. It reached the west of Taiwei and was extinguished 7 *chi* to the west of Youzhifa. It lasted for a total of more than 50 days.
[*Tang huiyao*] ch. 43

(b) 3rd year of the Qianyuan reign period of Emperor Suzong of Tang (i.e., 1st year of the Shangyuan reign period), 4th month, day *dingsi* [54] (27th day, i.e., May 16); a broom star was seen in the east. It was between Lou [LM 16] and Wei [LM 17]. Its color was white, and it was 4 *chi* long. It moved rapidly toward the east. It passed Mao [LM 18], Bi [LM 19], Zui [LM 20], Shen [LM 21], Dongjing [LM 22], Yugui [LM 23], Liu [LM 24], and Xuanyuan. It reached the west of Youzhifa. It lasted for a total of more than 50 days, then it disappeared.
[*Xin Tang shu*] ch. 32

NB: This is Comet Halley, return –16.

AD 761 May 9–Jun 7 [Korea] (Ho 275)

20th year of King Gyoengdoek of Silla, summer, 4th month; a broom star was seen.
[Samguk sagi] ch. 9/5

AD 767 Jan 22 [China] (Ho 277)

(a) 1st year of the Dali reign period of Emperor Daizong of the Tang Dynasty, 12th month, day *jihai* [36]; a broom star over 1 *chi* long arose at Paogua and trespassed on the Huanzhe star.
[*Jiu Tang shu: Daizong ji*] ch. 11

(b) 1st year of the Dali reign period of Emperor Daizong of the Tang Dynasty, 12th month, day *jihai* [36]; a broom star over 1 *chi* long emerged in PAOGUA and trespassed on the HUANZHE star.
[*Jiu Tang shu: tianwen zhi*] ch. 36

(c) 1st year of the Dali reign period of Emperor Daizong of the Tang Dynasty, 12th month, day *jihai* [36]; a broom star emerged at PAOGUA.
[*Xin Tang shu: Daizong ji*] ch. 6

(d) 1st year of the Dali reign period of Emperor Daizong of the Tang Dynasty, 12th month, day *jihai* [36]; a broom star over 1 *chi* long arose at PAOGUA and, after two 10-day weeks, was no longer seen; it trespassed on the HUANZHE star.
[*Xin Tang shu: tianwen zhi*] ch. 32

AD 768 spring [Korea] (Ho 278)

4th year of King Hyegong, spring; a broom star was seen in the east.
[Samguk sagi] ch. 9/5

AD 770 May 26–Jul 25 [China] (Ho 279)

(a) 5th year of the Dali reign period of Emperor Daizong of the Tang Dynasty, 4th month, day *jiwei* [56] (27th day) (May 26), at night; a broom star 3 *zhang* long arose at WUCHE...in the 5th month, day *jimao* [16], at night, a white broom star arose in the north...not until the 6th month, day *jiwei* [56] (Jul 25), was the broom star extinguished.
[*Jiu Tang shu: Daizong ji*] ch. 11

(b) 5th year of the Dali reign period of Emperor Daizong of the Tang Dynasty, 4th month, day *jiwei* [56] (27th day), at night; a broom star with bright rays 3 *zhang* long emerged at WUCHE and was luxuriant. In the 5th month, day *jimao* [16], at night, a white broom star arose in the north. On day *guiwei* [20] (Jun 19), at night, the broom star moved eastward with the heavens and approached BAGU...On day *guimao* [40] (Jul 9), the broom star was 2 *chi* away from SANGONG. On day *jiwei* [56] (Jul 25), the broom star was extinguished.
[*Jiu Tang shu: tianwen zhi*] ch. 36

(c) 5th year of the Dali reign period of Emperor Daizong of the Tang Dynasty, 4th month, day *jiwei* [56] (27th day); a broom star emerged in WUCHE.
[*Xin Tang shu: Daizong ji*] ch. 6

(d) 5th year of the Dali reign period of Emperor Daizong of the Tang Dynasty, 4th month, day *jiwei* [56] (27th day); there was a broom star 3 *zhang* long at WUCHE, its bright rays luxuriant. In the 5th month, day *jimao* [16], the white broom star appeared in the north. On day *guiwei* [20], it traveled eastward close to the middle star of BAGU. In the 6th month, on day *guimao* [40], the broom star was near SANGONG. On day *jiwei* [56], it was not seen.
[*Xin Tang shu: tianwen zhi*] ch. 32

(e) 5th year of the Dali reign period of Emperor Daizong of the Tang Dynasty, 4th month, day *jiwei* [56] (27th day); a broom star over 5 *zhang* long emerged at WUCHE; by the 6th month, 28th day, it was extinguished.
[*Tang huiyao*] ch. 43

AD 770 Jun 9 [Korea]

(a) 6th year of King Hyegong of Silla, summer, 5th month, 11th day; a broom star emerged north of WUCHE and lasted until the 12th day of the 6th month when it was extinguished.
[*Jeungbo munheon bigo*] ch. 6

(b) 6th year of King Hyegong of Silla, 5th month, 11th day; a broom star emerged north of WUCHE.
[*Samguk sagi*] ch. 9 (Ho 279)

AD 770 Jun 28–Aug 25 [Japan]

(a) 1st year of the Hōki reign period, 9th month, day *renwu* [17]; in this year, during the 6th and 7th months, a broom star entered BEIDOU.
[*Shoku Nihongi*]

(b) 1st year of the Hōki reign period, 7th month, day *guiwei* [20]; from the 6th and 7th months, a broom star entered BEIDOU.
[*Nihon kiryaku*] (Ho 279)

AD 773 Jan 17 [China] (Ho 280)

(a) 7th year of the Dali reign period of Emperor Daizong of the Tang Dynasty, 12th month, day *bingyin* [3]; a long star emerged at SHEN [LM 21].
[*Jiu Tang shu: Daizong ji*] ch. 11
[*Xin Tang shu: Daizong ji*] ch. 6

(b) 7th year of the Dali reign period of Emperor Daizong of the Tang Dynasty, 12th month, day *bingyin* [3]; there was a long star below SHEN [LM 21] that spanned the sky. Long stars belong to the category of comets.
[*Xin Tang shu: tianwen zhi*] ch. 32

(c) 7th year of the Dali reign period of Emperor Daizong of the Tang Dynasty, 12th month, day *bingyin* [3], at night; a long star emerged at SHEN [LM 21].
[*Jiu Tang shu: tianwen zhi*] ch. 36

AD 773 Jan 20 [Japan] (Ho 280)

3rd year of the Hōki reign period, 12th month, day *jisi* [6]; a broom star was seen in the south.
[*Shoku Nihongi*]; [*Nihon kiryaku*]

AD 815 Apr 13–May 12 [China] (Ho 282)

10th year of the Yuanhe reign period of Emperor Xianzong of the Tang Dynasty, 3rd month; there was a long star at TAIWEI whose tail reached XUANYUAN.
[*Xin Tang shu: tianwen zhi*] ch. 32

AD 815 Sep 7–Oct 6 [Korea]

7th year of King Heondoek of Silla, autumn, 8th month; a great star emerged between YI [LM 27] and ZHEN [LM 28], pointing toward azimuth *geng* (WSW); its rays were over 6 *chi* long and over 2 *cun* wide.
[*Jeungbo munheon bigo*] ch. 6
[*Samguk sagi*] ch. 10 (Ho 282)

AD 817 Feb 17 [China] (Ho 283)

(a) 12th year of the Yuanhe reign period of Emperor Xianzong of the Tang Dynasty, 1st month, day *wuzi* [25], at night; a broom star over 1 *zhang* long and pointing southwest emerged south of Bi [LM 19] for 3 days. It approached Shenqi in the south where it set.
[*Jiu Tang shu: Xianzong ji*] ch. 15

(b) 12th year of the Yuanhe reign period of Emperor Xianzong of the Tang Dynasty, 1st month, day *wuzi* [25]; a broom star over 2 *chi* long and pointing southwest emerged south of Bi [LM 19] for 3 days. It approached Shenqi where it set.
[*Jiu Tang shu: tianwen zhi*] ch. 36

(c) 12th year of the Yuanhe reign period of Emperor Xianzong of the Tang Dynasty, 1st month, day *wuzi* [25]; a broom star emerged at Bi [LM 19].
[*Xin Tang shu: Xianzong ji*] ch. 7

(d) 12th year of the Yuanhe reign period of Emperor Xianzong of the Tang Dynasty, 1st month, day *wuzi* [25]; there was a broom star at Bi [LM 19].
[*Xin Tang shu: tianwen zhi*] ch. 32

AD 821 Feb 26–27 [China] (Ho 284)

(a) 1st year of the Changqing reign period of Emperor Muzong of the Tang Dynasty, 1st month, day *wuwu* [55] (Feb 26), at night; a star became fuzzy in Yi [LM 27].
[*Jiu Tang shu: Muzong ji*] ch. 16

(b) 1st year of the Changqing reign period of Emperor Muzong of the Tang Dynasty, 1st month, day *jiwei* [56] (Feb 27); a star became fuzzy in Yi [LM 27].
[*Xin Tang shu: Muzong ji*] ch. 8
[*Xin Tang shu: tianwen zhi*] ch. 32

(c) 1st year of the Changqing reign period of Emperor Muzong of the Tang Dynasty, 1st month, day *jiwei* [56] (Feb 27), at night; a star became fuzzy in Yi [LM 27].
[*Jiu Tang shu: tianwen zhi*] ch. 36

AD 821 Mar 7 [China] (Ho 284)

(a) 1st year of the Changqing reign period of Emperor Mu of the Tang Dynasty, 1st month, day *dingmao* [4]; a star became fuzzy at azimuth *chen* [ESE], near the first star on the south in the western wall of TAIWEI.
[*Jiu Tang shu: Muzong ji*] ch. 16

(b) 1st year of the Changqing reign period of Emperor Muzong of the Tang Dynasty, 1st month, day *dingmao* [4]; a star became fuzzy at TAIWEI.
[*Xin Tang shu: Muzong ji*] ch. 8

(c) 1st year of the Changqing reign period of Emperor Muzong of the Tang Dynasty, 1st month, day *dingmao* [4]; a star became fuzzy by SHANGJIANG west of TAIWEI.
[*Xin Tang shu: tianwen zhi*] ch. 32

(d) 1st year of the Changqing reign period of Emperor Muzong of the Tang Dynasty, 1st month, day *dingmao* [4], at night; a star became fuzzy above azimuth *chen* [ESE], about 7 *cun* from the first star of the western enclosure of TAIWEI.
[*Jiu Tang shu: tianwen zhi*] ch. 36

AD 821 Jul 3–Aug 1 [China] (Ho 284)

(a) 1st year of the Changqing reign period of Emperor Muzong of the Tang Dynasty, 6th month; a broom star emerged in MAO [LM 18].
[*Xin Tang shu: Muzong ji*] ch. 8

(b) 1st year of the Changqing reign period of Emperor Muzong of the Tang Dynasty, 6th month; a broom star 1 *zhang* long emerged in MAO [LM 18]; after 10 days, it was no longer seen.
[*Xin Tang shu: tianwen zhi*] ch. 32

NB: Ho misdated this record to 822.

AD 823 Feb 19 [Japan] (Ho 286)

14th year of the Kōnin reign period, 1st month, day *xinyou* [58]; there was a fuzzy star in the southwest; after 3 days, it disappeared.
[*Nihon kiryaku*]

AD 828 Sep 3 [China] (Ho 287)

(a) 2nd year of the Dahe reign period of Emperor Wenzong of the Tang Dynasty, 7th month, day *jiachen* [41]; there was a broom star by YOUSHETI.
[*Xin Tang shu: Wenzong ji*] ch. 8

(b) 2nd year of the Dahe reign period of Emperor Wenzong of the Tang Dynasty, 7th month, day *jiachen* [41]; there was a broom star 2 *chi* long to the south of YOUSHETI.
[*Xin Tang shu: tianwen zhi*] ch. 32

AD 834 Oct 9 [China] (Ho 289)

(a) 8th year of the Dahe reign period of Emperor Wenzong of the Tang Dynasty, 9th month, day *xinhai* [48], at night; a broom star over 1 *zhang* long and pointing west arose in TAIWEI near LANGWEI. It traveled northwestward for 9 nights, passing over LANGWEI on the northwest by 5 *chi* where it was extinguished.
[*Jiu Tang shu: Wenzong ji*] ch. 17

(b) 8th year of the Dahe reign period of Emperor Wenzong of the Tang Dynasty, 9th month, day *xinhai* [48], during the fifth watch of the night; there was a broom star over 1 *zhang* long, pointing west in TAIWEIGONG near LANGWEI. It traveled northwestward for 9 nights, passing over the LANGWEI star on the northwest by 5 *chi* where it was extinguished.
[*Jiu Tang shu: tianwen zhi*] ch. 36

(c) 8th year of the Dahe reign period of Emperor Wenzong of the Tang Dynasty, 9th month, day *xinhai* [48]; a broom star emerged in TAIWEI.
[*Xin Tang shu: Wenzong ji*] ch. 8

(d) 8th year of the Dahe reign period of Emperor Wenzong of the Tang Dynasty, 9th month, day *xinhai* [48]; there was a broom star over 1 *zhang* in TAIWEI. It traveled northwestward surpassing LANGWEI; on day *gengshen* [57], it was no longer seen.
[*Xin Tang shu: tianwen zhi*] ch. 32

(e) 8th year of the Dahe reign period of Emperor Wenzong of the Tang Dynasty, 9th month, day *xinhai* [48]; a broom star arose in TAIWEI. It traveled northwestward passing above DIWEI (WUDI-ZUO) on the northwest; after 5 days, it was extinguished.
[*Tang huiyao*] ch. 34

AD 834 Oct 18 [China] (Ho 289)

(a) 8th year of the Dahe reign period of Emperor Wenzong of the Tang Dynasty, 9th month, day *gengshen* [57], at night; a broom star 3 *chi* long emerged in the east with a spectacular brilliance.
[*Jiu Tang shu: Wenzong ji*] ch. 17

(b) 8th year of the Dahe reign period of Emperor Wenzong of the Tang Dynasty, 9th month, day *gengshen* [57], at night; a broom star 3 *chi* long emerged in the east, its rays fiercely radiant.
[*Jiu Tang shu: tianwen zhi*] ch. 36

AD 836 Jul 17–Aug 15 [Korea] (Ho 290)

11th year of King Hoendoek of Silla, summer, 6th month; a fuzzy star was seen in the east.
[Samguk sagi] ch. 10

AD 837 Mar 22–Sep 9 [China] (Ho 291)

(a) 2nd year of the Kaicheng reign period of Emperor Wenzong of Tang, 2nd month, day *bingwu* [43] (Mar 22); at night, a broom star appeared in the east. It was more than 7 *chi* long and was situated in the 1st *du* of WEI [LM 12]; it pointed west. On the night of day *wushen* [45] (Mar 24), it was southwest of WEI [LM 12]. The length of the broom (star) was more than 7 *chi*, and the brightness of its rays was becoming fiercer. It again pointed west. On the night of day *guichou* [50] (Mar 29), the broom (star) was 8 *du* in WEI [LM 12]. On the night of day *gengshen* [57] (Apr 5), it was 3 and 1/2 *du* in XU [LM 11]. On the night of day *xinyou* [58] (Apr 6), the broom (star) was more than 1 *zhang* long. It moved westward in a straight line, pointing

slightly south. It was 1 and 1/2 *du* in Xu [LM 11]. On the night of day *renxu* [59] (Apr 7), it was more than 2 *zhang* long and 3 *chi* wide; it was 9 *du* in Nü [LM 10]. On the night of day *guihai* [60] (Apr 8), it increased in both length and width; it was 4 *du* in Nü [LM 10]. In the 3rd month, on the night of day *jiazi* [1] (Apr 9), its length was more than 5 *zhang*. It branched into two tails, one pointing toward Di [LM 3] and the other concealing Fang [LM 4]. It was 10 *du* in [Nan-]dou [LM 8]. On the night of day *bingyin* [3] (Apr 11), the length of the broom (star) was 6 *zhang*. The tail lacked branches and pointed north. It was 7 *du* in Kang [LM 2]. The emperor summoned the astronomer royal and asked him the reason for these star changes...That same night (i.e., still Apr 11) the broom (star) was 5 *zhang* long and 5 *chi* wide. It was moving toward the northwest and pointing east. On the night of day *wuchen* [5] (Apr 13], the length of the broom (star) was more than 8 *zhang*. It was still moving northwest and pointing east; it was 14 *du* in Zhang [LM 26]... On the night of day *guiwei* [20] (Apr 28), the broom star was 3 *chi* long. It appeared to the right of Xuanyuan and was pointing east. It was 7 *du* in Zhang [LM 26]...In the 8th month, on day *dingyou* [34] (Sep 9), the broom star appeared between Xu [LM 11] and Wei [LM 12].
[*Jiu Tang shu*] ch. 36

(b) 2nd year of the Kaichen reign period of Emperor Wenzong of Tang, 2nd month, day *bingwu* [43] (Mar 22). There was a broom star in Wei [LM 12]. It was more than 7 *chi* long and pointing west toward Nandou [LM 8]. On day *wushen* [45] (Mar 24), it was southwest of Wei [LM 12]; its rays had increased in brightness and had grown. On day *guichou* [50] (Mar 29), it was located in Xu [LM 11]. On day *xinyou* [58] (Apr 6), it was more than 1 *zhang* long, it moved west and pointed slightly toward the south. On the day *renxu* [59] (Apr 7), it was at Wunü [LM 10]. It was now more than 2 *zhang* long and 3 *chi* wide. On day *guihai* [60] (Apr 8), both it had further increased in length and width.

In the 3rd month, day *jiazi* [1] (Apr 9), it was at NANDOU [LM 8]. On day *yichou* [2] (Apr 10), it was 5 *zhang* long; its tail was in two branches, one pointing toward DI [LM 3] and one concealing FANG [LM 4]. On day *bingyin* [3] (Apr 11), it was 6 *zhang* long, but it was no longer branched. It pointed north and was situated 7 *du* in KANG [LM 2]. On day *dingmao* [4] (Apr 12), it moved northwestward but pointed east. On day *jisi* [6] (Apr 14), it was more than 8 *zhang* long, and it was located in ZHANG [LM 26]. On day *guiwei* [20] (Apr 28), its length was 3 *chi*. It disappeared to the right of XUANYUAN. All broom stars appearing in the morning point toward the west and, when appearing in the evening, point toward the east; this is normal. However, there had never been one pointing in all four directions and trespassing against so many (asterisms). On day *jiashen* [21] (Apr 29), a guest star appeared below DONGJING [LM 22]. On day *wuzi* [25] (May 3), another guest star appeared within DUANMEN near PINGXING. In the 4th month, day *bingwu* [43] (May 21), the guest star below DONGJING [LM 22] disappeared. In the 5th month, day *guiyou* [10] (Jun 17), the guest star within DUANMEN disappeared. On day *renwu* [19] (Jun 26), there was a guest star like a bushy star beside NANDOU [LM 8] and TIANYUE. In the 8th month, day *dingyou* [34] (Sep 9), there was a broom star at XU [LM 11] and WEI [LM 12].

[*Xin Tang shu*] ch. 32

NB: This is Comet Halley, return –15.

AD 837 Apr 4–17 [Japan] (Ho 291)

4th year of the Jōwa reign period, 3rd month, day *dingmao* [4] (Apr 12); a broom star was seen in the southeast. Its rays pointed east and reached the horizon. On day *renshen* [9] (Apr 17), the broom star was still visible, but the light of the moon reduced its brilliance; its rays had also diminished.

[*Zoku Nihon goki*] 6

NB: This is Comet Halley, return–15.

AD 837 Jun 26 [China] (Ho 291)

2nd year of the Kaicheng reign period of Emperor Wenzong of the Tang Dynasty, 5th month, day *renwu* [19]; a guest star like a fuzzy star was in NANDOU beside TIANYUE.
[*Xin Tang shu: tianwen zhi*] ch. 32

AD 837 Sep 9 [China] (Ho 291)

(a) 2nd year of the Kaicheng reign period of Emperor Wenzong of the Tang Dynasty, 8th month, day *dingyou* [34]; a broom star emerged between XU [LM 11] and WEI [LM 12].
[*Jiu Tang shu: Wenzong ji*] ch. 17
[*Jiu Tang shu: tianwen zhi*] ch. 36

(b) 2nd year of the Kaicheng reign period of Emperor Wenzong of the Tang Dynasty, 8th month, day *dingyou* [34]; a broom star emerged in XU [LM 11] and WEI [LM 12]; XU [LM 11] and WEI [LM 12] are [Jupiter Station] XUANXIAO.
[*Xin Tang shu: tianwen zhi*] ch. 32

AD 838 Nov 9 [China] (Ho 292)

(a) 3rd year of the Kaicheng reign period of Emperor Wenzong of the Tang Dynasty, 10th month, 19th day; a broom star over 2 *zhang* long appeared; on the night of the 20th day, it was 2 *zhang* 5 *chi* long; on the night of the 21st, it was 3 *zhang* long; on the night of the 22nd day, it was 3 *zhang* 5 *chi* long, side by side and above CHEN pointing west toward the box of ZHEN [LM 28].
[*Jiu Tang shu: tianwen zhi*] ch. 36

(b) 3rd year of the Kaicheng reign period of Emperor Wenzong of the Tang Dynasty, 10th month, 19th day; a broom star over 2 *zhang* long emerged above CHEN; on the night of the 20th day it appeared above CHEN and was 3 *zhang, 5 chi* long; on the night of the 21st day, it appeared at CHEN and was over 3 *zhang* long and pointing toward the southeastern star of ZHEN [LM 28]; on the night of the 22nd day, it appeared above azimuth *chen* [ESE] and

was over 3 *zhang* 5 *chi* long and pointing south toward the box of ZHEN [LM 28].
[*Tang huiyao*] ch. 43

AD 838 Nov 11 [China] (Ho 292)

(a) 3rd year of the Kaicheng reign period of Emperor Wenzong of the Tang Dynasty, 10th month, day *yisi* [42]; a broom star emerged in ZHEN [LM 28].
[*Xin Tang shu: Wenzong ji*] ch. 8

(b) 3rd year of the Kaicheng reign period of Emperor Wenzong of the Tang Dynasty, 10th month, day *yisi* [42]; there was a broom star over 2 *zhang* long in the box of ZHEN [LM 28] that gradually grew longer and pointed west.
[*Xin Tang shu: tianwen zhi*] ch. 32

(c) 3rd year of the Kaicheng reign period of Emperor Wenzong of the Tang Dynasty, 10th month, day *yisi* [42], at night; there was a broom star 3 *zhang* long in ZHEN [LM 28] pointing east to west.
[*Jiu Tang shu: Wenzong ji*] ch. 17

AD 838 Nov 12–AD 839 Feb 8 [Japan]

(a) 5th year of the Jōwa reign period, 10th month, day *bingwu* [43]; at night, a broom star was seen in the southeast, its vapor reddish-white; it stretched across the sky for more than several miles and then shortly disappeared. In the 11th month, day *xinwei* [8] (Dec 5), the broom star was seen in the east. The star arose on the 22nd day of the 10th month, and until the 17th day of the present month, it was seen in the east every night during the hour of *yin* [LT = 3–5 h], its length over 7 *chi*.
[*Zoku Nihon goki*] (Ho 292)

(b) 5th year of the Jōwa reign period, 10th month, day *bingwu* [43]; on this night, a broom star was seen in the southeast, its vapor reddish-white; it stretched across the sky for more than several miles and then shortly disappeared; on day *gengxu* [47] (Nov 16), it was seen again. In the 11th month, day *xinwei* [8] (Dec 5), the

broom star was seen in the east. The star arose on the 22nd day of the 10th month, and until the 17 day of the present month, it was seen every night in the east during the hour of *yin* [LT = 3–5 h], its length over 7 to 8 *zhang*. In the 6th year of the Jōwa reign period, 1st month, day *bingzi* [13] (Feb 8); the broom star was seen at azimuth *dui* [W], now over 1 *zhang* long.
[*Shodō kambun*]

AD 838 Nov 21 [China] (Ho 292)

(a) 3rd year of the Kaicheng reign period of Emperor Wenzong of the Tang Dynasty, 11th month, day *yimao* [52]; a broom star appeared in the east in Wei [LM 6] and Ji [LM 7], spanning the sky from east to west. By the 12th month, day *renchen* [29], it was no longer seen.
[*Xin Tang shu: tianwen zhi*] ch. 32

(b) 3rd year of the Kaicheng reign period of Emperor Wenzong of the Tang Dynasty, 11th month, day *yimao* [52], at night; a fuzzy broom star stretched across the sky from east to west.
[*Jiu Tang shu: Wenzong ji*] ch. 17

(c) 3rd year of the Kaicheng reign period of Emperor Wenzong of the Tang Dynasty, 11th month, day *yimao* [52], at night; a broom star emerged in the east and stretched across the sky from east to west.
[*Jiu Tang shu: tianwen zhi*] ch. 36

AD 839 Feb 7–17 [China] (Ho 292)

(a) 4th year of the Kaicheng reign period of Emperor Wenzong of the Tang Dynasty, 1st month, day *guiyou* [10] (Feb 7); a broom star emerged in Yulin. Intercalary month, day *bingwu* [43], it emerged in Juanshe.
[*Xin Tang shu: Wenzong ji*] ch. 8

(b) 4th year of the Kaicheng reign period of Emperor Wenzong of the Tang Dynasty, 1st month, day *guiyou* [10] (Feb 7); a broom star emerged in Yulin. Intercalary month, day *bingwu* [43], there was

a broom star to the northwest of JUANSHE. By the 2nd month, day *jimao* [16], it was no longer seen.
[*Xin Tang shu: tianwen zhi*] ch. 32

(c) 4th year of the Kaicheng reign period of Emperor Wenzong of the Tang Dynasty, 1st month, day *guiyou* [10] (Feb 7), at night; a broom star emerged in the west at 14 *du* in SHI [LM 13]. Intercalary month, 23rd day, it appeared again to the north of JUANSHE. In all, it appeared for 33 days; by the night of the 26th day (of the intercalary month), it was extinguished.
[*Jiu Tang shu: tianwen zhi*] ch. 36

NB: *Jiao kan ji* said, "the two characters '2nd month' should not appear before '26th day.' On checking, in this year the intercalary was the 1st month, so from day *guiyou* [10] in the 1st month to day *bingwu* [43], the 23rd day of the subsequent intercalary month, it was 33 days. There is no mistake in the text.

(d) 4th year of the Kaicheng reign period of Emperor Wenzong of the Tang Dynasty, 1st month, 30th day (Feb 17); a broom star appeared south of SHI [LM 13] and passed through lunar mansions BI [LM 14], KUI [LM 15], LOU [LM 16], and WEI [LM 17], and so forth. By the 23rd day of the intercalary month, it appeared again to the north of JUANSHE; in all, it appeared for 33 days. Subsequently, the 24th and 25th days were cloudy, but the 26th was clear and it had already extinguished.
[*Tang huiyao*] ch. 43

AD 840 Mar 20 [China] (Ho 293)

(a) 5th year of the Kaicheng reign period of Emperor Wenzong of the Tang Dynasty, 2nd month, day *gengshen* [57]; a broom star emerged in SHI [LM 13] and BI [LM 14].
[*Xin Tang shu: Wuzong ji*] ch. 8

(b) 5th year of the Kaicheng reign period of Emperor Wenzong of the Tang Dynasty, 2nd month, day *gengshen* [57]; a broom star emerged between YINGSHI [LM 13] and DONGBI [LM 14]; after 20, days it was extinguished.
[*Xin Tang shu: tianwen zhi*] ch. 32

AD 840 Dec 3 [China] (Ho 294)

(a) 5th year of the Kaicheng reign period of Emperor Wenzong of the Tang Dynasty, 11th month, day *wuyin* [15]; a broom star emerged in the east.
[*Xin Tang shu: Wuzong ji*] ch. 8

(b) 5th year of the Kaicheng reign period of Emperor Wenzong of the Tang Dynasty, 11th month, day *wuyin* [15]; there was a broom star in the east.
[*Xin Tang shu: tianwen zhi*] ch. 32

AD 841 Jul 22–Aug 19 [China] (Ho 295)

(a) 1st year of the Huichang reign period of Emperor Wuzong of the Tang Dynasty, 7th month; a broom star reappeared between SHI [LM 13] and BI [LM 14].
[*Jiu Tang shu: Wuzong ji*] ch. 18

(b) 1st year of the Huichang reign period of Emperor Wuzong of the Tang Dynasty, 7th month; a broom star emerged in YULIN.
[*Xin Tang shu: Wuzong ji*] ch. 8

(c) 1st year of the Huichang reign period of Emperor Wuzong of the Tang Dynasty, 7th month; there was a broom star at YULIN between YINGSHI [LM 13] and DONGBI [LM 14].
[*Xin Tang shu: tianwen zhi*] ch. 32

AD 841 Dec 22 [China] (Ho 296)

(a) 1st year of the Huichang reign period of Emperor Wuzong of the Tang Dynasty, 11th month, day *renyin* [39] (6th day), at night; a broom star arose in SHI [LM 13]; it appeared for 56 days in all and then was extinguished.
[*Jiu Tang shu: Wuzong ji*] ch. 18

(b) 1st year of the Huichang reign period of Emperor Wuzong of the Tang Dynasty, 11th month, day *renyin* [39] (6th day); a broom star appeared in the southwest in the initial *du* of SHI [LM 13]; it was visible for 56 days in all and then extinguished.
[*Jiu Tang shu: tianwen zhi*] ch. 36

(c) 1st year of the Huichang reign period of Emperor Wuzong of the Tang Dynasty, 11th month, day *renyin* [39] (6th day); a broom star emerged in YINGSHI [LM 13].
[*Xin Tang shu: Wuzong ji*] ch. 8

(d) 1st year of the Huichang reign period of Emperor Wuzong of the Tang Dynasty, 11th month, day *renyin* [39] (6th day); there was a broom star in BEILUOSHIMEN, in YINGSHI [LM 13], which entered ZIGONG. By the 12th month, day *xinmao* [28], it no longer appeared.
[*Xin Tang shu: tianwen zhi*] ch. 32

(e) 1st year of the Huichang reign period of Emperor Wuzong of the Tang Dynasty, 11th month, day *renyin* [39] (6th day); a broom star appeared in the southwest in the space of lunar mansion SHI [LM 13]; it was visible for 56 days in all and then was extinguished.
[*Tang huiyao*] ch. 43

AD 841 Dec 22–Jan 11 [Japan]

(a) 8th year of the Jōwa reign period, 11th month, day *renyin* [39]; a broom star was seen in the west; on day *dingsi* [54] (Jan 6), the broom star was seen again.
[*Zoku Nihon goki*]

(b) 8th year of the Jōwa reign period, 11th month, day *renyin* [39]; a broom star was seen through the 26th day (AD 842 Jan 11).
[*Dai Nihon shi*] (Ho 296)

AD 852 Mar 14 [Japan]

2nd year of the Ninju reign period, 2nd month, day *dingsi* [54]; on this night, a broom star about 5 *zhang* long was seen in the west.
[*Montoku jitsuroku*]; [*Nihon kiryaku*]
[*Dai Nihonshi*] (Ho 297)

AD 852 Mar 25–Apr 22 [China] (Ho 297)

6th year of the Dazhong reign period of Emperor Xuanzong of the Tang Dynasty, 3rd month; a broom star emerged in ZUI [LM 20] and SHEN [LM 21].

[*Xin Tang shu: Xuanzong ji*] ch. 8
[*Xin Tang shu: tianwen zhi*] ch. 32

AD 855 Feb 23 [Japan]
2nd year of the Saikō reign period, 2nd month, day *guichou* [50]; a long star was seen in the northeast.
[*Sandai jitsuroku*]
[*Nihon kiryaku*]
[*Dai Nihon shi*] (Ho 298)

AD 857 Sep 22 [China] Ho (299)

(a) 11th year of the Dazhong reign period of Emperor Xuanzong of the Tang Dynasty, 9th month, day *yiwei* [32]; a broom star 3 *chi* long emerged in the initial *du* of Fang [LM 4].
[*Jiu Tang shu: Xuanzong ji*] ch. 18

(b) 11th year of the Dazhong reign period of Emperor Xuanzong of the Tang Dynasty, 9th month, day *yiwei* [32]; a broom star emerged in Fang [LM 4].
[*Xin Tang shu: Xuanzong ji*] ch. 8

(c) 11th year of the Dazhong reign period of Emperor Xuanzong of the Tang Dynasty, 9th month, day *yiwei* [32]; there was a broom star 3 *chi* long in Fang [LM 4].
[*Xin Tang shu: tianwen zhi*] ch. 32

AD 864 Apr 10–May 9 [China]
5th year of the Xiantong reign period of Emperor Yizong of the Tang Dynasty, 3rd month; a broom star emerged in Lou [LM 16]; the Astronomical Bureau took it to be an auspicious star.
[*Shi shi zheng gang*] ch. 20

AD 864 Apr 23 [Japan]
6th year of the Jōgan reign period, 3rd month, 14th day *gengzi* [37]; a broom star over 4 *chi* long was seen in the east in lunar mansion Yingshi [LM 13].
[*Sandai jitsuroku*]

[*Nihon kiryaku*]
[*Dai Nihon shi*] ch. 359 (Ho 300)

AD 864 Jun 21 [China] (Ho 300)

(a) 5th year of the Xiantong reign period of Emperor Yizong of the Tang Dynasty, 5th month, day *jihai* [36]; a broom star emerged in LOU [LM 16].
[*Xin Tang shu: Yizong ji*] ch. 9

(b) 5th year of the Xiantong reign period of Emperor Yizong of the Tang Dynasty, 5th month, day *jihai* [36], at night; before one mark had elapsed on the clepsydra, a yellowish-white broom star 3 *chi* long emerged in the northeast in LOU [LM 16].
[*Xin Tang shu: tianwen zhi*] ch. 32

AD 868 Jan 21 [Japan] (Ho 301)

(a) 9th year of the Jōgan reign period, 12th month, 23rd day *wuzi* [25]; a broom star about 5 *chi* long was seen west of ZIWEIGONG and penetrated NEIJIE.
[*Sandai jitsuroku*]

(b) 9th year of the Jōgan reign period, 12th month, 23rd day *wuzi* [25]; during the hour of *xu* [LT = 19–21 h], a broom star 5 *chi* long was seen slightly west of the enclosure of ZIWEIGONG, which penetrated NEIJIE.
[*Shodō kanbun*]

NB: *Sandai jitsuroku* originally had "11th month, 23rd day *wuwu* [55]," which is a mistake for "12th month, 23rd day, *wuzi* [25]." Ho (1962) misdates as 867.

AD 868 Jan 29–Feb 26 [China] (Ho 303)

9th year of the Xiantong reign period of Emperor Yizong of the Tang Dynasty, 1st month; a broom star emerged in LOU [LM 16] and WEI [LM 17].
[*Xin Tang shu: Yizong ji*] ch. 9
[*Xin Tang shu: tianwen zhi*] ch. 32

AD 869 Sep 10–Oct 8 [China] (Ho 304)

(a) 10th year of the Xiantong reign period of Emperor Yizong of the Tang Dynasty, 8th month; a broom star emerged in DALING. [*Xin Tang shu: Yizong ji*] ch. 9

(b) 10th year of the Xiantong reign period of Emperor Yizong of the Tang Dynasty, 8th month; there was a broom star in DALING, pointing northeast. [*Xin Tang shu: tianwen zhi*] ch. 32

AD 875 Mar 12–Apr 9 [Korea] (Ho 305)

15th year of King Gyeongmun of Silla, spring, 2nd month; there was a fuzzy star in the east; after 20 days, it was extinguished.
[*Jeungbo munheon bigo*] ch. 6
[*Samguk sagi*] ch. 11

AD 875 Jun 5–23 [Japan]

17th year of the Jōgan reign period, 4th month, 28th day *gengchen* [17]; during the hour of *mao* [LT = 5–7 h], a white broom star was seen in the northeast, its color reddish and forming rays. By the 2nd day of the 5th month (Jun 9), when its body was about 1 *zhang* or more long, it first emerged from WUCHE and lightly brushed BAGU. Its vapors, though slightly dissipated, had not yet extinguished. On the 16th day of the 5th month, day *dingyou* [34] (Jun 22), at night a cloudy vapor shaped like a streamer stretched across the sky, its head resting on the western mountains and its tail on the eastern mountains. On the 18th day *jisi* [6] (Jun 23), at night there was a fuzzy star in the northeast.
[*Sandai jitsuroku*]
[*Dai Nihon shi*] (Ho 306)
NB: "On the 16th day of the 5th month" should read "on the 15th day of the 5th month."

AD 885 Jan 20–AD 886 Feb 7 [China] (Ho 309)

1st year of the Guangqi reign period of Emperor Xizong of the Tang Dynasty; there was a broom star between JISHUI and JIXIN.
[*Xin Tang shu: tianwen zhi*] ch. 32

AD 886 Jun 6–Jul 5 [China] (Ho 310)

(a) 2nd year of the Guangqi reign period of Emperor Xizong of the Tang Dynasty, 5th month, day *bingxu* [23] (Jun 13); a star became fuzzy in JI [LM 7] and WEI [LM 6].
[*Xin Tang shu: Xizong ji*] ch. 9

(b) 2nd year of the Guangqi reign period of Emperor Xizong of the Tang Dynasty, 5th month, day *bingxu* [23] (Jun 13); a star became fuzzy in JI [LM 7] and WEI [LM 6] that passed through BEIDOU and SHETI.
[*Xin Tang shu: tianwen zhi*] ch. 32

(c) 2nd year of the Guangqi reign period of Emperor Xizong of the Tang Dynasty, 5th month (Jun 6–Jul 5); a star became fuzzy in JI [LM 7] and WEI [LM 6] that passed through BEIDOU and SHETI.
[*Jiu Tang shu: Xizong ji*] ch. 19

AD 891 May 12 [China] (Ho 313)

(a) 2nd year of the Dashun reign period of Emperor Zhaozong of the Tang Dynasty, 4th month, day *gengchen* [17]; a broom star entered TAIWEI.
[*Xin Tang shu: Zhaozong ji*] ch. 10

(b) 2nd year of the Dashun reign period of Emperor Zhaozong of the Tang Dynasty, 4th month, day *gengchen* [17]; there was a broom star over 10 *zhang* long in SANTAI that traveled eastward and entered TAIWEI, sweeping DAJIAO and TIANSHI. In the 5th month, day *jiaxu* [11], it no longer appeared.
[*Xin Tang shu: tianwen zhi*] ch. 32

AD 892 Nov 23–Dec 22 [China] (Ho 314)

1st year of the Jingfu reign period of Emperor Zhaozong of the Tang Dynasty, 11th month; a star became fuzzy in DOU [LM 8] and NIU [LM 9].
[*Xin Tang shu: Zhaozong ji*] ch. 10
[*Xin Tang shu: tianwen zhi*] ch. 32

AD 892 Dec 28 [China] (Ho 314)

1st year of the Jingfu reign period of Emperor Zhaozong of the Tang Dynasty, 12th month, day *bingzi* [13]; a *Tiancen* (star) emerged in the southwest; on day *jimao* [16], it turned into a cloud and disappeared.
[*Xin Tang shu: tianwen zhi*] ch. 32

AD 893 May 6 [China] (Ho 315)

(a) 2nd year of the Jingfu reign period of Emperor Zhaozong of the Tang Dynasty, 4th month, day *yiyou* [22]; a broom star entered TAIWEI.
[*Xin Tang shu: Zhaozong ji*] ch. 10

(b) 2nd year of the Jingfu reign period of Emperor Zhaozong of the Tang Dynasty, 4th month, day *yiyou* [22] at night; there was a broom star over 10 *zhang* long in SANTAI that traveled eastward and entered TAIWEI, swept DAJIAO, and entered TIANSHI. For 37 days, it grew longer until it was over 20 *zhang* long and then was not seen due to clouds and overcast.
[*Xin Tang shu: tianwen zhi*] ch. 32

AD 894 Feb 10–Mar 10 [China] (Ho 316)

1st year of the Qianning reign period of Emperor Zhaozong of the Tang Dynasty, 1st month; a star became fuzzy in CHUNSHOU [i.e., JING (LM 22) and GUI (LM 23)].
[*Xin Tang shu: Zhaozong ji*] ch. 10
[*Xin Tang shu: tianwen zhi*] ch. 32

AD 894 Mar 11–Apr 9 [Japan] (Ho 316)

6th year of the Kampyō reign period, 2nd month; a broom star was seen.
[Dai Nihon shi]

AD 896 Nov 9–Dec 7 [China] (Ho 318)

3rd year of the Qianning reign period of Emperor Zhaozong of the Tang Dynasty, 10th month; there were three guest stars, one large

and two small, between XU [LM 11] and WEI [LM 12], now joining together, now separating, as if fighting, as they traveled eastward together. After 3 days, the two smaller stars disappeared; later, the large one disappeared.
[*Xin Tang shu: tianwen zhi*] ch. 32

AD 902 Feb 11–Mar 12 [China] (Ho 320)
2nd year of the Tianfu reign period of Emperor Zhaozong of the Tang Dynasty, 1st month; a guest star like a peach, gradually moved from below HUAGAI of ZIGONG to YUNÜ. On day *dingmao* [4], meteor showers emanated from WENCHANG and reached the guest star, which did not move; on day *jisi* [6], the guest star was at GANG and guarded it; by the following year, it still had not departed.
[*Xin Tang shu: tianwen zhi*] ch. 32

AD 905 May 18 [China] (Ho 321)
(a) 2nd year of the Tianyou reign period of Emperor Ai of the Tang Dynasty, 4th month, day *gengzi* [37] (May 18); a broom star emerged in the northwest; on day *jiachen* [41] (May 22), it emerged from BEIHE.
[*Xin Tang shu: Aidi ji*] ch. 10
(b) 2nd year of the Tianyou reign period of Emperor Ai of the Tang Dynasty, 4th month, day *gengzi* [37], evening; in the northwestern corner, there was a star like Venus with a gleam on the top resembling a broom; it was 3 to 4 *zhang* long and reddish brown in color. On the evening of day *xinchou* [38], its color was like white silk. All said, it was the Watery Star of WUCHE; one said it was the *Zhaoming* star.
[*Xin Tang shu: tianwen zhi*] ch. 32

AD 905 May 21–Jun 8 [Japan]
5th year of the Engi reign period, 4th month, 15th day *guimao* [40]… once again a broom star was seen at azimuth *qian* [NW]. Likewise, on the 16th, 18th, and 19th days, the broom star was seen; on the 24th

day *renzi* [49] (May 30), the broom star over 30 *zhang* long was seen at azimuth *qian* [NW], its rays pointing toward azimuth *xun* [SE]; on the 25th day *guichou* [50] (May 31), the broom star was seen at azimuth *qian* [NW] and stretched across the sky; on the 26th, 27th, 28th, and 29th, it likewise appeared. First day of the 5th month, *jiwei* [56] (Jun 6), on this night, the broom star gradually began to fade; on the 3rd day, *xinyou* [58] (Jun 8), from this night on, the broom star that had appeared in azimuth *qian* [NW] did not reappear.
[*Fusō ryakuki*]
NB: *Shodō kanmbun* had a similar record.
[*Dai Nihonshi*] (Ho 321)

AD 905 May 22 [China] (Ho 321)

(a) 2nd year of the Tianyou reign period of Emperor Ai of the Tang Dynasty, 4th month, day *jiachen* [41], at night; in the northwest, a broom star 3 *zhang* long arose in BEIHE and penetrated WENCHANG...5th month, day *yiyou* [22], at night, the broom star in the northwest was 60 to 70 *zhang* long, gleaming furiously and extending across the entire sky from DAJIAO of XUANYUAN to the western wall of TIANSHI.
[*Jiu Tang shu: Aidi ji*] ch. 20

(b) 2nd year of the Tianyou reign period of Emperor Ai of the Tang Dynasty, 4th month, day *jiachen* [41]; there was a broom star over 3 *zhang* long in BEIHE that approached ZHONGTAI and XIATAI and penetrated WENCHANG; 5th month, on day *yichou* [2], at night, from the left horn of XUANYUAN, it stretched across the sky to the western wall of TIANSHI, its bright rays gleaming furiously; on day *bingyin* [3], it was cloudy, and then on day *xinwei* [8], when it cleared, it was not seen.
[*Xin Tang shu: tianwen zhi*] ch. 32

AD 905 Jun 6–Jul 4 [China] (Ho 321)

2nd year of the Tianyou reign period of Emperor Ai of the Tang Dynasty, 5th month; in the northwest, a long star stretched across

the sky, sweeping TAIWEI, WENCHANG, DIZUO, and several lunar mansions.
[*Jiu Tang shu: liezhuan*] ch. 179

AD 906 Apr 26–May 25 [China]
3rd year of the Tianyou reign period of Emperor Ai of the Tang Dynasty, 4th month; a broom star emerged in the northwest, sweeping WENCHANG, XUANYUAN, and TIANSHI.
[*Xin Wudai shi: Tang liu chen zhuan*] ch. 35
NB: One suspects this should read '2nd year.'

AD 907 Apr 7–15 [Japan] (Ho 322)
7th year of the Engi reign period, 2nd month, 22nd day; on the 24th day of the 2nd month, *xinwei* [8], a broom star over 3 *zhang* long trespassed against Venus. In the 2nd month on the 22nd day, *jisi* [6] (Apr 8), during the hour of *xu* [LT = 19–21 h], a broom star was seen at azimuth *you* [W] and *xu* [WNW]; on the 29th day, the broom star was seen in the same direction; on the 30th day, *dingchou* [14] (Apr 15), it did not appear.
[*Fusō ryakuki*]
[*Dai Nihonshi*]
NB: The original "2nd month, 25th day *xinwei* [8]" is a mistake for "2nd month, 24th day *xinwei*."

AD 908 Mar 5–Apr 3 [Korea] (Ho 323)
12th year of King Hyogong of Silla, spring, 2nd month; a fuzzy star was seen in the east.
[Samguk sagi] ch. 11

AD 912 May 13–15 [China] (Ho 325)
(a) 2nd year of the Qianhua reign period Emperor Taizu of Hou Liang, 4th month, on the night of day *jiaxu* [11] (May 15); a broom star was seen to the west of LINGTAI.
[*Jiu wu dai shi*] ch. 139; [*Wu dai huiyao*] ch. 10

(b) 2nd year of the Qianhua reign period Emperor Taizu of Hou Liang, 4th month, day *renshen* [9] (May 13); a broom (star) appeared in ZHANG [LM 26]. On day *jiaxu* [11] (May 15), the broom (star) appeared in LINGTAI.
[*Xin Wudai shi*] ch. 59

NB: This is Comet Halley, return –14.

AD 912 Jul 19–28 [Japan] (Ho 325)

(a) 12th year of the Engi reign period, 6th month, 3rd day, *jimao* [16] (Jul 19); at night a broom star was seen in the northwest corner between *xu* [WNW] and *hai* [NNW]. On the 5th and 6th days (Jul 21 and Jul 22), it was the same. On the 8th day, *jiashen* [21] (Jul 24), it was seen in the southeast between *chen* [ESE] and *si* [SSE]. On the 9th day, *yiyou* [22] (Jul 25), it was seen in the northwest. On the 12th day, *wuzi* [25] (Jul 28), it was seen in the west.
[*Fusō ryakuki*] 23

(b) 12th year of the Engi reign period, 6th month, 3rd day (Jul 19). A broom star first appeared in the direction between *xu* [WNW] and *hai* [NNW]; it lasted seven nights.

NB: This is Comet Halley, return –14.

AD 918 Nov 7 [Japan] (Ho 326)

18th year of the Engi reign period, 10th month, 1st day; a broom star first was seen in the southwest and lasted for 3 nights.
[*Ichidai yōki*]; [*Shodō kanbun*]
[*Dai Nihonshi*]

AD 923 Nov 11–Dec 10 [China]

3rd year of the Tianzan reign period of Emperor Taizu of the Liao Dynasty (1st year of the Tongguang reign period), 10th month; a broom star over 1 *zhang* long appeared to emerge from YUGUI [LM 23].
[*Qidan guo zhi*] ch. 1

NB: The third year of the Tianzan reign period is not the first year of the Tongguang reign period; the date in the Western calendar is based on the first year of the Tongguang reign period.

AD 923 Dec 8 [China]

1st year of the Tongguang reign period of Emperor Zhuangzong of the Later Tang Dynasty, 10th month, day *wuxu* [35]; a broom star over 1 *zhang* long appeared in YUGUI [LM 23].
[*Yongle dadian*] ch. 7159

AD 928 Dec 13 [China] (Ho 328)

3rd year of the Tiancheng reign period of Emperor Mingzong of the Tang Dynasty, 10th month, on the night of day *gengwu* [7]; there was a fuzzy star in the southwest; it was over 1 *zhang* long and pointing southeast at 5 *du* in NIU [LM 9].
[*Jiu wu dai shi: tianwen zhi*] ch. 139

AD 930 May 31–Jul 27 [Japan] (Ho 329)

(a) 8th year of the Enchō reign period, from the 5th month to the 7th month; a guest star entered YULIN.
[*Meigetsuki*] Nonomiya ed.

(b) 8th year of the Enchō reign period, from the 5th month to the 7th month; a guest star entered YULIN, after first moving westward to YULIN.
[*Ichidai yōki*]

AD 934 Dec 19 [China] (Ho 331)

1st year of the Qingtai reign period of Emperor Mo of the Later Tang Dynasty, 11th month, day *dingwei* [44]; a broom star emerged in XU [LM 11] and WEI [LM 12] and swept TIANLEI and KUXING.
[*Xin Wudai shi: sitian kao*] ch. 59

AD 936 Sep 21 [China] (Ho 332)

3rd year of the Qingtai reign period of Emperor Mo of the Later Tang Dynasty, 9th month, day *jichou* [26]; a tiny, dim broom star

over 1 *chi* long emerged in Xu [LM 11] and Wei [LM 12] and passed Tianlei and Kuxing.
[*Jiu Wudai shi: tianwen zhi*] ch. 139

AD 936 Oct 19 [China] (Ho 332)
3rd year of the Qingtai reign period of Emperor Mo of the Later Tang Dynasty, 10th month, day *dingsi* [54], at night; a broom star over 1 *chi* long emerged in Xu [LM 11] and Wei [LM 12].
[*Jiu Wudai shi: Tang shu*] ch. 48
NB: Likely the same as the preceding.

AD 941 Mar 31–Apr 28 [Japan]
4th year of the Tengyō reign period, spring, 3rd month; there was a star at about azimuth *you* [W] with a light like a white rainbow, narrow at the base and gradually broadening toward the tip. It traveled over 10 *li* (miles) and lasted for 3 months.
[*Fusō ryakuki*]
[*Dai Nihonshi*] (Ho 334)

AD 941 Aug 25–Sep 23 [China] (Ho 335)
5th year of the Shengyuan reign period of Emperor Liezu of the Southern Tang Dynasty, 8th month; a star became fuzzy in Tianshi; it was several *chi* long, and after 70 days, it disappeared.
[*Nan Tang shu: Liezu ji*] ch. 1

AD 941 Oct 18 [China] (Ho 335)
6th year of the Tianfu reign period of Emperor Gaozu of the Later Jin Dynasty, 9th month, day *renwu* [19], at night; a broom star over 2 *zhang* long emerged in the west at 1 *du* in Fang [LM 4]. Its tail penetrated the wall of Tianshi as it traveled eastward. It lasted over a month before being extinguished.
[*Jiu Tang shu: Jin shu*] ch. 80

AD 943 Nov 5 [China] (Ho 336)
(a) 8th year of the Tianfu reign period of Emperor Gaozu of the Later Jin Dynasty, 10th month, day *gengxu* [47], at night; a broom star

with a tail over 1 *zhang* long and pointing west appeared in the east at 9 *du* in JIAO [LM 1].
[*Jiu Wudai shi: tianwen zhi*] ch. 139

(b) 8th year of the Tianfu reign period of Emperor Gaozu of the Later Jin Dynasty, 10th month, day *gengxu* [47]; on this night during the fifth watch, a broom star appeared in the east in JIAO [LM 1]; after one 10-day week, it was extinguished.
[*Jiu Wudai shi: Jin shu*] ch. 82

AD 946 Mar 2 [Japan] (Ho 339)

2nd year of the Tenryaku reign period, 1st month, 19th day; a broom star was seen.
[Dai Nihon shi]

AD 947 Sep 12 [China] (Ho 338)

(a) 12th year of the Tianfu reign period of Emperor Gaozu of the Later Han Dynasty, intercalary 7th month, day *dingchou* [14]; there was a broom star in ZHANG [LM 26]; after one 10-day week, it was extinguished.
[*Jiu Wudai shi: Han shu*] ch. 100

(b) 5th year of the Baoda reign period of Emperor Yuanzong of the Southern Tang Dynasty, intercalary 7th month, day *dingchou* [14], at night; a broom star emerged in the east in the obscurity near the horizon, its tail grazing SHAOWEI and CHANGYUAN from the side; it lasted until day *renchen* [29] in the 8th month before being extinguished.
[*Nan Tang shu: Yuanzong ji*] ch. 2

AD 956 Mar 13 [China] (Ho 340)

(a) 3rd year of the Xiande reign period of Emperor Taizu of the Zhou Dynasty, 1st month, day *renxu* [59], at night; there was a fuzzy broom star in the corner of SHEN [LM 21] whose rays pointed southeast.
[*Jiu Wudai shi: tianwen zhi*] ch. 139

(b) 3rd year of the Xiande reign period of Emperor Taizu of the Zhou Dynasty, 1st month, day *renxu* [59]; a star became fuzzy in SHEN [LM 21].
[*Xin Wudai shi: sitian kao*] ch. 59

AD 961 Mar 15 [Japan] (Ho 341)
1st year of the Ōwa reign period, 2nd month, 27th day; a broom star was seen in the southwest, its light resembling a wildfire.
[Dai Nihonshi]

AD 962 Jan 28 [China] (Ho 342)
2nd year of the Jianlong reign period of Emperor Taizu of the Song Dynasty, 12th month, day *jiyou* [46]; a guest star with tiny rays emerged east of the ZONGREN star of TIANSHIYUAN. In the 3rd year, 1st month, day *xinwei* [8] (Feb 19), it traveled southwestward and entered lunar mansion DI [LM 3]. In the 2nd month, day *guichou* [50](Apr 2), it reached QIXING [LM 25] and disappeared.
[*Song shi: tianwen zhi*] ch. 56

AD 967 Jan 8 [Japan]
(a) 3rd year of the Kōhō reign period, 11th month, 25th day *yimao* [52]; at night during the hour of *hai* [LT = 21–23 h], a broom star was seen in the south.
[*Shodō kanmon*]
[*Dai Nihonshi*] (Ho 344)
(b) 3rd year of the Kōhō reign period, 11th month, 25th day *yimao* [52]; during the hour of *hai* [LT = 21–23 h], a broom star was seen in the west.
[*Ichidai yōki*]

AD 975 [China]
Yihai [12] year of Emperor Houzhu of the Southern Tang Dynasty, 3rd month, day *dingsi* [54]; a white broom star 5 *chi* long emerged in WUCHE. In summer, the 6th month, it turned around and appeared

in the west where it trespassed on TAIWEI for 60 days when it was extinguished.
[*Nan Tang shu: Houzhu ji*] ch. 3
NB: There was no *dingsi* [54] day that month.

AD 975 Aug 2–3 [Japan]

(a) 3rd year of the Ten-en reign period, 6th month, 22nd day *guihai* [60]; at daybreak, a broom star shaped like a round fan 5 to 6 *chi* long was seen at azimuth *gen* [NE] and lasted for several nights; on the 23rd day, *jiazi* [1] (Aug 3), at daybreak, the Moon trespassed on BI [LM 14], and further, there was a guest star like a white rainbow.
[*Nihon ryakuki*]
[*Dai Nihonshi*] (Ho 346)

(b) 3rd year of the Ten-en reign period, 6th month, 22nd day *guihai* [60]; a white broom star 5 to 6 *zhang* long like a waving plume was seen and lasted for 5 days.
[*Shodō kammon*]

AD 975 Aug 3 [China] (Ho 346)

(a) 8th year of the Kaibao reign period of Emperor Taizu of Song, 6th month, day *jiazi* [1]; a broom star 4 *zhang* long visible at daybreak in the east emerged in LIU [LM 24].
[*Song shi: Taizu ji*] ch. 3

(b) 8th year of the Kaibao reign period of Emperor Taizu of Song, 6th month, day *jiazi* [1]; a broom star 4 *zhang* long emerged in LIU [LM 24] and was visible at daybreak in the east pointing southwest. In all, it passed through 11 lunar lodges, from YUGUI [LM 23] to DONGBI [LM 14], during 83 days, after which it was extinguished.
[*Song shi: tianwen zhi*] ch. 56

AD 977 Mar 16 [Japan] (Ho 347)

2nd year of the Tengen reign period, 2nd month, 22nd day; a broom star was seen.
[Dai Nihon shi]

AD 983 Apr 3 [China] (Ho 348)

8th year of the Taiping xingguo reign period of Emperor Taizong of the Song Dynasty, 6th month, day *jiachen* [41]; a guest star emerged east of DUANMEN of TAIWEIYUAN and approached PINGXING as it traveled northward.

[*Song shi: tianwen zhi*] ch. 56

AD 988 Jul 18 [China]

1st year of the Duangong reign period of Emperor Taizong of the Song Dynasty, 6th month, 2nd day; a broom star appeared in JING [LM 22] and GUI [LM 23]; after an uncounted number of days, it was extinguished.

[*Song huiyao jigao*] vol. 52

AD 989 Aug 12–Sep 10 [China] (Ho 349–350)

(a) 2nd year of the Duangong reign period of Emperor Taizong of Song, 7th month, day *dinghai* [24] (Aug 12); a guest star appeared to the northwest of the stars of BEIHE. It became slightly dimmer. It had short rays and a broom that pointed southwest. [*Song shi*] ch. 56

(b) 2nd year of the Duangong reign period of Emperor Taizong of Song, 7th month, day *wuzi* [25] (Aug 13); a broom star appeared at DONGJING [LM 22] to the west of JISHUI. Its color was greenish-white. Its rays gradually lengthened. In the morning, it was seen in the northeast for 10 days. In the evening, it was seen in the northwest. It passed YOUSHETI. After a total of 30 days, it reached KANG [LM 2] and disappeared. [*Song shi*] ch. 56

(c) 2nd year of the Duangong reign period of Emperor Taizong of Song, 7th month, day *wuzi* [25] (Aug 13); a broom star was seen in DONGJING [LM 22]...In the 8th month, day *bingchen* [53] (Sep 10), it was not seen at night. [*Song shi*] ch. 5

NB: This is Comet Halley, return –13.

AD 989 Aug 14–23 [Japan] (Ho 350)

(a) 1st year of the Eiso reign period, 6th month, 1st day, *gengxu* [47] (Jul 6); a broom star was seen in the east and west. During the middle 10-day week of the 7th month, on successive nights, a broom star was seen in the east and west. [*Nihon kiryaku*]

(b) 3rd year of the Ei-en reign period, 7th month, 13th day (Aug 16); a broom star was seen in the east for several nights. It was about 5 *chi* long. [*Shodō kanmon*]

NB: This is Comet Halley, return –13.

AD 989 Oct 18 [Korea] (Ho 350)

8th year of King Seongjong of Goryeo, 9th month, day *jiawu* [31] (Oct 18); a broom star was seen. [*Goryeosa*] ch. 47

NB: This is Comet Halley, return –13.

AD 990 Feb 2 [China] (Ho 351)

1st year of the Chunhua reign period of Emperor Taizong of the Song Dynasty, 1st month, day *xinsi* [18]; a guest star emerged in lunar mansion ZHEN [LM 28] and traveled retrograde to reach ZHANG [LM 26] in 70 days, traversing some 40 *du* before disappearing.
[*Song shi: tianwen zhi*] ch. 56

AD 998 Feb 12–23 [China] (Ho 352)

(a) 1st year of the Xianping reign year of Emperor Zhenzong of the Song Dynasty, 1st month, 13th day (Feb 12); a broom star with bright rays over 1 *chi* long emerged north of YINGSHI [LM 13]... on the 26th day, the broom star was extinguished.
[*Song huiyao jigao*] vol. 52

(b) 1st year of the Xianping reign year of Emperor Zhenzong of the Song Dynasty, 1st month, day *jiashen* [21] (Feb 23); a broom star emerged north of YINGSHI [LM 13].
[*Song shi: Zhenzong ji*] ch. 6

(c) 1st year of the Xianping reign year of Emperor Zhenzong of the Song Dynasty, 1st month, day *jiashen* [21] (Feb 23); a broom star with bright rays over 1 *chi* long emerged again north of YINGSHI [LM 13] until day *dingyou* [34], in all 14 days before being extinguished.
[*Song shi: tianwen zhi*] ch. 56

AD 998 Feb 25 [Japan]

(a) 4th year of the Chōtoku reign period, 1st month, 26th day, *bingxu* [23]; during the hour of *yin* [LT = 3–5 h], a broom star over 4 *chi* long was seen in the east.
[*Shodō kanmon*]
[*Dai Nihon shi*] (Ho 352)

(b) 4th year of the Chōtoku reign period, 1st month, 26th day; during the marks of *yin* [LT = 3–5 h], a broom star over 4 *chi* long was seen in the east.
[*Ichidai yōki*]

AD 1001 Feb 23 [Japan] (Ho 353)

3rd year of the Chōhō reign period, 1st month, 27th day; a broom star 2 *chi* long was seen north of TIANCHUAN and south of JUANSHE.
[*Dai Nihon shi*]

AD 1003 Dec 24 [China] (Ho 354)

(a) 6th year of the Xianping reign year of Emperor Zhenzong of the Song Dynasty, 11th month, day *jiayin* [51]; a star became fuzzy in JING [LM 22] and GUI [LM 23].
[*Song shi: Zhenzong ji*] ch. 7

(b) 6th year of the Xianping reign year of Emperor Zhenzong of the Song Dynasty, 11th month, day *jiayin* [51]; there was a greenish-white fuzzy broom star as big as a cup and with bright rays over 4 *chi* long in JING [LM 22] and GUI [LM 23]. It passed from WUZHUHOU to WUCHE and then entered SHEN [LM 21], in all lasting over 30 days before disappearing.
[*Song shi: tianwen zhi*] ch. 56

AD 1005 Oct 1–4 [China] (Ho 355)

(a) 2nd year of the Jingde reign year of Emperor Zhenzong of the Song Dynasty, 8th month, day *xinchou* [38] (Oct 1); a star became fuzzy in ZIWEI.
[*Song shi: Zhenzong ji*] ch. 7

(b) 2nd year of the Jingde reign year of Emperor Zhenzong of the Song Dynasty, 8th month, day *jiachen* [41] (Oct 4); a guest star emerged beside TIANBANG of ZIWEI, all fuzzy like white cotton wadding. It gradually entered the [ZIWEI] enclosure then passed YUNÜ and HUAGAI, in all lasting 11 days before disappearing.
[*Song shi: tianwen zhi*] ch. 56

AD 1014 Feb 12 [China] (Ho 359)
3rd year of the Kaitai reign period of Emperor Shengzong of the Liao Dynasty, 1st month, day *dingyou* [34]; this evening, a broom star appeared in the west.
[*Liao shi: Shengzong ji*] ch. 15

AD 1014 Feb 24 [China]
7th year of the Dazhong xiangfu reign period of Emperor Zhenzong of the Song Dynasty, 1st month, 22nd day, at night; a star emerged in the east with rays over 2 *chi* long; the director of astronomy said, "It is a HANYU star."
[*Song huiyao jigao*] vol. 52

AD 1014 Feb 27 [Korea] (Ho 359)
5th year of King Hyeonjong of Goryeo, 1st month, day *renzi* [49]; a broom star was seen in WUCHE; in the 2nd month, day *gengshen* [57], the broom star entered DALING.
[*Goryeosa]*] ch. 47

AD 1014 Mar 1 [Japan]
3rd year of the Chōwa reign period, 1st month, 27th day; from the 27th day of the 1st month past, during the hour of *xu* [LT = 19–21 h], a broom star 2 *chi* long was seen north of TIANCHUAN and south of JUANSHE.
[*Shō yūki*]
NB: "North" in the original is miswritten as "this."

AD 1014 Sep 29 [China]

7th year of the Dazhong xiangfu reign period of Emperor Zhenzong of the Song Dynasty, 9th month, day *bingxu* [23]; the HANYU star appeared again, like a broom; it had a tail, though not a long one.
[*Song shi: tianwen zhi*] ch. 56

AD 1018 Aug 2–23 [Japan]

(a) 2nd year of the Kannin reign period, 6th month, 18th day (Aug 2); 6th month, 23rd day *jiayin* [51] (Aug 7)...starting from the 18th day past, at the juncture of azimuths *xu* [WNW] and *hai* [NNW], a broom star was seen...having compared precedents in the literature, strange stars like this do not last for long before disappearing.
[*Sakei ki*]

(b) 2nd year of the Kannin reign period, 6th month, 19th day (Aug 3); a broom star over 2 *zhang* long was seen and lasted for several days.
[*Shodō kammon*]

(c) 2nd year of the Kannin reign period, 6th month, 19th day (Aug 3), a broom star over 2 *zhang* long was seen in the northwest. On the 20th day (Aug 4), it was likewise, and by the 29th day (Aug 13), its rays became more intense.
[*Dai Nihon shi*] (Ho 360)

AD 1018 Aug 3 [Korea] (Ho 360)

9th year of King Hyoenjong of Goryeo, 6th month, day *gengxu* [47]; a broom star emerged by the second star of BEIDOU, its gleam was over 4 *zhang* long, shooting into WENCHANG and TIANLAO.
[*Goryeosa]*] ch. 47

AD 1018 Aug 4 [China] (Ho 362)

(a) 2nd year of the Tianxi reign period of Emperor Zhenzong of the Song Dynasty, 6th month, day *xinhai* [48] (20th day); a broom

star emerged in the bowl of BEIDOU…7th month, day *dinghai* [24], the broom star disappeared.
[*Song shi: Zhenzong ji*] ch. 8

(b) 2nd year of the Tianxi reign period of Emperor Zhenzong of the Song Dynasty, 6th month, day *xinhai* [48] (20th day); a broom star about 3 *chi* long emerged northeast of the second star in the bowl of BEIDOU, on a level with the first star of BEIDOU. It traveled northward, passing TIANLAO and brushing WENCHANG, at which point it was over 3 *zhang* long. It passed ZIWEI, SANTAI, and XUANYUAN then traveled rapidly westward as far as QIXING [LM 25], in all for 37 days until it disappeared.
[*Song shi: tianwen zhi*] ch. 56

(c) 2nd year of the Tianxi reign period of Emperor Zhenzong of the Song Dynasty, 6th month, day *xinhai* [48] (20th day); a broom star emerged in the bowl of BEIDOU then traveled northward across the sky…on the evening of the 27th day, the broom star was first extinguished, after 37 days in all.
[*Song huiyao jigao*] ch. 52

NB: *Gujin tushu jicheng* and *Song huiyao jigao* made it "3rd year, 6th month, day *xinhai* [48]." *Song huiyao jigao*, "evening of the 27th day," probably omitted the two characters, "7th month." Ho gave the date as 1019.

AD 1019 Feb 6 [Korea] (Ho 361)

9th year of King Hyoenjong of Goryeo, 12th month, day *dingsi* [54]; a broom star was seen between TIANSHIYUAN, ZONGZHENG, ZONGREN, and SHILOU, pointing toward the west.
[*Goryeosa*] ch. 47

AD 1020 Jan 26 [Korea] (Ho 363)

10th year of King Hyoenjong of Goryeo, 12th month, day *xinhai* [48]; a broom star was seen between ZONGZHENG, ZONGREN, and SHILOU.
[*Goryeosa*] ch. 47

AD 1021 May 25 [China] (Ho 364)

(a) 5th year of the Tianxi reign period of Emperor Zhenzong of the Song Dynasty, 4th month, day *bingchen* [53]; a guest star emerged in Xuanyuan.
[*Song shi: Zhenzong ji*] ch. 8

(b) 5th year of the Tianxi reign period of Emperor Zhenzong of the Song Dynasty, 4th month, day *bingchen* [53]; a guest star large as a peach emerged northwest of the first star of Xuanyuan. Travelling rapidly, it passed the large star of Xuanyuan (Regulus) and entered Taiweiyuan; then it occulted Youzhifa, trespassed against Cijiang, passed northwest of Pingxing, lasting 75 days in all before disappearing in the horizon obscurity.
[*Song shi: tianwen zhi*] ch. 56

(c) 5th year of the Tianxi reign period of Emperor Zhenzong of the Song Dynasty, 4th month, day *bingchen* [53]; a star as large as a peach and like white cotton wadding in appearance emerged at the first star of Xuanyuan. It trespassed against Cijiang, entered Taiweiyuan, and passed northwest of Pingxing, lasting 75 days in all before disappearing in the horizon obscurity.
[*Song shi: tianwen zhi*] ch. 57

AD 1021 May 31 [Korea] (Ho 364)

12th year of King Hyoenjong of Goryeo, 4th month, day *renxu* [59]; a baleful star was seen in Taiwei.
[*Goryeosa*] ch. 47

AD 1032 Jul 15 [China] (Ho 366)

1st year of the Mingdao reign period of Emperor Renzong of the Song Dynasty, 6th month, day *yisi* [42]; a guest star with rayed tail emerged in the northeast near the horizon obscurity and lasted until day *dingsi* [54], in all for 13 days, before disappearing.
[*Song shi: tianwen zhi*] ch. 56

AD 1033 Mar 5 [China] (Ho 367)

2nd year of the Mingdao reign period of Renzong of the Song Dynasty, 2nd month, day *wuxu* [35]; a yellowish-white HANYU star appeared in the northeast with rays about 2 *chi* long.
[*Song shi: tianwen zhi*] ch. 56

AD 1033 Mar 5 [Japan]

(a) 6th year of the Chōgen reign period, 2nd month, 2nd day; a broom star 6 to 7 *chi* long was seen for 10 nights in a row. Likewise, in the 6th year, 2nd month, 3rd day *jihai* [36], at daybreak during the hour of *yin* [LT = 3–5 h], a white broom star over 6 *chi* long was seen in the east.
[*Shodō kanmon*]

(b) 6th year of the Chōgen reign period, 2nd month, 2nd day; a broom star 6 to 7 *chi* long was seen for 12 nights in a row. In the 6th year of the Chōgen reign period, 2nd month, 3rd day *jihai* [36]; at daybreak during the hour of *yin* [LT = 3–5 h], a white broom star over 6 *chi* long was seen in the east.
[*Ichidai yōki*]

NB: Both originals mistakenly wrote "2nd year of the Chōgen reign period" instead of "6th year"; a similar account in *Dai Nihon shi* has "6th year." (Ho 365)

AD 1034 Sep 20 [China] (Ho 368)

(a) 1st year of the Jingyou reign period of Emperor Renzong of the Song Dynasty, 8th month, day *renxu* [59]; a star became fuzzy in ZHANG [LM 26] and YI [LM 27].
[*Song shi: Renzong ji*] ch. 10

(b) 1st year of the Jingyou reign period of Emperor Renzong of the Song Dynasty, 8th month, day *renxu* [59], at night; a star became fuzzy and over 7 *chi* long by 5 *cun* wide in ZHANG [LM 26] and YI [LM 27], lasting 12 days before disappearing.
[*Song shi: tianwen zhi*] ch. 56

(c) 1st year of the Jingyou reign period of Emperor Renzong of the Song Dynasty, 8th month, day *renxu* [59]; a green and yellow and

white vapor like a broom star over 7 *chi* long emerged above ZHANG [LM 26] and YI [LM 27], for 33 days before disappearing.
[*Song shi: tianwen zhi*] ch. 60

AD 1034 Sep 28 [Japan] (Ho 368)
7th year of the Chōgen reign period, 8th month, 13th day; a broom star was seen.
[Dai Nihon shi]

AD 1037 Mar 19 [Korea] (Ho 370)
3rd year of King Jeongjong of Goryeo, 2nd month, day *guiyou* [10]; a broom star was seen five times, each time it was 5 to 6 *chi* long.
[*Goryeosa*] ch. 47

AD 1041 Aug 30–Sep 27 [Korea] (Ho 371)
7th year of King Jeongjong of Goryeo, 8th month; a broom star over 30 *chi* long was seen in the east and lasted more than 20 days before being extinguished.
[*Goryeosa*] ch. 47

AD 1041 Sep 1 [China]
1st year of the Qingli reign period of Emperor Renzong of the Song Dynasty, 8th month, day *gengchen* [17], at night; there was a white vapor over 10 *chi* long in the east in the space of lunar mansion XING [LM 25]. By the 10th day, it was over 1 *zhang* long and charged into TIANXIANG before coming to rest south of the large star of lunar mansion XING [LM 25]; it lasted for over 90 days in all before disappearing.
[*Song shi: tianwen zhi*] ch. 60

AD 1041 Oct 28–Nov 26 [Korea] (Ho 372)
7th year of King Jeongjong of Goryeo, 10th month; a broom star over 30 *chi* long emerged in the east for more than 10 days.
[*Goryeosa*] ch. 47

AD 1049 Mar 10 [China] (Ho 373)
1st year of the Huangyou reign period of Emperor Renzong of the Song Dynasty, 2nd month, day *dingmao* [4]; a broom star emerged in XU

[LM 11] and appeared at dawn in the east, pointing southwest; it passed ZIWEI as far as LOU [LM 16], for 114 days in all before disappearing.
[*Song shi: tianwen zhi*] ch. 56
NB: *Qidan guozhi* has "for 124 days in all."

AD 1053 Feb 25 [Korea] (Ho 374)

7th year of King Munjong of Goryeo, 2nd month, day *yihai* [12]; a broom star more than 1 *zhang* long emerged in KULOU and then entered YI [LM 27].
[*Goryeosa*] ch. 47

AD 1056 Aug 14–Sep 11 [China] (Ho 376)

(a) 1st year of the Jiayou reign period of Emperor Renzong of the Song Dynasty, 7th month; a broom star over 1 *zhang* long emerged in ZIWEIYUAN; 8th month, day *guihai* [60], this evening the broom star was extinguished.
[*Song shi: Renzong ji*] ch. 12

(b) 1st year of the Jiayou reign period of Emperor Renzong of the Song Dynasty, 7th month; a white broom star over 1 *zhang* long emerged in ZIWEI and passed by QIXING [LM 25], lasting until the 8th month, day *guihai* [60] when it was extinguished.
[*Song shi: tianwen zhi*] ch. 56

AD 1056 Sep 10 [Japan]

4th year of the Tengi reign period, 7th month, 28th day; at daybreak, a broom star was first seen in the east; by the 7th to 8th days of the 8th month, it was over 2 *zhang* long.
[*Shodō kammon*]
NB: According to Ho 376, *Dai Nihon shi* made it 2 *chi* long at first appearance.

AD 1056 Dec 10–AD 1057 Jan 7 [Korea]

10th year of King Munjong of Goryeo, 11th month; a broom star was seen in ZHEN [LM 28].
[*Jeungbo munheon bigo*] ch. 6

AD 1060 Dec 22 [Japan]

(a) 3rd year of the Kōhei reign period, 11th month, 27th day; a broom star over 5 *chi* long was seen in the south for 7 nights in succession.
[*Shodō kanmon*]

(b) 3rd year of the Kōhei reign period, 11th month, 27th day; a broom star over 5 *chi* long was seen in the south for 7 nights. [*Ichidai yōki*]
[*Dai Nihon shi*] (Ho 378)

AD 1063 Jun 8 [Korea]

17th year of King Munjong of Goryeo, 5th month, day *gengxu* [47]; there was a star that emerged at DAJIAO and became fuzzy in DI [LM 3].
[*Goryeosa*] ch. 47

AD 1063 Aug 22 [Korea]

17th year of King Munjong of Goryeo, 7th month, day *yichou* [2]; there was a star as big as a peck measure with a tail several *chi* long that emerged at azimuth *kan* [N] and became fuzzy in DI [LM 3].
[*Goryeosa*] ch. 47

AD 1066 Apr 2–Jul 2 [China] (Ho 380)

(a) 3rd year of the Zhiping reign period of Emperor Yingzong of Song, 3rd month, day *jiwei* [56] (Apr 2); a broom star appeared in YING-SHI [LM 13]; in the morning it was seen in the east. It was about 7 *chi* long. It pointed southwest toward WEI [LM 12] and reached FENMU. Gradually its speed increased as it moved eastward, and when near the Sun, it disappeared. On day *xinsi* [18] (Apr 24), in the evening, it was seen in the southwest. In the north, there was a star without rays or broom, advancing in an easterly direction. There was another white vapor about 3 *chi* wide. It penetrated ZIWEI and the polar asterism JIXING, joining FANG [LM 4]. Its head and tail went below the horizon. It advanced eastward, passed

WENCHANG and BEIDOU, and penetrated WEI [LM 6]. On day *renwu* [19] (Apr 25), the star again had its rays and a broom. It was more than 1 *zhang* long and more than 3 *chi* wide; it pointed northeast. It passed WUCHE. The white vapor became branched and stretched across the sky. It penetrated BEIHE, WU ZHUHOU, XUANYUAN, and also WUDIZUO and NEI WUZHUHOU of TAIWEI. It reached JIAO [LM 1], KANG [LM 2], DI [LM 3], and FANG [LM 4]. On day *guiwei* [20] (Apr 26), the broom star was 15 *chi* long. The star had a broom-like vapor and resembled a 1/10 peck measure. It passed from YINGSHI [LM 13] to *ZHANG* [LM 26], a total of 14 lunar mansions. After a total of 67 days, the star's vapor and "bushiness" all vanished.
[*Song shi*] ch. 56

(b) 3rd year of the Zhiping reign period of Emperor Yingzong of Song, spring, 3rd month, day *gengshen* [57] (Apr 3); a broom star was seen in the morning in [YING-]SHI [LM 13]...On day *xinsi* [18] (Apr 24), the broom star was seen in the morning in MAO [LM 18]. It was like Venus and 15 *chi* long. On day *renwu* [19] (Apr 25), it emerged fuzzy at BI [LM 19]. It was like the moon. In the 5th month, day *yichou* [2] (Jun 7), the broom star reached ZHANG [LM 26] and disappeared.
[*Song shi*] ch. 13

(c) 2nd year of the Xiaoyong reign period of Emperor Daozong of Liao, 3rd month, day *renwu* [19] (Apr 25); a broom star was seen in the west.
[*Liao shi*] ch. 22

NB: This is Comet Halley, return –12.

AD 1066 Apr 3–27 [Japan]

(a) 2nd year of the Jiryaku reign period, 3rd month, 6th day (Apr 3); at daybreak, a broom star was seen in the east. In the 4th month, 1st day (Apr 27), in the hour of *you* (5 p.m.–7 p.m.), the broom star was again seen in the west.
[*Jyōsandai yōryaku*]

(b) 3rd year (should be 2nd year) of the Jiryaku reign period, 3rd month (Mar 29–Apr 26); a broom star was seen in the east. After 20 days, it was no longer seen. Later it was seen in the west. [*Sankai ki*]

NB: This is Comet Halley, return –12. (Ho 380) misdates the '3rd month, 6th day' of the *Dai Nihon shi* record similar to *(a)* as Apr 22.

AD 1066 Apr 20 [Korea] (Ho 380)

20th year of King Munjong, 3rd month, day *dingchou* [14] (Apr 20); a star appeared in the northwest. It was as large as the Moon. Suddenly it changed into a fuzzy broom star (*hui bo*).

[*Goryeosa*] ch. 47

NB: This is Comet Halley, return –12.

AD 1075 Nov 17 [China] (Ho 385)

(a) 8th year of the Xining reign period of Emperor Shenzong of the Song Dynasty, 10th month, day *yiwei* [32]; a broom star emerged in ZHEN [LM 28]. On day *dingwei* [44], the broom star was not seen.
[*Song shi: Shenzong ji*] ch. 15

(b) 8th year of the Xining reign period of Emperor Shenzong of the Song Dynasty, 10th month, day *yiwei* [32]; a greenish-white broom star resembling Saturn emerged in the space of ZHEN [LM 28]. On day *bingshen* [33], it sprouted bright rays 3 *chi* long on the northwest, which pointed obliquely toward ZHEN [LM 28] like a broom. On day *dingyou* [34], the bright rays were 5 *chi* long. On day *wuxu* [35], the bright rays were 7 *chi* long and pointing obliquely toward ZUOXIA; it lasted until day *dingwei* [44] when it disappeared in the horizon obscurity.
[*Song shi: tianwen zhi*] ch. 56

AD 1075 Nov 18 [Korea] (Ho 385)

29th year of King Munjong of Goryeo, 10th month, day *bingshen* [33]; a broom star over 7 *chi* long was seen in ZHEN [LM 28].

[*Goryeosa*] ch. 47

AD 1075 Nov 19–Nov 24 [Japan]

(a) 2nd year of the Jōhō reign period, 10th month, 9th day; at daybreak, a broom star was seen for 5 nights.
[*Shodō kanmon*]

(b) 2nd year of the Jōhō reign period, 10th month, 9th day; a broom star was seen for 5 nights.
[*Ichidai yōki*]
[*Dai Nihon shi*] (Ho 385)

AD 1080 Jan 6 [China] (Ho 386)

5th year of the Dakang reign period of Emperor Daozong of the Liao Dynasty, 12th month, day *bingwu* [43]; a broom star trespassed on *Wei* [LM 6].
[*Liao shi: Daozong ji*] ch. 24

AD 1080 Aug 10 [China] (Ho 387)

(a) 3rd year of the Yuanfeng reign period of Emperor Shenzong of the Song Dynasty, 7th month, day *guiwei* [20]; a broom star with white vapor 1 *zhang* long emerged in the northwest, south of LANGWEI of TAIWEIYUAN, and pointing obliquely to the southeast, within the space of ZHEN [LM 28]. On day *bingxu* [23], it traveled toward the northwest into the space of YI [LM 27]. On day *wuzi* [25], it was 3 *chi* long and obliquely penetrated LANGWEI. On day *guimao* [40] (NB: Should be *guisi* [30]), it trespassed against XUANYUAN until day *dingyou* [34] when it entered the horizon obscurity and was no longer seen. At dawn on day *gengzi* [37], it reappeared in the space of ZHANG [LM 26], until day *wuzi* [25] (NB: Should be *wuwu* [55]), lasting 36 days in all, whereupon it disappeared from view.
[*Song shi: tianwen zhi*] ch. 56

(b) 3rd year of the Yuanfeng reign period of Emperor Shenzong of the Song Dynasty, 7th month, day *guiwei* [20]; a broom star

emerged in TAIWEIYUAN...In the 8th month, day *wuwu* [55], the broom star did not appear.
[*Song shi: Shenzong ji*] ch. 16

AD 1092 Jan 8 [China] (Ho 389)

6th year of the Yuanyou reign period of Emperor Zhezong of the Song Dynasty, 11th month, day *xinhai* [48]; a guest star emerged in the space of SHEN [LM 21], trespassed against and occulted the CE star, and on day *renzi* [49], it trespassed on the JIUYOU star. In the 12th month, day *guiyou* [10], it entered KUI [LM 15], not until the 7th year, 3rd month, day *xinhai* [48], did it disperse.
[*Song shi: tianwen zhi*] ch. 56

AD 1097 Sep 9–Oct 7 [China] (Ho 390)

(a) 4th year of the Shaosheng reign period of Emperor Zhezong of the Song Dynasty, 8th month, day *jiyou* [46] (Oct 6); a broom star emerged in the west...in the 9th month, on day *wuchen* [5], the broom star was extinguished.
[*Song shi: Zhezong ji*] ch. 18

(b) 4th year of the Shaosheng reign period of Emperor Zhezong of the Song Dynasty, 8th month, day *jiyou* [46] (Oct 6); a gleaming white broom star resembling Saturn with vapor 3 *zhang* long emerged in the space of DI [LM 3], and pointed obliquely toward the left star of TIANSHI. In the 9th month, day *renzi* [49], the bright rays were 5 *chi* long as it entered TIANSHIYUAN. On day *jiwei* [56], it trespassed against the HUANZHE star of TIANSHIYUAN. On day *gengshen* [57], it trespassed against DIZUO of TIANSHIYUAN. On day *wuchen* [5], it disappeared and was no longer seen.
[*Song shi: tianwen zhi*] ch. 56

(c) 3rd year of the Shouchang reign period of Emperor Daozong of the Liao Dynasty, 8th month (Sep 9–Oct 7); a broom star emerged in DI [LM 3], and pointed obliquely toward TIANSHIYUAN. Its

bright rays were over 3 *chi* long and over the course of 3 days grew to over 1 *zhang* and swept BA.
[*Qidan guozhi*] ch. 9

AD 1097 Oct 4 [Korea] (Ho 390)
2nd year of King Sukjong of Goryeo, 8th month, day *dingwei* [44]; a broom star was seen between DI [LM 3] and FANG [LM 4], its rays shooting into TIANSHIYUAN.
[*Goryeosa*] ch. 47

AD 1097 Oct 8 [Japan]
(a) 1st year of the Jōtoku reign period, 9th month, 1st day; this month, on the 1st day *xinhai* [48], at dusk during the hour of *you* [LT = 17–19 h], a white broom star over 1 *zhang* long emerged in the west, the vapors of its rays reaching TIANSHIYUAN; afterward, it moved eastward. Likewise, on the 2nd, 3rd, and 4th days, it was seen inside TIANSHIYUAN, pointing to azimuth *zhen* [E]; on the 5th day, the sky was overcast and it was not seen; on the 6th day (Oct 13), it moved northward and was seen north of TIANSHIYUAN, now 7 to 8 *zhang* long and its brightness slowly diminishing. Subsequently, the sky was overcast and it was not seen; on the 18th day, the sky was clear, but the star had gone and was not observed.
[*Shodō kammon*]
[*Dai Nihon shi*] (Ho 390)
(b) 1st year of the Jōtoku reign period, 9th month, 1st day; a broom star was seen in the west. It was likewise until the 18th day (25), after which it was not seen.
[*Ichidai yōki*]

AD 1106 Feb 9–15 [Japan]
(a) 1st year of the Kajō reign period, 1st month, 4th day; on the 4th day of this month, *dingyou* [34], at dusk during the hour of *you* [LT = 17–19 h], a white broom star over 10 *zhang* long emerged

at azimuth *kun* [SW], its rays pointing toward azimuth *zhen* [E] and reaching between asterisms TIANCANG and TIANYUAN. It was the same for 6 days and nights, by which time it was over 1 *zhang* long and its rays had dimmed. It was likewise for 7 days and nights, when it had moved considerably eastward and was over 3 to 4 *chi* long. On the nights of the 8th, 9th, and 10th, the sky was overcast and it was not seen; on the night of the 11th, it again moved eastward; on the nights of the 12th, 13th, and 14th, the sky was overcast and it was not seen; on the night of the 15th (Feb 15), it moved eastward and was over 2 *zhang* long.
[*Shodō kambun*]
[*Dai Nihon shi*] (Ho 391)

(b) 1st year of the Kajō reign period, 1st month, 4th day; a broom star over 10 *zhang* long emerged in azimuth *kun* [SW] for over 30 days when it was extinguished.
[*Hyaku ren shō*]

AD 1106 Feb 9 [Korea] (Ho 391)

1st year of King Yejong of Goryeo, 1st month, day *dingyou* [34]; a broom star over 10 *chi* long was seen in the southwest and lasted over a month before being extinguished.
[*Goryeosa*] ch. 47

AD 1106 Feb 10 [China] (Ho 391)

(a) 5th year of the Chongning reign period of Emperor Huizong of the Song Dynasty, 1st month, day *wuxu* [35]; a broom star emerged in the west and stretched across the sky…in the 3rd month, on day *bingshen* [33], the astral anomaly having vanished, by imperial edict the demand to straightforwardly report on it was rescinded.
[*Song shi: Huizong ji*] ch. 20

(b) 5th year of the Chongning reign period of Emperor Huizong of the Song Dynasty, 1st month, day *wuxu* [35]; a broom star emerged in the west resembling a cup with a wide mouth, bright rays emanating from it like fragmented stars. It was 6 *zhang* long and 3 *chi*

wide, and pointing obliquely toward the northeast. From lunar mansion KUI [LM 15] it penetrated LOU [LM 16], WEI [LM 17], MAO [LM 18], BI [LM 19], then entered the horizon obscurity and was no longer seen.
[*Song shi: tianwen zhi*] ch. 56

AD 1110 Jan 22–AD 1111 Feb 9 [China] (Ho 393)

4th year of the Daguan reign period of Emperor Huizong of the Song Dynasty; a star became fuzzy and several *zhang* long in the Central Palace. It emerged between WANGLIANG and ZAOFU, then passed by GEDAO, then retrograded to enter ZIGONG, repeatedly sweeping asterisms within and without the [ZIWEI]YUAN. Having retreated, it very soon advanced again to sweep DIZUO twice; from start to finish, it was over 30 days before it was extinguished.
Yuan Dynasty, Jiangsu ed. [*Jinling xin zhi*] ch. 3

AD 1110 May 31–Jun 29 [Korea] (Ho 393)

5th year of King Yejong of Koryŏ, 5th month, day *jiyou* [46] (May 31); a broom star entered ZIWEI; on day *yimao* [52] (Jun 21), during the second watch of the night, the broom star shone between the stars of TIANJIANGJUN and GEDAO until daybreak when it was extinguished. During the night of day *dingsi* [54] (Jun 23), the broom star shone by [FU-]LU and CE; on day *wuwu* [55] (Jun 24), it shone northwest of WANGLIANG; on day *jiwei* [56] (Jun 25), it shone between HUAGAI and CHUANSHE. During the night of *gengshen* [57] (Jun 26), it shone in the middle of HUAGAI; on day *xinyou* [58] (Jun 27), it shone north of the LIUJIA stars below HUAGAI. During the night of *guihai* [60] (Jun 29), it moved among the stars of NÜYUGONG.
[*Goryeosa*] ch. 47

AD 1110 Jun 1–21 [Japan]

(a) 1st year of the Ten-ei reign period, 5th month, 12 day; a broom star was seen in the east, its brightness over 5 to 6 *chi* long.
[*Tōji chōjya honin*]

(b) 1st year of the Ten-ei reign period, 5th month, 12th day...from the 12th day of the 5th month; [a broom star] emerged in the east, its brightness over 6 *chi* long...on the 9th day [of the 6th month], the broom star's brightness was extinguished; on the 10th day (Jun 11), the star's light was lost from sight.
[*Kujyakukyō misyuhō ki*]

(c) 1st year of the Ten-ei reign period, 5th month, 12th day; a broom star 5 *chi* long was seen in the east and lasted over 20 days before being extinguished.
[*Dai Nihon shi*] (Ho 393)

AD 1110 May 29 [China] (Ho 393)

(a) 4th year of the Daguan reign period of Emperor Huizong of the Song Dynasty, 5th month, day *dingwei* [44]; a broom star emerged in Kui [LM 15] and Lou [LM 16].
[*Song shi: Huizong ji*] ch. 20

(b) 4th year of the Daguan reign period of Emperor Huizong of the Song Dynasty, 5th month, day *dingwei* [44]; a broom star with bright rays 6 *chi* long emerged in Kui [LM 15] and Lou [LM 16]. It traveled northward and entered Ziweiyuan, reaching as far as the northwest where it entered the horizon obscurity and was no longer seen.
[*Song shi: tianwen zhi*] ch. 56

AD 1111 Aug 12 [China]

1st year of the Zhenghe reign period of Emperor Huizong of the Song Dynasty, 7th month, 6th day; a white broom star over 1 *zhang* long emerged in Ziweiyuan and passed by Qixing [LM 25], lasting from this until the night of the 14 day of the 8th month when it was extinguished.
[*Song huiyao jigao*] vol. 52

AD 1113 Aug 15 [Korea] (Ho 394)

8th year of King Yejong of Goryeo, 7th month, day *xinsi* [18]; there was a fuzzy star in Yingshi [LM 13].
[*Goryeosa*] ch. 47

AD 1123 Feb 2 [China]

5th year of the Xuanhe reign period of Emperor Huizong of the Song Dynasty, 1st month, 5th day; a broom star emerged in the west; from KUI [LM 15] it penetrated LOU [LM 16], WEI [LM 17], MAO [LM 18], BI [LM 19]…lasting until the 25th day when the broom star was extinguished.

[*Song huiyao jigao*] vol. 52

AD 1123 Aug 11 [Korea] (Ho 395)

1st year of King Injong of Goryeo, 7th month, day *jisi* [6]; there was a fuzzy star in BEIDOU.

[*Goryeosa*] ch. 47

AD 1126 Jul 19 [China] (Ho 396)

1st year of the Jingkang reign period of Emperor Qinzong of the Song Dynasty, 6th month, day *renxu* [59]; a broom star emerged in ZIWEIYUAN.

[*Song shi: Qinzong ji*] ch. 23

AD 1126 Jul 22 [Japan]

(a) 1st year of the Daiji reign period, 7th month, after the 1st day *yichou* [2]; a broom star was seen in the north, from 3 *chi* to over 1 *zhang* long.

[*Shodō kanmon*]

(b) 1st year of the Daiji reign period, 7th month, new moon day; a broom star over 3 *chi* long was seen in the north.

[*Ichidai yōki*]

(c) 1st year of the Daiji reign period, 7th month, 1st day; a broom star about 3 *chi* long was seen in the north.

[*Dai Nihon shi*] (Ho 396)

AD 1127 Jan 15 [China] (Ho 397)

1st year of the Jingkang reign period of Emperor Qinzong of the Song Dynasty, 12th month, day *renxu* [59]; a broom star several *zhang* long emerged northeast of ZIWEIYUAN, brushed DIZUO, and swept WENCHANG.

When the broom star was observed, there was a white vapor emerging from TAIWEIYUAN.
Yuan Dynasty, Jiangsu ed. [*Jinling xin zhi*] ch. 3

AD 1132 Jan 5 [China] (Ho 400)

(a) 1st year of the Shaoxing reign period of Emperor Gaozong of the Song Dynasty, 12th month, day *wuyin* [15]; a broom star appeared in WEI [LM 17].
[*Song shi: tianwen zhi*] ch. 56

(b) 1st year of the Shaoxing reign period of Emperor Gaozong of the Song Dynasty, 12th month, day *wuyin* [15]; in view of the emergence of the broom star, a straightforward report on it was demanded.
[*Song shi: Gaozong ji*] ch. 26

AD 1132 Oct 5–10 [Japan]

(a) 1st year of the Chōjō reign period, 8th month, 25th day; on the 28th day of the 8th month...last month on the night of the 25th day *renzi* [49], at daybreak during the hour of *yin* [LT = 3–5 h], a white broom star 3 *chi* long, its tail pointing westward, was seen in the north in the space of ZUI [LM 20] near BAGU; on the night of the 26th, the sky was overcast and it was not seen; on the 27th, at night during the hour of *hai* [LT = 21–23 h], it was seen at azimuth *yin* [ENE], close by the third star of LOU [LM 16]; it was over 3 *zhang* long, its intense rays and tail pointing toward azimuth *xu* [WNW] to *hai* [NNW]; on the night of the 28th, it had moved southward, its brightness slightly diminished but over 1 *zhang* long, its vaporous rays reaching lunar mansion KUI [LM 15]; on the night of the 29th, it was in the same space, still moving southward, in tandem with the asterism TUSIKONG; its sharp rays were diminished to about 2 to 3 *chi* long; on the 13th night and the night of the 1st day [of the 9th month], the whole sky was overcast and it was not seen; after the night of the 2nd day, it had gone and was no longer seen.

[*Chūyū ki*]
[*Dai Nihon shi*] (Ho 401)

NB: "Asterism BAGU" was erroneously written asterism "*ru ao*"; asterism "TUSIKONG" was mistakenly written asterism "*xia si kong*." *Dai Nihon shi* added that when first observed, the comet was trespassing on ZHEN [LM 28].

(b) 1st year of the Chōjō reign period, 8th month, 25th day; this daybreak a broom star was seen in the north; on the 27th (Oct 7), it moved to azimuth *xun* [SE] and was over 3 *zhang* long.
[*Jyūsandai yōryaku*]

AD 1132 Oct 5 [Korea] (Ho 401)

10th year of King Injong of Goryeo, 8th month, day *renzi* [49]; a broom star was seen at BAGU pointing southeast; on day *jiayin* [51] (Oct 7), it pointed northwest and was 3 *chi* long.
[*Goryeosa*] ch. 47

AD 1132 Oct 6–7 [China] (Ho 401)

(a) 2nd year of the Shaoxing reign period of Emperor Gaozong of the Song Dynasty, 8th month, 26th day (Oct 6); the grand scribe said, "Tonight during the fourth watch, a broom star resembling Mars or Jupiter emerged in the space of lunar mansion WEI [LM 17]"...reported during the initial watch of the next night; then it again trespassed against TUSIKONG...until the 17th day of the 9th month when the broom star disappeared.
[*Song huiyao jigao*] vol. 52

(b) 2nd year of the Shaoxing reign period of Emperor Gaozong of the Song Dynasty, 8th month, day *jiayin* [51] (Oct 7); a broom star emerged in WEI [LM 17]...in the 9th month, day *jiaxu* [11], the broom star disappeared.
[*Song shi: Gaozong ji*] ch. 27

(c) 2nd year of the Shaoxing reign period of Emperor Gaozong of the Song Dynasty, 8th month, day *jiayin* [51] (Oct 7); a broom star appeared in WEI [LM 17]; on day *bingchen* [53], it moved

and trespassed on TUSIKONG; not until the 9th month, day *jiaxu* [11], did the broom star disappear.
[*Song shi: tianwen zhi*] ch. 56

AD 1138 Aug 27 [Japan]
4th year of the Hōen reign period, 7th month, 20th day; a white broom star 5 to 6 *chi* long was seen at azimuth *xu* [WNW]–*hai* [NNW]; after 2 to 3 days, it gradually faded and disappeared.
[*Ichidai yōki*]
[*Dai Nihon shi*] (Ho 402)

AD 1138 Sep 3 [China] (Ho 403)
8th year of the Shaoxing reign period of Emperor Gaozong of the Song Dynasty, 7th month, day *xinhai* [48]; a broom star emerged in the east...in the 8th month, day *dingchou* [14], the broom star was extinguished.
[*Song shi: Gaozong ji*] ch. 29

AD 1142 Aug 8 [China]
12th year of the Shaoxing reign period of Emperor Gaozong of the Song Dynasty, 7th month, day *bingwu* [43]; a broom star appeared in the northeast in the space of lunar mansion JING [LM 22].
Qing Dynasty Kangxi period ed. [*Shaanxi tongzhi*] ch. 30

AD 1145 Apr 26–Jul 9 [China] (Ho 406)
(a) 15th year of the Shaoxing reign period of Emperor Gaozong of Southern Song, 4th month, day *wuyin* [15] (Apr 26); a broom star was seen in the east. On day *bingshen* [33] (May 14), it was again seen in the space of SHEN [LM 21]. In the 5th month, day *dingsi* [54] (Jun 4), it changed into a guest star. Its color was greenish-white. On day *renxu* [59] (Jun 9), it was beside ZHANG [LM 26] and guarding it. In the 6th month, day *dinghai* [24] (Jul 4), it dispersed.
[*Song shi*] ch. 56

(b) 15th year of the Shaoxing reign period of Emperor Gaozong of Southern Song, 4th month, day *wuyin* [15] (Apr 26); a broom star was seen in the east. On day *dinghai* [24] (May 5), there was a general amnesty on account of the broom star. On day *guisi* [30] (May 11), the broom star disappeared...In the 5th month, day *bingchen* [53] (Jun 3), a guest star was seen...In the 6th month, day *dinghai* [24] (Jul 6), the guest star disappeared.
[*Song shi*] ch. 30

(c) 5th year of the Huangtong reign period of Emperor Xizong of Jin, 4th month, day *bingshen* [33] (May 14); a broom star appeared in the northwest. It was more than 1 *zhang* long. In the 5th month, day *renxu* [59] (Jun 9), it was extinguished.
[*Jin shi*] ch. 20

NB: This is Comet Halley, return –11.

AD 1145 May 3–Jun 27 [Japan] (Ho 406)

(a) 1st year of the Kyūan reign period, 4th month, 10th day, *yiyou* [22] (May 3). The erudite of astronomy said that a star had appeared fuzzy in Kui [LM 15]. Two others stated that on Apr 23 or Apr 27 the broom star was first seen. On the 16th day, *xinmao* [28] (May 9), in the morning, its rays pointed west; they were about 2 *zhang* long. On the 21st day, *bingshen* [33] (May 14), in the morning, the broom star disappeared. On the 23rd day, *wuxu* [35] (May 16), the broom star was seen again in azimuth *xin* (WNW). Its rays were about 5 *chi* long and pointed east; they were partly covered by clouds. On the 24th day, *jihai* [36] (May 17), the sky was clear and the broom star was seen. Its rays were about 2 *zhang* long. On the 25th day, *gengzi* [37] (May 18), the broom star was in the east; [later] it disappeared. In the 5th month, 7th day, *renzi* [49] (May 30), the light of the broom star was gradually fading, possibly on account of the moon's light. On the 12th day, *dingsi* [54] (Jun 4), the broom star was gradually fading. On the 13th day (Jun 5), the broom star was close to the moon and its rays were not seen. On the 16th day (Jun 8), at night it was again seen;

its rays were 3 *chi* (long). On the 20th day, *yichou* [2] (Jun 12), the broom star still had not dispersed. On the 26th day (Jun 18), the sky was clear and the broom star was seen. After this, it was cloudy every night until Jun 27. On that evening, it was clear but the broom star was not seen.
[*Tai ki*]

(b) 1st year of the Kyūan reign period, 4th month, 5th day, *gengchen* [17] (Apr 28), at the hour of *yin* (LT 3–5); a broom star was seen in the east. (Commentary: Its length was 1 *zhang* and its color was white)…It penetrated and trespassed against Kui [LM 15]… On the 20th day (May 13), the rays and vapor were not seen…On the 23rd day (May 16), it was seen in the west again.
[*Honchō seiki*]

NB: This is Comet Halley, return –11.

AD 1145 May 14 [Korea] (Ho 406)

23rd year of King Injong, 4th month, day *bingshen* [33] (May 14); a broom star was seen in the northwest for 15 days. It was more than 1 *zhang* long.
[*Goryeosa*] ch. 47
NB: This is Comet Halley, return –11.

AD 1146 Dec 29 [China] (Ho 407)

16th year of the Shaoxing reign period of Emperor Gaozong of the Southern Song Dynasty, 11th month, day *gengyin* [27]; a broom star appeared in the southwest in lunar mansion Wei [LM 12].
[*Song shi: tianwen zhi*] ch. 56

AD 1147 Jan 4–Feb 14 [Japan]

(a) 2nd year of the Kyūan reign period, 12th month, 3rd day *wuxu* [35] (Jan 6); there was a broom star between Wei [LM 12] and Shi [LM 13], its light over 1 *zhang* long; on the 19th day, *jiayin* [51] (Jan 22), on this night the broom star faded, as if to disappear.
[*Yasuchika ason ki*]

(b) 2nd year of the Kyūan reign period, 1st month, 12th day *bingzi* [13] (Feb 13); today at daybreak during the hour of *yin* [LT = 3–5 h], a broom star was seen in the east between Nü [LM 10] and Xu [LM 11], its light 1 *zhang* long and coming close to asterism Paogua; on the 13th day, *dingchou* [14] (Feb 14), today at daybreak the broom star's location is the same as yesterday but inclining rather northward; on the 24th day, *wuzi* [25], on this night the broom star was not seen.
[*Honchō seiki*]
[*Dai Nihon shi*] (Ho 408)

NB: The original miswrote "asterism Paogua" as "asterism Paoyou."

(c) 2nd year of the Kyūan reign period, 12th month, 1st day (Jan 4); a broom star was seen at azimuth *kun* [SW]; 3rd year, 1st month, 12th day (Feb 13); the broom star was seen in the east.
[*Hyaku ren shō*]

(d) 2nd year of the Kyūan reign period, 12th month, 1st day (Jan 4); a broom star about 2 to 3 *zhang* long was seen in the west; on the 3rd day (Jan 6), it was about 1 *zhang* long and was seen between Wei [LM 12] and Yingshi [LM 13]; on the 6th day (Jan 9), it was seen in the southwest, now about 10 *zhang* long; after over 10 days, it faded away.
[*Dai Nihon shi*] (Ho 407)

AD 1147 Jan 6 [China] (Ho 408)

(a) 16th year of the Shaoxing reign period of Emperor Gaozong of the Song Dynasty, 12th month, day *wuxu* [35] (3rd day); a broom star appeared in the southwest; on day *yisi* [42] it was extinguished.
[*Song shi: Gaozong ji*] ch. 30

(b) 16th year of the Shaoxing reign period of Emperor Gaozong of the Song Dynasty, 12th month, day *wuxu* [35] (3rd day); on this night, a broom star appeared in the southwest…until the night of the 9th day of this month when it vanished.
[*Song huiyao jigao*] vol. 52

AD 1147 Feb 8 [China] (Ho 408)

7th year of the Huangtong reign period of Emperor Xizong of the Jin Dynasty, 1st month, day *xinwei* [8]; a broom star over 1 *zhang* long emerged in the east for 15 days in all before being extinguished.

[*Jin shi: tianwen zhi*] ch. 20

AD 1147 Feb 12 [China]

17th year of the Shaoxing reign period of Emperor Gaozong of the Song Dynasty, 1st month, day *yihai* [12]; a baleful star as small as Jupiter and with bright rays 5 *zhang* long emerged in the northeast within lunar mansion Nü [LM 10]; not until the 2nd month, day *bingyin* [3], was it was extinguished.

[*Song shi: tianwen zhi*] ch. 60

AD 1156 Jul 23–Aug 4 [Japan]

(a) 1st year of the Hōgen reign period, 7th month, 11th day; at daybreak a white broom star over 6 *chi* long was seen at azimuth *yin* [ENE]; between the 12th and 13th days, it moved northeastward, and by the 15th (Aug 4) at daybreak, it was trespassing against asterisms Wuzhuhou and Sangong, from which it was separated by over 3 *cun*.
[*Ichidai yōki*]

(b) 1st year of the Hōgen reign period, 7th month, 5th day (Jul 23); a broom star was seen in the east; on the 11th day (Jul 29), it was about 6 *chi* long and white; on the 12th day (Jul 30), it moved northeastward; on the 15th day (Aug 4), it trespassed against Wuzhuhou and Sangong.
[*Dai Nihon shi*] (Ho 409)

AD 1156 Jul 25–26 [China] (Ho 409)

(a) 26th year of the Shaoxing reign period of Emperor Gaozong of the Song Dynasty, 7th month, day *bingwu* [43] (Jul 25); a broom star about 1 *zhang* long with bright rays 2 *chi* long appeared in

DONGJING [LM 22]; on day *guichou* [50], it further trespassed on WUZHUHOU.
[*Song shi: tianwen zhi*] ch. 56

(b) 26th year of the Shaoxing reign period of Emperor Gaozong of the Song Dynasty, 7th month, day *bingwu* [43] (Jul 25); a broom star appeared in the northeast in lunar mansion JING [LM 22]; on day *dingwei* [44], the star was like Jupiter, with bright rays 2 *chi* long.
[*Wenxian tongkao*] ch. 286

(c) 26th year of the Shaoxing reign period of Emperor Gaozong of the Song Dynasty, 7th month, day *dingwei* [44] (Jul 26); a broom star emerged in JING [LM 22]; on day *bingchen* [53], the broom star was extinguished.
[*Song shi: Gaozong ji*] ch. 31

(d) 26th year of the Shaoxing reign period of Emperor Gaozong of the Song Dynasty, 7th month, 8th day (Jul 26); a broom star emerged in lunar mansion JING [LM 22]...the 17th day of this month when it set and vanished.
[*Song huiyao jigao*] vol. 52

AD 1156 Aug 14–25 [Korea] (Ho 409)
10th year of King Uijong of Goryeo, 7th month, day bingyin [3]; a broom star was seen in the east and was not extinguished until day dingchou [14] (Aug 25) in the 8th month.
[Goryeosa] ch. 47

AD 1161 Jul 22 [China] (Ho 410)
31st year of the Shaoxing reign period of Emperor Gaozong of the Song Dynasty, 6th month, day *jisi* [6]; a broom star appeared northeast of BEIDOU and the star TIANQUAN, and the Prefect Grand Astrologer mistakenly called it a HANYU star.
[*Song shi: tianwen zhi*] ch. 56

AD 1166 Apr 23–30 [Japan]

(a) 1st year of the Ninnan reign period, 3rd month, 22nd day, *yichou* [2]; at night during the hour of *hai* [LT = 21–23 h], a white broom

star over 3 *chi* long was seen within TAIWEIGONG at the gap in the asterism [NEI-]PING, its rays pointing toward YOUZHIFA; after the 23rd day, the broom star was in the north; on the 29th day, it was near WUDIZUO, more than 1 *chi* away, its vaporous rays pointing toward HUANGDI and BAIDIZUO; on the thirtieth day (May 1), the sky was overcast and it was not seen. After this, it was consistently overcast. In the 4th month, on the 11th day (May 13), the sky was clear, but the broom star had vanished and was not observed.
[*Yasuchika ason ki*]

(b) 1st year of the Ninnan reign period, 3rd month, 22nd day; a white broom star 3 *chi* long was seen in TAIWEIGONG, within YOUYE-MEN. It emerged west of [NEI-]PING and penetrated ZHIFA. In the 4th month, on the 11th day (May 13), it vanished and was not observed, having been seen for 20 days.
[*Ichidai yōki*]

(c) *1st year of the Ninnan reign period, 4th month, 22nd day; a guest star was seen at TAIWEIYUAN.*
[*Dai Nihon shi*] (Ho 412)

NB: In the *Dai Nihon shi* record (Ho 412), "4th month" should be "3rd month."

AD 1166 May 1 [China] (Ho 412)

2nd year of the Qiandao reign period of Emperor Xiaozong of the Song Dynasty, 3rd month, day *guiyou* [10]; [a guest star] appeared within TAIWEIYUAN, west of the large star of WUDIZUO. It was small and dim, and greenish-white in color.
[*Song shi: tianwen zhi*] ch. 56

AD 1178 Jan 14–27 [Japan]

(a) 1st year of the Jijō reign period, 12th month, 24th day: Jijō reign period, 2nd year 1st month day *renyin* [39]...this daybreak at azimuth *xun* [SE] a broom star was seen...last year on the 24th day of the 12th month it was seen, but afterward did not appear [again]; this daybreak it was seen again.
[*Sankaiki*]

(b) 1st year of the Jijō reign period, 12th month, 24th day: Jijō reign period, 2nd year 1st month, 7th day (Jan 27); during the hour of *yin* [LT = 3–5 h], a broom star was seen at about azimuth *xun* [SE]. The courtier Yasuchika memorialized about the report, saying that it was seen last year in the 12th month, 24th day.
[*Hyaku ren shō*]

(c) 1st year of the Jijō reign period, 12th month, 24th day: a broom star was seen in the at azimuth *xun* [SE]; on the 28th day (Jan 18) its rays became more intense...2nd year of the Jijō reign period, 1st month, 7th day (Jan 27); the broom star was seen.
[*Dai Nihon shi*] (Ho 414)

AD 1185 Feb 2 [Japan] (Ho 416)

4th year of the Juei reign period, 1st month, 1st day; CHIYOU'S banner [i.e., comet with a bent tail] over 1 zhang long was seen in the southeast.
[Dai Nihon shi]

AD 1189 Mar 16 [Japan]

5th year of the Bunji reign period, 2nd month, 28th day; a reddish-white broom star over 1 *zhang* long was seen in the east, beside LANGWEI at the north of TAIWEIYUAN.
[*Ichidai yōki*]
[*Dai Nihon shi*] (Ho 417)

AD 1210 Oct 19 [Japan]

(a) 4th year of the Jōgen reign period, 9th month, thirtieth day, *yimao* [52], clear skies; during the hour of *xu* [LT = 19–21 h], a strange star was seen in the west beside the third star of TIANSHIYUAN; its brightness over 3 *chi* long and pointing east, and its vaporous rays about 1 *zhang* long were exceptionally profuse; based on the records this star belongs to the broom stars, and so it was reported.
[*Azuma kagami*]
[*Dai Nihon shi*] (Ho 421)

(b) 4th year of the Jōgen reign period, 9th month, 30th day; during the hour of *xu* [LT = 19–21 h], a broom star was seen at azimuth *kun* [SW].
[*Ichidai yōki*]

AD 1220 Jan 25 [Japan]

(a) 1st year of the Jōkyū reign period, 12th month, 28th day; during the hour of *xu* [LT = 19–21 h], a red broom star 1 *chi* long was observed at azimuth *qian* [NW], east of the TENGSHE star.
[*Ichidai yōki*]

(b) 1st year of the Jōkyū reign period, 12th month, 28th day; a reddish broom star was seen in the northwest 1 *chi* east of TENGSHE.
[*Dai Nihon shi*] (Ho 422)

AD 1220 Feb 6 [Korea] (Ho 423)

7th year of King Gojong of Goryeo, 1st month, day *renchen* [29]; a broom star over 3 *chi* long emerged at GOUXING, its tail pointing northwest.
[*Goryeosa*] ch. 47

AD 1220 Mar 21 [Korea] (Ho 423)

7th year of King Gojong of Goryeo, 2nd month, day *bingzi* [13]; there was a fuzzy star in XUANYUAN.
[*Goryeosa*] ch. 47

AD 1220 Dec 27–AD 1221 Jan 24 [Korea] (Ho 424)

7th year of King Gojong of Goryeo, 12th month; there was a fuzzy star in BEIDOU.
[*Goryeosa*] ch. 48

AD 1222 Sep 10–Oct 23 [China] (Ho 425)

(a) 15th year of the Jiading reign period of Emperor Ningzong of Southern Song, 8th month, day *jiawu* [31] (Sep 25); a broom star was seen at YOUSHETI; its rays were more than 3 *chi* (long).

Its body was similar to Jupiter. It lasted for 2 months. It passed DI [LM 3], FANG [LM 4] and XIN [LM 5] and then disappeared.
[*Song shi*] ch. 56

(b) 15th year of the Jiading reign period of Emperor Ningzong of Southern Song, 8th month, day *jiawu* [31] (Sep 25); a broom star appeared in DI [LM 3]...In the 9th month, day *renxu* [59] (Oct 23), the broom star disappeared.
[*Song shi*] ch. 40

(c) 6th year of the Xingding reign period of Emperor Xuanzong of Jin, (i.e., 1st year of the Yuanguang reign period), 8th month, day *jimao* [16] (Sep 10); a broom star appeared in KANG [LM 2], between YOUSHETI and ZHOUDING. It pointed toward DAJIAO (Arcturus)...In the 9th month, day *dingwei* [44] (Oct 8), it was extinguished.
[*Jin shi*] ch. 20

NB: This is Comet Halley, return –10.

AD 1222 Sep 8–Oct 5 [Japan] (Ho 425)

(a) 1st year of the Jōō reign period, 8th month, 2nd day, *dingchou* [14] (Sep 8), the hour of *xu* (LT 19–21h); a broom star was seen in the northwest. The center of the star was as large as a half-moon. Its color was white but its rays were red. Its length was more than 17 *chi*...On the 15th day, *gengyin* [27] (Sep 21) it was clear; on this night the rays of the broom star were more than 1 *chi* (long). From the 2nd day (Sep 8) on, it was seen every night. On the 22nd day, *dingyou* [34] (Sep 29), the broom star was again seen in the southwest. It was 5 *chi* long. On the 29th day, *jiachen* [41] (Oct 5), it was clear but the broom star was not seen.
[*Azuma kagami*] Kitsugawa edition

(b) 1st year of the Jōō reign period, 8th month, 1st day; from this evening a broom star was seen at azimuth *qian* [NW].
[*Hyaku ren shō*]

NB: This is Comet Halley, return –10.

AD 1222 Sep 3–9 [Korea] (Ho 425)

9th year of King Gojong, 7th month, day *renshen* [9] (Sep 3); a broom star appeared within SANTAI. Its tail pointed west and it was more than 3 *chi* long…On day *yihai* [12] (Sep 6), the broom star was seen in the northwest. It was about 3 *chi* long. In the 8th month, day *dingchou* [14] (Sep 8), it was seen in the northwest. Its length was about 20 *chi*. On day *wuyin* [15] (Sep 9), it was seen during the day, and Venus was also seen during the day across the sky.

[*Goryeosa*] ch. 48

NB: This is Comet Halley, return –10.

AD 1223 Aug 28–29 [Japan]

2nd year of the Jōō reign period, 8th month, 1st day; a white broom star was seen at azimuth *qian* [NW]; on the 2nd day (Aug 29), its bright rays intensified and it grew to over 2 *zhang* in length.

[*Dai Nihon shi*] (Ho 426)

NB: Probably a misdated reference to the former.

AD 1230 Dec 4–30 [Japan]

2nd year of the Kangi reign period, 10th month, 28th day; a guest star was seen in the west…on the 3rd day *gengyin* [27]…on the 4th day *xinmao* [28] (Dec 6), the sky was clear…at night the sky was clear and a strange star was seen. The star was indistinct and glowed faintly, but it was not small. Two days ago the courtier Yasutoshi reported that since the previous night, which was moonless, a strange star was seen at azimuth *xin* [W by N], east of ZHINÜ and beside XIZHONG, northeast of TIANJIN. Its body showed signs of being either a guest star or a broom star, without rays…in the 11th month, 25th day (Dec 30) the guest star returned to the south at daybreak.

[*Meigetsuki*] Nonomiya ed.

NB: According to Ho 428, *Dai Nihon shi* concluded, "until the 8th day in the 11th month (Dec 13) it was seen in the west in the evening and in the northeast in the morning."

AD 1230 Dec 15 [China] (Ho 428)

3rd year of the Shaoding reign period of Emperor Lizong of the Song Dynasty, 11th month, day *dingyou* [34]; a star became fuzzy below TIANSHIYUAN and the TUSI star; the following year, 2nd month, day *renwu* [19], it vanished.

[*Song shi: tianwen zhi*] ch. 56

AD 1231 [China]

7th year of the Zhengda reign period of Emperor Aizong of the Jin Dynasty, 12th month, day *gengyin* [27]; a star the size of Saturn and of indistinct color emerged below TIANJIN and initially trespassed on NIANDAO. On the 2nd day it appeared in the northeast, south of ZHINÜ. On day *yiwei* [32], it entered TIANSHIYUAN, not emerging until day *wushen* [45]. On day *guichou* [50], it passed north of FANG [LM 4], once again traveling southeastward and entering JIXIN, in all for 25 days, whereupon it was extinguished.

[*Jin shi: tianwen zhi*] ch. 20

NB: There was no day *gengyin* [27] in the said month.

AD 1232 Oct 16–Nov 13 [China] (Ho 429)

5th year of the Shaoding reign period of Emperor Lizong of the Song Dynasty, intercalary 9th month; a broom star appeared in the east; not until the 10th month, day *jiwei* [56], was it was extinguished.

[*Song shi: tianwen zhi*] ch. 56

AD 1232 Oct 17–18 [China] (Ho 429)

(a) 1st year of the Tianxing reign period of Emperor Aizong of the Jin Dynasty, intercalary 9th month, day *jiyou* [46] (Oct 17); a white broom star over 1 *zhang* long appeared in the east, bent like an elephant's tusk. It emerged in JIAO [LM 1] and ZHEN [LM 28], traveling southward, and by the 12th day it was 2 *zhang* long. On the 16th day it was obscured by the moonlight. On the 27th day, during the fifth watch, it again emerged in the southeast, over 4 *zhang* long, until the 1st day of the 10th month when it was extinguished, in all, after 48 days.

[*Jin shi: tianwen zhi*] ch. 20

(b) 5th year of the Shaoding reign period of Emperor Lizong of the Song Dynasty, intercalary month, day *gengxu* [47] (Oct 18); a broom star emerged in JIAO [LM 1].
[*Song shi: Lizong ji*] ch. 41

AD 1232 Oct 19–Nov 19 [Japan]

1st year of the Jōei reign period, intercalary 9th month, 4th day *xinhai* [48]; clearing skies; during the hour of *yin* [LT = 3–5 h], a broom star's vapor was seen at azimuth *yi* [E by S], pointing to azimuth *geng* [W by S]; it was 2 *chi* long, 8 *cun* wide, and reddish-white in color… the 8th day *yimao* [52] (Oct 23); for the past 2 to 3 days, it has either been cloudy or raining, but today at dawn, when it was possible to see blue sky, the broom star still was seen, now increased in brightness. Its rayed vapors were 2 *zhang* long, over 1 *chi* wide, and pointing in the same direction. It retrograded about 1 *zhang*, and moved southward 4 *chi*, so it was about 5 *chi* from the eastern mountains…on the 15th day *renxu* [59] (Oct 30), the sky was clear and the broom star was fainter, its rayed vapor not visible…10th month, 5th day *xinsi* [18] (Nov 19), clear sky; from the 28th of last month until today at dawn the broom star was seen nightly. Its rays, though faint, were 4 *zhang* long, and pointed toward azimuth *qian* [NW] from the south.
[*Azuma kagami*] Yoshikawa ed.
NB: (Ho 429) gives a much abbreviated record from *Dai Nihon shi*.

AD 1240 Jan 27–Mar 2 [Japan]

(a) 1st year of the Ninji reign period, 1st month, 2nd day *dingmao* [4] (Jan 27); the sky is clear…during the hour of *xu* [LT = 19=21 h], a reddish-white broom star was seen at azimuth *shen* [WSW], pointing to azimuth *xun* [SE], its rayed vapor 3 *chi* long…the 7th day *renshen* [9] (Feb 1), clear sky; during the hour of *xu* [LT 19–21 h], the broom star was seen beside Jupiter at a distance of about 3 *chi*, its rays 5 *chi* long and pointing toward azimuth *gen* [NE], the star at its hub as big as Venus; on the 8th day *guiyou* [10], during the hour of *xu* [LT = 19=21 h], the broom star approached Jupiter to a distance of about 2 *chi* or more…on the 11th day *bingzi* [13]

(Feb 5), during the hour of *xu* [LT = 19=21 h], the broom star trespassed on the 1st star of BI [LM 14] at a distance of about 1 *chi* or more; at the same time the Moon entered DONGJING [LM 22]; on the 18th day *guiwei* [20], the broom star approached KUI [LM 15]; on the 19th day *jiashen* [21], the broom star entered KUI [LM 15]…for the past 17 days until tonight its rays have successively grown; on the 26th day *xinmao* [28] (Feb 20), during the hour of *xu* [LT = 19=21 h], the broom star was seen and trespassed on the fifth star of WANGLIANG, its rays now fading; in the 2nd month, 6th day *xinchou* [38] (Mar 2)…after nightfall, the broom star was seen; from the 4th day of the 1st month until today, it has not disappeared.
[*Azuma kagami*]

(b) 1st year of the Ninji reign period, 1st month, 4th day *jisi* [6] (Jan 29); during the hour of *xu* [LT = 19=21 h], a broom star was seen at azimuth *kun* [SW], its rays about 3 *zhang* long and dazzling the eyes of a myriad people. For 5 or 6 days while the sky was cloudy, the broom star was not seen; on the 8th day, it was seen again, its rayed vapor touching the east, sometimes over 2 *zhang* long, sometimes 1 *zhang* or more.
[*Hyaku ren shō*]

(c) 1st year of the Ninji reign period, 1st month, 2nd day *jisi* [6] (Jan 27); a reddish-white broom star 3 *chi* long and pointing southeast was seen in the southwest; on the 4th day (Jan 29) it reappeared, now the same size as Saturn, with rays up to 4 *chi* long; on the 4th and 5th days (Jan 29–30), it did not appear due to cloudiness; on the 7th day (Feb 1), it was seen alongside Jupiter and was as big as Venus, its rays 5 *chi* long and pointing northeast; on the 8th day (Feb 2), it was facing Jupiter and was visible all night; on the 11th day (Feb 5), it trespassed against the first star of BI [LM 14]; on the 18th day (Feb 12), the broom star approached KUI [LM 15] which it entered on the 19th day (Feb 13); on the 27th day of the month (Feb 21), its rays were still faintly visible.
[*Dai Nihon shi*] (Ho 432)

AD 1240 Jan 31 [China] (Ho 432)

(a) 4th year of the Jiaxi reign period of Emperor Lizong of the Song Dynasty, 1st month, day *xinwei* [8]; a broom star emerged in YINGSHI [LM 13]. On day *jiawu* [31], the broom star trespassed against the second star of WANGLIANG. In the 3rd month, day *xinwei* [8], the broom star vanished from sight.
[*Song shi: Lizong ji*] ch. 42

(b) 4th year of the Jiaxi reign period of Emperor Lizong of the Song Dynasty, 1st month, day *xinwei* [8]; a broom star appeared in SHI [LM 13], until the 3rd month, day *xinwei* [8], whereupon it vanished.
[*Song shi: tianwen zhi*] ch. 56

AD 1241 Feb 17 [Japan] (Ho 434)

2nd year of the Ninji reign period, 1st month, 5th day; a broom star was seen.
[Dai Nihon shi]

AD 1245 Feb 24–Apr 4 [Japan]

(a) 3rd year of the Kangen reign period, 1st month, 26th day *renxu* [59] (Feb 24); clear sky; today at dawn a guest star was seen…3rd month, 1st day *bingshen* [33] (Mar 30), clear sky; today at dawn a broom star was seen in the east…6th day *xinchou* [38] (Apr 4); from dawn today the broom star has not been seen.
[*Hyaku ren shō*]

(b) 3rd year of the Kangen reign period, 1st month, 27th day *guihai* [60] (Feb 25); clear sky; during the hour of *chou* [LT 1–3 h], a guest star was seen in TIANSHIYUAN at azimuth *xun* [SE] in the space of DOU [LM 8]; 28th day *jiazi* [1], clear sky; during the hour of *yin* [LT = 3–5 h], the guest star still was seen south of lunar mansion NIU [LM 9]…2nd month, 1st day *bingyin* [3], clear sky; the guest star was seen in the space of QIANNIU [LM 9], in two nights covering a span of about 3 *zhang*…3rd month, 1st day *bingshen* [33] (Mar 30), the sky was clear; during the hour of

yin [LT 3–5 h], a broom star was seen between SHI [LM 13] and BI [LM 14], with a length of 2 *chi*.
[*Azuma kagami*]

(c) 3rd year of the Kangen reign period, 1st month, 26th day (Feb 24); a guest star was seen in the southeast; on the 27th day (Feb 25), it was seen southeast of TIANSHIYUAN within the space of DOU [LM 8]; on the 28th day (Feb 26), it was seen to the south of NIU [LM 9]; on the 29th day (Feb 27), it did not appear due dark clouds; on the 1st day of the 2nd month (Feb 28), it was seen within the space of QIANNIU [LM 9], and resembled cotton wadding…on the 1st day of the 3rd month of the 3rd year of the Kangen reign period (Mar 30); a broom star 2 *chi* long was seen between SHI [LM 13] and BI [LM 14]; it disappeared on the 6th day (Apr 4).
[*Dai Nihon shi*] (Ho 435)

AD 1264 Jul 21–Sep 18 [Japan]

(a) 1st year of the Bun-ei reign period, 6th month, 26th day (Jul 21). In the 7th month, 4th day *bingzi* [13] (Jul 28); today at dawn a white broom star 4 to 5 *chi* long was seen at azimuth *yin* [ENE]. Last month on the 26th, it was first seen in azimuth *xu* [WNW]; on the 5th day *dingchou* [4] (Jul 29), during the hour of *mao* [LT = 5–7 h], the broom star was seen in the east, its rays exceptionally large… on the 11th day, *guiwei* [20] (Aug 4), the broom star still was seen, about 6 to 7 *zhang* long…on the 17th day *jichou* [2] (Aug 10), the broom star still was seen, its rays gradually fading…on the 21st day *guisi* [30] (Aug 14), the broom star still was seen, its rays once again increasing…9th month, 2nd day *guiyou* [10] (Aug 23), the broom star was seen, its rays over 10 *zhang* long.
[*Geki nikki*]

(b) 1st year of the Bun-ei reign period, 6th month, 26th to 27th days (Jul 21–22); a broom star was seen at azimuth *gen* [NE]; 7th month, 4th day (Jul 28); at dawn the broom star about 3 *chi* long was seen in the east; on the 7th day, at dawn its rays was large; in

the 8th month, 22nd day (Sep 13), the broom star was not seen; on the 26th day (Sep 17), the broom star was seen again.
[*Ichidai yōki*]

(c) 1st year of the Bun-ei reign period, 6th month, 26th day (Jul 21); a broom star was seen in the northwest on the 27th day (Jul 22) it was the same; on the 4th day in the 7th month (Jul 28), it was seen in the east with a length of about 3 *chi*; on the 7th day (Jul 31), its rays stretched across the sky; from this time on it was seen at intervals until it was extinguished on the 27th day [of the 8th month] (Sep 18).
[*Dai Nihon shi*] (Ho 436)

AD 1264 Jul 26 [China] (Ho 436)

(a) 5th year of the Jingding reign period of Emperor Lizong of the Song Dynasty, 7th month, day *jiaxu* [11]; a broom star emerged in Liu [LM 24]. On day *jimao* [16], the broom star retreated to Gui [LM 23]. On day *xinsi* [18], the broom star retreated to Jing [LM 22]. On day *wuxu* [35], the broom star retreated to Shen [LM 21]. In the 8th month, day *wuwu* [55], the broom star vanished from sight. On day *jiazi* [1], the broom star reappeared in Shen [LM 21]. On day *xinwei* [8], the broom star changed into colorful twilight vapors.
[*Song shi: Lizong ji*] ch. 45

(b) 5th year of the Jingding reign period of Emperor Lizong of the Song Dynasty, 7th month, day *jiaxu* [11]; a broom star over 10 *zhang* long appeared in Liu [LM 24], its rayed horns lighting up the sky for over a month, being attenuated only when the sun was high. On day *jimao* [16], the broom star retreated and appeared in Gui [LM 23]. On day *xinsi* [18], the broom star was in Jing [LM 22]. On day *bingshen* [33], it appeared in Shen [LM 21]. On day *wuxu* [35], it was within the space of Shen [LM 21]. At the end of the 8th month, the bright rays diminished somewhat, but it was not extinguished for 4 months.
[*Song shi: tianwen zhi*] ch. 56

AD 1264 Jul 26 [China] (Ho 436)

(a) 1st year of the Zhiyuan reign period of Emperor Shizu of the Yuan Dynasty, 7th month, day *jiaxu* [11]; a broom star appeared at dusk in the northwest in YUGUI [LM 23], then penetrated SHANGTAI, swept ZIWEI, WENCHANG and reached BEIDOU; then at dawn it appeared in the northeast, in all for over 40 days.
[*Yuan shi: Shizu ji*] ch. 5
[*Yuan shi: tianwen zhi*] ch. 48

(b) 1st year of the Zhiyuan reign period of Emperor Shizu of the Yuan Dynasty, 7th month, day *jiaxu* [11]; a broom star emerged in GUI [LM 23] and LIU [LM 24]. It appeared at dusk in the northwest, several tens of *zhang* long, its gleam lighting up the sky. It penetrated SHANGTAI, swept ZIWEI, WENCHANG, and BEIDOU; then at dawn it appeared in the northeast, in all for over 40 days.
[*Yuan shi lei bian*] ch. 2

AD 1264 Jul 26–Oct 5 [Korea] (Ho 436)

5th year of King Wonjong of Goryeo, 7th month, day *jiaxu* [11]; a broom star first was seen at azimuth *gen* [NE], its tail 7 to 8 *chi* long. Gradually it split into five and headed northwest…8th month, day *renyin* [39] (Aug 23), its rays joined again to become one and gradually lengthened; on day *renxu* [59] (Sep 12), the broom star's rays blazed with greater intensity, spanning the sky…in the 9th month, day *yiyou* [22] (Oct 5), the broom star was extinguished; from the 7th month until this it was 72 days in all.
[*Goryeosa*] ch. 48

AD 1266 Jan 17–Feb 15 [Japan]

(a) 2nd year of the Bun-ei reign period, 12th month, 11th day *yihai* [12] (Jan 18); today at dawn a broom star was seen at azimuth *shen* [WSW], its rays over 1 *chi* long…3rd year, 1st month, 1st day *yiwei* [32] (Feb 7); on this night the broom star still was seen in the west, its rays about 1 *zhang* long…9th day *guimao* [40] (Feb 15); tonight the broom star was seen.
[*Geki nikki*]

(b) 2nd year of the Bun-ei reign period, 12th month, 14th day *wuyin* [15] (Jan 21), clearing; today at dawn a broom star was seen in the east…27th day *xinmao* [28]; clearing, today the white broom star was seen in the west, in lunar mansion SHI [LM 13], its rayed vapor over 2 *chi* long. In the 3rd year, 1st month, 1st day *yiwei* [32] (Feb 7); the sky was clear and the wind still…as dusk turned to dark the broom star was seen in the west, at 8 *du* in BI [LM 14].
[*Azuma kagami*]

NB: The original miswrote "in lunar mansion SHI" as "there is lunar mansion SHI" and "at 8 *du* in BI" as "at 8 *du* in *bi*."

(c) 2nd year of the Bun-ei reign period, 12th month, 10th day (Jan 17); a broom star was seen in the east and lasted until the 1st month of the next year (Feb 7–Mar 8).
[*Dai Nihon shi*] (Ho 437)

AD 1268 Aug 27 [Japan] (Ho 438)

5th year of the Bun-ei reign period, 7th month, 18th day (Aug 27); a broom star was seen.
[Dai Nihon shi]

AD 1273 Feb 5–6 [Japan]

(a) 10th year of the Bun-ei reign period, 1st month, 16th day (Feb 5); during about the hour of *xu* [LT = 19–21 h], a white broom star over 1 *zhang* long was seen in the west; after this it was seen nightly.
[*Moromori ki*]

NB: (Ho 439) provides a similar account from *Dai Nihon shi*.

(b) 10th year of the Bun-ei reign period, 1st month, 17th day (Feb 6); during the hour of *you* [LT = 17–19 h], a broom star over 1 *chi* long was seen and lasted for several days.
[*Kantō hyōjōden*]

AD 1273 Feb 17 [Korea] (Ho 439)

14th year of King Wonjong of Goryeo, *1st month, day renwu [19]; a broom star was seen.*
[Goryeosa] ch. 47

AD 1273 Apr 9 [China] (Ho 439)

10th year of the Zhiyuan reign period of Emperor Shizu of the Yuan Dynasty, 3rd month, day *guiyou* [10]; a greenish-white guest star like cotton wadding arose in BI [LM 19], then paused north of WUCHE before penetrating the ladle of DOU via WENCHANG. It passed GENGHE as far as ZUOSHETI, lasting in all for 21 days.
[*Yuan shi: Shizu ji*] ch. 8
[*Yuan shi: tianwen zhi*] ch. 48

AD 1277 Mar 9 [Japan]

3rd year of the Kenji reign period, 2nd month, 4th day; tonight a broom star was seen, its rays about 3 *chi* long; last night it likewise was seen. After this, it was seen repeatedly.
[*Moromori ki*]

AD 1277 Apr 1 [Korea] (Ho 440)

3rd year of King Chungnyeol of Goryeo, 2nd month, day bingxu [23]; Chi You's banner was seen.
[Goryeosa] ch. 47

AD 1293 Nov 7 [China] (Ho 441)

30th year of the Zhiyuan reign period of emperor Shizu of the Yuan Dynasty, 10th month, day *gengyin* [27]; a broom star with bright rays about 1 *chi* long entered ZIWEIYUAN and reached the bowl of DOU, in all for 1 month before being extinguished.
[*Yuan shi: Shizu ji*] ch. 17
[*Yuan shi: tianwen zhi*] ch. 48
NB: *Xu wenxian tongkao* has "reached BEIDOU."

AD 1293 Nov 13–23 [Japan]

1st year of the Einin reign period, 10th month, 14th day; during the hour of *mao* [LT = 5–7h], a guest star was seen in the east; for 10 days, its rayed vapor was seen—it was a broom star.
[*Hōjō kudai ki*]

AD 1293 Nov 21–Dec 23 [Korea] (Ho 441)

19th year of King Chungnyeol of Goryeo, 10th month, day *jiachen* [41]; a broom star was seen in the east. On day *wushen* [45] (Nov 25), the broom star, over 1 *chi*, 5 *cun* long, was seen at the ZUOYEMEN of TAIWEI...in the 11th month, on day *bingzi* [13] (Dec 23), the broom star trespassed on ZIWEI, and also trespassed on BEIDOU.

[*Goryeosa*] ch. 49

AD 1297 Mar 12–25 [Korea] (Ho 442)

23rd year of King Chungnyeol of Goryeo, 2nd month, day *xinhai* [48]; a broom star was seen in DONGJING [LM 22]; after 6 days, it was extinguished. In the 3rd month, on day *jiazi* [1] (Mar 25), the broom star was seen in DONGJING [LM 22].

[*Goryeosa*] ch. 49

AD 1297 Mar 13 [Japan]

(a) 5th year of the Einin reign period, 2nd month, 19th day; tonight a broom star was seen in the east.
[*Moromori ki*]

(b) 5th year of the Einin reign period, 2nd month, 19th day; during the hour of *xu* [LT = 19–21 h], a broom star was seen with rayed vapor 5 to 6 *chi* long.
[*Hōjō kudai ki*]

AD 1297 Sep 14–21 [Japan]

(a) 5th year of the Einin reign period, 8th month, 26th day (Sep 14); during the hour of *chou* [LT = 1–3 h], a guest star was seen in the east, its body like white fluff.
[*Hōjō kudai ki*]

(b) 5th year of the Einin reign period, 9th month, 3rd day (Sep 21); during the hour of *xu* [LT = 19–21 h], a white broom star with rays over 1 *chi* long was seen at azimuth *yi* [E by S] in the east; after this it was seen nightly.
[*Moromori ki*]

AD 1299 Jan 6–Feb 3 [Japan]

(a) 6th year of the Einin reign period, 12th month, 3rd day; today at dawn a white broom star was seen in the south in the space of lunar mansion JIAO [LM 1]; its rays were about 1 *chi* 7 *cun* long and pointing toward azimuth *qian* [NW]; after this it was seen nightly. 7th year, 1st month, 1st day (Feb 3); the broom star was seen; after this it was seen repeatedly.
[*Moromori ki*]

(b) 6th year of the Einin reign period, 12th month, 3rd day; today at dawn a broom star was seen in the south, its rays over 1 *chi* long.; on the 20th day (Jan 23); at night the broom star was seen in the south, its rays 7 *chi* long.
[*Hōjō kudai ki*]

(c) 6th year of the Einin reign period, 12th month (Jan 4–Feb 1); a broom star was seen.
[Dai Nihon shi] (Ho 444)

AD 1299 Jan 24 [Korea] (Ho 444)
24th year of King Chungnyeol of Goryeo, 12th month, day jiaxu [11]; a broom star was seen in the south.
[Goryeosa] ch. 47

AD 1299 Oct 23 [Japan]
1st year of the Shōan reign period, 9th month, 28th day; during the hour of *yin* [LT = 3–5 h], a broom star was seen at azimuth *chen* [ESE], its rays 2 *zhang* long.
[*Hōjō kudai ki*]

AD 1299 Jan 24 [China] (Ho 444)
2nd year of the Dade reign period of Emperor Chengzong of the Yuan Dynasty, 12th month, day *jiaxu* [11]; a broom star emerged beneath ZI and SUN.
[*Yuan shi: Chengzong ji*] ch. 19
[*Yuan shi: tianwen zhi*] ch. 48

AD 1301 Sep 16–Nov 1 [China] (Ho 446)

5th year of the Dade reign period of Emperor Chenzong of Yuan, 8th month, day *gengchen* [17] (Sep 16); a broom star appeared at 24 *du* 40 *fen* within [DONG-]JING [LM 22]. It was like the large star of NANHE. Its color was white and it was 5 *chi* long. It formed a straight line toward the northwest. Later it passed WENCHANG and the head of [BEI-]DOU. It swept to the south of TAIYANG and then swept BEIDOU, TIANJI, the wall of ZIWEI, SANGONG and GUANSUO. The length of the star was more than 1 *zhang*. It reached the east of BASHU at the TIANSHI wall, [continued to] the south of LIANG and ZHU and above the star SONG. Its length was rather more than 1 *chi*. After a total of 46 days, it was extinguished.

[*Yuan shi*] ch. 48

NB: This is Comet Halley, return –9.

AD 1301 Sep 15–23 [Japan]

(a) 3rd year of the Shōan reign period, 8th month, 13th day (Sep 15), at the hour of *yin* (LT 3–5h); a broom star appeared in the east. Its rays were more than 3 *chi* long. Over several days they increased to 1 *zhang* in length.
[*Moromori ki*]

(b) 3rd year of the Shōan reign period, 8th month, 14th day (Sep 16); today at daybreak, a broom star appeared in the east. After this it appeared nightly; 21st day (Sep 23), on this night, at the hour of *xu* (LT 19=21h), the broom star again appeared in the northwest, its rays more than 1 *zhang* long.
[*Hōdō kyūdai ki*]

NB: This is Comet Halley, return –9.

AD 1301 Sep 14–Oct [Korea] (Ho 446)

27th year of King Chungnyeol of Goryeo, 8th month, day *wuyin* [15] (Sep 14); a broom star was seen at BEIDOU and ZIWEI. On day *renwu* [19] (Sep 18), it was seen in BEIDOU. On day *yiwei* [32] (Oct 1), it was seen in SHANGTAI and entered the wall of TIANSHI…In the 9th

month, day *guimao* [40] (Oct 9), the broom star was seen at the wall of TIANSHI.
[*Goryeosa*] ch. 49
NB: This is Comet Halley, return –9.

AD 1303 Jul 17–27 [Japan]
1st year of the Kagen reign period, 6th month, 13th day; a white broom star with a star at its hub was seen at azimuth *gen* [NE], its rayed vapor over 1 *chi* long; it has was seen each day at dawn for the past 10 days.
[*Moromori ki*]

AD 1304 Feb 3–7 [Korea] (Ho 447)
29th year of King Chungnyeol of Goryeo, 12th month, day *gengxu* [47]; a broom star was seen in the west. In the 30th year, 1st month, day *jiayin* [51] (Feb 7), the broom star was seen in KUI [LM 15].
[*Goryeosa*] ch. 49

AD 1304 Feb 4 [Japan]
1st year of the Kagen reign period, 12th month, 28th day; a broom star was seen. 2nd year, 1st month, 1st day; during the hour of *xu* [LT 19–21 h], a broom star was seen at azimuth *wei* [SSW]–*shen* [WSW], its rays over 2 *chi* long; after this it was seen nightly.
[*Moromori ki*]

AD 1304 Apr 18 [China] (Ho 447)
(a) 8th year of the Dade reign period of Emperor Chengzong of the Yuan Dynasty, 3rd month, day *yichou* [2]; since day *gengxu* [47] (Feb 2) of the 12th month of last year a broom star over 1 *chi* long has appeared at the 11th *du* of SHI [LM 13]. It entered ZIWEIYUAN whereupon it was extinguished, in all after 74 days.
[*Yuan shi: Chengzong ji*] ch. 21
(b) 8th year of the Dade reign period of Emperor Chengzong of the Yuan Dynasty, 3rd month, day *yichou* [2]; since day *gengxu* [47]

(Feb 3) of the 12th month of last year a white broom about fully 1 *chi* long has appeared pointing southeast, observation placing it at the 11th *du* of Shi [LM 13]. It gradually grew to over 1 *chi* long, and pointing northwest swept Tengshe then entered Ziweiyuan; on reaching this point it was extinguished, in all after 74 days.
[*Yuan shi: tianwen zhi*] ch. 48

AD 1304 Dec 24–30 [Korea] (Ho 448)
30th year of King Chungnyeol of Goryeo, 11th month, day *yihai* [12]; a broom star was seen between Xu [LM 11] and Wei [LM 12]. In the 12th month, on day *wuyin* [15] (Dec 27), the broom star trespassed on Xu [LM 11]; on day *gengchen* [17] (Dec 29), it penetrated Xu [LM 11]; on day *xinsi* [18] (Dec 30), it entered Wei [LM 12].
[*Goryeosa*] ch. 49

AD 1304 Dec 26 [Japan]
2nd year of the Kagen reign period, 11th month, 29th day; tonight a white broom star was seen in the west at azimuth *geng* [W by S], south of the southernmost star of lunar mansion Xu [LM 11], its rayed vapor over 1 *chi* long and pointing east; after this it was seen nightly.
[*Moromori ki*]

AD 1305 Jun 19 [Japan]
3rd year of the Kagen reign period, 5th month, 27th day; during the hour of *yin* [LT =3–5 h], a broom star was seen in azimuth *gen* [NE]; after the 9th day of the 6th month, it was no longer seen.
[*Buke nendai ki*] *ka*

AD 1307 Aug 24 [Korea] (Ho 449)
33rd year of King Chungnyeol of Goryeo, 7th month, day *wuzi* [25]; a broom star was seen in Wei [LM 6].
[*Goryeosa*] ch. 49

AD 1313 Apr 13 [China] (Ho 450)

2nd year of the Huangqing reign period of Emperor Renzong of the Yuan Dynasty, 3rd month, day *dingwei* [44]; a broom star emerged in DONGJING [LM 22].
[*Yuan shi: Renzong ji*] ch. 24
[*Yuan shi: tianwen zhi*] ch. 48

AD 1313 Apr 13 [Korea] (Ho 450)

5th year of King Chungseon of Goryeo, 3rd month, day *dingwei* [44]; there was a broom star in DONGJING [LM 22].
[*Goryeosa*] ch. 49

AD 1313 Apr 14–25 [Japan]

(a) 2nd year of the Shōwa reign period, 3rd month, 18th day; tonight during the hour of *xu* [LT = 19–21 h], a broom star was seen in the west, its rayed vapor pointing east; on the 19th, it was likewise.
[*Moromori ki*]

(b) 2nd year of the Shōwa reign period, 3rd month, 18th day. 3rd month, 19th day *jiyou* [46]; clear sky…last night a broom star was seen in the west…tonight the broom star reappeared; 29th day *jiwei* [56] (Apr 25), clear sky, tonight the broom star was seen.
[*Hanazonoin shinki*]

(c) 2nd year of the Shōwa reign period, 3rd month, 17th day (Apr 13); a broom star was seen.
[Dai Nihon shi] (Ho 450)

AD 1315 Oct 29 [China] (Ho 451)

2nd year of the Yanyou reign period of Emperor Renzong of the Yuan Dynasty, 10th month, new moon day *bingzi* [13]; a guest star appeared in TAIWEIYUAN. In the 11th month, day *bingwu* [43], the guest star changed into a broom star and trespassed on ZIWEIYUAN, then traversed 15 lunar mansions from ZHEN [LM 28] as far as BI [LM 19]; it was not extinguished until day *gengyin* [27] (1316 Mar 11) in the 2nd month of the following year.

[*Yuan shi: Renzong ji*] ch. 25
[*Yuan shi: tianwen zhi*] ch. 48

AD 1337 May 4 [China] (Ho 454)

3rd year of the Zhiyuan reign period of Emperor Shun of the Yuan Dynasty, 4th month, day *jiaxu* [11]; a star became fuzzy in WANGLIANG until day *renyin* [39] in the 7th month when it disappeared in GUANSUO.
[*Yuan shi: Shundi ji*] ch. 39

AD 1337 Jun 25–Sep 7 [Korea] (Ho 452 [misdated 1319], 454)

Latter 6th year of King Chungsuk of Goryeo, 5th month, day *bingyin* [3]; a broom star over 1 *zhang* long was seen, reaching from north of TIANCHUAN to WANGLIANG and GEDAO. In the 6th month, on the day of the new moon *gengwu* [7] (Jun 29), the broom star was seen at azimuth *gen* [NE]; on day *guiyou* [10] (Jul 2), it was seen again at ZIWEI's western wall in HUAGAI, GOUZHEN, and BEIJI. In the 7th month, on day *guisi* [30] (Aug 1), it was seen again at ZIWEI's eastern wall; on day *dingyou* [34] (Aug 7), it further trespassed on GUANSUO. In the 8th month, on day *gengzi* [37] (Aug 28), the broom star was seen in TIANSHIYUAN, and after 40 days it was extinguished.
[*Goryeosa*] ch. 49
[*Jeungbo munheon bigo*] ch. 6
NB: [*Goryeosa*] mistook "7th month, day *guisi*" for "7th month, day *guimao*"; mistook "day *dingyou,* further trespassed on GUANSUO" for "day *dingwei,* further trespassed on GUANSUO"; and mistook "8th month, day *gengzi*" for "8th month, day *gengchen.*"

AD 1337 Jun 26 [China] (Ho 455)

(a) 3rd year of the Zhiyuan reign period of Emperor Shun of the Yuan Dynasty, 5th month, day *dingmao* [4]; a white broom star as large as the TIANCHUAN star and more than about 1 *chi* long appeared in the northeast. The broom star was pointing southwest and was not extinguished until day *gengwu* [7] in the 8th month…on day *gengwu* [7] in the 8th month the broom star did not appear; from day *dingmao* [4] in the 5th month when it first appeared until now, in all it was visible for 63 days until it was extinguished, passing by 15 lunar mansions from MAO [LM 18] to FANG [LM 4].

[*Yuan shi: Shundi ji*] ch. 39

(b) 3rd year of the Zhiyuan reign period of Emperor Shun of the Yuan Dynasty, 5th month, day *dingmao* [4]; a white broom star as large as the TIANCHUAN star and more than about 1 *chi* long appeared in the northeast. The broom star was pointing southwest, measurement placing it at 5 *du* in MAO [LM 18]...on day *gengwu* [7] in the 8th month the broom star did not appear; from its first appearance on day *dingmao* [4] in the 5th month the star headed southwest on day *wuchen* [5], daily picking up speed until day *xinwei* [8] in the 6th month its rayed broom increased in length to over about 2 *chi*; on day *dingchou* [14] it swept SHANGCHENG; on day *jimao* [16], its bright rays increased greatly to over 3 *chi* and it entered HUANWEI; on day *renwu* [19], it swept HUAGAI and GANGXING; on day *yiyou* [22], it swept the large star of GOUCHEN and TIANHUANGDADI; on day *bingxu* [23], it penetrated SIFU and passed SHUXIN; on day *jiawu* [31], it emerged from HUANWEI; on day *dingyou* [34], it emerged from ZIWEIYUAN; on day *wuxu* [35], it trespassed against GUANSUO, and swept TIANJI; on day *gengzi* [37], in the 7th month, it swept HEJIAN; on day *guimao* [40], it passed ZHENG and JIN and entered TIANSHIYUAN; on day *bingwu* [43], it swept LIESI; by day *jiyou* [46], the moon's glare had increased so that its rayed tail was barely discernible. It emerged from TIANSHIYUAN and swept the LIANG star; by day *xinyou* [58], its bright rays were short and dim, and it was observed above JIANBI and lunar mansion FANG [LM 4], due west of the middle star of FA, but difficult to measure as it daily traveled southward. By this time it had been visible for 63 days in all, until it was extinguished after having traversed 15 lunar mansions from MAO [LM 18] to FANG [LM 4].

[*Yuan shi: tianwen zhi*] ch. 49

AD 1337 Jun 27–Jul 19 [Japan]

(a) 4th year of the Kemmu reign period, 5th month, 28th day (Jun 27); tonight during the hour of *yin* [LT = 3–5 h], a broom star was seen in the east.
[*Moromori ki*]

(b) 4th year of the Kemmu reign period, 6th month, 21st day (Jul 19); a broom star invaded BEIJI.
[*Kō nendai ryakuki*]; [*Kō nendai shiki*]
[*Dai Nihon shi*] (Ho 455)

AD 1340 Mar 24 [China] (Ho 456)

(a) 6th year of the Zhiyuan reign period of Emperor Shun of the Yuan Dynasty, 2nd month, day *jiyou* [46]; a white broom star shaped like cotton wadding and as big as the FANG star, the tail more than about 5 *cun* long. The broom star pointed southwest as it gradually travelled northwest. In the 3rd month, day *xinsi* [18], the broom star did not appear; from day *jiyou* [46] in the 2nd month until day *gengchen* [17] in the 3rd month, it was observed for 32 days in all.
[*Yuan shi: Shundi ji*] ch. 40

(b) 6th year of the Zhiyuan reign period of Emperor Shun of the Yuan Dynasty, 2nd month, day *jiyou* [46]; a white broom star shaped like cotton wadding and as big as the FANG star, the tail more than about 5 *cun* long. The broom star pointed southwest and measurement placed it at 7 *du* in FANG [LM 4], then it gradually travelled northwest. In the 3rd month, day *xinsi* [18], on this night the broom star did not appear. From day *jiyou* [46] in the 2nd month until day *gengchen* [17] in the 3rd month, it was observed for 32 days in all.
[*Yuan shi: tianwen zhi*] ch. 49

(c) 6th year of the Zhiyuan reign period of Emperor Shun of the Yuan Dynasty, 2nd month, day *jiyou* [46]; a star became fuzzy in FANG [LM 4]. The star was large and white, shaped like cotton wadding,

the tail over 5 *cun* long. The broom star pointed southeast as it gradually travelled northwest, for 32 days in all before being extinguished.
[*Yuan shi lei bian*] ch. 10

AD 1340 Mar 25–Mar 29 [Korea] (Ho 453)
Latter 1st year of King Chunghye of Goryeo, 2nd month, day *gengxu* [47]; a broom star was seen in the east and entered TAIWEI. On day *jiayin* [51] (Mar 29), it reappeared in the east.
[*Goryeosa*] ch. 49

AD 1340 Apr 1 [Japan]
3rd year of the Ryakuō reign period [of the Northern Dynasty], 3rd month, 4th day…left the court to pay homage to the broom star…tonight it already trespassed on the DONGCIXIANG asterism of the eastern wall of TAIWEIGONG.
[*Gyokuei kishō*]

AD 1345 Jul 31 [Korea] (Ho 457)
1st year of King Chungmok of Goryeo, 7th month, day *jiashen* [21]; a broom star was seen in ZIWEIYUAN; on day *dinghai* [24], the broom star was seen north of BEIHE.
[*Goryeosa*] ch. 49

AD 1345 Aug 2 [Japan]
6th year of the Kōkoku reign period, 7thmonth, 4th day…today at dawn during the hour of *yin* [LT = 3–5 h], a baleful star was seen…5th day…in the northeast its rays were already about 4 *chi* long, its white color unusually intense.
[*Entai ryaku*]
[*Dai Nihon shi*] (Ho 457)

AD 1351 Nov 24–26 [China] (Ho 459)
(a) 11th year of the Zhizheng reign period of Emperor Shun of the Yuan Dynasty, 11th month, day *xinhai* [48] (Nov 24); a fuzzy

star appeared in lunar mansion KUI [LM 15]; on day *guiychou* [50], the fuzzy star appeared in lunar mansion LOU [LM 16]; on day *jiayin* [51], the fuzzy star appeared in lunar mansion WEI [LM 17]; on day *yimao* [52], it was likewise; on day *bingchen* [53], the fuzzy star appeared in lunar mansion MAO [LM 18]; on day *dingsi* [54], the fuzzy star was barely visible in lunar manison BI [LM 19].
[*Yuan shi: tianwen zhi*] ch. 49

(b) 11th year of the Zhizheng reign period of Emperor Shun of the Yuan Dynasty, 11th month, day *guichou* [50] (Nov 26); a fuzzy star appeared in lunar mansion LOU [LM 16]; on day *jiayin* [51], the fuzzy star appeared in lunar mansion WEI [LM 17]; on days *yimao* [52] and *bingchen* [53], it was likewise; on day *dingsi* [54], the fuzzy star was barely visible in lunar manison BI [LM 19].
[*Yuan shi: Shundi ji*] ch. 42

AD 1356 Sep 21 [China] (Ho 461)

(a) 16th year of the Zhizheng reign period of Emperor Shun of the Yuan Dynasty, 8th month, day *jiaxu* [11]; a greenish-white broom star over 1 *chi* long appeared in lunar mansion ZHANG [LM 26] pointing southwest, until day *wuwu* [55] in the 10th month when it was extinguished.
[*Yuan shi: Shundi ji*] ch. 44

(b) 16th year of the Zhizheng reign period of Emperor Shun of the Yuan Dynasty, 8th month, day *jiaxu* [11]; a greenish-white broom star over 1 *chi* long and as big as the left horn of XUANYUAN appeared due east, and pointing southwest. Measurement placed it at 17 *du*, 10 *fen* in lunar mansion ZHANG [LM 26]; by day *wuwu* [55] in the 10th month when traces were extinguished, it had traveled northwestward for over 40 days.
[*Yuan shi: tianwen zhi*] ch. 49

NB: *Yuan shi lei bian* has “by new moon day *wuwu* [55] in the 12th month when it was extinguished.”

AD 1362 Feb 25–Mar 26 [Japan] (Ho 463)

17th year of the Shōhei reign period, 2nd month; a broom star was seen.

[Dai Nihon shi]

AD 1362 Mar 5 [China] (Ho 463)

(a) 22nd year of the Zhizheng reign period of Emperor Shun of the Yuan Dynasty, 2nd month, day *yiyou* [22]; a greenish-white broom star with bright rays over 1 *zhang* long appeared in lunar mansion WEI [LM 12]; on day *dingyou* [34], the broom star trespassed on the western star of LIGONG; by the end of the 2nd month, the bright rays were over 2 *zhang* long. In the 3rd month, day *wushen* [45], the body of the broom star was not seen, only the white vapor, its curving shape stretching across the sky and pointing west as it swept DAJIAO. On day *renzi* [49], the broom star passed before the sun and there was only the form of the star without rays; it was located in lunar mansion MAO [LM 18]; not until day *wuwu* [55], was it was extinguished.
[*Yuan shi: Shundi ji*] ch. 46

(b) 22nd year of the Zhizheng reign period of Emperor Shun of the Yuan Dynasty, 2nd month, day *yiyou* [22]; a greenish-white broom star with bright rays over 1 *zhang* long appeared. Observation placed it at 7 *du*, 20 *fen* in lunar mansion WEI [LM 12]; on day *dingyou* [34], the broom star trespassed on the western star of LIGONG; by the end of the 2nd month, the bright rays were over 2 *zhang* long. In the 3rd month, day *wushen* [45], the body of the broom star was not seen, only the white vapor, its curving shape stretching across the sky and pointing west as it swept DAJIAO. On day *renzi* [49], the broom star passed before the sun and there was only the form of the star without rays, as big as a wine cup, indistinct, and white in color; measurement placed it at 6 *du* in lunar mansion MAO [LM 18]; not until day *wuwu* [55], was it was extinguished.
[*Yuan shi: tianwen zhi*] ch. 49

AD 1362 Apr 25 [China] (Ho 464)

22nd year of the Zhizheng reign period of Emperor Shun of the Yuan Dynasty, 4th month, new moon day *bingzi* [13]; a long star several tens of *zhang* long and shaped like white silk appeared between XU [LM 11] and WEI [LM 12]; it was not extinguished until after more than 40 days.

[*Yuan shi: Shundi ji*] ch. 46

AD 1362 Jun 29 [China] (Ho 465)

(a) 22nd year of the Zhizheng reign period of Emperor Shun of the Yuan Dynasty, 6th month, day *xinsi* [18]; a broom star with bright rays over 1 *chi* long appeared in ZIWEIYUAN, pointing southeast as it traveled southwest; on day *wuzi* [25], the bright rays of the broom star swept SHANGZAI; on day *yimao* [52] in the 7th month, the traces of the broom star were extinguished.

[*Yuan shi: Shundi ji*] ch. 46

(b) 22nd year of the Zhizheng reign period of Emperor Shun of the Yuan Dynasty, 6th month, day *xinsi* [18]; a white broom star with bright rays over 1 *chi* long appeared in ZIWEIYUAN, pointing southeast as it traveled southwest; observation placed it at 2 *du*, 90 *fen* in NIU [LM 9]; on day *wuzi* [25], the bright rays of the broom star swept SHANGZAI; on day *yimao* [52] in the 7th month, all trace of the broom star was extinguished.

[*Yuan shi: tianwen zhi*] ch. 49

AD 1362 Jun 29–Jul 2 [Korea] (Ho 465)

11th year of King Gongmin of Goryeo, 6th month, day *xinsi* [18]; a broom star over 1 *chi* long was seen below ZIWEIYUAN and HUAGAI for 3 days.

[*Goryeosa*] ch. 49

AD 1363 Mar 16 [China] (Ho 466)

23rd year of the Zhizheng reign period of Emperor Shun of the Yuan Dynasty, 3rd month, new moon day *xinchou* [38]; a

broom star appeared in the east for a month whereupon it was extinguished.
[*Yuan shi: Shundi ji*] ch. 46

AD 1364 Mar 30 [Korea] (Ho 467)

13th year of King Gongmin of Goryeo, 2nd month, day *xinyou* [58]; red broom stars over 1 *chi* long were seen, one south of TAIWEI, one beside DAJIAO, one northeast of BEIDOU, one north of DI [LM 3].
[*Goryeosa*] ch. 49

AD 1366 Oct 25 [China] (Ho 468)

26th year of the Zhizheng reign period of Emperor Shun of the Yuan Dynasty, 9th month, day *gengzi* [37]; a fuzzy star about as big as a peck measure and resembling white cotton wadding in color appeared in ZIWEIYUAN, alongside the star QUAN of BEIDOU; it traveled southeastward, passing by and trespassing on TIANBANG; on day *xinchou* [38], measurement placed the fuzzy star at 18 *du*, 50 *fen* in WEI [LM 6]; on day *renyin* [39], measurement placed the fuzzy star at 2 *du*, 50 *fen* in Nü [LM 10]; on day *guimao* [40], measurement placed the fuzzy star at 9 *du*, 90 *fen* in Nü [LM 10]; on day *jiachen* [41], measurement placed the fuzzy star at 80 *fen* of the initial *du* of XU [LM 11]; on day *yisi* [42], it emerged from between ZIWEIYUAN, BEIDOU, QUANXING, and YUHENG; it was in lunar mansion ZHEN [LM 28] and moving southeastward; it passed by and trespassed on TIANBANG, passed JIANTAI, NIANDAO, then departed lunar mansion XU [LM 11], and the western asterisms LEI, BI [LM 14], and ZHEN before being extinguished and vanishing there.
[*Yuan shi: tianwen zhi*] ch. 49

AD 1366 Oct 26–27 [Korea] (Ho 468)

15th year of King Gongmin of Goryeo, 9th month, day *xinchou* [38]; there was a fuzzy star above the stars of FANG [LM 4] and XU [LM 11]; on day *renyin* [39] (Oct 27), it reappeared in the south.
[*Goryeosa*] ch. 49

AD 1368 Feb 7 [China] (Ho 469; Ho & Ang 1)

(a) 28th year of the Zhizheng reign period of Emperor Shun of the Yuan Dynasty, 1st month, day *gengyin* [27]; a broom star appeared between MAO [LM 18] and BI [LM 19].
[*Yuan shi: Shundi ji*] ch. 47

(b) 1st year of the Hongwu reign period of Emperor Taizu of the Ming Dynasty, 1st month, day *gengyin* [27]; a broom star appeared in MAO [LM 18] and BI [LM 19].
[*Ming shi: tianwen zhi*] ch. 27

AD 1368 Mar 7–Apr 21 [Korea] (Ho 470)

17th year of King Gongmin of Goryeo, 2nd month, day *jiwei* [56]; a broom star over 1 *zhang* long was seen in the west. In the 3rd month, day *jiashen* [21] (Apr 4), it was seen in the west; on day *gengyin* [27] (Apr 10), it emerged between DALING and JISHI; on day *xinmao* [28] (Apr 11), it emerged between DALING and JUANSHE, shooting to the ninth star of TIANCHUAN; on day *bingshen* [33] (Apr 16), it emerged above JUANSHE; on day *jihai* [36] (Apr 19), it emerged above DALING. In the 4th month, on day *xinchou* [38] (Apr 21), it was seen again.
[*Goryeosa*] ch. 49

AD 1368 Mar 8–Apr 9 [Japan]

23rd year of the Shōhei reign period, 2nd month, 19th day. 3rd month 22nd day *renchen* [29] (Apr 9), clear sky; last night during the hour of *xu* [LT = 19–21 h] at azimuth *qian* [NW], there was a star whose rays were extremely long…it was seen in the west on the 19th of last month (Mar 8) during the hour of *xu* [LT 19–21 h]…4th month…12th day *renzi* [49] (Apr 29), clear sky; the broom star was not seen at this time.
[*Gu kan ki*]
[*Dai Nihon shi*] (Ho 470)

AD 1368 Apr 8 [China] (Ho 470; Ho & Ang 2)

(a) 1st year of the Hongwu reign period of Emperor Taizu of the Ming Dynasty, 3rd month, day *xinmao* [28], at night; a broom

star emerged…4th month, day *jiyou* [46], at night, the broom star disappeared. Prior to this, on day *xinmao* [28] in the 3rd month, the broom star with rays more than about 8 *chi* long emerged north of MAO [LM 18], between DALU (should be DALING), and TIANCHUAN; on day *xinchou* [38], it pointed toward WENCHANG and approached WUCHE; on this night it disappeared north of WUCHE.
[*Ming Taizu shilu*] ch. 31

(b) 1st year of the Hongwu reign period of Emperor Taizu of the Ming Dynasty, 3rd month, day *xinmao* [28]; a broom star with rays over 8 *chi* long emerged north of MAO [LM 18], between DALING and TIANCHUAN; it pointed toward WENCHANG and approached WUCHE; in the 4th month, on day *jiyou* [46], it disappeared north of WUCHE.
[*Ming shi: tianwen zhi*] ch. 27

(c) 1st year of the Hongwu reign period of Emperor Taizu of the Ming Dynasty, 3rd month, day *xinmao* [28], at night; a broom star emerged north of MAO [LM 18]; on the night of day *xinchou* [38], in the 4th month, the broom star pointed toward WENCHANG and approached WUCHE; on the night of day *jiyou* [46], the broom star was extinguished.
[*Guo que*] ch. 3

AD 1370 Jan 31 [Korea] (Ho 471)
19th year of King Gongmin of Goryeo, 1st month, day jiawu [31]; a broom star appeared in the northeast.
[Goryeosa] ch. 49

AD 1373 Apr 23–May 22 [China] (Ho 472) (Ho & Ang 3)
6th year of the Hongwu reign period of Emperor Taizu of the Ming Dynasty, 4th month; a broom star entered ZIWEIYUAN 3 times.
[*Ming shi: tianwen zhi*] ch. 27

AD 1374 Mar 11–30 [Japan]

(a) 3rd year of the Bunchū reign period, 1st month, 28th day *jiawu* [31] (Mar 11); tonight a fuzzy star emitting bright rays in all

directions was seen in the west…13th day *bingshen* [33] (Mar 13); from tonight on there was a broom star seen in the south.
[*Zoku shigushō*]

(b) 3rd year of the Bunchū reign period, 2nd month, 7th day *guimao* [40] (Mar 20);…on the 13th of last month during the hour of *xu* [LT = 19–21 h], a broom star was seen in the south; on the 1st and 2nd, it was cloudy and so it did not appear; since the 3rd day it has appeared every night…on the 17th day *guichou* [50] (Mar 30)… when questions were exchanged about the matter of the broom star and I personally reported it to the assembled courtiers, it had still not disappeared.
[*Gu kan ki*]

(c) 3rd year of the Bunchū reign period, 1st month, 25th day (Mar 8); a broom star was seen; it reappeared on the 26th day (Mar 9).
[Dai Nihon shi] (Ho 473)

AD 1374 Mar 15–Apr 29 [Korea] (Ho 473)

23rd year of King Gongmin of Goryeo, 2nd month, day *wuxu* [35]; a broom star over 1 *zhang* long was seen in the east for 45 days before being extinguished.
[*Goryeosa*] ch. 49

AD 1376 Jun 22–Aug 8 [China] (Ho 474; Ho & Ang 4)

(a) 9th year of the Hongwu reign period of Emperor Taizu of the Ming Dynasty, 6th month, day *wuzi* [25] (Jun 22); there was a white star as large as a crossbow bolt. It stopped at TIANCANG, then passed WAIPING, JUANSHE, and entered ZIWEIYUAN; it swept WENCHANG and pointed toward NEICHU, then entered ZHANG [LM 26]; on day *yihai* [12] in the 7th month it was extinguished.
[*Ming shi: tianwen zhi*] ch. 27

(b) 9th year of the Hongwu reign period of Emperor Taizu of the Ming Dynasty, 7th month, day *yihai* [12] (Aug 8); the guest star disappeared. Prior to this, on day *wuzi* [25] in the 6th month, the guest star was as large as a crossbow bolt, white in color and stopped at TIANCANG; on day *guisi* [30], it increased in brightness; on day

jiawu [31], it passed WAIPING; on day *renyin* [39], it passed JUANSHE and entered ZIWEIYUAN; on day *gengxu* [47], it swept WENCHANG and pointed toward NEICHU; on day *renzi* [49], it swept WENCHANG; on day *guihai* [60] of this month it entered ZHANG [LM 26], and on this evening it was extinguished.
[*Ming Taizu shilu*] ch. 107

AD 1376 Jul 9–23 [Japan]

(a) 2nd year of the Tenju reign period, 6th month, 23rd day (Jul 10); during the hour of *yin* [LT = 3–5 h], a broom star was seen at azimuth *chou* [NNE]. Tonight at the beginning of the hour of *hai* [LT = 21–23 h] the broom star was seen at azimuth *qian* [NW]; during the dawn watch, it was again like this; it appeared nightly at azimuth *chou* [NNE]. In the 7th month…6th day *wuwu* [55] (Jul 23)…the broom star still was seen, at azimuth *qian* [NW], its movement strongly westward. As for its almost entering ZIWEIGONG, since the Tennan reign period (AD 857), this has occurred 5 or 6 times, but there is no precedent for its trespassing on TIANHUANGDADI as in this instance.
[*Gu kan ki*]

(b) 2nd year of the Tenju reign period, 6th month, 22nd day (Jul 9); during the hour of *yin* [LT = 3–4 h], a white broom star was seen at azimuth *chou* [NNE] its brightness over 3 *chi* long; likewise on the 27th (Jul 14), during the hour of *xu* [LT = 19–21 h], it was seen at azimuth *hai* [NNW], with its brightness over 1 *zhang* long; afterward, it was seen again at dawn at azimuth *chou* [NNE] as in the beginning.
[*Kyūrei zōji ki*]
[*Dai Nihon shi*] (Ho 474)

AD 1376 Jul 12–18 [Korea] (Ho 474)

2nd year of King U of Goryeo, 6th month, day *wushen* [45]; a broom star was seen in WENCHANG, its rays shooting toward ZIWEIYUAN…in the 7th month, on day *jiayin* [51] (Jul 18), the broom star was seen

4 to 5 *chi* west of WENCHANG, its gleam shooting toward the bowl of [BEI-]DOU.
[*Goryeosa*] ch. 49

AD 1376 Jul 21 [China]
Bingchen [53] year of the Hongwu reign period of Emperor Taizu of the Ming Dynasty, 7th month, 5th day; a star became fuzzy in BEIDOU.
[*Er shen ye lu*] ch. 1

AD 1378 Sep 22–Oct 21 [China] (Ho 475; Ho & Ang 5)
Wuwu (11th) year of the Hongwu reign period of Emperor Taizu of the Ming Dynasty, 9th month; a star became fuzzy in TIANSHI.
[*Er shen ye lu*] ch. 1

AD 1378 Oct 11–Nov 10 [China] (Ho 475)
11th year of the Hongwu reign period of Emperor Taizu of Ming, 9th month, day *jichou* [26] (Oct 11); a guest star swept TIANBIAN, moved southwest and entered WEI [LM 6]. Before this, on day *jiaxu* [11] (Sep 26), it was seen to the northeast of WUCHE. On day *dingchou* [14] (Sep 29), it entered ZIWEI. On day *wuyin* [15] (Sep 30), it emitted rays more than 1 *zhang* long; it swept NEIJIE. On day *gengchen* [17] (Oct 2), it entered ZIWEIGONG and swept the five stars of BEIJI. On day *renwu* [19] (Oct 4), it trespassed against SHAOZAI at the eastern wall (of ZIWEI). On day *jiashen* [21] (Oct 6), it entered the wall of TIANDI and on day *dinghai* [24] (Oct 9), it entered TIANSHI itself. From then (i.e., after Oct 11) until the 10th month, day *jiwei* [56] (Nov 10), it was cloudy and the comet was not seen.
[*Ming shilu*] ch. 119
NB: This is Comet Halley, return –8.

AD 1378 Oct 1 [Japan]
4th year of the Tenju reign period, 9th month, 9th day, *jimao* [16] (Oct 1). The officials reported that a comet appeared last night. From the

15th day *jiayin* [51] to the 25th day *jiazi* [1] the Emperor offered sacrifices to the comet.
[*Gu kan ki*] 22
NB: This is Comet Halley, return –8.

AD 1378 Oct 1 [Korea]
4th year of the King U of Goryeo, 9th month, day *jimao* [16] (Oct 1); a star appeared fuzzy at the western wall of ZIWEI. It trespassed against SIFU at the north pole and emerged at the eastern wall [of ZIWEI]. It trespassed against TIANBEI.
[*Goryeosa*] ch. 49
NB: This is Comet Halley, return –8.

AD 1381 Nov 7–22 [Korea] (Ho 477)
7th year of King U of Goryeo, 10th month, day *renshen* [9]; a broom star over 1 *zhang* long was seen in DI [LM 3] for 15 days before being extinguished.
[*Goryeosa*] ch. 49

AD 1381 Nov 8–Dec 1 [Japan]
1st year of the Kōwa reign period, 10th month, 22nd day; a broom star was seen in the east, its rays about 1 *zhang*, 5 to 6 *chi* long; on the 27th day *wuyin* [15], clear sky; today at dawn the broom star was seen, its rays truly long…11th month, 15th day (Dec 1); at nightfall the courtier Tameyo brought in a report on the broom star saying that for the past 2 to 3 nights it had disappeared; [he] says that he will submit the confidential report today, and that the broom star this time lacks a star at its hub, for which there is precedent.
[*Gu kan ki*]

AD 1382 Mar 11 [Korea] (Ho 478)
8th year of King U of Goryeo, 2nd month, day bingzi [13]; a fuzzy star was seen in the north.
[Goryeosa] ch. 49

AD 1382 Sep 19 [Korea] (Ho 480)

8th year of King U of Goryeo, 8th month, day *wuzi* [25]; a broom star over 1 *zhang* long was seen by TAIWEI's eastern wall.
[*Goryeosa*] ch. 49

AD 1385 May 10–Jun 7 [China]

Yichou [2] year of the Hongwu reign period of Emperor Taizu of the Ming Dynasty, 4th month; a broom star appeared and swept YI [LM 27], and then swept TIANMIAO.
[*Er shen ye lu*] ch. 1

AD 1385 Oct 23 [China] (Ho 481; Ho & Ang 7)

(a) 18th year of the Hongwu reign period of Emperor Taizu of the Ming Dynasty, 9th month, day *wuyin* [15]; a guest star entered TAIWEIYUAN; on the night of day *xinsi* [18], the guest star trespassed on YOUZHIFA and emerged from DUANMEN; on the night of day *yiyou* [22], the guest star entered YI [LM 27], its broom over 1 *zhang* long. On the night of day *dinghai* [24], the guest star was southwest of lunar mansion ZHEN [LM 28], its broom sweeping lunar mansion YI [LM 27]; on day *wuzi* [25], the guest star re-entered TAIWEI; in the 10th month, on the night of day *gengyin* [27], the guest star trespassed on JUNMEN and its broom swept TIANMIAO.
[*Ming Taizu shilu*] ch. 175

(b) 18th year of the Hongwu reign period of Emperor Taizu of the Ming Dynasty, 9th month, day *wuyin* [15]; a star appeared in TAIWEIYUAN, trespassed on YOUZHIFA, then emerged from DUANMEN; on day *yiyou* [22], it entered YI [LM 27], its broom over 1 *zhang* long. By day *gengyin* [27] in the 10th month, it trespassed on JUNMEN, and its broom swept TIANMIAO.
[*Ming shi: tianwen zhi*] ch. 27

AD 1385 Nov 10 [China]

18th year of the Hongwu reign period of Emperor Taizu of the Ming Dynasty, 10th month, day *bingshen* [33]; a star became fuzzy in DOU.
[*Guo que*] ch. 8

AD 1389 Aug 6 [China]

(a) 22nd year of the Hongwu reign period of Emperor Taizu of the Ming Dynasty, 7th month, day *xinsi* [18]; a broom star appeared beside ZIWEI; on day *yimao* [52] it was extinguished.
[*Guo que*] ch. 9

(b) 22nd year of the Hongwu reign period of Emperor Taizu of the Ming Dynasty, 6th month, day *xinsi* [18]; a white broom star with brightness more than about 1 *zhang* long appeared alongside ZIWEI; it was at 90 *fen* in NIU [LM 9] and pointing southeast as it traveled northwestward; on day *wuzi* [25] the broom's brightness swept SHANGZAI; on day *yimao* [52] in the 7th month it was extinguished.
Ming Dynasty Wanli period ed. Zhejiang [*Xiushui xianzhi*] ch. 10

NB: The original's "6th month" is a mistake for "7th month," as the 6th month lacked day *xinsi* [18]; the 7th month lacked day *yimao* [52].

AD 1391 May 23 [China] (Ho 483; Ho & Ang 9)

(a) 24th year of the Hongwu reign period of Emperor Taizu of the Ming Dynasty, 4th month, day *bingzi* [13], at night; two broom stars: 1 entered CHANGHE gate of ZIWEIYUAN and trespassed on TIANCHUANG; 1 trespassed on LIUJIA and swept WUDIZUO.
[*Ming Taizu shilu*] ch. 208

(b) 24th year of the Hongwu reign period of Emperor Taizu of the Ming Dynasty, 4th month, day *bingzi* [13]; two broom stars: one entered CHANGHE gate of ZIWEIYUAN and trespassed on TIANCHUANG; one trespassed on LIUJIA and swept WUDIZUO.
[*Ming shi: tianwen zhi*] ch. 27

AD 1392 Mar 18 [Korea] (Ho 484)

4th year of King Gongyang of Goryeo, 2nd month, day bingzi [13]; a broom star stretched across the sky.
[Goryeosa] ch. 49

AD 1402 Feb 20–Mar 19 [Korea] (Ho 487)

2nd year *renwu* of King Taejo of the Joseon Dynasty, 1st month, day *renyin* [39]; a broom star over 6 *chi* long was seen in the east of KUI

[LM 15], its rays pointing eastward…on day *jiachen* [41] (Feb 22), the star over 1 *zhang* long was seen in the east, its rays radiating in all directions…in the 2nd month, on day *jisi* [6] (Mar 19), the broom star disappeared. [It was visible] from its first appearance on the previous *renyin* [29] day until this day when it was extinguished.
[*Joseon wangjo sillok: T'aejong*] ch. 3

AD 1402 Feb 21–Mar 16 [Japan]

(a) 9th year of the Ōei reign period, 1st month, 19th day (Feb 21); at dusk during the hour of *xu* [LT = 19–21 h], a white broom star was seen in the west in the space of Kui [LM 15], its rays about 3 *chi* long…[2nd month], 16th day *jisi* [6] (Mar 19), clear sky; the broom star was seen through the night, but after the 17th day, it was no longer seen.
[*Yoshidake hitsujiki*]
[*Dai Nihon shi*] (Ho 487)

(b) 9th year of the Ōei reign period, 2nd month (Mar 3–Apr 1); a broom star was seen at Kamakura.
[*Yoshidake hitsujiki*]

AD 1403 Dec 30 [Korea] (Ho 488)

3rd year of King T'aejong of the Joseon Dynasty, 11th month, day wushen [45]; a broom star was seen in the northeast.
[Joseon wangjo sillok: T'aejong] ch. 3

AD 1410 Oct 28–Nov 25 [China]

8th year of the Yongle reign period of Emperor Taizong of the Ming Dynasty, 10th month; a star became fuzzy in Nandou.
[*Ming Taizong shilu*] ch. 167

AD 1421 Jan 9 [Japan]

27th year of the Ōei reign period, 12th month, 6th day; during the hour of *xu* [LT = 19–21 h], a broom star was seen at azimuth *qian* [NW] its rays about 5 *chi* long.
[*Kōyō ki*]
NB: The original miswrote "about 5 *chi*" as "about 5 seen."

AD 1430 Nov 14 [China] (Ho 501; Ho & Ang 14)

(a) 5th year of the Xuande reign period of Emperor Xuanzong of the Ming Dynasty, 10th month, day *bingshen* [33], at night; a fluffy white star like cotton wadding appeared south of WAIPING. It gradually traveled southeastward, passing TIANCANG and TIANYU in all for 8 days before being extinguished.
[*Ming Xuanzong shilu*] ch. 71

(b) 5th year of the Xuande reign period of Emperor Xuanzong of the Ming Dynasty, 10th month, day *bingshen* [33]; a fluffy star appeared south of WAIPING It traveled southeastward, passing TIANCANG and TIANYU. After 8 days, it was extinguished.
[*Ming shi: tianwen zhi*] ch. 27

(c) 5th year of the Xuande reign period of Emperor Xuanzong of the Ming Dynasty, 10th month, day *bingshen* [33]; a white star fluffy like cotton wadding appeared at WAIPING. It traveled southward, passing TIANCANG and TIANYU. In 8 days, it was extinguished.
[*Guo que*] ch. 21

AD 1431 May 12–Jun 9 [China] (Ho 503; Ho & Ang 16)

(a) 6th year of the Xuande reign period of Emperor Xuanzong of the Ming Dynasty, 4th month, day *wuxu* [35] (May 15); at dusk a broom star emerged in DONGJING [LM 22], fluffy, with horn-like rays over 5 *chi* long.
[*Ming Xuanzong shilu*] ch. 78

(b) 6th year of the Xuande reign period of Emperor Xuanzong of the Ming Dynasty, 4th month, day *wuxu* [35] (May 15); a star became fuzzy and over 5 *chi* long in DONGJING [LM 22].
[*Ming shi: tianwen zhi*] ch. 27

(c) 6th year of the Xuande reign period of Emperor Xuanzong of the Ming Dynasty, 4th month; a star became fuzzy in DONGJING [LM 22].
Ming Dynasty Wanli period ed. [*Sichuan zongzhi*] ch. 27

AD 1432 Feb 3 [China] (Ho 504; Ho & Ang 17)

(a) 7th year of the Xuande reign period of Emperor Xuanzong of the Ming Dynasty, 1st month, day *renxu* [59], at night; a broom star with bright rays over 1 *zhang* long emerged in the east; its tail swept TIANJIN, then it gradually traveled southeastward for 10 days before being extinguished...on day *wuzi* [25], at night, the broom star re-emerged in the west for 17 days before it was extinguished. [*Ming Xuanzong shilu*] ch. 86

(b) 7th year of the Xuande reign period of Emperor Xuanzong of the Ming Dynasty, 1st month, day *renxu* [59]; a broom star over 1 *zhang* long emerged in the east, its tail sweeping TIANJIN as it traveled southeastward for 10 days before being extinguished...on day *wuzi* [25] of this month, it reemerged in the west for 17 days before being extinguished.
[*Ming shi: tianwen zhi*] ch. 27

NB: The original (b) "for 10 months before being extinguished" is a mistake for "for 10 days before being extinguished."

AD 1433 Sep 15 [China] (Ho 505; Ho & Ang 18)

(a) 8th year of the Xuande reign period of Emperor Xuanzong of the Ming Dynasty, intercalary 8th month, day *renzi* [49]; at dusk a broom star with bright rays 1 *zhang* long emerged beside TIANCANG; on day *jisi* [6], it entered GUANSUO and swept QIGONG; on day *jimao* [16], it entered TIANSHIYUAN, swept the JIN star, and then after another 24 days it was extinguished.
[*Ming Xuanzong shilu*] ch. 105

(b) 8th year of the Xuande reign period of Emperor Xuanzong of the Ming Dynasty, intercalary 8th month, day *renzi* [49]; a broom star with bright rays about 1 *zhang* long emerged beside TIANCANG; on day *jisi* [6], it entered GUANSUO and swept QIGONG; on day *jimao* [16], it entered TIANSHIYUAN, swept the JIN star, then after 24 days, it was extinguished.
[*Ming shi: tianwen zhi*] ch. 27

AD 1433 Oct 2–Nov 4 [Korea] (Ho 505)

15th year *guichou* of King Sejong of the Joseon Dynasty, intercalary 8th month, day *jisi* [6]; a broom star first was seen within the space of GUANSUO; on day *jiaxu* [11] (Oct 7), it was seen at the western wall

of TIANSHI; in the 9th month, on day *wuzi* [25] (Oct 21), the broom star grew faint; on day *gengzi* [37] (Nov 2), the broom star was on the western horizon and very faint; on day *renyin* [39] (Nov 4), it went out of sight in the west.
[*Joseon wangjo sillok: Sejong*] ch. 61

AD 1433 Oct 3 [China]

(a) 8th year of the Xuande reign period of Emperor Xuanzong of the Ming Dynasty, intercalary 8th month, day *gengwu* [7]; at dusk, a broom star with a tail over 5 *chi* long emerged from within GUANSUO, then gradually traveled southwestward; it was not extinguished until after 34 days.
[*Ming Xuanzong shilu*] ch. 105

(b) 8th year of the Xuande reign period of Emperor Xuanzong of the Ming Dynasty, intercalary 8th month, day *gengwu* [7]; at dusk, a broom star emerged at GUANSUO and was extinguished after 34 days.
[*Guo que*] ch. 22

(c) 8th year of the Xuande reign period of Emperor Xuanzong of the Ming Dynasty, intercalary 8th month, day *renzi* [49]; a broom star about 1 *zhang* long emerged beside TIANCANG; on day *jisi* [6] it entered GUANSUO and swept QIGONG; on day *jimao* [16] it further entered TIANSHIYUAN and swept the JIN star; after 24 days it was extinguished.
[*Ming shi: tianwenzhi*] ch. 27 (Ho & Ang 18)

NB: there was no day *renzi* [49] in the intercalary 8th month. Ho & Ang misdate the report to Sep 15.

AD 1433 Oct 8–Nov 2 [Japan]

(a) 5th year of the Eikyō reign period, 8th month, 25th day; at dusk during the hour of *xu* [LT = 19–21 h], a white broom star was seen between azimuth *you* [W] and *xu* [WNW] near GUANSUO in the space of WEI [LM 6].
[*Kammon gyoki*]

(b) 5th year of the Eikyō reign period, 8th month, 25th day, clear sky; at the beginning of the hour of *xu* [LT = 19–21 h] in the west,

a star's brightness pointed east...9th month...on the 19th of this month (Oct 31), its rays gradually dimmed, clearly in process of disappearing; likewise on the 20th and 21st (Nov 1 and Nov 2), this process continued—it is extinguishing.
[*Manzai jugō nikki*]
[*Dai Nihon shi*] (Ho 505)

AD 1434 Apr 14 [Japan] (Ho 506)
6th year of the Eikyō reign period, 3rd month, 5th day; a broom star was seen in the east.
[Kanda]

AD 1434 Sep 11 [Japan] (Ho 507)
6th year of the Eikyō reign period, 8th month, 9th day; a broom star was seen in the east.
[Kanda]

AD 1439 Mar 15–Apr 8 [Japan]

(a) 11th year of the Eikyō reign period, 2nd month, 15th day (Mar 29); a broom star was seen.
[*Daijyōin nikki mokuroku*]

(b) 11th year of the Eikyō reign period, 2nd month (Mar 15–Apr 13); on the 23rd day (Apr 6) the Chief Astrologer reported that a broom star was seen in the north; from the 18th to 19th days on, it had been barely discernible, appearing in lunar mansion Gui [LM 23].
[*Kennai ki*]

(c) 11th year of the Eikyō reign period, 2nd month, 23rd day *xinchou* [38] (Apr 6); the broom star which was seen on this night in Ziwei, on the 25th (Apr 8)...memorialized about it.
[*Zoku shigushō*]

AD 1439 Mar 25 [China] (Ho 511; Ho & Ang 19)

(a) 4th year of the Zhengtong reign period of Emperor Yingzong of the Ming Dynasty, intercalary 2nd month, day *jichou* [26], at

night; a white broom star as big as a crossbow bolt and moving westward was produced in the southeast beside lunar mansion ZHANG [LM 26]; on day *dingyou* [34], at night, the rays of the broom star were over 5 *chi* long as it traveled northwestward sweeping JIUQI; on the night of day *gengzi* [37], the broom star was traveling northwestward and was over 7 *chi* long when it trespassed on GUI [LM 23].
[*Ming Yingzong shilu*] ch. 52

(b) 4th year of the Zhengtong reign period of Emperor Yingzong of the Ming Dynasty, intercalary 2nd month, day *jichou* [26]; a broom star as big as a crossbow bolt and moving westward appeared beside lunar mansion ZHANG [LM 26]; on day *dingyou* [34], it was over 5 *zhang* long as it traveled westward sweeping JIUQI; it meandered northward and trespassed on GUI [LM 23].
[*Ming shi: tianwen zhi*] ch. 27

AD 1439 Apr 1–16 [Korea]

(a) 21st year *jiwei* of King Sejong of the Joseon Dynasty, intercalary 2nd month, day *bingshen* [33]; a broom star was seen; on day *dingyou* [34] (Apr 2), it reappeared; on day *dingwei* [44] (Apr 12), it was seen in JING [LM 22]; on days *wushen* [45] (Apr 13) it was seen, and on day *xinhai* [48] (Apr 16) in the 3rd month it reappeared, then disappeared after 12 days in all.
[*Joseon wangjo sillok: Sejong*] ch. 84 (Ho 511)

(b) 21st year *jiwei* of King Sejong of the JoseonDynasty, intercalary 2nd month, day *bingshen* [33], a broom star was seen in DONGJING [LM 22] for 22 days in all.
[*Jeungbo munheon bigo*] ch. 6

AD 1439 Jul 12 [China] (Ho 511)

(a) 4th year of the Zhengtong reign period of Emperor Yingzong of the Ming Dynasty, 6th month, day *wuyin* [15]; a broom star with bright rays over 1 *zhang* long and pointing southwest appeared

in lunar mansion Bi [LM 19], until the 25th day of the 7th month when it disappeared.
[*Ming Yingzong shilu*] ch. 56

(b) 4th year of the Zhengtong reign period of Emperor Yingzong of the Ming Dynasty, 6th month, day *wuyin* [15]; a broom star over 1 *zhang* long and pointing southwest appeared beside lunar mansion Bi [LM 19] for about 55 days, then it was extinguished.
[*Ming shi: tianwen zhi*] ch. 27

AD 1444 Aug 6 [China] (Ho 513; Ho & Ang 21)

(a) 9th year of the Zhengtong reign period of Emperor Yingzong of the Ming Dynasty, 7th month, day *gengwu* [7], at dusk; a broom star with bright rays 1 *zhang* long at the eastern wall of Taiwei gradually grew longer. By the 2nd day of the intercalary 7th month, it entered lunar mansion Jiao [LM 1] then disappeared.
[*Ming Yingzong shilu*] ch. 118

(b) 9th year of the Zhengtong reign period of Emperor Yingzong of the Ming Dynasty, 7th month, day *gengwu* [7]; a broom star about 1 *zhang* long appeared at the eastern wall of Taiwei and gradually lengthened day by day. By day *jimao* [16] of the intercalary 7th month, it entered Jiao [LM 1] then disappeared.
[*Ming shi: tianwen zhi*] ch. 27

AD 1444 Aug 7–23 [Japan]

(a) 1st year of the Bunnan reign period, intercalary 6th month, 23rd day; it was subsequently reported that tonight a broom star was seen. Once it was learned that a broom star had appeared, at nightfall it was observed there in the west, with rays about 2 *chi* long; it was a small star.
[*Morosato ki*]

(b) 1st year of the Bunnan reign period, intercalary 6th month, 23rd day *xinwei* [8]; there was a broom star observed at azimuth *xu*

[WNW] through the first 10 days of the next month after which it disappeared.
[*Zoku shigushō*]
[*Dai Nihon shi*] (Ho 513)

AD 1444 Aug 12 [Korea] (Ho 513)

26th year of King Sejong of the Joseon Dynasty, 7th month, day bingzi [13], a broom star was seen.
[Joseon wangjo sillok: Sejong] ch. 105

AD 1449 Dec 20 [China] (Ho 514; Ho & Ang 22)

(a) 14th year of the Zhengtong reign period of Emperor Yingzong of the Ming Dynasty, 12th month, day *renzi* [49], at daybreak; a broom star appeared beside the SHILOU star of TIANSHIYUAN; at daybreak on day *guichou* [50], the greenish-white broom star appeared at 12 *du* in WEI [LM 6], with rays over 1 *chi* long and its sweep pointing due west; at daybreak on day *yimao* [52], the broom star appeared at 11 *du* in WEI [LM 6], now over 2 *chi* long. At daybreak on day *bingchen* [53], the broom star appeared at 11 *du* in WEI [LM 6], traveling westward, until the night of day *yihai* [12] when it was no longer seen.
[*Ming Yingzong shilu*] ch. 186

(b) 14th year of the Zhengtong reign period of Emperor Yingzong of the Ming Dynasty, 12th month, day *renzi* [49]; a broom star appeared beside the SHILOU star of TIANSHIYUAN; passing through the space of Wei [LM 6] it was over 2 *chi* long, until day *yihai* [12] when it disappeared.
[*Ming shi: tianwen zhi*] ch. 27

AD 1449 Dec 26–AD 1450 Jan 30 [Korea] (Ho 514)

31st year *jisi* of King Sejong of the Joseon Dynasty, 12th month, 12th day *wuwu* [55]; a broom star was seen within TIANSHIYUAN in the east, after which the sky was overcast and it could not be observed. Coming to the 21st day *dingmao* [4] (Jan 4), the broom star was again seen

in TIANSHI, its tail over 5 to 6 *chi* long...32nd year, 1st month, day *guisi* [30] (Jan 30), for the first time the broom star was not seen.
[*Joseon wangjo sillok: Sejong*] ch. 126–127

AD 1450 Jan 19 [China] (Ho 514; Ho & Ang 22)

(a) 1st year of the Jingtai reign period of Emperor Yingzong of the Ming Dynasty, 1st month, day *renwu* [19]; a broom star emerged outside TIANSHIYUAN and swept the asterism TIANJI.
[*Ming Yingzong shilu*] ch. 187
[*Ming shi: tianwen zhi*] ch. 27

(b) 1st year of the Jingtai reign period of Emperor Yingzong of the Ming Dynasty, 1st month, day *renwu* [19], at night; a broom star appeared outside TIANSHIYUAN and swept TIANJI.
[*Guo que*] ch. 29

NB: Ho considered this a report of the same comet as the preceding.

AD 1452 Mar 21 [China] (Ho 515; Ho & Ang 23)

3rd year of the Jingtai reign period of Emperor Yingzong of the Ming Dynasty, 3rd month, new moon day *jiawu* [31], at night; a fuzzy star appeared in BI [LM 19].
[*Ming Yingzong shilu*] ch. 214

AD 1456 May 27 [China] (Ho 517; Ho & Ang 27)

(a) 7th year of the Jingtai reign period of Emperor Yingzong of the Ming Dynasty, 4th month, day *renxu* [59], at night; a broom star with bright rays 2 *zhang* long appeared in the northeast in lunar mansion WEI [LM 17], its tail pointing southwestward... in the 5th month, day *guiyou* [10], at night, the bright rays of the broom star gradually lengthened to over 1 *zhang*, pointing southwest...at dusk on day *wuzi* [25], the broom star appeared in the northwest in lunar mansion LIU [LM 24], its tail over 9 *chi* long, sweeping and trespassing on XUANYUAN...at dusk on day *jiawu* [31], the broom star appeared in lunar mansion *Zhang* [LM 26], its bright rays over 7 *chi* long, and its tail sweeping the western wall of TAIWEI as it traveled southwestward…in the 6th month, on the night of day *renyin* [39], the broom star entered within the western wall of TAIWEI beside SHANGJIANG, its rays over 5 *chi* long.

[*Ming Yingzong shilu*] ch. 265–267

(b) 7th year of the Jingtai reign period of Emperor Yingzong of the Ming Dynasty, 4th month, day *renxu* [59]; a broom star 2 *chi* long appeared in the northeast in WEI [LM 17], pointing southwest; by day *guiyou* [10] in the 5th month, it had gradually lengthened to over 1 *zhang*; on day *wuzi* [25], it appeared in the northwest in LIU [LM 24], now over 9 *chi* long, sweeping and trespassing on XUANYUAN; on day *jiawu* [31], it appeared in ZHANG [LM 26], now over 7 *chi* long, and its tail sweeping the north of TAIWEI as it traveled southwestward; in the 6th month, on day *renyin* [39], it entered TAIWEIYUAN, now over 1 *chi* long.

[*Ming shi: tianwen zhi*] ch. 27

NB: This is Comet Halley, return –7.

AD 1456 May 31–Jul 3 [Japan]

2nd year of the Kōshō reign period, 4th month, 27th day; in the 5th month, 7th day *yihai* [12] (Jun 9), after it rained late at night, it was subsequently learned that today at dawn a broom star was seen at azimuth *gen* [NE], its rays about 3 *chi* long; some said they had seen it on the 27th of last month…6th month, 1st day *jihai* [36] (Jul 3); tonight the broom star was seen again.

[*Morosato ki*]

[*Dai Nihon shi*] (Ho 517)

NB: This is Comet Halley, return –7.

AD 1456 Jun 6–29 [Korea] (Ho 517)

2nd year of King Sejo of the Joseon Dynasty, 5th month, day *renshen* [9]; a broom star was seen; on days *guiyou* [10] (Jun 7), *jiaxu* [11] (Jun 8), *yihai* [12] (Jun 9), *wuyin* [15] (Jun 12), *jimao* [16] (Jun 13), *yiyou* [22] (Jun 19), *wuzi* [25] (Jun 22), *xinmao* [28] (Jun 25),

jiawu [31] (Jun 28), and *yiwei* [32] (Jun 29), the broom star was seen.
[*Joseon wangjo sillok: Sejo*] ch. 4
NB: This is Comet Halley, return –7

AD 1456 Aug 1–30 [China]

(a) 7th year of the Jingtai reign period of Emperor Yingzong of the Ming Dynasty, during the 7th month; during daylight hours corresponding to the end of the hour of *shen* [LT 15–17 h], a broom star shaped like a washing brush was faintly visible in the west; after the hour of *you* [LT 17–19 h], it gradually lengthened to become like the sweep of a broom and people called it a broom star; it set daily after stretching across half the sky; it was like this for 2 months before being extinguished.
[*Xian zhong gu jin lu*]

(b) 7th year of the Jingtai reign period of Emperor Yingzong of the Ming Dynasty, during the 7th month; a broom star appeared during daylight in the west from the hour of *shen* [LT 15–17 h] until sunset; it stretched across half the sky and lasted for 2 months before being extinguished.
[*Er shen ye lu*] ch. 2

AD 1457 Jan 14 [China] (Ho 518; Ho & Ang 26)

(a) 7th year of the Jingtai reign period of Emperor Yingzong of the Ming Dynasty, 12th month, day *jiayin* [51], at night; a broom star with bright rays 5 *cun* long reappeared in lunar mansion Bi [LM 19], slowly traveling southeastward. Its bright rays gradually lengthened from this day through day *guihai* [60].
[*Ming Yingzong shilu*] ch. 273

(b) 7th year of the Jingtai reign period of Emperor Yingzong of the Ming Dynasty, 12th month, day *jiayin* [51]; a broom star with bright rays 5 *cun* long reappeared in lunar mansion Bi [LM 19], traveling southeastward. It gradually lengthened until day *guihai* [60] when it disappeared.
[*Ming shi: tianwen zhi*] ch. 27

AD 1457 Jun 15 [China] (Ho 519; Ho & Ang 27)

(a) 1st year of the Tianshun reign period of Emperor Yingzong of the Ming Dynasty, 5th month, day *bingxu* [23], at night; a greenish-white broom star like cotton wadding appeared to vibrate in lunar mansion WEI [LM 12]; by the night of day *dinghai* [24] it had traveled eastward 1 *du*, its faint rays 5 *cun* long and pointing southwest...in the 6th month, on the night of day *guisi* [30], the broom star appeared in lunar mansion SHI [LM 13], its horn-like rays over 1 *zhang* long...on the night of day *jihai* [36], the broom star trespassed on lunar mansion BI [LM 14], its tail pointing to the second star of DONGBI [LM 14]...on the night of day *yisi* [42], the broom star trespassed against TIANDAJIANGJUN...on the night of day *renzi* [49], the broom star passed by and trespassed on the third star of JUANSHE...in the 7th month, on the night of day *wuyin* [15], the broom star trespassed on lunar mansion JING [LM 22]...on the night of day *xinmao* [28], the broom star trespassed on the second star on the south of SHUIWEI...in the 8th month, on day *jiawu* [31], the Emperor personally offered a prayer to the Supreme Lord of Boundless Heaven, with the words, "At present the broom star has appeared for 2 months, with bright rays still persisting, a suspended anomaly which does not disappear."

[*Ming Yingzong shilu*] ch. 278–281

(b) 1st year of the Tianshun reign period of Emperor Yingzong of the Ming Dynasty, 5th month, day *bingxu* [23]; a broom star with bright rays 5 *cun* long and pointing southwest appeared in WEI [LM 12], seemingly vibrating as it traveled eastward 1 *du*; in the 6th month, on new moon day *guisi* [30], it appeared in SHI [LM 13], over 1 *zhang* long and extending from WEI [LM 6] to DONGBI [LM 14], trespassing on TIANDAJIANGJUN, the third star of JUANSHE, and the second star on the south of SHUIWEI in lunar mansion JING [LM 22].

[*Ming shi: tianwen zhi*] ch. 27

AD 1457 Jun 23–Jul 16 [Korea] (Ho 519)

3rd year of King Sejo of the Joseon Dynasty, 6th month, day *jiawu* [31]; a broom star was seen. On days *yiwei* [32] (Jun 24, *bingshen* [33] (Jun 25), *dingyou* [34] (Jun 26), *wuxu* [35] (Jun 27), *renyin* [39] (Jul 1)), *dingwei* [44] (Jul 6), *jiyou* [46] (Jul 8), *gengxu* [47] (Jul 9), *xinhai* [48] (Jul 10), *renzi* [49] (Jul 11), *jiayin* [51] (Jul 13), *bingchen* [53] (Jul 15), and *dingsi* [54] (Jul 16), the broom star was seen.

[*Joseon wangjo sillok: Sejo*] ch. 4

AD 1457 Jul 30–Aug 19 [Japan]

(a) 1st year of the Chōroku reign period, 7th month, 9th day *xinwei* [8] (Jul 30); today is the beginning of autumn. At dawn, a broom star began to appear; it emerged during the hour of *mao* [LM = 5–7 h] at azimuth *yin* [ENE]; its color was pale, and it had a tail about 5 *chi* long.

[*Kyōgaku shi yōshō*]

(b) 1st year of the Chōroku reign period, 7th month, 29th day (Aug 19); the broom star has not yet disappeared; it is in the east and I saw it at dawn, with rays about 7 to 8 *chi* long.

[*Munekata kyōki*] Yanagiharake kiroku

NB: A record in Kanda states that the comet was first observed at the beginning of the 6th month; see (Ho 519).

AD 1457 Oct 26 [China] (Ho 520; Ho & Ang 28)

(a) 1st year of the Tianshun reign period of Emperor Yingzong of the Ming Dynasty, 10th month, day *jihai* [36], at night; a broom star with rays over 5 *cun* long reappeared in lunar mansion JIAO [LM 1], its tail pointing northward…on the night of day *jiyou* [46], toward daybreak, the broom star trespassed against the north star of lunar mansion JIAO [LM 1]…on the night of day *renzi* [49], the broom star trespassed on the eastern star of PINGDAO.

[*Ming Yingzong shilu*] ch. 283

(b) 1st year of the Tianshun reign period of Emperor Yingzong of the Ming Dynasty, 10th month, day *jihai* [36]; a broom star over 5 *cun* long appeared in JIAO [LM 1], pointing northward and trespassing against the north star of lunar mansion JIAO [LM 1] and the eastern star of PINGDAO.
[*Ming shi: tianwen zhi*] ch. 27

AD 1458 Dec 24 [China] (Ho 521; Ho & Ang 29)

(a) 2nd year of the Tianshun reign period of Emperor Yingzong of the Ming Dynasty, 11th month, day *guimao* [40], at night; a white guest star appeared in lunar mansion XING [LM 25] traveling westward; by the night of day *bingwu* [43], its body was faintly larger, like cotton wadding in appearance, and next to XUANYUAN…on the night of day *gengxu* [47], the guest star sprouted rays about 5 *cun* long, then passed by and trespassed on the northwest star of GUANWEI…in the 12th month, on the night of day *renxu* [59], the guest star disappeared in lunar mansion JING [LM 22].
[*Ming Yingzong shilu*] ch. 297–298

(b) 2nd year of the Tianshun reign period of Emperor Yingzong of the Ming Dynasty, 11th month, day *guimao* [40]; a white guest star appeared in lunar mansion XING [LM 25] traveling westward; by day *bingwu* [43], its body was faint, like cotton wadding in shape, and next to XUANYUAN; on day *gengxu* [47], it sprouted rays 5 *cun* long, then trespassed on the northwest star of GUANWEI; by the 12th month, day *renxu* [59], it disappeared in DONGJING [LM 22].
[*Ming shi: tianwen zhi*] ch. 27

AD 1458 Dec 29–AD 1459 Apr 8 [Korea]

4th year of King Sejo of the Joseon Dynasty, 11th month, day *wushen* [45]; a broom star was seen; on day *gengxu* [47] (Dec 31) through day *yimao* [52] (Jan 5), the 1st day in the 12th month, the broom star was seen; on day *bingchen* [53] (Jan 6) through day *guihai* [60] (Jan 13), the broom star was seen but was faint; on day *renwu* [19] (Feb 1) through *bingxu* [23] (Feb 5), in the 1st month of the 5th year, and on day *wuzi*

[25] (Feb 7), the broom star was seen; on day *jiawu* [31] (Feb 13), the broom star was north of Lou [LM 16], but its rays were faint; on day *yiwei* [32] (Feb 14) through day *jiayin* [51] (Mar 5), the 1st day in the 2nd month, the broom star was seen; on day *yimao* [52] (Mar 6) through day *guiwei* [20] (Apr 3), the 1st day in the 3rd month, the broom star was seen; on day *wuzi* [5] (Apr 8), the broom star was seen, but its light was faint.
[*Joseon wangjo sillok: Sejo*] ch. 14–15 (Ho 521)

AD 1459 Apr 8 [Korea]
5th year *jimao* of King Sejo of the Joseon Dynasty, 1st month, day *bingxu* [23]; a broom star was seen in Lou [LM 16], its rays faint, until the 3rd month, day *wuzi* [25] (Apr 8).
[*Jeungbo munheon bigo*] ch. 6

AD 1461 Jul 30 [China] (Ho 522; Ho & Ang 31)
5th year of the Tianshun reign period of Emperor Yingzong of the Ming Dynasty, 6th month, day *renchen* [29], at night; an anomalous powdery-white star appeared beside Zongzheng of Tianshiyuan; by the night of day *yiwei* [32], it changed into a white vapor and vanished.
[*Ming Yingzong shilu*] ch. 329

AD 1461 Aug 5 [China] (Ho 523; Ho & Ang 32)
(a) 5th year of the Tianshun reign period of Emperor Yingzong of the Ming Dynasty, 6th month, day *wuxu* [35], at night; a broom star with bright rays over 3 *chi* long appeared in the east, its tail pointing southwest…7th month, toward daybreak of the night of day *gengzi* [37], the broom star was at 4 *du* in Jing [LM 22], its bright rays over 1 *chi* long, and its tail pointing southwest…at daybreak of the night of day *bingwu* [43], the broom star still appeared in Jing [LM 22] until the night of day *bingyin* [3] when it was extinguished.
[*Ming Yingzong shilu*] ch. 329–330

(b) 5th year of the Tianshun reign period of Emperor Yingzong of the Ming Dynasty, 6th month, day *wuxu* [35]; a broom star appeared in the east, pointing southwest; it entered the space of JING [LM 22]; it was not extinguished until day *bingyin* [3] in the 7th month.
[*Ming shi: tianwen zhi*] ch. 27

(c) 5th year of the Tianshun reign period of Emperor Yingzong of the Ming Dynasty, 6th month, day *wuxu* [35], at night; a broom star appeared in the east, its rays over 2 *chi* long; on the night of day *gengzi* [37], in the 7th month, the broom star appeared in lunar mansion JING [LM 22]; on the night of day *bingwu* [43], the broom star appeared in lunar mansion JING [LM 22]; on the night of day *bingyin* [3], it was extinguished.
[*Guo que*] ch. 33

AD 1462 Jun 29 [China] (Ho 524; Ho & Ang 33)

(a) 6th year of the Tianshun reign period of Emperor Yingzong of the Ming Dynasty, 6th month, day *bingyin* [3], at night; a grayish-white guest star appeared next to CEXING…on the night of day *jisi* [6], the guest star entered ZIWEIYUAN…on the night of day *yihai* [12], the guest star entered and trespassed on TIANLAO…on the night of day *guiwei* [20], the guest star stayed below ZHONGTAI where it grew progressively fainter.
[*Ming Yingzong shilu*] ch. 341

(b) 6th year of the Tianshun reign period of Emperor Yingzong of the Ming Dynasty, 6th month, day *bingyin* [3]; a grayish-white star appeared next to CEXING then entered ZIWEIYUAN and trespassed on TIANLAO; by day *guiwei* [20], the guest star stayed below ZHONGTAI where it grew progressively fainter.
[*Ming shi: tianwen zhi*] ch. 27

AD 1465 Mar 27–Apr 24 [China] (Ho 525; Ho & Ang 34)

1st year of the Chenghua reign period of Emperor Xianzong of the Ming Dynasty, 3rd month; a broom star over 3 *zhang* long appeared again in the northwest, lasting over 3 months before disappearing.

[*Ming shi: tianwen zhi*] ch. 27
Ming Dynasty Wanli reign period ed. Hubei [*Xiangyang fu zhi*] ch. 33
NB: The Shanxi (Kangxi period ed.) of *Linfen xianzhi* had "2nd month" with remaining text the same.

AD 1468 Jul 19–Aug 17 [China]

4th year of the Chenghua reign period of Emperor Xianzong of the Ming Dynasty, 7th month; a star became fuzzy in [SAN-]TAI and DOU.
Ming Dynasty Jiajing ed. Hubei [*Sui zhi*] ch. *shang*

AD 1468 Sep 14–Oct 10 [Korea]

(a) 14th year *wuzi* of King Sejo of the Joseon Dynasty, 9th month, day *guihai* [60] (Sep 22); a broom star was seen in TIANSHI, its tail over 3 *zhang* long.
[*Jeungbo munheon bigo*] ch. 6

(b) 14th year of King Sejo of the Joseon Dynasty, 9th month, day *jiwei* [56]; a broom star was seen. On days *gengshen* [57] (Sep 14) and *renxu* [59] (Sep 16), the broom star was seen. The comet was seen in the accession year of King Yejong of the Joseon Dynasty, 9th month, days *jiazi* [1] (Sep 18), *bingyin* [3] (Sep 20), *wuchen* [5] (Sep 22), *jisi* [6] (Sep 23), *gengwu* [7] (Sep 24), *xinwei* [8] (Sep 25), *renshen* [9] (Sep 26), *guiyou* [10] (Sep 27), *jiaxu* [11] (Sep 28), *yihai* [12] (Sep 29), *dingchou* [14] (Oct 1), *wuyin* [15] (Oct 2), *jimao* [16] (Oct 3), *gengchen* [17] (Oct 4), *xinsi* [18] (Oct 5), *renwu* [19] (Oct 6), *guiwei* [20] (Oct 7), *jiashen* [21] (Oct 8), *yiyou* [22] (Oct 9), *bingxu* [23] (Oct 10).
[*Joseon wangjo sillok: Sejo*] ch. 47; [*Joseon wangjo sillok: Yejong*] ch. 1 (Ho 528)

NB: Ho mistakenly dated record (b) to 1469.

AD 1468 Sep 18 [China] (Ho 526; Ho & Ang 35)

(a) 4th year of the Chenghua reign period of Emperor Xianzong of the Ming Dynasty, 9th month, day *jiwei* [56], at night; a guest star appeared at 5 *du* in XING [LM 25] traveling northeastward…on the night of day *guihai* [60], the guest star was grayish-white with bright rays over 3 *zhang* long, its tail pointing southwest, having changed into a broom star…on day *wuchen* [5], the broom star appeared at dawn in the northeast…on day *jisi* [6], the broom star appeared at dusk in the southwest…on day *gengwu* [7] at dusk, the broom star trespassed on SANGONG…on day *xinwei* [8] at dusk, the broom star trespassed on the YAOGUANG star of BEIDOU…on day *dingchou* [14] at dusk, the broom star trespassed on the fourth star on the west of QIGONG…on day *renwu* [19] at dusk, the broom star entered TIANSHIYUAN…in the 10th month, on day *yisi* [42], the broom star emerged from TIANSHIYUAN, its body gradually shrinking…on day *jiayin* [51], the broom star trespassed on the first star on the west of TIANPING…at night on day *gengshen* [57], in the 11th month, the broom star was extinguished.

[*Ming Xianzong shilu*] ch. 58–60

(b) 4th year of the Chenghua reign period of Emperor Xianzong of the Ming Dynasty, 9th month, day *jiwei* [56]; a star appeared at 5 *du* in XING [LM 25] traveling northeastward. After 5 days, its rays were over 3 *zhang* long, its tail pointing southwest, having changed into a broom star. Subsequently, it appeared at dawn in the east and at dusk appeared in SHI [LM 13], and then to the south it trespassed on SANGONG, BEIDOU, YAOGUANG, and QIGONG; then it turned and entered TIANSHIYUAN. When it emerged from [TIANSHI]YUAN, its body gradually diminished as it trespassed on the first star on the west of TIANPING; not until the 11th month, day *gengchen* [17], was it extinguished.

[*Ming shi: tianwen zhi*] ch. 27

AD 1468 Sep 22–Oct 6 [Japan]

(a) 2nd year of the Ōnin reign period, 9th month, 6th day (Sep 22); a broom star was seen for 30 days.

[*Daijyōin nikki mokuroku*]

(b) 2nd year of the Ōnin reign period, 9th month, 20th day *dingchou* [4] (Oct 6); a broom star was seen in the west with rays over

1 *zhang* long; all said that it had been observed from the first 10-day week [of the month].
[*Hekizan nichiroku*]
[*Dai Nihon shi*] (Ho 526)

AD 1468 Sep 27 [China]
4th year of the Chenghua reign period of Emperor Xianzong of the Ming Dynasty, 9th month, day *wuchen* [5]; [according to *Ming Zhao dai dian*] a broom star appeared at dawn in the northeast; on day *jisi* [6], the broom star appeared at dawn in the southwest; in the 9th month, day *dingchou* [14], at dusk, the broom star trespassed on the four stars on the west of QIGONG; on day *renwu* [19], at dusk, the broom star entered TIANSHIYUAN; in the 10th month, on day *jiayin* [51], the broom star trespassed on the first star on the west of TIANPING; in the 11th month, on day *wuwu* [55], at night, the broom star was extinguished.
[*Gujin tushu jicheng*: *shuzheng dian*: *xing bian bu*] ch. 55

AD 1468 Oct 2 [China]
4th year of the Chenghua reign period of Emperor Xianzong of the Ming Dynasty, 9th month, day *guiyou* [10]; [according to *Dazheng ji*] at night, a grayish-white guest star with bright rays over 3 *zhang* long, its tail pointing southwest, changed into a broom star; on day *dingchou* [14], at dusk, the broom star trespassed on the four stars of QIGONG; on day *renwu* [19], at dusk, the broom star entered TIANSHIYUAN; in the 10th month, on day *jiayin* [51], the broom star trespassed on the second star on the west of TIANPING; in the 11th month, on day wuwu [55], the broom star was extinguished.
[*Gujin tushu jicheng*: *shuzheng dian*: *xing bian bu*] ch. 55

AD 1469 Sep 3 [Japan] (Ho 528)
1st year of the Bummei reign period, 7th month, 27th day; a broom star was seen in the east.
[Kanda]

AD 1471 Dec 12–AD 1472 Jan 9 [China]

(a) 7th year of the Chenghua reign period of Emperor Xianzong of the Ming Dynasty, 11th month; a broom star appeared, emerging from TIANTIAN and entering TAIWEIYUAN.
Ming Dynasty Wanli reign period ed. Shanxi [*Hongdongxian zhi*] ch. 8

(b) 7th year of the Chenghua reign period of Emperor Xianzong of the Ming Dynasty, 11th month; a broom star emerged at XUANYUAN; in the 12th month, on day *dingchou* [14], at night, the broom star traveled northward, laterally sweeping LANGWEI of TAIWEIYUAN; on day *jimao* [16], the bright rays of the broom star stretched across the sky from east to west; from the 11th day it traveled 28 *du* northward, trespassing on TIANQIANG, its tail sweeping BEIDOU, TAIYANG and SANGONG; on day *bingxu* [23], the Beginning of Spring, at dusk, the broom star trespassed on TIANHE.
[*Er shen ye lu*] ch. 2

(c) 7th year of the Chenghua reign period of Emperor Xianzong of the Ming Dynasty, 12th month, day *jiaxu* [11]; a broom star appeared in TIANTIAN, pointing west; it traveled northward for 8 *chi*, trespassing on YOUSHETI, and sweeping TAIWEIYUAN, SHANGJIANG, and reached XINGCHEN, TAIZI, CONGGUAN, its tail pointing due west and sweeping horizontally across TAIWEIYUAN and LANGWEI. On day *jimao* [16], its bright rays were big and long, stretching across the sky from east to west. It traveled northward over 28 *du*, trespassing on TIANQIANG, and sweeping BEIDOU, SANGONG, TAIYANG, and entering ZIWEIYUAN, being visible in broad daylight. From DI [LM 3] through BEIDOU, SHUZI, HOUGONG, GOUCHEN, TIANSHU, SANSHI, TIANLAO, ZHONGTAI, TIANHUANG DADI, SHANGWEI, GEDAO, WENCHANG, SHANGTAI, there were none it did not trespass against. On day *yiyou* [22], it traveled southward, trespassing on LOU [LM 16], TIANHE, TIANYIN, WAIPING, TIANJUN."
[*Ming shi: tianwenzhi*] ch. 27 (Ho 530)

NB: Ho 530 adds: "On day *bingwu* [43], in the 1st month of the 8th year it passed KUI [LM 15] and WAIPING. It gradually diminished in size but took a long time to disappear altogether."

AD 1472 Jan 7–Feb 17 [Korea]

(a) 2nd year of King Seongjong of the Joseon Dynasty, 11th month, night of day *yichou* [2] (Jan 7); a star with faint rays was seen in JIAO [LM 1], above the TIANTIAN star…in the 12th month, day *yihai* [12], the broom star was seen in JIAO [LM 1], below the ZHOUDING star, its tail over 2 *zhang* long…on day *jimao* [16] (Jan 21), the broom star was seen in TIANCANG stars of ZIWEI-YUAN, its tail over 2 *zhang* long…on day *gengchen* [17] (Jan 22), the broom star was seen at HUAGAI, and the GANG star, its tail over 2 *zhang* long….. on day *xinsi* [18] (Jan 23), the broom star was seen north of WANGLIANG…on day *jiashen* [21] (Jan 26), the broom star was seen south of KUI [LM 15], its tail over 1 *zhang* long…in the 3rd year, 1st month, on day *jihai* [36] (Feb 10), the broom star was seen south of the WAIPING star, its tail over 5 *chi* long…on day *bingwu* [43] (Feb 17), the broom star was faintly visible.
[*Joseon wangjo sillok: Seongjong*] ch. 13–14 (Ho 530)

(b) 2nd year *xinmao* of King Seongjong of the Joseon Dynasty, 11th month, day *yichou* [2] (Jan 7); a broom star was seen in JIAO [LM 1], passed by ZHOUDING and reached HUAGAI, its tail over 2 *zhang* long. In the 12th month, day *gengyin* [27] (Feb 1), [the broom star] traversed WANGLIANG and moved south of KUI [LM 15]. The next year, in the 1st month, on day *bingwu* [43] (Feb 17), it was extinguished.
[*Jeungbo munheon bigo*] ch. 6

AD 1472 Jan 13–Feb 9 [Japan]

(a) 3rd year of the Bummei reign period, 12th month, 3rd day (Jan 13); on the 3rd of this month, during the hour of *zi* [LT = 23–1 h],

a white broom star with rays about 3 *chi* long was seen in the east, penetrating the third star of TAIWEI.
[*Chikanaga kyōki*]

NB: The original miswrote "penetrating TAIWEI" as "penetrating *dawei*."

(b) 3rd year of the Bummei reign period, 1st month, 1st day (Feb 9), clear sky; today at dawn in the western sky a broom star was seen which did not disappear; when it first was seen it was in the east, then it circled northward, then turned to the west where it ended up.
[*Chikanaga kyōki*]

(c) 3rd year of the Bummei reign period, 12th month, 26th day (Feb 5); since the beginning of this month a broom star has appeared, its brightness about 10 *zhang* long; in the beginning it was seen at dawn in the east, lately at the beginning of night it has appeared in the west.
[*Daijyōin nikki mokuroku*]

NB: The report in Kanda states that as of the 2nd day in the 1st month of the following year—Feb 10—it was still being seen; see (Ho 530).

AD 1472 Jan 16 [China] (Ho 530; Ho & Ang 37)

(a) 7th year of the Chenghua reign period of Emperor Xianzong of the Ming Dynasty, 12th month, day *jiaxu* [11]; a broom star appeared in TIANTIAN, its bright rays pointing west...on day *yihai* [12], the broom star traveled northward, its bright rays gradually brightening, trespassing on YOUSHETI, and sweeping SHANGJIANG of TAIWEIYUAN, and XINGCHEN, TAIZI, CONGGUAN...on the night of day *dingchou* [14], the broom star traveled northward for over 5 *du*, its tail pointing due west, and its rays increasingly bright, as it laterally swept LANGWEI and TAIWEIYUAN...on day *jimao* [16], at night, the brightness of the broom star grew longer and larger as it stretched across the sky from east to west, from the 11th day it traveled northward for over 28 *du*, trespassing on TIANQIANG, and its tail sweeping BEIDOU, SANGONG, and TAIYANG...on day *gengchen* [17], the broom star traveled northward and entered ZIWEIYUAN, being still visible in broad daylight it passed by and trespassed on DIXING, the second star of the bowl of BEIDOU, SHUZI, HOUGONG, the lower star of GOUCHEN, the first star of the bowl of BEIDOU, the third star of GOUCHEN, TIANSHU, SANSHI, TIANLAO, ZHONGTAI, SHANGWEI, TIANHUANGDADI...on day *xinsi* [18], at dusk, the broom star entered ZIWEIYUAN, passed by and trespassed on GEDAO, WENCHANG, and SHANGTAI...on day *yiyou* [22], at dusk, the broom star traveled southward and trespassed on lunar mansion LOU [LM 16]; on day *bingxu* [23], at dusk, the broom star trespassed on TIANHE...on day *jichou* [26], at dusk, the broom star trespassed on TIANYIN...on day *guisi* [30], at dusk, the broom star trespassed on WAIPING...on day *yiwei* [32], at dusk, the broom star trespassed on TIANQUN...in the 8th year, 1st month, on the night of day *bingwu* [43], the broom star moved beneath lunar mansion KUI [LM 15] and WAIPING, growing progressively smaller.

[*Ming Xianzong shilu*] ch. 99–100

(b) 7th year of the Chenghua reign period of Emperor Xianzong of the Ming Dynasty, 12th month, day *jiaxu* [11]; a broom star appeared in TIANTIAN, pointing west as it slowly traveled northward, trespassing on YOUSHETI, and sweeping SHANGJIANG of TAIWEIYUAN, XINGCHEN, TAIZI, and CONGGUAN, its tail pointing due west as it laterally swept LANGWEI of TAIWEIYUAN; on day *jimao* [16], its bright rays grew longer and larger as it stretched across the sky from east to west, traveling northward over 28 *du*, and trespassing on TIANQIANG, sweeping BEIDOU, SANGONG, and TAIYANG, and entering ZIWEIYUAN, being still visible in broad daylight; from DIXING, to BEIDOU, KUI, SHUZI, HOUGONG, GOUCHEN, TIANSHU, SANSHI, TIANLAO, ZHONGTAI, TIANHUANGDADI, SHANGWEI, GEDAO, WENCHANG and SHANGTAI, there was none which was not trespassed against; on day *yiyou* [22], it traveled southward and trespassed on LOU [LM 16], TIANHE, TIANYIN, WAIPING, and TIANQUN; in the 8th year, 1st month, day *bingwu* [43], it traveled

by lunar mansion KUI [LM 15] and WAIPING, gradually growing fainter and not being extinguished for a long time.
[*Ming shi: tianwen zhi*] ch. 27

AD 1472 Jan 24 [China] (Ho 530)

7th year of the Chenghua reign period of Emperor Xianzong of the Ming Dynasty, 12th month, day *renwu* [19]; a broom star entered ZIWEIYUAN.
[*Ming shi: Xianzong ji*] ch. 13
NB: Probably the same comet as the preceding.

AD 1480 May 17 [Japan] (Ho 532)

12th year of the Bummei reign period, 2nd month, 6th day; a broom star was seen in the east.
[Kanda]

AD 1490 Dec 31 [China] (Ho 534; Ho & Ang 39)

(a) 3rd year of the Hongzhi reign period of Emperor Xiaozong of the Ming Dynasty, 11th month, day *wuxu* [35], at dusk; a broom star with rays over 1 *chi* long, appeared south of TIANJIN, its tail pointing northeast…on day *xinchou* [38], at dusk, the broom star trespassed on RENXING…in the 12th month, on day *wushen* [45], the broom star entered lunar mansion SHI [LM 13]…on day *gengshen* [57], at dusk, the broom star trespassed on TIANCANG.
[*Ming Xiaozong shilu*] ch. 45–46

(b) 3rd year of the Hongzhi reign period of Emperor Xiaozong of the Ming Dynasty, 11th month, day *wuxu* [35]; a broom star appeared south of TIANJIN, its tail pointing northeast; it trespassed on RENXING, and passed by CHU and JIU; in the 12th month, on new moon day *wushen* [45], it entered lunar mansion YINGSHI [LM 13]; on day *gengshen* [57], it trespassed on TIANCANG.
[*Ming shi: tianwen zhi*] ch. 27

(c) 3rd year of the Hongzhi reign period of Emperor Xiaozong of the Ming Dynasty, 11th month, day *wuxu* [35], at dusk; a broom

star with rays over 1 *chi* long appeared at TIANJIN; on day *xinchou* [38], at dusk, the broom star trespassed on RENXING; on day *jiwei* [56], at dusk, the broom star trespassed on TIANCANG.
[*Guo que*] ch. 42

AD 1490 Dec 31 [Japan] (Ho 534)
1st year of the Entoku reign period, 11th month, 24th day (Jan 4); a broom star 5 chi long was seen in the west among the stars of SHI [LM 13].
[Kanda]

AD 1491 Jan 3–Feb 6 [Korea]
(a) 21st year of King Seongjong of the Joseon Dynasty, 11th month, day *xinchou* [38] (Jan 3); during the first watch of the previous night, there was a faint glow 3 to 4 *chi* long within the range of XU [LM 11]...day *guimao* [40]; last night a faintly glowing star moved into the space of WEI [LM 12]...day *yisi* [42] (Jan 7); last night the broom star moved to the 6th *du* of WEI [LM 12], at a polar distance of 65 *du*...on day *dingwei* [44] (Jan 9); last night the broom star moved to the 11th *du* of WEI [LM 12], at a polar distance of 76 *du*, its gleam over 1 *zhang* long...in recent days the broom star left XU [LM 11] and WEI [LM 12]. 12th month, day *jiyou* [46] (Jan 11); last night the broom star moved into the 2nd *du* of SHI [LM 13], at a polar distance of 81 *du*; the notation of the official in charge of celestial observations says, "last night during the first watch, the broom star entered to the east of the southernmost large star of SHI [LM 13], its gleam more than 1 *zhang* long"...day *gengxu* [47] (Jan 12); last night the broom star moved to 4.5 *du* of SHI, at a polar distance of 84 *du*...the notation of the official in charge of celestial observations [says], "last night during the first watch, the broom star over 1 *zhang* long was east of the LEIDIAN star in the range of SHI [LM 13]"...day *xinhai* [48] (Jan 13); tonight it was overcast and at first the broom star was not visible, then during the

fourth watch it was between the second and third stars of the western wall of TAIWEI...day *jiayin* [51] (Jan 16); last night, the broom star 3 to 4 *chi* long moved south of the first star on the east of YUNYU, within the range of SHI [LM 13]...day *yimao* [52] (Jan 17); last night the broom star moved south of the stars of BI [LM 14], day *wuwu* [55] (Jan 20); last night the broom star moved among the stars of LEI, BI [LM 14], and TIANHUN...day *jiwei* [56] (Jan 21); last night the broom star moved above the first star of TIANCANG...day *gengshen* [57] (Jan 22); last night the broom star moved to the west of the second star on the west of TIANCANG...day *yihai* [12] (Feb 6); last night during the first watch, the broom star moved east of the first star on the east of TIANCANG, it lacked rays and its body was extremely faint.
[*Joseon wangjo sillok: Seongjong*] ch. 247–248 (Ho 534)

(b) 21st year *gengxu* of King Sŏngjong of the Joseon Dynasty, 11th month, day *xinchou* [38] (Jan 3); a broom star was seen in XU [LM 11], gradually moving eastward into WEI [LM 12], its tail over 1 *zhang* long. It passed DONGBI [LM 14] and reached TIANCANG; the next year, in the 1st month, it was extinguished.
[*Jeungbo munheon bigo*] ch. 6

AD 1491 Jan 4–8 [Japan]

(a) 2nd year of the Entoku reign period, 11th month, 24th day (Jan 4), clear sky; subsequently reported: tonight a broom star was seen in the west, its brightness about 5 *chi* long; on the 27th day (Jan 7), clear sky; I saw the broom star there in the west, its brightness about 3 *chi* long.
[*Chikanaga kyōki*]

(b) 2nd year of the Entoku reign period, 11th month, 28th day (Jan 8); during the hour of *xu* [LT 19–21 h], a white broom star was seen in the west in lunar mansion SHI [LM 13], its length about 5 *chi*.
[*Go hōkōin masaie ki*]

AD 1491 Jan 19 [China] (Ho 535; Ho & Ang 40)

(a) 3rd year of the Hongzhi reign period of Emperor Xiaozong of the Ming Dynasty, 12th month, day *dingsi* [54], at night; a grayish-white guest star appeared within TIANSHIYUAN and slowly traveled southeastward; on day *wuchen* [5], at dusk, the grayish-white guest star appeared below TIANCANG, gradually moving in the direction of lunar mansion BI [LM 14].
[*Ming Xiaozong shilu*] ch. 46

(b) 3rd year of the Hongzhi reign period of Emperor Xiaozong of the Ming Dynasty, 12th month, day *dingsi* [54]; a star appeared within TIANSHIYUAN traveling southeastward; on day *wuchen* [5], the guest star appeared below TIANCANG, gradually moving in the direction of BI [LM 14].
[*Ming shi: tianwen zhi*] ch. 27

NB: Possibly the same as the preceding.

AD 1495 Jan 7 [China] (Ho 536; Ho & Ang 41)

(a) 7th year of the Hongzhi reign period of Emperor Xiaozong of the Ming Dynasty, 12th month, day *bingyin* [3], at daybreak; a guest star appeared beside TIANJIANG in lunar mansion WEI [LM 6]; from *gengwu* [7] to *gengchen* [17] it gradually approached lunar mansion DOU [LM 8]; in the 8th year, 1st month, day *gengxu* [47], at daybreak, the guest star entered lunar mansion WEI [LM 12] and trespassed on the upper star.
[*Ming Xiaozong shilu*] ch. 95–96

(b) 7th year of the Hongzhi reign period of Emperor Xiaozong of the Ming Dynasty, 12th month, day *bingyin* [3]; a star appeared beside TIANJIANG and slowly approached DOU [LM 8]; by the 8th year, 1st month, day *gengxu* [47], it entered WEI [LM 12].
[*Ming shi: tianwen zhi*] ch. 27

AD 1499 Aug 16 [China] (Ho 537; Ho & Ang 42)

(a) 12th year of the Hongzhi reign period of Emperor Xiaozong of the Ming Dynasty, 7th month, day *wuchen* [5], at dusk; a guest star appeared beside the Zong star of Tianshiyuan...on day *bingzi* [13], at dusk, the brightness of the guest star pointed southeast as it moved to a point beside Shaozai of the eastern enclosure of Ziweiyuan...on day *dingchou* [14], at dusk, the guest star entered Ziweiyuan beside Shangshu on the eastern enclosure...on the night of day *wuyin* [15], the guest star traveled to a point beside Taizi within Ziweiyuan, and gradually faded...on the night of day *jimao* [16], the guest star was next to Hou within Ziweiyuan...at dusk on day *jiashen* [21], the guest star traveled to a point beside Shaofu within the western enclosure of Ziweiyuan, then gradually grew fainter...on day *dinghai* [24], at dusk, the guest star emerged from Ziweiyuan via the western enclosure...in the 8th month, on the night of day *jichou* [26], the guest star was not seen.

[*Ming Xiaozong shilu*] ch. 152–153

(b) 12th year of the Hongzhi reign period of Emperor Xiaozong of the Ming Dynasty, 7th month, day *wuchen* [5], at dusk; a star appeared beside the Zong star of Tianshiyuan, then entered Ziweiyuan via the eastern enclosure and passed by Shaozai and Shangshu, reaching Taizi in Hougong. It emerged from the western enclosure at a point next to Shaofu, until day *jichou* [26], in the 8th month when it was extinguished.

[*Ming shi: tianwen zhi*] ch. 27

AD 1500 May 8 [China] (Ho 538; Ho & Ang 43)

(a) 13th year of the Hongzhi reign period of Emperor Xiaozong of the Ming Dynasty, 4th month, day *jiawu* [31]; a broom star appeared above lunar mansion Shi [LM 13] and Leibizhen...on the night of day *bingshen* [33], the broom star appeared northwest of lunar mansion Shi [LM 13]...on the night of day *jihai* [36], the broom star appeared between Shi [LM 13] and Bi [LM 14], its rays over 1 *chi* long...on day *yisi* [42], the broom star's rays were over 3 *chi* long, its tail pointing toward Ligong...on the night of day *jiayin* [51], in the 5th month, the broom star swept Zaofu...on the night of day *dingmao* [4], the broom star passed

by ZIWEIYUAN then gradually faded…on the night of day *xinsi* [18], the broom star entered ZIWEIYUAN and approached NÜSHI… in the 6th month, on day *dinghai* [24], the broom star further trespassed on SHANGSHU…on the night of day *wuxu* [35], the broom star did not appear.
[*Ming Xiaozong shilu*] ch. 161–163

(b) 13th year of the Hongzhi reign period of Emperor Xiaozong of the Ming Dynasty, 4th month, day *jiawu* [31]; a broom star appeared above lunar mansion SHI [LM 13] and LEIBIZHEN, then entered between SHI [LM 13] and BI [LM 14], gradually lengthening to over 3 *chi*. It pointed toward LIGONG, swept ZAOFU, and passed by TAIWEIYUAN, as it gradually faded. It entered ZIWEIYUAN, approached NÜSHI, and trespassed on SHANGSHU; on day *dingyou* [34], in the 6th month, it disappeared.
[*Ming shi: tianwen zhi*] ch. 27

(c) 13th year of the Hongzhi reign period of Emperor Xiaozong of the Ming Dynasty, 4th month, day *jiawu* [31]; a broom star appeared in lunar mansion SHI [LM 13]; on day *jihai* [36], the broom star appeared between SHI [LM 13] and BI [LM 14], its rays over 1 *chi* long; on the night of day *yisi* [42], the broom star was over 3 *chi* long; on the night of day *jiayin* [51], in the 5th month, the broom star swept WENCHANG; on the night of day *dingmao* [4], the broom star passed the asterism ZIWEI then gradually faded; on the night of day *xinsi* [18], the broom star entered ZIWEIYUAN and approached NÜSHI; on day *dingyou* [34], in the 6th month, it was extinguished; on day *dinghai* [24], in the 6th month, the broom star further trespassed on SHANGSHU; on the night of day *wuxu* [35], the broom star was extinguished.
[*Guo que*] ch. 44

AD 1500 May 18–Jul 1 [Korea]

(a) 6th year *gengshen* of King Yeonsangun of the Joseon Dynasty, 4th month, day *jiachen* [41]; a broom star was seen with a tail 5 to 6 *chi* long, until the 6th month, day *wuzi* [25] (Jul 1).
[*Jeungbo munheon bigo*] ch. 6 (Ho 538)

(b) 6th year *gengshen* of King Yeonsangun of the Joseon Dynasty, 4th month, day *jiachen* [41]; a broom star 2 to 3 *chi* long was seen, its tail pointing between south and west.
[*Joseon wangjo sillok: Yeonsangun*] ch. 37

AD 1500 May 24–Jun 11 [Japan]

(a) 9th year of the Meiō reign period, 4th month, 26th day *gengxu* [47] (May 24), clear sky; in the past few days a broom star has been seen in the sky; the head of the star was in the northeast and its tail pointing southwest.
[*Kazunaga kyōki*]

(b) 9th year of the Meiō reign period, 5th month, 14th day *dingmao* [4] (Jun 11); clear sky; during this time, a broom star appeared...On the 4th day (Jun 1) of the present month, at the hour of *zi* (LT = 23–1 h), the broom star was seen in the north-northeast in NEICHU; it was about 2 *chi* long, pointing south-southwest, with white rays.
[*Go hōkōin masaie ki*]

NB: (Ho 538) gives an abbreviated account from Kanda.

AD 1502 Nov 28 [China] (Ho 539; Ho & Ang 44)

(a) 15th year of the Hongzhi reign period of Emperor Xiaozong of the Ming Dynasty, 10th month, day *wuchen* [5]; a guest star appeared beside TIANMIAO in lunar mansion ZHANG [LM 26]...on day *jisi* [6], the guest star moved to lunar mansion YI [LM 27]...11th month, day *guiyou* [10], the guest star was again in lunar mansion ZHANG [LM 26]; on the night of day *wuyin* [15], the guest star was extinguished.
[*Ming Xiaozong shilu*] ch. 192–193

(b) 15th year of the Hongzhi reign period of Emperor Xiaozong of the Ming Dynasty, 10th month, day *wuchen* [5]; a star appeared beside TIANMIAO; from ZHANG [LM 26] it reached as far as YI [LM 27], then retreated again to ZHANG [LM 26]; on day *wuyin* [15], it was extinguished.
[*Ming shi: tianwen zhi*] ch. 27

AD 1506 Jul 31 [China] (Ho 540; Ho & Ang 45)

(a) 1st year of the Zhengde reign period of Emperor Wuzong of the Ming Dynasty, 7th month, day *jichou* [26], at night; a grayish-white star like a crossbow bolt appeared in the north outside the western enclosure of ZIWEI; on the night of day *renchen* [29], the star appeared again with faint rays for 3 days in succession and was repeatedly observed between SHEN [LM 21] and JING [LM 22]; by the night of day *wuxu* [35], the rays gradually lengthened to 2 *chi*, sweeping sideways like a broom as it slowly traveled in the northwest as far as WENCHANG…on the night of day *renyin* [39], the broom star's bright rays were about 5 *chi* long as it swept and trespassed on the uppermost star of XIATAI…at dusk on day *guimao* [40], the broom star entered TAIWEIYUAN.
[*Ming Wuzong shilu*] ch. 15

(b) 1st year of the Zhengde reign period of Emperor Wuzong of the Ming Dynasty, 7th month, day *jichou* [26]; a grayish-white star like a crossbow bolt appeared outside the western enclosure of ZIWEI; over the course of several days it appeared with faint rays between SHEN [LM 21] and JING [LM 22], the rays gradually lengthening to 2 *chi* like a broom as it traveled in the northwest as far as WENCHANG; in the 7th month, on day *gengzi* [37], the broom star appeared with a gleam gliding southeastward, now 3 *chi* long; over the course of 3 days it grew to about 5 *chi* in length, sweeping the uppermost star of XIATAI, then entering TAIWEIYUAN.
[*Ming shi: tianwen zhi*] ch. 27

(c) 1st year of the Zhengde reign period of Emperor Wuzong of the Ming Dynasty, 7th month, day *jichou* [26]; a broom star appeared in SHEN [LM 21] and JING [LM 22], then swept TAIWEIYUAN until day *wuxu* [35] when it was extinguished.
Ming Dynasty Wanli period ed. [*Shaanxi tongzhi*] ch. 4

AD 1506 Aug 7–10 [Japan]

(a) 3rd year of the Eishō reign period, 7th month, 18th day (Aug 7); 7th month, 21st day, clear sky; a broom star was seen; on the

18th last numerous people saw it, and today the noble Arinobu submitted a report.
[*Nobutane kyō ki*]

(b) 3rd year of the Eishō reign period, 7th month, 21st day (Aug 10); during the hour of *xu* [LT = 19–21 h], a broom star [was seen] at azimuth *hai* [NNW] in NEIJIE; the star was about 1 *zhang* long, pointing east, its rays white in color.
[*Go hōjyōji hisamichi kō ki*]

NB: (Ho 540) reproduces the same record from Kanda.

(c) 3rd year of the Eishō reign period, 7th month, 21st day *jihai* [36] (Aug 10), clear sky; a broom star was seen; tonight everyone saw it…last month on the 19th (Jul 9), it was seen at azimuth *qian* [NW] during the hour of *xu* [LT =19–21 h], with rays over 1 *zhang* long.
[*Sanetaka kō ki*]

AD 1518 Sep 5–Oct 4 [China]

13th year of the Zhengde reign period of Emperor Wuzong of the Ming Dynasty, 8th month; a star became fuzzy in NANDOU [LM 8].
Qing Dynasty Daoguang ed. Guangdong [*Enpingxian zhi*] ch. 1

AD 1523 Jul 23–Aug 10 [China] (Ho 543; Ho & Ang 48)

2nd year of the Jiajing reign period of Emperor Shizong of the Ming Dynasty, 6th month; a star became fuzzy in TIANSHI.
[*Ming shi: tianwen zhi*] ch. 27

AD 1529 Sep 1–18 [Korea] (Ho 545)

24th year *jichou* of King Jungjong of the Joseon Dynasty, 7th month, day *renxu* [59]; a white broom star 4 to 5 *chi* long was seen in the west; on day *jimao* [16] (Sep 18), it had moved and was seen at azimuth *geng* [WSW].
[*Jeungbo munheon bigo*] ch. 6

AD 1530 Jul 11 [China]

9th year of the Jiajing reign period of Emperor Shizong of the Ming Dynasty, 6th month, day *yihai* [12]; a broom star emerged in DONGJING [LM 22].

Qing Dynasty Guangxu ed. Shaanxi [*Fengxian zhi*] ch. 9

AD 1530 autumn [China]

9th year of the Jiajing reign period of Emperor Shizong of the Ming Dynasty, autumn; a broom star appeared, stopping at the tail of BI [LM 19] for a month before being extinguished.

Ming Dynasty Wanli period ed. Shandong [*Qingchengxian zhi*] ch. 2

AD 1531 Feb 5 [Japan] (Ho 547)

4th year of the Kyōroku reign period, 1st month, 18th day; a broom star was seen; it was not extinguished until the 3rd month (Mar 19–Apr 16). [Kanda]

AD 1531 Aug 5 [China] (Ho 547; Ho & Ang 50)

(a) 10th year of the Jiajing reign period of Emperor Shizong of the Ming Dynasty, intercalary 6th month, day *yisi* [42]; a broom star with bright rays over 1 *chi* long appeared in DONGJING [LM 22], pointing southwest...on the night of day *gengxu* [47], the broom star moved and swept the first star on the north of XUANYUAN, its bright rays gradually lengthening...in the 7th month, day *yimao* [52], the broom star moved into the space of YI [LM 27], its bright rays over 7 *chi* long and its tail pointing northeast as it swept TAIZUN...on day *wuwu* [55], it entered TAIWEIYUAN and swept LANGWEI...on day *bingzi* [13], the broom star moved into the space of JIAO [LM 1], its tail pointing southeast, as it swept the second star on the north of KANG [LM 2]; its rays gradually diminishing over the course of 34 days as it disappeared.

[*Ming Shizong shilu*] ch. 127–128

NB: This is Comet Halley, return –6.

(b) 10th year of the Jiajing reign period of Emperor Shizong of the Ming Dynasty, intercalary 6th month, day *yisi* [42]; a broom star with bright rays over 1 *chi* long appeared in DONGJING [LM 22], and swept the first star of XUANYUAN. Its bright rays gradually lengthened until it reached YI [LM 27], when it was over 7 *chi* long. It swept TAIZUN in the northeast and entered TAIWEIYUAN, sweeping LANGWEI. It moved into the space of JIAO [LM 1], and in the southeast it swept the second star on the north of KANG [LM 2], its rays gradually diminishing; it lasted a total of 34 days before disappearing.
[*Ming shi: tianwen zhi*] ch. 27

NB: This is Comet Halley, return –6.

AD 1531 Aug 8–9 [Japan] (Ho 547)

(a) 4th year of the Kyōroku reign period, 6th month, 26th day *wushen* [45] (Aug 8); tonight during the hour of *yin* [LT = 3–5 h], a greenish-white broom star 5 *chi* long was seen at azimuth *gen* [NE]; it was seen for several days…7th month, 10th day *renxu* [59]; from today on for 7 days at numerous shrines, there have been imperial prayers to the broom star.
[*Zoku shigushō*]

(b) 4th year of the Kyōroku reign period, 6th month, 27th day (Aug 9); during the hour of *mao* [LT = 5–7 h], a greenish-white broom star 5 *chi* long was seen at azimuth *gen* [NE]. During the 7th month, it shifted to azimuth *qian* [NW].
[*Kanda*] NB: This is Comet Halley, return –6.

AD 1531 Aug 10 [Korea]

(a) 26th year *xinmao* of King Jungjong of the Joseon Dynasty, intercalary 6th month, day *gengxu* [47]; a white broom star with a tail over 10 *chi* long was seen outside the eastern wall of TAIWEI.
[*Jeungbo munheon bigo*] ch. 6 (Ho 547)

(b) 26th year *xinmao* of King Jungjong of the Joseon Dynasty, intercalary 6th month, day *gengxu* [47]; at azimuth *gen* (NE) there was a broom star with a tail over 10 *chi* long.

[*Joseon wangjo sillok: Jungjong*] ch. 71
NB: This is Comet Halley, return –6.

AD 1532 Feb 6–AD 1533 Jan 24 [China]

(a) 11th year of the Jiajing reign period of Emperor Shizong of the Ming Dynasty; a broom star appeared in the east, pointing toward the southwest for over 1 *zhang*, and not disappearing until the end of winter.
Ming Dynasty Jiajing period ed. Fujian [*Longxi xianzhi*] ch. 5

(b) 11th year of the Jiajing reign period of Emperor Shizong of the Ming Dynasty; at the beginning of the Jiajing reign period of the Ming Dynasty, He Dong submitted a memorial, "In the 8th month of this year a broom star emerged in the space of JING [LM 22], and was visible at dawn in the east as it traversed GUI [LM 23], LIU [LM 24], XING [LM 25], ZHANG [LM 26], and entered the space of ZHEN [LM 28], passing by TAIWEIYUAN and trespassing on SHANGJIANG, positions which could be ascertained. When it first emerged, it was several *chi* long, and as it gradually traveled eastward, it grew to several *zhang* in length, and its shape was seen. It moved back and forth in the space of JING [LM 22], at first traveling quite slowly, then after passing several lunar mansions it gradually sped up; from the 5th day of the 8th month to the 18th day of the 9th month, in all 43 days, it traveled over 80 *du*, which motion again could be investigated. Initially it was grayish-white, later it turned reddish; colors which could be seen. In the 8th month, the 5th day of the month, on day *gengchen* [17], it followed the ecliptic in as it traveled prograde, until the 15th day of the 9th month, on day *gengshen* [57], when it trespassed on SHANGJIANG of TAIWEI most rapidly, which dates and timing can again be investigated."
Qing Dynasty Jiaqing period ed. Shaanxi [*Chang'anxian zhi*] ch. 27

NB: Checking according to the dates "8th month, 5th day *gengchen* [17]" and "9th month, 15th day *gengshen* [57]" shows that this was the 11th year of the Jiajing reign period.

AD 1532 Mar 9 [China] (Ho 548; Ho & Ang 51)

(a) 11th year of the Jiajing reign period of Emperor Shizong of the Ming Dynasty, 2nd month, day *renwu* [19], fifth watch of the night; a grayish-white star with faint rays appeared in the southeast; it was not extinguished for 19 days.
[*Ming Shizong shilu*] ch. 135

(b) 11th year of the Jiajing reign period of Emperor Shizong of the Ming Dynasty, 2nd month, day *renwu* [19]; a grayish-white star with rays appeared in the southeast; it lasted 19 days in all and was extinguished.
[*Ming shi: tianwen zhi*] ch. 27

AD 1532 Jun 7 [China]

Renchen year of the Jiajing reign period of Emperor Shizong of the Ming Dynasty, 5th month, 5th day; a broom star with rays over 1 *chi* long appeared in the east, then traveled northeastward, passing TIANJIN and watching lunar mansion JING [LM 22] from a distance, as its rays gradually grew to over 1 *zhang* in length. Then it swept TAIWEIYUAN and TIANMEN in lunar mansion JIAO [LM 1]; in the 8th month, day *jimao* [16], the broom star appeared in DONGJING [LM 22].
[*Er shen ye lu*] ch. 4

AD 1532 Jun 21–Aug 10 [Japan] (Ho 549)

1st year of the Tembun reign period, 5th month, 18th day…a broom star was seen at azimuth *gen* [NE] and increased in brightness; in the 7th month, 10th day (Aug 10); until now it has been there, after this it disappeared.
[*Kaigen sōzu ki*]

AD 1532 Sep 2 [China] (Ho 550; Ho & Ang 52)

(a) 11th year of the Jiajing reign period of Emperor Shizong of the Ming Dynasty, 8th month, day *jimao* [16], at night; a broom star over 1 *chi* long appeared in DONGJING [LM 22], and later traveled northeastward, passing asterisms at TIANJIN as its rays gradually grew to over 1 *zhang* in length; it swept several asterisms in TAIWEIYUAN together with TIANMEN of lunar mansion JIAO [LM 1]; by day *jiaxu* [11] in the 12th month, it had lasted 115 days in all when it was extinguished.

[*Ming Shizong shilu*] ch. 141

(b) 11th year of the Jiajing reign period of Emperor Shizong of the Ming Dynasty, 8th month, day *jimao* [16]; a broom star about 1 *chi* long appeared in DONGJING [LM 22]; later it traveled northeastward, passing TIANJIN and gradually growing to over 1 *zhang* in length; it swept several asterisms of TAIWEIYUAN together with lunar mansion JIAO [LM 1] and TIANMEN; by day *jiaxu* [11] in the 12th month, it had lasted 115 days in all when it was extinguished.

[*Ming shi: tianwen zhi*] ch. 27

NB: *Ming hui yao* mistakenly wrote "155 days and then was extinguished."

AD 1532 Sep 14–Oct 31 [Korea] (Ho 550)

27th year *renchen* of King Jungjong of the Joseon Dynasty, 8th month, day *xinmao* [28]; a broom star was seen in LIU [LM 24]; 9th month, *jiyou* [46] (Oct 2) it moved and was seen at azimuth *mao* (E), its tail over 1 *zhang* long and its color white; by the 12th month, day *wuyin* [15] (Oct 31), its gleam was faint.

[*Jeungbo munheon bigo*] ch. 6

AD 1532 Sep 25–Nov 6 [Japan]

(a) 1st year of the Tembun reign period, 8th month, 26th day; a broom star was seen in the east until after the 10th day of the 10th month (Nov 6).

[*Kugyō bunin*]

(b) 1st year of the Tembun reign period, 8th month, *renyin* [39]; a broom star was seen in the east and lasted until the 10th month before disappearing.

[*Zoku honchō tsūgan*] ch. 183 (Ho 550)

AD 1533 Jun 27–Aug 26 [Korea] (Ho 551)

28th year *guisi* of King Jungjong of the Joseon Dynasty, 6th month, day *dingchou* [14]; a white broom star with a tail 7 to 8 *chi* long was seen north of WUCHE. It entered TIANCHUAN, passed WANGLIANG, traversed TENGSHE, and reached TIANJIN; by the 8th month, day *dingchou* [14] (Aug 26), it gradually diminished in size.
[*Jeungbo munheon bigo*] ch. 6

AD 1533 Jul 1 [China] (Ho 551; Ho & Ang 53)

(a) 12th year of the Jiajing reign period of Emperor Shizong of the Ming Dynasty, 6th month, day *xinsi* [18]; a broom star with rays over 5 *chi* long appeared in WUCHE, its tail pointing toward the southwest…on day *renwu* [19], the broom star reemerged between BI [LM 19] and MAO [LM 18], its bright rays barely visible…on the night of day *jihai* [36], the broom star swept DALING together with TIANDAJIANGJUN, its rays now over 1 *zhang* long…in the 7th month, day *jiayin* [51], the broom star swept GEDAO and moved to TENGSHE; by the 28th day of the 8th month, it was extinguished.
[*Ming Shizong shilu*] ch. 151–152

(b) 12th year of the Jiajing reign period of Emperor Shizong of the Ming Dynasty, 6th month, day *xinsi* [18]; a broom star with rays over 5 *chi* long appeared in WUCHE; it swept DALING together with TIANDAJIANGJUN, gradually growing to over 1 *zhang* in length; it swept GEDAO and trespassed on TENGSHE; by the 8th month, day *wuxu* [35], it was extinguished.
[*Ming shi: tianwen zhi*] ch. 27

AD 1533 Jul 15–24 [Japan]

2nd year of the Tenbun reign period, 6th month, 23rd day; a broom star was in the north, fuzzy and over 1 *zhang* long…7th month, 3rd day (Jul 24); the broom star was seen in the north during the second watch, its fuzziness much dimmed.
[*Sanetaka kōki*]
NB: (Ho 551) reproduces an abbreviated record from Kanda.

AD 1533 Aug 20–Sep 17 [China]

12th year of the Jiajing reign period of Emperor Shizong of the Ming Dynasty, 8th month; the broom star re-emerged in DONGJING [LM 22] and swept TAIWEIYUAN until the 12th month.
[*Er shen ye lu*] ch. 4

AD 1534 Jun 12 [China] (Ho 552; Ho & Ang 54)

(a) 13th year of the Jiajing reign period of Emperor Shizong of the Ming Dynasty, 5th month, new moon day *dingmao* [4], during the fourth watch of the night; a guest star appeared in TENGSHE, then passed TIANJIU and entered GEDAO; by the 24th day, it was extinguished.
[*Ming Shizong shilu*] ch. 163

(b) 13th year of the Jiajing reign period of Emperor Shizong of the Ming Dynasty, 5th month, new moon day *dingmao* [4]; a star appeared in TENGSHE, passed TIANJIU and entered GEDAO; on the 24th day, it was extinguished.
[*Ming shi: tianwen zhi*] ch. 27

AD 1536 Mar 24–Apr 27 [China] (Ho 553; Ho & Ang 55)

(a) 15th year of the Jiajing reign period of Emperor Shizong of the Ming Dynasty, 3rd month, day *wuwu* [55] (Mar 24), at night; a guest star appeared beside the asterism TIANBANG.
[*Ming Shizong shilu*] ch. 185

(b) 15th year of the Jiajing reign period of Emperor Shizong of the Ming Dynasty, 3rd month, day *wuwu* [55] (Mar 24); a star appeared beside TIANBANG then moved eastward, passing TIANCHU and entering the Milky Way from the west; by the 4th month, day *renchen* [29], it disappeared.
[*Ming shi: tianwen zhi*] ch. 27

(c) 15th year of the Jiajing reign period of Emperor Shizong of the Ming Dynasty, 3rd month, day *wuwu* [55] (Mar 24), at night; a guest star appeared in ZIWEIYUAN; in the 4th month, on day *renchen* [29], it was extinguished.
[*Guo que*] ch. 56

(d) 15th year of the Jiajing reign period of Emperor Shizong of the Ming Dynasty, 3rd month, day *renchen* [29] (Apr 27); a guest star appeared beside TIANBANG then gradually traveled eastward, passing TIANCHU on the west and entering the Milky Way; not until this date [original has "first day"] did it disappear.
[*Ming Shizong shilu*] ch. 186

AD 1537 Mar 8 [Japan] (Ho 544)
6th year of the Tenbun reign period, 1st month, 27th day; a broom star was seen at azimuth qian [NW].
[Kanda]

AD 1537 Sep 4–Oct 3 [China]
16th year of the Jiajing reign period of Emperor Shizong of the Ming Dynasty, 8th month; a broom star appeared in the southwest and was not extinguished for over a month.
Qing Dynasty Shunzhi period ed. Henan [*Wenxiangxian zhi*] ch. 1

AD 1538 Jan 9 [Japan] (Ho 555)
6th year of the Tenbun reign period, 12th month, 9th day; a broom star was seen.
[Kanda]

AD 1538 Jan 21–29 [Korea] (Ho 555)
32nd year *dingyou* of King Jungjong of the Joseon Dynasty, 12th month, day *bingyin* [3]; a broom star over 3 *zhang* long was seen at azimuth *you* (W), its color white; by day *jiaxu* [11] (Jan 29), it had a faint vapor.
[*Jeungbo munheon bigo*] ch. 6

AD 1539 Apr 20–May 24 [Korea]
(a) 34th year of King Jungjong of the Joseon Dynasty, 4th month, day *gengzi* [37]; a star with a tail over 1 *chi* long was seen below the SHANGTAI star of SANTAI, its form and body like a

broom star...on the night of day *wushen* [45], the broom star was seen by the second star of the tail of XUANYUAN...on the night of day *xinyou* [58] (May 11), the broom star was seen west of the 10th star of XUANYUAN, its tail over 5 to 6 *chi* long, and its color white...in the 5th month, on the night of day *jisi* [6], the broom star was seen at the western corner of XUANYUAN, its tail over 4 to 5 *chi* long, and its color white... on the night of day *jiaxu* [11] (May 24), the broom star was faintly visible.
[*Joseon wangjo sillok: Chungjong*] ch. 90

(b) 34th year *jihai* of King Jungjong of the Joseon Dynasty, 4th month, day *gengzi* [37]; a broom star with a tail 5 to 6 *chi* long was seen in XUANYUAN, its color was white. In the 5th month it passed JIUQI, to a point south of XING [LM 25].
[*Jeungbo munheon bigo*] ch. 6 (Ho 556)

AD 1539 Apr 30 [China] (Ho 556; Ho & Ang 56)

(a) 18th year of the Jiajing reign period of Emperor Shizong of the Ming Dynasty, 4th month, day *gengxu* [47], at night; a broom star with rays about 3 *chi* long appeared, its brightness pointing southeastward; it swept the 8th star on the north of XUANYUAN, and was not extinguished for 10 days.
[*Ming Shizong shilu*] ch. 223

(b) 18th year of the Jiajing reign period of Emperor Shizong of the Ming Dynasty, 4th month, day *gengxu* [47]; a broom star with rays about 3 *chi* long appeared, its brightness pointing southeastward; it swept the 8th star on the north of XUANYUAN, and was not extinguished for 10 days.
[*Ming shi: tianwen zhi*] ch. 27

AD 1540 Apr 30 [China]

19th year of the Jiajing reign period of Emperor Shizong of the Ming Dynasty, 3rd month, day *bingchen* [53]; there was a star like a broom over 1 *zhang* long that passed by WEI [LM 17], SHI [LM 13], and

BI [LM14], and on new moon day *bingchen* [53] in the intercalary month, entered LOU [LM 16].
Qing Dynasty Tongzhi ed. Shandong [*Jimoxian zhi*] ch. 11
NB: The Kangxi period edition and the Qianlong period edition lack this record.

AD 1545 Dec 26 [China] (Ho 557; Ho & Ang 57)

(a) 24th year of the Jiajing reign period of Emperor Shizong of the Ming Dynasty, 11th month, day *renwu* [19], at night; a guest star emerged at TIANBANG and entered lunar mansion JI [LM 7]; over the course of 3 days it turned northeastward and did not disappear for a month.
[*Ming Shizong shilu*] ch. 305

(b) 24th year of the Jiajing reign period of Emperor Shizong of the Ming Dynasty, 11th month, day *renwu* [19]; a star emerged at TIANBANG and entered JI [LM 7], then turned northeastward and disappeared after a month.
[*Ming shi: tianwen zhi*] ch. 27

AD 1549 Mar 7 [Korea] (Ho 558)

4th year *jiyou* of King Myeongjong of the Joseon Dynasty, 2nd month, day *jiyou* [46]; a broom star was seen at azimuth *gen* (NE); by the 3rd month (Mar 29–Apr 26), it had moved and was seen at azimuth *mao* (E).
[*Jeungbo munheon bigo*] ch. 6

AD 1546 Mar 2–31 [China]

25th year of the Jiajing reign period of Emperor Shizong of the Ming Dynasty, 2nd month; a broom star over 1 *zhang* long appeared in the north and lasted over 30 days before being extinguished.
Qing Dynasty Qianlong ed. Fujian [*Longxixian zhi*] ch. 20

AD 1554 May 2–30 [China]

33rd year of the Jiajing reign period of Emperor Shizong of the Ming Dynasty, 4th month; a broom star appeared; it emerged from the

mouth of BEIDOU, its tail so long it stretched across the sky; after 5 days, it was extinguished.
Ming Dynasty Longqing period ed. Jiangsu [*Yizhenxian zhi*] ch. 13

AD 1554 Jun 23 [China] (Ho 559; Ho & Ang 58)

(a) 33rd year of the Jiajing reign period of Emperor Shizong of the Ming Dynasty, 5th month, day *guihai* [60], at night; a broom star appeared next to TIANQUAN of BEIDOU…on the night of day *xinwei* [8] in the 6th month, the broom star gradually traveled northwestward, trespassing on WENCHANG…on the night of day *jichou* [26], the broom star entered the obscurity at the horizon and only then disappeared.
[*Ming Shizong shilu*] ch. 410–411

(b) 33rd year of the Jiajing reign period of Emperor Shizong of the Ming Dynasty, 5th month, day *guihai* [60]; a broom star appeared next to TIANQUAN and trespassed on WENCHANG; then it entered the obscurity at the horizon, and after 27 days, it disappeared.
[*Ming shi: tianwen zhi*] ch. 27

AD 1554 Jun 26–Jul 11 [Korea]

(a) 9th year of King Myeongjong of the Joseon Dynasty, 5th month, day *bingyin* [3]; there was a star like a broom star over 1 *zhang* long that emerged in front of the bowl of BEIDOU…in the 6th month, day *gengwu* [7] (Jun 30), the broom star was seen in front of the bowl of BEIDOU and behind WENCHANG, more than 2 *du* from ZHANG [LM 26]; its color was white and its tail over 3 to 4 *cun* long…on the night of *xinsi* [18] (Jul 11), the broom star was not seen.
[*Joseon wangjo sillok: Myoengjong*] ch. 16

(b) 9th year *jiayin* of King Myeongjong of the Joseon Dynasty, 5th month, day *bingyin* [3]; a broom star was seen in BEIDOU, its tail was 1 *chi*, 4 to 5 *cun* long, and its color white; by the 6th month, day *xinsi* [18] (Jul 11), it no longer was seen.
[*Jeungbo munheon bigo*] ch. 6 (Ho 559)

AD 1555 Oct 15–Nov 13 [China]

34th year of the Jiajing reign period of Emperor Shizong of the Ming Dynasty, 10th month; a broom star emerged in MAO [LM 18] and gradually traveled northward, reaching DOU.

Qing Dynasty Shunzhi period ed. Shanxi [*Lu'anfu zhi*] ch. 15

AD 1556 Jan 9 [Japan] (Ho 560)

2nd year of the Kōji reign period, 1st month, 23rd day; a broom star was seen in the east.

[Kanda]

AD 1556 Jan 31 [China]

34th year of the Jiajing reign period of Emperor Shizong of the Ming Dynasty, 12th month, 20th day; a broom star initially appeared in the southeast, then swerved and entered the northwest, lasting over a month before disappearing, its rays several tens of *zhang* long.

Qing Dynasty Shunzhi period ed. Gansu [*Chongxinxian zhi*] ch. *xia*

AD 1556 Mar 1 [China] (Ho 560; Ho & Ang 59)

(a) 35th year of the Jiajing reign period of Emperor Shizong of the Ming Dynasty, 1st month, day *gengchen* [17], at night; a broom star with rays 1 *chi* long and pointing southwest appeared beside JINXIAN; it gradually lengthened to over 3 *chi* as it passed by and swept CIXIANG of TAIWEIYUAN, then traveled northeastward and entered ZIWEIYUAN where it trespassed on TIANCHUANG; not until the 2nd day of the 4th month was it was extinguished.

[*Ming Shizong shilu*] ch. 431

(b) 35th year of the Jiajing reign period of Emperor Shizong of the Ming Dynasty, 1st month, day *gengchen* [17]; a broom star with rays about 1 *chi* long and pointing southwest appeared beside JINXIAN; it gradually grew to over 3 *chi* as it passed by and swept TAIWEIYUAN and northeast of CIXIANG, then entered ZIWEIYUAN where it trespassed on TIANCHUANG; on the 2nd day of the 4th month, it was extinguished.

[*Ming shi: tianwen zhi*] ch. 27

AD 1556 Mar 1–Apr 21 [Korea]

(a) 11th year of King Myeongjong of the Joseon Dynasty, 1st month, day *gengchen* [17], at night; a star was seen over 5 to 6 *du* north of ZHEN [LM 28], pointing west and with rayed horns...on the night of day *renwu* [19] (Mar 3)...the broom star was seen over 8 to 9 *du* northeast of ZHEN [LM 28], outside the eastern wall of TAIWEI, its tail pointing southwest; it was over 1 *chi* long and its color faintly white...2nd month, new moon day *gengyin* [27] (Mar 11); during the first watch of the night, the broom star was seen at azimuth *yin* (ENE), from the second watch through the fifth watch the broom star was seen above the ZHAOYAO star, in the initial *du* of lunar lodge KANG [LM 2], at a polar distance of 44 *du*, its tail pointing southwest and over 4 *chi* long, its color white...on day *yiwei* [32] (Mar 16), the broom star was seen in front of CHUANSHE north of ZIWEIGONG, at a polar distance of over 23 *du*; its vapor reached HUAGAI, its tail was over 2 *chi* long, pointing southwest, its color white...on the night of *dingsi* [54] (Apr 7), the broom star was seen at azimuth *yin* (ENE), north of BI [LM 14] at a polar distance of 60 *du*, its tail pointing southwest and over 2 *chi* long, its color white...in the 3rd month, on the night of *xinwei* [8] (Apr 21), the broom star was seen at azimuth *mao* (E).
[*Joseon wangjo sillok: Myeongjong*] ch. 20

(b) 11th year *bingchen* of King Myeongjong of the Joseon Dynasty, 1st month, day *gengchen* [17] (Mar 1); a broom star was seen in ZHEN [LM 28], traversed JIAO [LM 1] and KANG [LM 2], and entered ZIWEI; it was white and its tail over 5 *chi* long. By the 3rd month, day *gengwu* [7] (Apr 20), it reached DONGBI [LM 14], its form and color gradually growing fainter.
[*Jeungbo munheon bigo*] ch. 6 (Ho 560)

AD 1556 Mar 11–Apr 10 [China]

35th year of the Jiajing reign period of Emperor Shizong of the Ming Dynasty, 2nd month; a broom star over 1 *zhang* long appeared in the north and lasted over 30 days before being extinguished.
Ming Dynasty Jiajing ed. Fujian [*Changtaixian zhi*] ch. *xia*

AD 1557 Oct 10 [China] (Ho 561; Ho & Ang 60)

(a) 36th year of the Jiajing reign period of Emperor Shizong of the Ming Dynasty, 9th month, day *wuchen* [5]; a broom star over 1 *chi* long and pointing northeast appeared next to LIESI within TIANSHIYUAN; not until the 23rd day of the 10th month was it was extinguished.
[*Ming Shizong shilu*] ch. 451

(b) 36th year of the Jiajing reign period of Emperor Shizong of the Ming Dynasty, 9th month, day *wuchen* [5]; a broom star pointing northeast appeared next to LIESI within TIANSHIYUAN; by the 23rd day of the 10th month, it was extinguished.
[*Ming shi: tianwen zhi*] ch. 27

AD 1557 Oct 13–26 [Korea]

(a) 12th year of King Myeongjong of the Joseon Dynasty, 9th month, day *xinwei* [8], second watch of the night; a broom star was seen in the west between HUANZHE, DIZUO, CHESI, and DOU [LM 8], its tail over 1 *chi* long and pointing toward HOU, its color faintly white...on day *jiaxu* [11] (Oct 16), during the first watch of the night, the broom star was seen above the [SHI-]LOU star of TIANSHI, its tail over 2 *chi* long and pointing toward the ZONGREN star, its color faintly white. At the first point of the second watch, it entered the horizon in the west...on day *jimao* [16] (Oct 21), the broom star was seen in the Milky Way west of the first star in the handle of NANDOU, its glow obscured by the Milky Way, its body dim...in the 10th month, day *guiwei* [20] (Oct 25), during the first watch of the night, the broom star was seen above BIE, its body dim, at 5 points of the first watch it entered the horizon in the west...on day *jiashen* [21] (Oct 26), during the first watch of the night, the broom star was seen at the first star in the bowl of NANDOU, its body somewhat faint, and at 5 points of the first watch it entered the horizon in the west.
[*Joseon wangjo sillok: Myeongjong*] ch. 23

(b) 12th year *dingsi* of King Myoengjongof the Joseon Dynasty, 9th month, day *xinwei* [8]; a broom star was seen within TIANSHI,

its tail over 1 *chi* long and its color white. In the 10th month (Oct 22–Nov 20), it moved by NANDOU, its color and body dim.
[*Jeungbo munheon bigo*] ch. 6 (Ho 561)

AD 1558 Aug 8–Sep 13 [Korea]

(a) 13th year of King Myoengjong of the Joseon Dynasty, 7th month, day *gengwu* [7], at night; a broom star was seen in TAIWEIYUAN, its tail over 4 to 5 *chi* long and its color white. Its vaporous rays pointed toward the zenith, and it set on the horizon at azimuth *qian* (NW)...intercalary 7th month, the night of day *gengchen* [17] (Aug 18), the broom star was seen within TAIWEIYUAN, the tail over 4 *chi* long and pointing to the south of the LANGJIANG star, its color was white, and it entered the horizon at azimuth *qian* (NW)...on the night of day *yiwei* [32] (Sep 3), the broom star was seen north of TIANSHIYUAN, its tail over 1 *chi* long and pointing south of the GUANSUO star, its color faintly white, until the third watch when it entered the western horizon...in the 8th month, new moon day *yisi* [42] (Sep 13), at night, the broom star was seen within TIANSHIYUAN above SHILOU, its tail over 4 to 5 *chi* long and pointing south of ZONGREN, its color faintly white, until the third watch when it entered the western horizon.
[*Joseon wangjo sillok: Myeongjong*] ch. 24

(b) 13th year *wuwu* of King Myoengjongof the Joseon Dynasty, 7th month, day *gengwu* [7], a broom star was seen within TAIWEI, its tail 4 to 5 *chi* long, and its color white. During the intercalary [7th] month (Aug 14–Sep 11), it moved within TIANSHI, its body dim.
[*Jeungbo munhon bigo*] ch. 6 (Ho 562)

AD 1558 Aug 9–19 [Japan]

1st year of the Eiroku reign period, intercalary 6th month, 25th day; in the sky a broom star was seen straight through to the 5th to 6th days of the 7th month (Jul 19–Jul 20).
[*Hosokawa ryōke ki*]
NB: (Ho 562) from Kanda is the same.

AD 1567 Jan 10 [Korea]

21st year *bingyin* of King Myoengjong of the Joseon Dynasty, 12th month, day *dinghai* [24]; in the south there was a star shaped like a long star, 3 to 4 *chi* wide and 5 to 6 *chi* long, broader above and narrower below; it lasted a good while before it was extinguished.
[*Joseon wangjo sillok: Myoengjong*] ch. 33
[*Jeungbo munheon bigo*] ch. 6 (Ho 563)

AD 1567 Oct 27 [China]

46th year of the Jiajing reign period of Emperor Shizong of the Ming Dynasty, 9th month, 26th day, morning twilight; a white vapor appeared in the southeast and lasted half the month before being extinguished; on searching, a star became fuzzy in the east that gradually traveled northward, its brightness several *zhang* long and stretching right across the middle of the sky.
Qing Dynasty Kangxi period ed. Jiangsu [*Changzhoufu zhi*] ch. 3

AD 1567 Nov 1–30 [China]

46th year of the Jiajing reign period of Emperor Shizong of the Ming Dynasty, 10th month, at midnight; there was a white vapor several *zhang* long arising in the southeast for over a month; a broom star reached GUI [LM 23] and DOU [LM 8], both of them appearing on *mao* days and passing by the two asterisms LING and LIAN, the allocated field of Zhe.
Qing Dynasty Qianlong period ed. Henan [*Zhechengxian zhi*] ch. 18
NB: LIAN, 链, is probably a mistake for JIAN, 键.

AD 1569 Nov 9 [China] (Ho 564; Ho & Ang 61)

(a) 3rd year of the Longqing reign period of Emperor Muzong of the Ming Dynasty, 10th month, day *xinchou* [38]; a grayish-white broom star over 2 *chi* long appeared in TIANSHIYUAN, its rays pointing northeast, until the 20th day when it was extinguished.
[*Ming Muzong shilu*] ch. 38

(b) 3rd year of the Longqing reign period of Emperor Muzong of the Ming Dynasty, 10th month, day *xinchou* [38]; a broom star

appeared in TIANSHIYUAN pointing northeast until day *gengshen* [57], when it was extinguished.
[*Ming shi: tianwen zhi*] ch. 27

AD 1573 Sep 26–Oct 25 [China]
1st year of the Wanli reign period of Emperor Shenzong of the Ming Dynasty, 9th month, at night; a star became fuzzy in the southwest whose vapors spread upward to a height of more than about 10 *zhang*, pointed at the top and wide at the bottom; this lasted for 3 months before being extinguished.
Qing Dynasty Kangxi ed. Fujian [*Anhai zhi*] ch. 8

AD 1574 Dec 13–AD 1575 Jan 11 [China]
2nd year of the Wanli reign period of Emperor Shenzong of the Ming Dynasty, 12th month; a fiery star with a tail and bright rays over 1 *zhang* long like a broom star guarded lunar mansion KUI [LM 15] for a month before dissipating.
Ming Dynasty Tianqi reign period ed. Jiangsu [*Huai'anfu zhi*] ch. 24

AD 1577 Jul 15–Aug 13 [Korea] (Ho 567)
10th year of King Seonjo of the Joseon Dynasty, 7th month; a broom star was seen.
[Jeungbo munheon bigo] ch. 6

AD 1577 Aug 10 [China]
5th year of the Wanli reign period of Emperor Shenzong of the Ming Dynasty, 7th month, 27th day; there was a star like a white vapor several *zhang* long that appeared in the southwest and pointed straight northeast; by the 11th month it gradually grew smaller and then disappeared.
Qing Dynasty Kangxi period ed. Fujian [*Shaxian zhi*] ch. 11

AD 1577 Aug 13 [China]
5th year of the Wanli reign period of Emperor Shenzong of the Ming Dynasty, the end of the 7th month; a broom star appeared in the southwest

and trespassed on the space of DOU [LM 8] until the 12th month when it disappeared.
Qing Dynasty Kangxi period ed. Zhejiang [*Jinyunxian zhi*] ch. 4

AD 1577 Sep 11 [China]

5th year of the Wanli reign period of Emperor Shenzong of the Ming Dynasty, 8th month, thirtieth day; a broom star appeared in the southwest with bright rays like hairpins shooting straight toward CHE; it did not dissipate until the first 10-day week of the 12th month.
Qing Dynasty Kangxi period ed. Anhui [*Zhixian zhi*] ch. 5

AD 1577 Oct 14 [China]

(a) 5th year of the Wanli reign period of Emperor Shenzong of the Ming Dynasty, 9th month, 4th day; a broom star over about 10 *zhang* long and shaped like a silk train appeared in the west; from WEI [LM 6] it passed JI [LM 7] and surpassed the space of DOU [LM 8] and NIU [LM 9]; by the 19th day of the 11th month, it ceased.
Qing Dynasty Qianlong period ed. Jiangxi [*Longquanxian zhi*] ch. *mo*

(b) 5th year of the Wanli reign period of Emperor Shenzong of the Ming Dynasty, 9th month, 4th day; a broom star appeared in the west; it was about 10 *zhang* long, 3 *zhang* wide, with a root over 5 *zhang* long and shaped like white cloud with silk train; from WEI [LM 6] it passed JI [LM 7] and surpassed the space of DOU [LM 8] and NIU [LM 9]; by the 29th day of the 11th month it ceased.
Qing Dynasty Qianlong period ed. Jiangxi [*Dayuxian zhi*] ch. 1

AD 1577 Nov 7 [China]

(a) 5th year of the Wanli reign period of Emperor Shenzong of the Ming Dynasty, 9th month, 28th day; in the southwest, an extremely large star burst into view, its purplish vapors filling the sky.

Ming Dynasty Tianqi reign period ed. Zhejiang [*Pinghuxian zhi*] ch.18

(b) 5th year of the Wanli reign period of Emperor Shenzong of the Ming Dynasty, 9th month, 28th day, during the hour of *you* [LT 17–19 h]; in the southwest sky, a star with bright rays over 1 *chi* long appeared, pointing upwards, and gradually growing to several *zhang* in length. After several days, it moved to the zenith and its traces gradually were extinguished.
Qing Dynasty Kangxi period ed. Zhejiang [*Wenzhoufu zhi*] ch. 30

(c) 5th year of the Wanli reign period of Emperor Shenzong of the Ming Dynasty, 9th month, 28th day; a broom star appeared in the southwest and swept the allocated field of Chu [LM 27–28]; on the 2nd day of the 10th month, its bright rays grew longer, like a piece of silk 100 *chi* long stretching across the western sky.
Qing Dynasty Daoguang ed. Jiangsu [*Huangjing zhi gao*] ch. 7

AD 1577 Nov 8 [China]

(a) 5th year of the Wanli reign period of Emperor Shenzong of the Ming Dynasty, 9th month, 29th day; a broom star with bright rays over 3 *zhang* long appeared in the southwest pointing eastward for half the day; it was not extinguished until the 15th day in the 11th month.
Qing Dynasty Jiaqing period ed. Jiangsu [*Dongtaixian zhi*] ch.7

(b) 5th year of the Wanli reign period of Emperor Shenzong of the Ming Dynasty, 9th month, 29th day; a broom star over 10 *zhang* long appeared in Wei [LM 6] and Ji [LM 7], its gleam obscuring Nandou [LM 8], its flaming rays like banners; in all it lasted over 40 days before being extinguished. This star was red and exceptionally long; hence it was a "silk train" star.
[*Cui Mingwu jishi*]

AD 1577 Nov 8–9 [Japan]

(a) 5th year of the Tenshō reign period, 9th month, 28th day *renwu* [19] (Nov 8); a broom star 5 *zhang* long was observed at azimuth *wei* [SSW]–*shen* [WSW]; it lasted until the following year before disappearing.

[*Zoku shigushō*]

(b) 5th year of the Tenshō reign period, 9th month, 29th night (Nov 9); a broom star was seen for over 100 days until the second 10-day week of the 1st month of the following year, its gleam like a bright star. From the 12th month its shape was like the character *dai* [i.e., like a spread-eagled stick figure of a man], its head to the east and tail to the west and about 20 *jian* long [1 *jian* = 6 ft].

[*I hon tōji chōchō*]

NB: (Ho 568) reproduced from Kanda is a composite of these two records, but says the comet 'was as bright as the moon.'

AD 1577 Nov 10–14 [China] (Ho 568; Ho & Ang 64)

(a) 5th year of the Wanli reign period of Emperor Shenzong of the Ming Dynasty, 10th month, new moon day (Nov 10); a broom star with a tail several tens of *zhang* long appeared between DOU [LM 8] and NIU [LM 9], its bright rays stretching across the sky like a piece of silk, its vapor forming a white arc.

Qing Dynasty Kangxi period ed. Anhui [*Taipingfu zhi*] ch. 3

(b) 5th year of the Wanli reign period of Emperor Shenzong of the Ming Dynasty, 10th month, new moon day (Nov 10); a broom star several tens of *zhang* long, its tail pointing toward WUNÜ [LM 10], appeared between DOU [LM 8] and NIU [LM 9], its bright rays stretching across the sky like a piece of silk, and its vapor forming a white arc.

[*Er shen ye lu*] ch. 5

(c) 5th year of the Wanli reign period of Emperor Shenzong of the Ming Dynasty, 10th month, new moon day (Nov 10); a broom star several tens of *zhang* long and pointing toward lunar mansion WUNÜ [LM 10] appeared between DOU [LM 8] and NIU [LM 9].

Qing Dynasty Kangxi period ed. Hebei [*Changpingzhou zhi*] ch. 26

(d) 5th year of the Wanli reign period of Emperor Shenzong of the Ming Dynasty, 10th month, day *wuzi* [25] (Nov 14); a broom star several *zhang* long, with a bright gleam shaped like a small cup and greenish-white rays, appeared in the southwest; from WEI [LM 6] and JI [LM 7] it surpassed DOU [LM 8] and NIU [LM 9] and directly pressed on toward lunar mansion NÜ [LM 10].
[*Ming Shenzong shilu*] ch. 68

(e) 5th year of the Wanli reign period of Emperor Shenzong of the Ming Dynasty, 10th month, day *wuzi* [25] (Nov 14); a greenish-white broom star several *zhang* long appeared in the southwest, its vapors forming a white arc; from WEI [LM 6] and JI [LM 7] it surpassed DOU [LM 8] and NIU [LM 9] and pressed on toward lunar mansion NÜ [LM 10]; after a month it was extinguished.
[*Ming shi: tianwen zhi*] ch. 27

AD 1577 Nov 10–Dec 8 [Korea] (Ho 568)

10th year of King Seonjo of the Joseon Dynasty, 10th month; a baleful star was seen in the west with rays tens of feet long. It appeared like a broom star, but was not a broom star; people called it Chiyou's banner.
[*Jeungbo munheon bigo*] ch. 6
NB: Chinese and Japanese records make it clear this was a comet.

AD 1578 autumn [Korea] (Ho 569)

11th year of King Soenjo of the Joseon Dynasty, autumn; a long star stretched across the sky, its shape like a white silk; it lasted several months before being extinguished.
[*Jeungbo munheon bigo*] ch. 6

AD 1579 Sep 21–Oct 19 [China]

7th year of the Wanli reign period of Emperor Shenzong of the Ming Dynasty, 9th month; a broom star appeared between YI [LM 27] and ZHEN [LM 28].
Qing Dynasty Kangxi period ed. Zhejiang [*Wuqing wenxian*] ch. 3

AD 1580 Oct 9–Nov 6 [China]

(a) 8th year of the Wanli reign period of Emperor Shenzong of the Ming Dynasty, 9th month; a broom star appeared between YI [LM 27] and ZHEN [LM 28].
Ming Dynasty Wanli period ed. Zhejiang [*Chongdexian zhi*] ch. 11

(b) 8th year of the Wanli reign period of Emperor Shenzong of the Ming Dynasty, 9th month; a broom star several *zhang* long emerged in the southeast; by the end of the 10th month it disappeared.
Ming Dynasty Tianqi reign period ed. Anhui [*Feng shu*] ch. 4

NB: (Ho & Ang 66) provide a similar record from *Ming shi* dated Oct 1.

AD 1580 Oct 9 [Japan] (Ho 570)
8th year of the Tenshō reign period, 9th month, 1st day; from this day a broom star was seen for several days.
[Kanda]

AD 1580 Oct 15 [Korea] (Ho 570)
13th year of King Seonjo of the Joseon Dynasty, 9th month, day jiaxu [11]; a broom star was seen.
[Jeungbo munheon bigo] ch. 6

AD 1581 Apr 3–May 2 [China] (Ho & Ang 67)
9th year of the Wanli reign period of Emperor Shenzong of the Ming Dynasty, 3rd month; a broom star appeared in ZIWEIYUAN, the bright rays of its tail shooting northwestward.
Ming Dynasty Tianqi period ed. Jiangsu [*Huai'anfu zhi*] ch. 24

AD 1581 winter [China]
9th year of the Wanli reign period of Emperor Shenzong of the Ming Dynasty, winter; a "silk train" star several *chi* long emerged between NIU [LM 9] and DOU [LM 8], gradually growing longer until it reached several *zhang*; only after more than a month was it was extinguished.
Qing Dynasty Kangxi period ed. Henan [*Fugouxian zhi*] ch. 4
NB: "Silk train" is a common name for a broom star.

AD 1582 May 13–15 [Japan]

(a) 10th year of the Tenshō reign period, 4th month, 23rd day; a broom star 10 *zhang* long was seen at azimuth *qian* [NW], its head at *xu* [WNW]–*hai* [NNW] and its tail at *chen* [ESE]–*si* [SSE].
[*Tabunin nikki*]

(b) 10th year of the Tenshō reign period, 4th month, 21st day; a broom star 10 *zhang* long was seen in the evening in the northwest like a white cloud or a rainbow.
[*Kanda*] (Ho 571)

AD 1582 May 20 [China] (Ho 571; Ho & Ang 68)

(a) 10th year of the Wanli reign period of Emperor Shenzong of the Ming Dynasty, 4th month, day *bingchen* [53], at night; a broom star appeared in the northwest, its tail pointing toward WUCHE, not until after more than 20 days was it was extinguished.
[*Ming Shenzong shilu*] ch. 123

(b) 10th year of the Wanli reign period of Emperor Shenzong of the Ming Dynasty, 4th month, day *bingchen* [53]; a broom star shaped like a bolt of silk cloth appeared in the northwest, its tail pointing toward WUCHE, after more than 20 days, it was extinguished.
[*Ming shi: tianwen zhi*] ch. 27

(c) 10th year of the Wanli reign period of Emperor Shenzong of the Ming Dynasty, 4th month, day *bingchen* [53], at night; a broom star appeared in the northwest with a tail pointing toward WUCHE, not until after an entire 10-day week was it extinguished.
[*Guo que*] ch. 71

AD 1582 Jun 20 [China]

10th year of the Wanli reign period of Emperor Shenzong of the Ming Dynasty, 6th month, day *dinghai* [24], 3rd day of the new moon; a broom star emerged at the mouth of WUCHE, south of ZHU "Pillar."
[*Er shen ye lu*] ch. 5

AD 1585 Oct 13 [China] (Ho 573; Ho & Ang 70)

(a) 13th year of the Wanli reign period of Emperor Shenzong of the Ming Dynasty, 9th month, day *wuzi* [25]; a grayish-white broom star shaped like a crossbow bolt with a tail about 1 *chi* long and pointing northeast emerged beside YULIN; afterward it traveled eastward each evening as its body and rays gradually diminished in size…in the 10th month, day *guiyou* [10], the broom star was extinguished and did not appear.
[*Ming Shenzong shilu*] ch. 165–167

(b) 13th year of the Wanli reign period of Emperor Shenzong of the Ming Dynasty, 9th month, day *wuzi* [25]; a broom star with a tail about 1 *chi* long pointing northeast emerged beside YULIN; each evening it traveled eastward, gradually growing smaller, until day *guiyou* [10] in the 10th month when it was extinguished.
[*Ming shi: tianwen zhi*] ch. 27

(c) 13th year of the Wanli reign period of Emperor Shenzong of the Ming Dynasty, 9th month, day *wuzi* [25], at night; a bluish broom star shaped like a crossbow bolt with a tail about 1 *chi* long and pointing northeast emerged beside YULIN; each evening it gradually diminished in size on the west as it traveled eastward…on the night of day *guiyou* [10] in the 10th month, the broom star was extinguished.
[*Guo que*] ch. 73

AD 1585 Oct 13 [Japan] (Ho 573)

13th year of the Tenshō reign period, intercalary 8th month, 25th day; a broom star was seen at azimuth xun [SE] and azimuth gen [NE].
[Kanda]

AD 1585 Oct 16–20 [Korea]

(a) 18th year of King Seonjo of the Joseon Dynasty, 9th month, day *xinmao* [28]; during the first watch of the night, a star was seen at YULIN, seemingly with vaporous rays, strangely out of the ordinary...on day *yiwei* [32] (Oct 20), during the first watch of the night, the broom star was seen within the first star on the west of TIANCANG, at 7 *du* in lunar mansion BI [LM 14], at a polar distance of 101.5 *du*, its tail about 1 *chi* long and pointing toward azimuth *gen* (NE), its color greenish-white; by the first point of the fifth watch it entered the haze in the west and could not be observed further.

[*Joseon wangjo sillok: Sŏnjo*] ch. 19

(b) 18th year *yiyou* of King Sŏnjo of the Joseon Dynasty, 9th month, day *xinmao* [28]; a broom star was seen in YULIN, its tail over 1 *chi* long.

[*Jeungbo munheon bigo*] ch. 6 (Ho 573)

AD 1587 Oct 2–30 [Korea] (Ho 575)

20th year of King Seonjo of the Joseon Dynasty, 9th month; a broom star was seen in the west, its curved handle was 30–40 *zhang* long with glowing rays that lit up the ground; it lasted for 3 months before being extinguished.

[*Jeungbo munheon bigo*] ch. 6

AD 1593 Aug 6–Sep 19 [Korea]

(a) 26th year of King Seonjo of the Joseon Dynasty, 7th month, day *renxu* [59]; during the fifth watch, there was a star like a broom that emerged outside but close to the western wall of ZIWEI...its color was grayish-white, and its tail more than 3 to 4 *chi* long... on day *yichou* [2], during the fifth watch of the night, the broom star gradually moved northeastward, to a position below BAGU and in front of NEIJIE...on day *xinsi* [18], from the third watch until the fifth watch, the broom star gradually moved upward outside HUAGAI and within CHUANSHE, its body temporarily dimmed, its color grayish-white and its tail more than 2 to 3 *chi* long...in the 8th month, day *dinghai* [24], the broom star moved to the outside of the fifth and sixth stars on the west of HUAGAI, and within the eighth star on the west of CHUANSHE, its body small and dim,

its color blue-white, with no tail…on day *bingwu* [43], the broom star vanished and was not seen.
[*Joseon wangjo sillok: Seonjo*] ch. 41

(b) 26th year *guisi* of King Seonjo of the Joseon Dynasty, 7th month, day *renxu* [59]; a broom star was seen outside ZIWEI. On day *xinsi* [18] (Aug 25), it reached CHUANSHE, its body gradually growing dimmer, until the 8th month, day *bingwu* [43] (Sep 19) when it was extinguished.
[*Jeungbo munheon bigo*] ch. 6 (Ho 578)

AD 1589 Jan 2 [China]

16th year of the Wanli reign period of Emperor Shenzong of the Ming Dynasty, 10th month, day *yichou* [2]; a broom star over 1 *zhang* long pointing southeast, emerged in DI [LM 3]; it gradually pointed northwest, entering the space of KANG [LM 2] and sweeping BEIDOU, WENCHANG, WUCHE, and pressing on the right of ZIGONG; in the 11th month, day *jiachen* [41], it was extinguished.
Qing Dynasty Tongzhi period ed. Hunan [*Guiyang zhili zhou zhi*] ch. 22
NB: The "10th month" should be "11th month," and "11th month" should be "12th month."

AD 1591 Apr 13–23 [China] (Ho 576; Ho & Ang 71)

(a) 19th year of the Wanli reign period of Emperor Shenzong of the Ming Dynasty, 3rd month, day *bingchen* [53] (20th day) (Apr 13); in the northwest there was a star like a broom over 1 *chi* long; it passed WEI [LM 17], SHI [LM 13], BI [LM 14], with a length of 2 *chi*; in the intercalary 3rd month, new moon day *bingyin* [3], it entered LOU [LM 16].
[*Ming shi: tianwen zhi*] ch. 27

(b) 19th year of the Wanli reign period of Emperor Shenzong of the Ming Dynasty, 3rd month, day *bingchen* [53] (20th day) (Apr 13), at night; a broom star with a tail 2 *chi* long appeared between WEI [LM 17] and SHI [LM 13]; in the intercalary 3rd month, on the night of day *bingyin* [3], it entered lunar mansion LOU [LM 16].
[*Guo que*] ch. 75

(c) 19th year of the Wanli reign period of Emperor Shenzong of the Ming Dynasty, 3rd month, day *bingchen* [53] (20th day) (Apr 13); a broom star with a tail about 1 *chi* long appeared in the northwest in the allocated space of lunar mansion Wei [LM 17]; it then moved northeastward into the allocated space of Shi [LM 13] and Bi [LM 14], with a tail 2 *chi* long.
Qing Dynasty Kangxi ed. Hebei [*Changpingzhou zhi*] ch. 25

(d) 19th year of the Wanli reign period of Emperor Shenzong of the Ming Dynasty, intercalary 3rd month, new moon day *bingyin* [3] (Apr 23); a broom star entered Lou [LM 16]...on day *guiyou* [10], a scholar of the Imperial Academy Shen Shixing wrote, "on the 20th day of the 3rd month, after the first watch, in the northwest there was a star like a broom with a tail over 1 *chi* long, which the Imperial Directorate of Astronomy and the Calendar said was in the space of lunar mansion Wei [LM 17]; at present, it is in the northeast in the space of lunar mansions Shi [LM 13] and Bi [LM 14], with a tail about 2 *chi* long; most alarming!"
[*Ming Shenzong shilu*] ch. 234

AD 1591 Apr 23–May 22 [China]

19th year of the Wanli reign period of Emperor Shenzong of the Ming Dynasty, intercalary 3rd month; a broom star with a tail 2 *chi* long was in the northwest in the space of Shi [LM 13] and Bi [LM 14].
Ming Dynasty Chongzhen period ed. Shanghai [*Songjiangfu zhi*] ch. 47

AD 1591 Aug 19–Sep 16 [China]

19th year of the Wanli reign period of Emperor Shenzong of the Ming Dynasty, 7th month; a broom star appeared in Dongjing [LM 22], then retrograded into Ziwei.
Qing Dynasty Qianlong period ed. Guangxi [*Quanzhou zhi*] ch. 11

AD 1593 Jul 30 [China] (Ho 578; Ho & Ang 72)

(a) 21st year of the Wanli reign period of Emperor Shenzong of the Ming Dynasty, 7th month, day *yimao* [52], fifth drum of the night; a broom star over 3 *chi* long appeared in the space of Jing

[LM 22]...on day *yihai* [12], the broom star retrograded and entered ZIWEIYUAN and trespassed on HUAGAI.
[*Ming Shenzong shilu*] ch. 262

(b) 21st year of the Wanli reign period of Emperor Shenzong of the Ming Dynasty, 7th month, day *yimao* [52]; a broom star appeared in DONGJING [LM 22]; on day *yihai* [12], it retrograded and entered ZIWEIYUAN, trespassing on HUAGAI.
[*Ming shi: tianwen zhi*] ch. 27

(c) 21st year of the Wanli reign period of Emperor Shenzong of the Ming Dynasty, 7th month, day *yimao* [52], at night; a broom star with rays over 1 *chi* long appeared in lunar mansion JING [LM 22]; on the night of day *yihai* [12], it retrograded and entered ZIWEIYUAN, trespassing on HUAGAI.
[*Guo que*] ch. 76

AD 1593 Aug 3–16 [Japan]

(a) 2nd year of the Bunroku reign period, 7th month, 7th day (Aug 3); a broom star was seen at daybreak during the hour of *chou* [LT = 1–3 h]...20th day (Aug 16); the broom star disappeared.
[*Tamon-in nikki ryaku*]

(b) 2nd year of the Bunroku reign period, 7th month, 8th day (Aug 4), clear sky; past midnight, a broom star 1 *zhang* long was seen at azimuth *chou* [NNE]–*yin* [ENE].
[*Tokiyoshi kyōki*]

(c) 2nd year of the Bunroku reign period, 7th month, 7th day (Aug 3); at midnight a broom star 1 *zhang* long was seen in the northwest.
[*Kanda*] (Ho 578)

AD 1593 Aug 6 [Korea]

(a) 26th year of King Seonjo of the Joseon Dynasty, 7th month, day *renxu* [59]; during the fifth watch there was a star like a broom star that emerged from just outside the west wall of ZIWEI...it was white with a tail about three or four *chi* long...on day *yichou*, during the fifth watch of the night, the broom star gradually moved northwestward, to a position below BAGU and in front of

NEIJIE...on day *xinsi* [18], from the third watch to the fifth watch, the broom star gradually moved upward outside HUAGAI within CHUANSHE, its form and shape temporarily diminished, its color grayish-white, with a tail about two or three *chi* long...in the 8th month, on day *dinghai* [24], the broom star moved outside HUAGAI between the fifth and sixth stars, within CHUANSHE at the eighth star on the west, its form and shape tiny, its color grayish-white, and with no tail...on day *bingwu* [43], the broom star disappeared and was no longer seen.
[*Joseon wangjo sillok: Seonjo*] ch. 41

(b) 26th year of King Seonjo of the Joseon Dynasty, 7th month, day *renxu* [59]; a broom star was seen outside ZIWEI; on day *xinsi* [18], it reached CHUANSHE, its form gradually tinier; on day *bingwu* [43] in the 8th month it was extinguished.
[*Jeungbo munheon bigo*] ch. 6

AD 1594 Jan 21–Feb 19 [China]

21st year of the Wanli reign period of Emperor Shenzong of the Ming Dynasty, 12th month; a broom star emerged in TIANJIN.
Qing Dynasty Kangxi period ed. Jiangsu [*Huaianfu zhi*] ch. 1

AD 1596 May 27–Jul 31 [Japan]

(a) 1st year of the Keichō reign period, 5th month (May 27–Jun 25); a broom star appeared.
[*Zoku shigushō*]

(b) 1st year of the Keichō reign period, 6th month, 29th day (Jul 24), clear sky; tonight a broom star was seen during the hours of *you* and *xu* [LT = 17–21 h] at azimuth *xu* [WNW] to *hai* [NNW]. 7th month, 2nd day (Jul 26), clear sky; tonight the broom star was seen about 5 *chi* long...7th day (Jul 31), clear sky; the broom star was seen.
[*Takasue sukune hinami ki*]

NB: (Ho 579) is a confused composite of these two records which implies the comet was seen at the same hour throughout its period of visibility.

AD 1596 Jul 19–Aug 22 [Korea]

(a) 29th year of King Seonjo of the Joseon Dynasty, 6th month, day *gengshen* [57] (Jul 19); a broom star was seen in JING [LM 22], its body was like the big star of WUCHE; it lasted until the 7th month, day *jiawu* [31] (Aug 22), when it was extinguished.
[*Jeungbo munheon bigo*] ch. 6 (Ho 579)

(b) 29th year of King Seonjo of the Joseon Dynasty, 7th month, new moon day *bingyin* [3] (Jul 25); during the first watch of the night, a broom star was seen east of ZHONGTAI…on day *guiyou* [10], during the first watch of the night, the broom star was seen at XIATAI…on day *renwu* [19] (Aug 10), during the first watch of the night, the broom star was seen below CHANGZHEN…on day *jiawu* [31] (Aug 22), the broom star was extinguished.
[*Joseon wangjo sillok: Seonjo*] ch. 77

AD 1596 Jul 25–Aug 23 [China]

(a) 24th year of the Wanli reign period of Emperor Shenzong of the Ming Dynasty, 7th month; a broom star appeared in the northwest.
Qing Dynasty Kangxi period ed. Guangxi [*Hepuxian zhi*] ch. 1
Qing Dynasty Kangxi period ed. Guangxi [*Lianzhoufu zhi*] ch. 1

(b) 24th year of the Wanli reign period of Emperor Shenzong of the Ming Dynasty, 7th month; a broom star appeared in BEIDOU.
Qing Dynasty Qianlong period ed. Fujian [*Shaowufu zhi*] ch. 24

AD 1596 Aug 5 [China] (Ho 579; Ho & Ang 73)

(a) 24th year of the Wanli reign period of Emperor Shenzong of the Ming Dynasty, 7th month, day *dingchou* [14]; a greenish-white broom star like a crossbow pellet appeared in the north-west with rays pointing southeast; it entered the space of lunar mansion YI [LM 27], at about 1 *chi* in length, then traveled northwestward and entered [the horizon].
[*Ming Shenzong shilu*] ch. 299

(b) 24th year of the Wanli reign period of Emperor Shenzong of the Ming Dynasty, 7th month, day *dingchou* [14]; a broom star like a

crossbow bolt appeared in the northwest and entered Yi [LM 27] when it was over 1 *chi* long; then it traveled northwestward.
[*Ming shi: tianwen zhi*] ch. 27

AD 1600 Sep 2–27 [Korea] (Ho 580)
33rd year *gengzi* of King Seonjo of the Joseon Dynasty, 7th month, day *bingyin* [3]; a broom star was seen at Zhongtai. On day *xinmao* [28] (Sep 27), it moved inside Taiwei, its glowing rays gradually dimming.
[*Jeungbo munheon bigo*] ch. 6

AD 1601 Dec 20 [Korea]
34th year *xinchou* of King Seonjo of the Joseon Dynasty, 11th month, 26th day; at Xiongchuan County in Qingshang Circuit a broom star was seen between azimuths *zi* and *chou* [i.e., N to NNE], its tail several *chi* long.
[*Jeungbo munheon bigo*] ch. 6

AD 1602 [Korea]
35th year *renyin* of King Seonjo of the Joseon Dynasty, 7th month, day *yisi* [42]; a broom star was seen between Santai and Beiji, then moved below Langwei, its tail 7 to 8 *chi* long; it lasted a long time without being extinguished.
[*Jeungbo munheon bigo*] ch. 6
NB: The new moon day of the 7th month was *gengshen* [57]; hence there was no day *yisi* [42].

AD 1604 autumn [China]
32nd year of the Wanli reign period of Emperor Shenzong of the Ming Dynasty, autumn; a broom star about 2 to 3 *chi* long appeared in the middle of the sky, its color dim and indistinct; only after about 6 days did it vanish.
Qing Dynasty Qianlong period ed. Guangdong [*Chaozhoufu zhi*] ch. 11
NB: This may be a record of Kepler's supernova.

AD 1607 Sep 20 [China]

(a) 35th year of the Wanli reign period of Emperor Shenzong of the Ming Dynasty, 7th month, day *gengshen* [57], at night; a broom star over 2 *chi* long appeared in DONGJING [LM 22]; on the night of day *dingchou* [14] in the 8th month, the broom star passed FANG [LM 4]; on the night of day *renwu* [19], the broom star passed from FANG [LM 4] to XIN [LM 5].
[*Guo que*] ch. 80

(b) 35th year of the Wanli reign period of Emperor Shenzong of the Ming Dynasty, 7th month, day *gengshen* [57], at the fifth drum of the night; a broom star over 1 *chi* long emerged in the northeast in lunar mansion JING [LM 22], its tail pointing southwest.
Ming Dynasty Chongzhen period ed. Shanghai [*Songjiangfu zhi*] ch. 47

NB: This is Comet Halley, return –5.

AD 1607 Sep 21 [China] (Ho & Ang 75)

(a) 35th year of the Wanli reign period of Emperor Shenzong of the Ming Dynasty, 8th month, new moon day *xinyou* [58]; a grayish-white broom star about 2 *chi* long appeared in the space of JING [LM 22], its tail pointing southwest, and gradually traveled northwestward…on day *dingchou* [14], the broom star passed FANG [LM 4]…on day *renwu* [19], the broom star passed XIN [LM 5].
[*Ming Shenzong shilu*] ch. 437

(b) 35th year of the Wanli reign period of Emperor Shenzong of the Ming Dynasty, 8th month, new moon day *xinyou* [58]; a broom star appeared in DONGJING [LM 22], pointing southwest, and gradually traveled northwestward; on day *renwu* [19], the broom star passed from FANG [LM 4] to XIN [LM 5] and was extinguished.
[*Ming shi: tianwen zhi*] ch. 27

(c) 35th year of the Wanli reign period of Emperor Shenzong of the Ming Dynasty, 8th month, new moon day *xinyou* [58]; a broom star 2 *chi* long appeared in the space of JING [LM 22] and gradually traveled northwestward; on day *renwu* [19], it was extinguished.

Qing Dynasty Yongzheng period ed. [*Shaanxi tongzhi*] ch. 47

NB: This is Comet Halley, return –5.

AD 1607 Sep 23–Oct 24; Nov 19–Dec 18 [Korea]

(a) 40th year of King Seonjo of the Joseon Dynasty, 8th month, day *guihai* [60]; during the fifth watch of the night, through a break in the clouds a broom star was seen south of SHANGTAI, its color was grayish-white and it appeared to have an obliquely pointing tail…in the 9th month, on day *jiawu* [31] (Oct 24), the broom star was seen by the first star of TIANSHIYUAN.

[*Joseon wangjo sillok: Seonjo*] ch. 215–216

(b) 40th year *dingwei* of King Sŏnjo of the Joseon Dynasty, 8th month, day *guihai* [60]; a broom star was seen in SANTAI, its tail 7 to 8 *chi* long and grayish-white; it moved to TIANSHI and by the 18th day of the 9th month, it was extinguished. In the 10th month (Nov 19–Dec 18) it was seen within SANTAI, its tail over 10 *zhang* long and pointing toward WENCHANG and the bowl of BEIDOU, its color was grayish-white; it lasted over a month before being extinguished.

[*Jeungbo munheon bigo*] ch. 6

NB: This is Comet Halley, return –5.

AD 1607 Sep 25–Nov 10 [Japan]

(a) 12th year of the Keichō reign period, 8th month, 14th day *jiaxu* [11] (Oct 4); in the east (at the juncture of azimuth *yin* [ENE] and *chou* [NNE]) and the west a broom star was seen…9th month, 11th day *xinchou* [38] (Nov 12); clear sky; for two nights now the broom star has not been seen.

[*Takasue sukune hinami ki*]

(b) 12th year of the Keichō reign period, 8th month, 5th day *yichou* [2] [Sep 25]; during the hour of *chou* [LT = 1–3 h], a white broom star 8 *chi* long was seen at azimuth *gen* [NE], its bright vapor pointing toward azimuth *wei* [SSW]…11th day *xinwei* [8] (Oct 1),

during the hour of *xu* [LT = 19–21 h], the white broom star 25 *cun* long appeared at azimuth *dui* [W], its bright vapor pointing toward azimuth *zhen* [E]; 9th month, 5th day *yiwei* [32] (Nov 6); during the hour of *xu* [LT = 19–21 h], it moved from azimuth *gen* [NE] to *kun* [SW], its length over 2 *zhang* and its color red as fire.
[*(Tsuchimikado) kahi yōroku*]

NB: This is Comet Halley, return –5.

AD 1607 Oct 21–Nov 18 [China]

35th year of the Wanli reign period of Emperor Shenzong of the Ming Dynasty, 9th month; a broom star appeared in ZHEN [LM 28].
Qing Dynasty Qianlong period ed. Guangxi [*Lianzhoufu zhi*] ch. 5

AD 1614 Mar 14 [Japan]

(a) 19th year of the Keichō reign period, 2nd month, 4th day; a broom star was seen in the east; the court and the land verged on confusion, taking it to be inauspicious.
[*Isshi*]

(b) 19th year of the Keichō reign period, 2nd month, a certain day; someone recorded that there was a broom star several *zhang* in length whose rays pointed to the northwest, which lased until summer before disappearing.
[*Zoku shigushō*]

AD 1616 winter [China]

44th year of the Wanli reign period of Emperor Shenzong of the Ming Dynasty, winter; a broom star appeared; from DONGJING [LM 22] it entered ZIWEIYUAN, pointing south with a gleam.
Qing Dynasty Kangxi period ed. Shanxi [*Yucixian xuzhi*] ch. 12

AD 1617 Jun 3–Jul 2 [China]

45th year of the Wanli reign period of Emperor Shenzong of the Ming Dynasty, 5th month; a broom star over 3 *zhang* long emerged in NANDOU [LM 8] for more than a month.
Qing Dynasty Kangxi period ed. Zhejiang [*Changxingxian zhi*] ch. 4

AD 1618 Jan 26–Feb 24 [China]

46th year of the Wanli reign period of Emperor Shenzong of the Ming Dynasty, 1st month; a white broom star emerged in DI [LM 3] and KANG [LM 2].

[*Er shen ye lu*] ch. 6

AD 1618 Mar 26–Apr 24 [China]

46th year of the Wanli reign period of Emperor Shenzong of the Ming Dynasty, 3rd month; a broom star emerged in XU [LM 11].

Qing Dynasty Qianlong ed. [*Shandong tongzhi*] ch. 33

AD 1618 spring [China]

46th year of the Wanli reign period of Emperor Shenzong of the Ming Dynasty, spring; a broom star emerged in the southeast, thrust toward ZIWEIYUAN, and after 50 days it was extinguished.

Qing Dynasty Qianlong period ed. Shanxi [*Jiezhou Ruichengxian zhi*] ch. 11

AD 1618 Aug 28–Dec 20 [Japan]

(a) 4th year of the Genna reign period, 7th month, 9th day (Aug 28); at dawn a broom star was seen in the northeast, its tail 3 *chi* long… 11th month, 2nd (Dec 20) day; at dawn the broom star was seen, moving from the east to the north.

[*Takasue nichi jiki*]

(b) 4th year of the Genna reign period, 7th month, 9th day *yiwei* [32] (Aug 28), clear sky; this dawn a broom star 3 *chi* long was seen between azimuth *chou* [NNE] and *yin* [ENE]…10th month, 28th day *guiwei* [20] (Dec 14), clear sky; after midnight a broom star was seen at the juncture of azimuths *chou* [NNE] and *yin* [ENE].

[*Takasue sukune hinami ki*]

(c) 4th year of the Genna reign period, 8th month, 8th day (Sep 26); a broom star was seen.

[*Sōto nikki*]

(d) 4th year of the Genna reign period, 10th month, 11th day *bingyin* [3] (Nov 27); a broom star was seen in the east (at daybreak); by sunrise it was over 1 *zhang* long.
[*Zoku shigushō*]

AD 1618 Sep 19–Oct 17 [China]

(a) 46th year of the Wanli reign period of Emperor Shenzong of the Ming Dynasty, 8th month; a broom star emerged in the southeast, its tail sweeping the northwest for over 80 days when it was extinguished.
Qing Dynasty Shunzhi period ed. Hebei [*Raoyangxian zhi*] ch. 5

(b) 46th year of the Wanli reign period of Emperor Shenzong of the Ming Dynasty, 8th month; a broom star entered LOU [LM 16].
Qing Dynasty Kangxi period ed. Guangdong [*Chaoyangxian zhi*] ch. 3

(c) 46th year of the Wanli reign period of Emperor Shenzong of the Ming Dynasty, 8th month; a white vapor extended from east to west; after several evenings there was a giant star over 1 *zhang* long that appeared every night in the east, while suspended in the air there was a white vapor shaped like long-handled sword, or elongated rectangular serpentine spear, or again like white cotton cloth; emerging at sector *chen* (ESE = LM 1–2) in the space of JIAO [LM 1] and KANG [LM 2], its tail thrusting toward the space of KUI [LM 15], LOU [LM 16], and BI [LM 14]; it spread across the whole sky from the north to the south, and was not extinguished until the 10th month.
Qing Dynasty Kangxi period ed. Guangxi [*Pinglexian zhi*] ch. 6

AD 1618 Nov 14 [China] (Ho & Ang 77)

(a) 46th year of the Wanli reign period of Emperor Shenzong of the Ming Dynasty, 9th month, day *guichou* [50]; in Guangzhou a broom star appeared, emerging in sector *chen* (ESE = LM 1–2) in the space of JIAO [LM 1] and KANG [LM 2], its tail thrusting toward the space of KUI [LM 15], LOU [LM 16], and BI [LM 14];

several nights previously a white vapor extended from east to west like a sword, then appeared together with the broom star, its pointed rays like a broom; after 2 months, it was extinguished.
[*Er shen ye lu*] ch. 6

(b) 46th year of the Wanli reign period of Emperor Shenzong of the Ming Dynasty, 9th month, day *guichou* [50]; a broom star several *zhang* long and shaped like an arrow appeared in the east; emerging at sector *chen* (ESE = LM 1–2) in the space of JIAO [LM 1] and KANG [LM 2], its tail thrusting toward the space of KUI [LM 15], LOU [LM 16], and BI [LM 14].
Qing Dynasty Kangxi period ed. Guangdong [*Sanshuixian zhi*] ch. 1
Qing Dynasty Kangxi period ed. Guangxi [*Zhaoqingfu zhi*] ch. 3

AD 1618 Nov 15–16 [China]

(a) 3rd year of the Tianming reign period of Emperor Taizu of the Qing Dynasty, 9th month, day *jiayin* [51] (Nov 15) at night; a white long star shaped like a bolt of cloth and over 1 *chi* wide by over 2 *zhang* long appeared in YI [LM 27] and ZHEN [LM 28] for 19 days in all before being extinguished.
[*Guo que*] ch. 83

(b) 46th year of the Wanli reign period of Emperor Shenzong of the Ming Dynasty, 9th month, day *yimao* [52] (Nov 16); a long star appeared in the southeast, 1 swath of white vapor shaped like a bolt of cloth, over 1 *chi* wide by over 2 *zhang* long; on the east it reached ZHEN [LM 28] and on the west entered YI [LM 27]; it lasted 19 days in all before being extinguished.
[*Ming Shenzong shilu*] ch. 574

(c) 46th year of the Wanli reign period of Emperor Shenzong of the Ming Dynasty, 9th month, day *yimao* [52] (Nov 16); in the southeast there was a swath of white vapor over 1 *chi* wide by over 2 *zhang* long; on the east it reached ZHEN [LM 28] and on the west entered YI [LM 27]; it lasted 19 days and then was extinguished.
[*Ming shi: tianwen zhi*] ch. 27

AD 1618 Nov 26 [China] (Ho & Ang 77)

(a) 46th year of the Wanli reign period of Emperor Shenzong of the Ming Dynasty, 10th month, day *yichou* [2]; a grayish-white broom star as big as a hen's egg and over 1 *zhang* long emerged in DI [LM 3], its tail pointing southeast; after 10-odd days it turned and pointed northwest, sweeping and trespassing against TAIYANGSHOU, then entered KANG [LM 2] at 7 *du*, 10 *fen*; after another 3 days, it gradually traveled northwestward, its tail sweeping BEIDOU, TIANXUAN, TIANJI, WENCHANG, WUCHE, and so forth, and bore down on ZIYUAN, at a location 6 *du* in KANG [LM 2]. Not until the 19th day of the following month was it extinguished.

[*Ming Shenzong shilu*] ch. 575

(b) 46th year of the Wanli reign period of Emperor Shenzong of the Ming Dynasty, 10th month, day *yichou* [2]; a broom star over 1 *zhang* long emerged in DI [LM 3], pointing southeast, then gradually pointed northwest. It swept and trespassed against TAIYANGSHOU then entered the space of KANG [LM 2]; then to the northwest it swept BEIDOU, TIANJI, WENCHANG, WUCHE, and bore down on the right side of ZIWEIYUAN; by day *jiachen* [41] of the 11th month it was extinguished.

[*Ming shi: tianwen zhi*] ch. 27

(c) 46th year of the Wanli reign period of Emperor Shenzong of the Ming Dynasty, 10th month, day *yichou* [2], at night; a greenish-white broom star as big as a hen's egg and over a *zhang* long emerged in lunar mansion DI [LM 3], its tail pointing southeast; after several evenings it turned and pointed northwest, trespassing against TAIYANGSHOU; after another 3 days it gradually headed northwestward, its tail sweeping TIANXUAN of BEIDOU, TIANJI, WENCHANG, WUCHE, and so forth, and bore down on ZIWEI; by the 19th day of the following month it was extinguished.

[*Guo que*] ch. 83

(d) 46th year of the Wanli reign period of Emperor Shenzong of the Ming Dynasty, 10th month, day *yichou* [2]; a broom star appeared, emerging in the division of *mao* [E], in the space of Di [LM 3] and Fang [LM 4]; its tail thrusting up and pointing toward the division of *shen* [WSW], the spaces of Zui [LM 20] and Shen [LM 21]; observing from below it was over 1 *zhang* long and shaped like a sword; the flames of its rays were not extinguished for over a month.
Qing Dynasty Kangxi period ed. Guangdong [*Sanshuixian zhi*] ch. 1
Qing Dynasty Kangxi period ed. Guangdong [*Zhaoqingfu zhi*] ch. 3

AD 1618 Nov 26–AD 1619 Jan 13 [Korea]

(a) 10th year of Gwanghaegun of the Joseon Dynasty, 10th month, day *yichou* [2]; a broom star over 1 *zhang* long was seen in the east within the space of Di [LM 3], its color yellowish-red; when the sky brightened, it was no longer seen…on day *xinsi* [18] (Dec 12), during the third watch, when the broom star first was seen, it was about 20 *zhang* long and over 4 to 5 *chi* wide; it trespassed close to the second and third stars of and went straight through the stars of Wenchang, its size somewhat reduced compared to before and its color gradually fading…in the 11th month, on day *dingyou* [34] (Dec 28), the broom star's vaporous rays dimmed…on the night of day *guichou* [50] (Jan 13), the body of the broom star was faintly visible, its rays almost completely extinguished.
[*Joseon wangjo sillok: Kwanghaegun*] ch. 133–134

(b) 10th year *wuwu* of Kwanghaegun of the Joseon Dynasty, 10th month, day *yichou* [2]; a broom star was seen in Di [LM 3], its tail 20 *zhang* long and 4 to 5 *chi* wide and its color red. It trespassed on Beidou and penetrated Wenchang. In the 11th month…on day *guichou* [50] (Jan 13), the broom star had indeed vanished.
[*Jeungbo munheon bigo*] ch. 6

AD 1618 Nov 27 [China] (Ho & Ang 77)

(a) 3rd year of the Tianming reign period of Emperor Taizu of the Qing Dynasty, 10th month, 11th day (*bingyin* [3]); at the time of the fifth watch, in the southeast sector a white vapor 5 *chi* wide emerged from a star and surged straight at the Ming state; after the 14th day it no longer appeared. The star that produced it gradually moved toward BEIDOU each night, until the 29th day when it went right over the handle of BEIDOU, after which it was no longer seen. [*Da Qing Manzhou shilu*]

(b) 3rd year of the Tianming reign period of Emperor Taizu of the Qing Dynasty, 10th month, 11th day (*bingyin* [3]); a broom star appeared in the east, to the east of the previously sighted white vapor which was 5 *chi* wide, with its tail pointing right at the Ming state; each night it gradually moved toward BEIDOU in all for 19 days until it passed over the tail of BEIDOU, and only then was it was extinguished.
[*Qing Taizu Gao huangdi shilu*] ch. 5

(c) 3rd year of the Tianming reign period of Emperor Taizu of the Qing Dynasty, 10th month, 11th day (*bingyin* [3]); a broom star with a tail 5 *zhang* long appeared in the east; each night it gradually moved toward BEIDOU; on the 19th day it disappeared.
[*Qing shi gao: tianwen zhi*] ch. 39

AD 1618 Dec 17–AD 1619 Jan 15 [China]

(a) 46th year of the Wanli reign period of Emperor Shenzong of the Ming Dynasty, 11th month; a star became fuzzy in ZHEN [LM 28], red like a vapor or a sword and over 1 *zhang* long; it was extinguished in the 1st month of the next year.
Ming Dynasty Chongzhen period ed. Guangdong [*Nanhaixian zhi*] ch. 2

(b) 46th year of the Wanli reign period of Emperor Shenzong of the Ming Dynasty, 11th month; a broom star 5 *chi* long appeared in the east, below it broke into JUN, its bright rays brushing BEIDOU.
Qing Dynasty Jiaqing period ed. Jiangsu [*Yangzhoufu zhi*] ch. 70

AD 1619 mid-Feb [China] (Ho & Ang 78)

(a) 47th year of the Wanli reign period of Emperor Shenzong of the Ming Dynasty, end of the 1st month; a broom star several [*sic*] hundred *chi* long appeared in the east, shooting by beneath the moon, its tip curved and pointed; it appeared a while in the north-east, and then again in the west.
[*Ming Shenzong shilu*] ch. 578

(b) 47th year of the Wanli reign period of Emperor Shenzong of the Ming Dynasty, end of the 1st month; a broom star several [sic] hundred *chi* long appeared in the southeast, its tip curved and sharp pointed, and its bright rays shooting downward; it appeared a while in the northeast, and then again in the west.
[*Ming shi: tianwen zhi*] ch. 27

AD 1619 spring [China]

47th year of the Wanli reign period of Emperor Shenzong of the Ming Dynasty, spring; a broom star appeared at sector *wu* (S = LM 24–26) and later pointed north, then it entered BEIDOU; it was not extinguished until after 20 days.
Qing Dynasty Kangxi period ed. Guangdong [*Wenchangxian zhi*] ch. 9

AD 1622 Sep 7 [Japan]

8th year of the Genna reign period, 8th month, 2nd day *bingyin* [3]; light rain…a broom star was seen at azimuth *qian* [NW], its length measured at 1 *jian* [i.e., 6 ft.], it was seen around the hours of *zi–chou* [LT = 23–3 h].
[*Takasue sukune hinami ki*]

AD 1624 Jul 27 [Japan]

1st year of the Kan-ei reign period, 7th month, 1st day *guichou* [50], clear sky; some people said that on the 13th and 14th of the 6th month past (Jul 27–28), on the night of the full moon, a broom star about 5 *chi* long was seen in the south.
[*Takasue sukune hinami ki*]

AD 1627 summer [Korea]

5th year of King Injo of the Joseon Dynasty, summer; there was a fuzzy star by the Moon.
[*Jeungbo munheon bigo*] ch. 6

AD 1628 Aug 13 [Japan]

5th year of the Kan-ei reign period, 7th month, 14th day *guiyou* [10], clear sky; during the hour of *yin* [LT = 3–5 h], a broom star was seen at azimuth *chen* [ESE]; 16th day *yihai* [12], clear sky; the broom star was seen during the hour of *yin* [LT = 3–5 h].
[*Takasue sukune hinami ki*]

AD 1630 Feb 1 [Japan]

6th year of the Kan-ei reign period, 12th month, 19th day *gengwu* [7], clear sky; a broom star was seen at azimuth *qian* [NW].
[*Takasue sukune hinami ki*]

AD 1638 Jun 12–Jul 10 [Korea]

16th year of King Injo of the Joseon Dynasty, 5th month, a broom star was seen at azimuth *xun* (SE), its glowing rays extremely reddish; it lasted a whole month before being extinguished.
[*Jeungbo munheon bigo*] ch. 6

AD 1639 Aug 29–Sep 26 [China]

12th year of the Chongzhen reign period of Emperor Yizong of the Ming Dynasty, 8th month; a broom star entered BEIDOU.
Qing Dynasty Qianlong period ed. Shanxi [*Jiexiuxian zhi*] ch. 1

AD 1639 Sep 27–Oct 25 [China] (Ho & Ang 80)

12th year of the Chongzhen reign period of Emperor Yizong of the Ming Dynasty, 9th month; a broom star appeared in the space of SHEN [LM 21].
[*Guoque*] ch. 97

AD 1639 autumn [China]

12th year of the Chongzhen reign period of Emperor Yizong of the Ming Dynasty, autumn; a broom star appeared in the space of SHEN [LM 21].

[*Ming shi: tianwen zhi*] ch. 27

NB: Same as the preceding.

AD 1639 Nov 25–Dec 23 [China]

12th year of the Chongzhen reign period of Emperor Yizong of the Ming Dynasty, 11th month, at night; a broom star entered BEIDOU.

Qing Dynasty Kangxi period ed. Shanxi [*Jiexiuxian zhi*] ch. 1

AD 1652 Nov 1–30 [China]

(a) 9th year of the Shunzhi reign period of Emperor Shizu of the Qing Dynasty, 10th month; a broom star appeared with a tail 2 *zhang* long.

Qing Dynasty Qianlong period ed. Guangdong [*Shundexian zhi*] ch. 16

(b) 9th year of the Shunzhi reign period of Emperor Shizu of the Qing Dynasty, 10th month; in the south a swarm of stars clustered together, one as large as the moon gleamed with rays; it was evidently a fuzzy star. From JING [LM 22] it retrograded as far as BI [LM 19] and MAO [LM 18] and then disappeared.

Qing Dynasty Daoguang period ed. Jiangsu [*Huangjing zhi gao*] ch. 7

AD 1652 Dec 16–24 [Japan]

1st year of the Jōō reign period, 11th month, 16th day; at night a guest star was seen, above the star was a halo like joined hat strings about 5 *chi* long and pointing northeast; after the 24th day, the halo was dimmer and went away day by day.

[*Yamaga sokō kafu nenpu*]

AD 1652 Dec 22 [China] (Ho & Ang 82)

9th year of the Shunzhi reign period of Emperor Shizu of the Qing Dynasty, 11th month, day *gengyin* (27); an anomalous star with grayish-white vapor appeared in SHEN [LM21], it traveled northwestward and entered BI [LM 19].

[*Qing shi gao:tianwen zhi*] ch. 39

AD 1653 Sep 20 [Korea]

4th year *guisi* of King Hyojong of the Joseon Dynasty, 7th month, *renxu* [59]; a broom star was seen outside the western wall of ZIWEI, its body smaller than BEIHE, gradually it entered below GU, its tail 7 to 8 *chi* long and pointing southwest, its color greenish-white; by the 25th day of the 8th month (Oct 20–Nov 18), it was no longer seen.

[*Jeungbo munheon bigo*] ch. 6

AD 1656 Jul 5? [Korea]

7th year *bingshen* of King Hyojong of the Joseon Dynasty, 6th month, day *gengshen* [57]; a broom star was seen in JING [LM 22], its body like the large star of WUCHE, its tail 2 to 3 *chi* long, and its color greenish-white; by the 29th day of the 7th month it was extinguished.

[*Jeungbo munheon bigo*] ch. 6

NB: The new moon day of the 6th month was *wuyin* [15]; hence there was no day *gengshen* [57].

AD 1659 winter [China]

16th year of the Shunzhi reign period of Emperor Shizu of the Qing Dynasty, winter; a broom star trespassed on BEIDOU.

Qing Dynasty Shunzhi period ed. Shandong [*Dengzhoufu zhi*] ch. 1

AD 1661 Feb 4 [Korea]

(a) 2nd year of King Heonjong of the Joseon Dynasty, 1st month, day *bingchen* [53]; during the fifth watch of the night, a broom star was seen in the east within the space of lunar mansion NIU [LM 9], between HEGU and BAIGUA; its body resembled the small

star of Hegu, its tail over 2 *chi* long and pointing toward azimuth *qian* (NW), its color was white; it was not extinguished for 10 or more days.
[*Joseon wangjo sillok: Heonjong*] ch. 4

(b) 2nd year *xinchou* of King Heonjong of the Joseon Dynasty, 1st month, day *bingchen* [53]; a broom star was seen in Niu [LM 9], its tail over 2 *chi* long and its color white.
[*Jeungbo munheon bigo*] ch. 6

AD 1663 Oct 31–Nov 29 [China]

(a) 2nd year of the Kangxi reign period of Emperor Shengzu of the Qing Dynasty, 10th month; a broom star appeared in the space of Yi [LM 27] and Zhen [LM 28] with rays like a broom over 2 *zhang* long; in the 12th month it disappeared.
Qing Dynasty Kangxi period ed. Guangdong [*Haiyangxian zhi*] *zaixiang*

(b) 2nd year of the Kangxi reign period of Emperor Shengzu of the Qing Dynasty, 10th month; a broom star over 1 *zhang* long appeared, arising in the southeast, its brightness lit up the northwest; it lasted for 10 days and nights before dissipating, then in the 3rd month of the next year it appeared again.
Qing Dynasty Kangxi period ed. Hebei [*Jizhou zhi*] ch. 1

(c) 2nd year of the Kangxi reign period of Emperor Shengzu of the Qing Dynasty, 10th month; a broom star appeared in the southeast for over 50 days before disappearing; in the 3rd year, 2nd month, the broom star reappeared.
Qing Dynasty Kangxi period ed. Anhui [*Huoqiuxian zhi*] ch. 10

(d) 2nd year of the Kangxi reign period of Emperor Shengzu of the Qing Dynasty, 10th month; a broom star appeared in the southeast for over 50 days; then in the 2nd month of the next year the broom star reappeared.
Qing Dynasty Kangxi period ed. Jiangsu [*Jiangningfu zhi*] ch. 3
Qing Dynasty Kangxi period ed. Jiangsu [*Shangyuanxian zhi*] ch. 13

AD 1664 Mar 28 [Japan]

4th year of the Kambun reign period, 3rd month, from new moon day and for 7 days, a broom star was seen at azimuth *wei* [SSW]–*shen* [WSW]. [*Tōsho nendaiki*] *Hirayama nikki*

AD 1664 Oct 19–28 [China]

3rd year of the Kangxi reign period of Emperor Shengzu of the Qing Dynasty, 9th month, first 10-day week; there was a star over 1 *chi* long with bright vapors like a candle that followed the star as it ascended; it appeared during the fifth drum, and every night it was like a broom until the end of winter, when its brightness gradually grew shorter, though still about 4 to 5 *cun* long; it gradually turned and became earlier; subsequently, it appeared in the evening until the spring of the following year when it was no longer seen.
Qing Dynasty Kangxi period ed. Anhui [*Zhaoxian zhi*] ch. 4

AD 1664 Oct 19–Nov 17 [China]

(a) 3rd year of the Kangxi reign period of Emperor Shengzu of the Qing Dynasty, 9th month; a broom star appeared in YI [LM 27] and ZHEN [LM 28] and reached 17 lunar mansions, until the 11th month when it stopped.
Qing Dynasty Yongzheng period ed. Hubei [*Mianyangzhou zhi*] ch. 13

(b) 3rd year of the Kangxi reign period of Emperor Shengzu of the Qing Dynasty, 9th month; a broom star appeared in the space allocated to YI [LM 27] and ZHEN [LM 28] its tail toward the northwest; after several days it again headed northeastward when it was over 3 *zhang* long and dusky white; it was not extinguished for over 50 days.
Qing Dynasty Jiaqing period ed. Hunan [*Chenzhou zong zhi*] ch. 41

AD 1664 Nov 18 [China] (Ho & Ang 83–84)

(a) 3rd year of the Kangxi reign period of Emperor Shengzu of the Qing Dynasty, 10th month, new moon day *jiwei* [56]; a broom star appeared in the space of lunar mansion Zhen [LM 28]; on day *dinghai* [24], the broom star appeared in the space of lunar mansion Yi [LM 27], pointing northeast; on day *dingyou* [34] in the 11th month, the broom star trespassed on lunar mansion Zhang [LM 26]; on day *gengzi* [37], the broom star trespassed on lunar mansion Jing [LM 22]; on day *bingwu* [43], the broom star trespassed on lunar mansion Wei [LM 17], its tail pointing northwest; on day *renxu* [59] in the 12th month, the broom star was in the space of lunar mansion Kui [LM 15], its size gradually diminishing; on day *dingmao* [4], Supervising Secretary of the Office of Scrutiny for Justice, Yang Yongjian, memorialized, "In the first 10-day week of the 10th month of the present year a broom star appeared. To date it has passed 13 lunar mansions in over 50 days"...on day *jisi* [6] in the 2nd month of the 4th year, the broom star appeared in the space of lunar mansion Nü [LM 10]; on day *yiyou* [22], the broom star was at 1 *du* and 30 *fen* in lunar mansion Bi [LM 14], its tail over 5 *chi* long; on day *jichou* [26] in the 3rd month, the broom star moved and entered lunar mansion Kui [LM 15].

[*Da Qing Shengzu renhuangdi shilu*] ch. 13

(b) 3rd year of the Kangxi reign period of Emperor Shengzu of the Qing Dynasty, 10th month, new moon day *jiwei* [56]; there was a fluffy star in the east in Zhen [LM 28]; on day *dingmao* [4], its tail was 7 to 8 *cun* long, grayish-white, and pointing southwest; on day *dinghai* [24], its tail was over 3 *chi* long and pointing northwest as it retrograded as far as Yi [LM 27]; on day *wuxu* [35] in the 11th month, its tail was over 5 *chi* long and pointing north when it reached Zhang [LM 26]; on day *gengzi* [37], it was in Jing [LM 22]; on day *guimao* [40], it traveled northwestward, reaching Mao [LM 18]; on day *yisi* [42], its tail pointed northeast as it reached Wei [LM 17]; on day *gengxu* [47], it reached Lou [LM 16], its tail pointing east and grayish in color; on day *renxu* [59] in the 12th month, when it reached Kui [LM 15], its body was smaller and its tail over 2 *chi* long; on day *jisi* [6] in the 2nd month of the 4th year, an anomalous star appeared in the southeast in NÜ [LM 10]; on day *jiaxu* [11], its tail was 7 *cun* long, pointing southwest and grayish-white in color; on day *dingchou* [14], its tail was over 1 *chi* long, as it advanced northeastward to XU [LM 11]; on day *xinsi* [18], it reached SHI [LM 13], its body gradually increasing in size, and its tail over 80 *chi* long; on day *yiyou* [22], it reached BI [LM 14], its tail over 5 *chi* long. [*Qing shi gao: tianwen zhi*] ch. 39

(c) 3rd year of the Kangxi reign period of Emperor Shengzu of the Qing Dynasty, 10th month, new moon day *jiwei* [56]; a broom star with a tiny body appeared in the southeast, next to YOUXIA in lunar mansion ZHEN [LM 28]; on day *dingmao* [4], it appeared in the east, its tail over 7 *cun* long, grayish in color, and pointing southwest; on day *dinghai* [24], it retrograded southwestward into lunar mansion YI [LM 27], its body gradually increasing in size, its tail over 3 *chi* long and pointing northwest; on day *wuxu* [35] in the 11th month, it was in lunar mansion ZHANG [LM 26], its tail over 5 *chi* long and pointing northward; on day *gengzi* [37], it was in lunar mansion JING [LM 22]; on day *guimao* [40], it retrograded northwestward into lunar mansion MAO [LM 18]; on day *yisi* [42], it was in lunar mansion WEI [LM 17], its tail pointing east; on day *gengxu* [47], it was in lunar mansion LOU [LM 16]; on day *renxu* [59] in the 12th month, the broom star moved into lunar mansion KUI [LM 15], its body gradually growing smaller; on day *guisi* [30] in the 1st month of the 4th year, it was no longer seen; on day *jisi* [6] in 2nd month, it reappeared in lunar mansion NÜ [LM 10] in the southeast, its body having gradually diminished in size; on day *jiaxu* [11], it appeared in the northeast, its tail over 7 *chi* long, grayish-white, and pointing southwest; on day *dingchou* [14], it was in lunar mansion XU [LM 11], its tail over 1 *chi* long; on day *xinsi* [18], it was in lunar mansion SHI [LM 13], its body having gradually increased in size and its tail over 8 *chi* long;

on day *yiyou* [22], it was in lunar mansion Bi [LM 14], its tail over 5 *chi* long.
[*Qing ding huang chao wenxian tongkao*: *xiangwei*] ch. 267

AD 1664 Nov 26–Feb 20 [Korea]

(a) 5th year of King Heonjong of the Joseon Dynasty, 10th month, day *dingmao* [4], at night; there was a star that was seen on the east side of lunar mansion Zhen [LM 28], its tail pointed obliquely, and its color was white, so it was a broom star…on the night of day *yiyou* [22] (Dec 14), the broom star was seen west of the Youxia star, its body gradually growing larger…in the 11th month, on the night of new moon day *wuzi* [25] (Dec 17), the broom star was seen below Yi [LM 27]…on the night of *wuxu* [35] (Dec 27), the broom star was seen east of the Hu star and below the Waichu star…in the 12th month, on the night of day *wuwu* [55] (Jan 16), the broom star was seen within the space of lunar mansion Lou [LM 16] and above Tianqun…on the night of day *yihai* [12] (Feb 2), the broom star was seen within the space of lunar mansion Kui [LM 15]…in the 6th year, 1st month, day *guisi* [30] (Feb 20), on this night the broom star was extinguished.
[*Joseon wangjo sillok: Heonjong*] ch. 9

(b) 5th year *jiachen* of King Heonjong of the Joseon Dynasty, 10th month, day *dingmao* [4]; a broom star was seen at Zhen [LM 28], seemingly with a tail; it traversed nine lunar mansions and reached Lou [LM 16]; by the 1st month of the following year it was extinguished.
[*Jeungbo munheon bigo*] ch. 6

AD 1664 Nov 27–AD 1665 Feb 26 [Japan]

(a) 4th year of the Kambun reign period, 10th month, 10th day (Nov 27); a broom star was seen starting from the left linchpin of Zhen [LM 28] and pointing west, its white vapor over 2 *jian* [= 6’]

long; it lasted until the end of the 1st month of the next year (Jan 28–Feb 26) before disappearing.
[*Yamaga sokō kafu nenpu*]

(b) 4th year of the Kambun reign period, 11th month, 29th day (Dec 16); after having first retired at midnight, before the hour of *yin* [LT 3–5 h] I went out of the inner court and looked in the distance, and in the southeast there was a strange star like a broom star. Its rays were bluish-red and several paces long, pointing toward the northwest. The rays were narrow at the base and large at the tip, where they divided it was as if sinewy; certainly this is an anomaly, perhaps a broom star? Prior to this, those who saw it said that, during the first third of the month, this star was in the space of lunar mansion SHEN [LM 21]. In a matter of days it retreated through several lodges, then from the hour of *chou* [LT = 1–3 h] its brightness was seen to fade, then to increase again by the hour; after the hour of *yin* [LT = 3–5 h] its brightness intensified greatly, then from the middle of the hour of *yin* [LT = 3–5 h] its brightness again faded.
[*Kokushikan nichiroku*]

AD 1665 Mar 13 [Japan]

5th year of the Kambun reign period, 1st month, 27th day *bingyin* [3]; in the west there was a white vapor like a pillar about 2 *zhang* long; it was seen nightly from right after sunset until the hour of *xu* [LT = 19–21 h] when it dissipated, and lasted until the first 10-day week of the 2nd month when it was extinguished.
[*Zoku shigushō*]

AD 1665 Apr 1 [Japan]

5th year of the Kambun reign period, 2nd month, 16th day *guiyou* [10]; for days a broom star was seen at azimuth *yin* [ENE] (dawn observation) until the first 10-day week of the 3rd month when it was extinguished.
[*Zoku shigushō*]

AD 1667 Mar 11 [China]

6th year of the Kangxi reign period of Emperor Shengzu of the Qing Dynasty, 2nd month, 17th day; a broom star appeared between Jiao [LM 1] and Kang [LM 2].

Qing Dynasty Yongzheng period ed. Guangxi [*Qinzhou zhi*] ch. 1

AD 1668 Mar 3–12 [China] (Ho & Ang 85)

7th year of the Kangxi reign period of Emperor Shengzu of the Qing Dynasty, 1st month, last 10-day week; a broom star like a bolt of cloth lit up the sky. It appeared in lunar mansion Zhen [LM 28], then for over 3 months it passed through 15 lunar mansions before disappearing.

Qing Dynasty Kangxi period ed. Shandong [*Huangxian zhi*] ch. 7

AD 1669 Feb 1–AD 1670 Jan 20 [China]

8th year of the Kangxi reign period of Emperor Shengzu of the Qing Dynasty; the rays of a broom star entered the mouth of Dou.

Qing Dynasty Kangxi period ed. Shaanxi [*Hannanjun zhi*] ch. 2

AD 1670 Jan 21–AD 1671 Feb 8 [China]

(a) 9th year of the Kangxi reign period of Emperor Shengzu of the Qing Dynasty; a broom star spitting out rays stretched across the sky and entered the mouth of Dou.

Qing Dynasty Kangxi period ed. Shaanxi [*Yangxian zhi*] ch. 1

(b) 9th year of the Kangxi reign period of Emperor Shengzu of the Qing Dynasty; a broom star over 1 *zhang* long appeared for several months before disappearing.

Qing Dynasty Qianlong period ed. Shaanxi [*Changzhixian zhi*] ch. 21

AD 1673 May 9 [China] (Ho & Ang 86)

12th year of the Kangxi reign period of Emperor Shengzu of the Qing Dynasty, 3rd month, day *guisi* [30]; an anomalous white star, as big as a walnut and with a tail over 1 *chi* long pointing east, appeared in Lou [LM 16]; on day *jiawu* [31], it still appeared.

[*Qing shi gao: tianwen zhi*] ch. 39

NB: The original has "2nd month," but the 2nd month had no day *guisi* [30]; if it was the 3rd month, then it was May 9.

AD 1676 Nov 15 [Japan]

4th year of the Empō reign period, 10th month, 10th day; a broom star was seen at the left linchpin of ZHEN [LM 28], its vapor gradually reaching 2 to 3 *chi* in length.
[*Yamaga sokō kafu nenpu*]

AD 1677 Apr 28 [China]

16th year of the Kangxi reign period of Emperor Shengzu of the Qing Dynasty, 3rd month, day *guimao* [40]; an anomalous star appeared in the northeast in LOU [LM 16], its body glossily radiant and its tail over 1 *chi* long and pointing southwest; the prognostication called it a HANYU.
[*Qing shi gao: tianwen zhi*] ch. 39

AD 1678 Jan 23–AD 1679 Feb 10 [China]

17th year of the Kangxi reign period of Emperor Shengzu of the Qing Dynasty; in the southwest a single star, bearing a white vapor above over 1 *zhang* long like the sweep of a broom, charged into BEIDOU; after over a month it disappeared.
Qing Dynasty Shunzhi period ed. Shandong [*Dengzhoufu zhi*] ch. 1

AD 1679 Sep 2 [China]

(a) 18th year of the Kangxi reign period of Emperor Shengzu of the Qing Dynasty, 7th month, 28th day; a star appeared in the west with a white vapor that stretched across the sky; it passed through WEI [LM 6] and JI [LM 7] as far as NIU [LM 9] and NÜ [LM 10]; after over a month it disappeared.
Qing Dynasty Kangxi period ed. Anhui [*Tongchengxian zhi*] ch. 1

(b) 18th year of the Kangxi reign period of Emperor Shengzu of the Qing Dynasty, 7th month, 28th day; a broom star appeared.
Qing Dynasty Daoguang period ed. Hebei [*Baoanzhou zhi*] ch. 1

AD 1679 Dec 3–AD 1680 Jan 1 [China]

18th year of the Kangxi reign period of Emperor Shengzu of the Qing Dynasty, 11th month; in the west there was a star like a broom, a small white star with a black halo, with vapors spurting out from inside; it swept as far as KUI [LM 15] and LOU [LM16], lasting for a month before being extinguished.

Qing Dynasty Kangxi period ed. Guangdong [*Yangchunxian zhi*] ch. 1

AD 1680 Nov 14–AD 1681 Jan 19 [Japan]

(a) 8th year of the Empō reign period, 9th month, 23rd day (Nov 14); at night a broom star was seen in the southwest through the first 10-day week of the 12th month when it was no longer seen.
[*Tokugawa jiki*] *Jōken-inden gojikki ni*

(b) 8th year of the Empō reign period, 10th month, 9th day (Nov 29); today at dawn, past the 7th hour, a broom star was seen to rise in the east in company with JIAO [LM 1] and SHANGXIANG, the vapor of its broom reaching across half the sky. Typically [a broom] of this extent precludes its being a guest star. It was seen at the start of the month, moving slowly, and after 15 days, it had descended very much eastward, the vapor it was emitting also becoming very tenuous.
[*Yamaga sokō kafu nenpu*]

(c) 8th year of the Empō reign period, 11th month, 3rd day (Dec 23); a long star was seen in the west; at first, the star was not seen, but was like a stretched white silk; by the 7th and 8th nights the star was seen, its rays reaching halfway across the sky; by degrees it was extinguished until the end of the month when it disappeared.
[*Bunrui honchō nendaiki*]

(d) 8th year of the Empō reign period, 11th month (AD 1680 Dec 21–AD 1681 Jan 19); a long star was seen in the west, its rays pointing eastward, 3 *chi* wide and several tens of *zhang* long, until the 12th month when it disappeared.
[*Zoku kōnendai ryakuki*]

AD 1680 Nov 21–Dec 20 [China] (Ho & Ang 88)

(a) 19th year of the Kangxi reign period of Emperor Shengzu of the Qing Dynasty, 10th month; there was a star emitting a white gleam, so long as to stretch straight across the sky from the northwest to the southeast; at dusk it was not yet high, but by the middle of the night it had moved between DOU [LM 8] and NIU [LM 9]; it disappeared only after 2 months.
Qing Dynasty Kangxi period ed. Zhejiang [*Dinghaixian zhi*]

(b) 19th year of the Kangxi reign period of Emperor Shengzu of the Qing Dynasty, 10th month; a star like silk cloth appeared in the west; it arose in lunar mansion NÜ [LM 10] and entered XU [LM 11], lasting over 40 days.
Qing Dynasty Qianlong period ed. Anhui [*Dongliuxian zhi*] ch. 7

(c) 19th year of the Kangxi reign period of Emperor Shengzu of the Qing Dynasty, 10th month; a long star like silk cloth appeared in the west; it arose in lunar mansion NÜ [LM 10] and entered XU [LM 11], lasting over 40 days in all.
Qing Dynasty Jiaqing period ed. Jiangxi [*Jiujiangfu zhi*] ch. 30

(d) 19th year of the Kangxi reign period of Emperor Shengzu of the Qing Dynasty, 10th month; a long star with a broom 2 to 3 *zhang* long appeared in the southwest, pointing to the Milky Way in the northeast; it was not extinguished for over a month.
Qing Dynasty Kangxi period ed. [*Shanxi tong zhi*] ch. 30

AD 1680 Nov 21–Dec 21 [Korea]

(a) 6th year of King Sukjong of the Joseon Dynasty, 10th month, new moon day *bingxu* [23]; during the fifth watch of the night, a star was seen outside the west wall of TAIWEI, seemingly with a tail...on the night of day *dinghai* [24] (Nov 22), the star that was seen last night moved to a position 6 *du* in lunar mansion YI [LM 27], close by the YOUZHIFA star of TAIWEI, its body was like the great star of ZHINÜ, its color greenish-white, seemingly with a tail...on day *guisi* [30] (Nov 28), the broom star was seen above the right JIAO [LM 1] star, its appearance and color unchanged

from before, its tail almost reaching 1 *zhang* in length…in the 11th month, new moon day *bingchen* [53] (Dec 21), at dusk, a white vapor shaped like a broom emanated from the west and pointed to the zenith; it persisted for several days…the broom star was seen and disappeared, emerged and set above NIU [LM 9], DOU [LM 8], SHEN [LM 21], XU [LM 11], WEI [LM 12], SHI [LM 13], BI [LM 14], KUI [LM 15], LOU [LM 16], and WEI [LM 17].
[*Joseon wangjo sillok: Sukjong*] ch. 10

(b) 6th year *gengshen* of King Sukjong of the Joseon Dynasty, 10th month, day *bingxu* [23]; a broom star was seen outside the western wall of TAIWEI, its tail 3 to 4 *chi* long; it traversed JIAO [LM 1] and KANG [LM 2], but was not seen in DI [LM 3]. In the 11th month, on day *bingchen* [53] (Dec 21), at the beginning of dusk twilight [it was] a white vapor in the west 3 to 4 *zhang* long, the star's body faintly visible, but then it was no longer observable in the haze.
[*Jeungbo munheon bigo*] ch. 6

AD 1680 Nov 23 [China] (Ho & Ang 88)

(a) 19th year of the Kangxi reign period of Emperor Shengzu of the Qing Dynasty, 10th month, 3rd day (*wuzi* [25]), at night; in the southwest a white vapor stretched across the sky for half a month before dissipating; each night it shifted until it reached due west and then vanished.
Qing Dynasty Kangxi period ed. Zhejiang [*Linhaixian zhi*] ch. 11

(b) 19th year of the Kangxi reign period of Emperor Shengzu of the Qing Dynasty, 10th month, 3rd day (*wuzi* [25]); a broom star appeared in the space of lunar mansion YI [LM 27]; on day *bingchen* [53] in the 11th month, the broom star appeared in the evening in the west.
[*Da Qing Shengzu renhuangdi shilu*] ch. 92–93

(c) 19th year of the Kangxi reign period of Emperor Shengzu of the Qing Dynasty, 10th month, 3rd day (*wuzi* [25]); a white broom star appeared by YOUZHIFA with a tail over 1 *chi* long pointing

west and traveling eastward very fast; on day *renyin* [39], it approached the sun and was lost from sight; on new moon day *bingchen* [53] in the 11th month, the tail appeared in the evening in the west; on day *renxu* [59], the body of the star appeared, its color grayish-white, and with a tail over 6 *zhang* long, 2 *chi* wide, pointing northeast.
[*Qing shi gao: tianwen zhi*] ch. 39

AD 1680 Dec 21 [China]

(a) 19th year (*gengshen*) of the Kangxi reign period of Emperor Shengzu of the Qing Dynasty, 11th month, new moon day *bingchen* [53]; in Cangzhou there was a white vapor like the sweep of a broom from the southwest to the northeast for a whole 10-day week before being extinguished; in Lulong there was a white vapor like a cloud long enough to stretch eastward across the sky; after several evenings its color faded but its rays rose up high like a broom; in Jiangxian a white vapor like a rainbow appeared at night.
[*Qing shi gao: zaiyi si*] ch. 43

(b) 19th year (*gengshen*) of the Kangxi reign period of Emperor Shengzu of the Qing Dynasty, 11th month, new moon day *bingchen* [53]; from west to west-northwest a white vapor like a broom and shaped like a large sash stretched across the sky; from where it arose in the southwest it reached straight over to the Milky Way, gradually streaming westward until it reached the north where it transformed; on the 4th day of the month, its base was thick and its tip fine, with a yellow gleam and a greenish trace through the middle; on the 5th day of the month, the base linked up with a small dusky star; on the 9th day of the month, its color was pale white and its tip again atop the star; by the 15th day it was even thinner and smaller and about to reach the zenith, while the star was still bright; afterward it was hidden from view, repeatedly changing from long to short for over 50 days before transforming.

Qing Dynasty Kangxi period ed. Shanxi [*Wuxiangxian zhi*] ch. 4

(c) 19th year (*gengshen*) of the Kangxi reign period of Emperor Shengzu of the Qing Dynasty, 11th month, new moon day *bingchen* [53]; just after sunset, a baleful star emitting a white gleam arose in the southwest and pointed northeast; its circumference was over 1 *chi* and it stretched across the sky, its top and bottom both pointed, forceful and straight in appearance, it lasted the season before being extinguished; by the first 10-day week in the 12th month, it was it hidden from view and no longer observed. [*Sangang shi lue*] ch. 8

AD 1680 Dec 21–AD 1681 Jan 19 [China]

(a) 19th year of the Kangxi reign period of Emperor Shengzu of the Qing Dynasty, 11th month; due west there was a star spitting out white vapor that surged straight northeastward until it reached YOUYEWEI of ZIWEIYUAN; its bright rays were shaped like a bolt of silk cloth, the tail opening out several tens of *zhang*; it lasted until the spring of the following year when it began to gradually diminish and then disappeared.
Qing Dynasty Kangxi period ed. Jiangxi [*Ganzhoufu zhi*] ch. 18

(b) 19th year of the Kangxi reign period of Emperor Shengzu of the Qing Dynasty, 11th month; a white vapor surged up from the southwest to penetrate DOU [LM 8] and NIU [LM 9], the vaporous tail having a star.
Qing Dynasty Kangxi period ed. Hebei [*Yutianxian zhi*] ch. 8

(c) 19th year of the Kangxi reign period of Emperor Shengzu of the Qing Dynasty, 11th month; a broom star appeared in the southwest, its long streaks changing into a white rainbow several *zhang* long with sharp ends; on the 13th to 14th day and so forth, a single star again spit out white rays that stretched horizontally across the zenith for a long time without being extinguished.
Qing Dynasty Qianlong period ed. Guangdong [*Chaozhoufu zhi*] ch. 11

AD 1680 Dec 22 [China]

(a) 19th year of the Kangxi reign period of Emperor Shengzu of the Qing Dynasty, 11th month, 2nd day; west of Zhenyang a white vapor several *zhang* long stretched across the sky until the next season when it was extinguished; in Linzi, a white vapor appeared from the west to the east.
[*Qing shi gao: zaiyi*] ch. 43

(b) 19th year of the Kangxi reign period of Emperor Shengzu of the Qing Dynasty, 11th month, 2nd day; there was a streak of white vapor several tens of *zhang* long; it appeared in the southwest, like a bolt of white silk cloth, and penetrated straight through SANTAI until the 10th day of the 12th month when it was no longer seen.
Qing Dynasty Kangxi period ed. Sichuan [*Chengdufu zhi*] ch. 25

(c) 19th year of the Kangxi reign period of Emperor Shengzu of the Qing Dynasty, 11th month, 2nd day; a broom star appeared in NÜ [LM 10] and XU [LM 11] for over a month before being extinguished.
Qing Dynasty Kangxi period ed. Hunan [*Linglingxian zhi*] ch.14

AD 1682 Aug 24–25 [Korea]

(a) 8th year of King Sukjong of the Joseon Dynasty, 7th month, day *dingmao* [4]; there was a star like a broom that was seen at azimuth *gen* (NE) above BEIHE…on day *wuchen* [5] (Aug 25), the broom star gradually moved to south of the SHANGJIE star of SANTAI; after this it repeatedly was seen above JIAO [LM 1], KANG [LM 2] and SHETI; in the 9th year, for the first time it was not seen.
[*Joseon wangjo sillok: Sukjong*] ch. 13

(b) 8th year *renxu* of King Sukjong of the Joseon Dynasty, 7th month, day *dingmao* [4]; a broom [star] was seen in SANTAI, its tail 3 to 4 *chi* long; it was as large as the ZHINÜ star, its color white, and it traversed the YOUJIAO star as far as SHETI; not until the 9th month was it extinguished.
[*Jeungbo munheon bigo*] ch. 6

NB: This is Comet Halley, return –4.

AD 1682 Aug 26 [China] (Ho & Ang 89)

(a) 21st year of the Kangxi reign period of Emperor Shengzu of the Qing Dynasty, 7th month, day *jisi* [6]; a broom star appeared in the space of lunar mansion Jing [LM 22] with a tail over 2 *chi* long.
[*Da Qing Shengzu renhuangdi shilu*] ch. 103

(b) 21st year of the Kangxi reign period of Emperor Shengzu of the Qing Dynasty, 7th month, day *jisi* [6]; a white broom star appeared north of Beihe with a tail over 2 *chi* long, pointing southwest and moving swiftly northeastward; on day *renshen* [9], it entered zodiacal sign *wu* (Gemini) at which time its tail was over 6 *chi* long.
[*Qing shi gao: tianwenzhi*] ch. 39

(c) 21st year of the Kangxi reign period of Emperor Shengzu of the Qing Dynasty, 7th month, day *jisi* [6]; a white broom star appeared north of Beihe in lunar mansion Jing [LM 22] in the northeast, its tail over 2 *chi* long and pointing southwest; on day *renshen* [9], it traveled northeastward, its tail over 6 *chi* long.
[*Qing ding huang chao wenxian tongkao: xiangwei*] ch. 267

NB: This is Comet Halley, return –4.

AD 1682 Sep 3–21 [Japan]

(a) 2nd year of the Tenna reign period, 8th month, 2nd day (Sep 3); today at night a broom star was seen at azimuth *xu* [WNW], its rays over 1 *zhang* long; during this time, it was seen for a long time; 20th day; tonight after sunset, the broom star was seen in the west.
[*Yamaga sokō kafu nenpu*]

(b) 2nd year of the Tenna reign period, 8th month, 3rd day (Sep 4); tonight a broom star was seen in the west between azimuth *xu* [WNW] and *you* [W], its rays about 6 *chi* long. Today I left the capital to go to the palace of the Great Lord of the Left, and on the way back, I saw it from the road.
[*Gyōjo hoshinnō nikki*]

NB: This is Comet Halley, return –4.

AD 1682 Sep 9 [China]

21st year of the Kangxi reign period of Emperor Shengzu of the Qing Dynasty, 8th month, 8th day; a broom star appeared in JIAO [LM 1] and subsequently in FANG [LM 4] lasting a month before being extinguished.

Qing Dynasty Kangxi period ed. Jiangxi [*Pengzexian zhi*] ch. 14

NB: The same appears recorded as 8th month only in Qing Dynasty Qianlong period ed. Anhui [*Dongliuxian zhi*] ch. 7 and Qing Dynasty Jiaqing period ed. Jiangxi [*Jiujiangfu zhi*] ch. 30.

AD 1683 Aug 2 [China] (Ho & Ang 90)

22nd year of the Kangxi reign period of Emperor Shengzu of the Qing Dynasty, intercalary 6th month, day *gengxu* [47]; an anomalous white star appeared north of WUCHE and east of BAGU, then retrograded southwestward; on day *wuchen* [5], it entered WUCHE.

[*Qing shi gao: tianwenzhi*] ch. 39

AD 1684 Jun 26 [China] (Ho & Ang 91)

23rd year of the Kangxi reign period of Emperor Shengzu of the Qing Dynasty, 5th month, day *jiashen* [21]; an anomalous star appeared in TAIWEIYUAN within ZHEN [LM 28] in the east, large, bright and white in color it advanced northeastward; on day *yiyou* [22], it traveled over 4 *du*, reaching a point below YOUSHETI.

[*Qing shi gao: tianwenzhi*] ch. 39

NB: Ho and Ang made this July 1.

AD 1686 Aug 26 [China] (Ho & Ang 92)

25th year of the Kangxi reign period of Emperor Shengzu of the Qing Dynasty, 7th month, day *gengyin* [27]; an anomalous white star appeared in the east near the horizon moving eastward at a moderate rate; on day *dingyou* [34], it traveled a total of 16 *du* as far as LIU [LM 24], now with a trace of a tail; on day *renyin* [39], it reached XING [LM 25], then gradually disappeared.

[*Qing shi gao: tianwenzhi*] ch. 39

AD 1686 Sep 14–26 [Korea]

12th year *bingyin* of King Sukjong of the Joseon Dynasty, 7th month, day *jiyou* [46]; a broom star was seen in ZHANG [LM 26] and was not extinguished until the 8th month, day *xinyou* [58] (Sep 26).
[*Jeungbo munheon bigo*] ch. 6

AD 1689 Dec 6 [Japan]

(a) 2nd year of the Genroku reign period, 10th month, 25th day *wuzi* [25]; a strange star was seen at azimuth *xun* [SE].
[*Zoku shigushō*]

(b) 2nd year of the Genroku reign period, 10th month, 25th day; at night, a strange star was seen at azimuth *xun* [SE].
[*Bukō nempyō*]

AD 1695 after Oct 22 [China]

34th year of the Kangxi reign period of Emperor Shengzu of the Qing Dynasty, 9th month, after the 15th day; during the fifth watch, a broom star appeared, arising from the Milky Way in the eastern sky, its ghost lasting until the sky was already bright before fading from view; as of the 8th day of the 10th month, it had still not vanished.
Qing Dynasty Kangxi period ed. Shanghai [*Linian ji*]

AD 1695 Nov 1 [Japan]

8th year of the Genroku reign period, 9th month, 25th day; at dawn in the southeast, there was a white vapor with its root at ZHEN [LM 28] and its tip reaching to Yi [LM 27]. In the second 10-day week of the 10th month it disappeared; this was in the astral field allocated to Suruga [Shizuoka].
[*Jinki roku*]

AD 1701 Mar 28–Apr 10 [Korea]

27th year *xinsi* of King Sukjong of the Joseon Dynasty, 2nd month, day *dingchou* [14]; a broom star was seen in WEI [LM 12] and SHI [LM 13], its tail over 2 *zhang* long, its body was like the big star of

WUCHE and its color white; it moved and was seen in KUI [LM 15], until the 4th day of the 3rd month (Apr 10) when it was no longer seen. [*Jeungbo munheon bigo*] ch. 6

AD 1701 Oct 28–Nov 15 [Korea]

(a) 27th year of King Sukjong of the Joseon Dynasty, 9th month, day *xinhai* [48]; [a broom star] was seen outside the DUANMEN of TAIWEI, passed by YI [LM 27] and reached ZHANG [LM 26], its tail 2 to 3 *chi* long; by the 10th month, day *jisi* [6] (Nov 15), it was no longer seen.
[*Jeungbo munheon bigo*] ch. 6

(b) 27th year of King Sukjong of the Joseon Dynasty, 9th month, day *xinhai* [48], at night; a star was seen outside the DUANMEN of TAIWEIYUAN, seemingly with a tail…day *renzi* [49] (Oct 29), the body of the star that was seen outside the DUANMEN of TAIWEIYUAN is like the big star of KUI [LM 15] (β And), its tail now somewhat longer; it seems to be a broom star…day *guichou* [50] (Oct 30), at night, the broom star was seen at 16 *du* in lunar mansion YI [LM 27], outside the DUANMEN of TAIWEIYUAN, at a polar distance of 93 *du*, its color greenish-white, and about 2 to 3 *chi* long, its body unchanged compared to yesterday…10th month, day *renxu* [59] (Nov 8), the broom star was seen within lunar mansion ZHANG [LM 26], its body somewhat fainter than yesterday, its tail almost gone…day *jiazi* [1], the broom star was seen within lunar mansion ZHANG [LM 26], its body even fainter than yesterday.
[*Joseon wangjo sillok: Sukjong*] ch. 35

AD 1702 Feb 25–Mar 8 [Japan]

(a) 15th year of the Genroku reign period, 1st month, 29th day (Feb 25); from this evening, a broom star was seen in the west.
[*Arai hakuseki nikki*] *jō*

(b) 15th year of the Genroku reign period, last quarter of the 1st month to 1st quarter of the 2nd month (Feb 17–Mar 8); at the beginning of dusk at azimuth *shen* [WSW]–*you* [W], there was

a white vapor. Its root emerged at the 14th *du* of Kui [LM 15] within Tiancang, and its tip reached the initial *du* of Mao [LM 18] within Tianyuan; the astral field [allocated to] the Chūgoku and Saigoku regions.
[*Jinki roku*]

AD 1702 Feb 28 [China]
41st year of the Kangxi reign period of Emperor Shengzu of the Qing Dynasty, 2nd month, day *jiayin* [51]; during the hour of *you* [LT 17–19 h], in the southwest there was a streak of white cloud over 2 *zhang* long and over 1 *chi* wide; it pierced Tiancang and Tianyuan and entered the horizon; by day *dingsi* [54], it was over 3 *zhang* long.
[*Qing shi gao: tianwenzhi*] ch. 39

AD 1723 Oct 11 [China] (Ho & Ang 95)
1st year of the Yongzheng reign period of Emperor Shizong of the Qing Dynasty, 9th month, day jichou [26]; an anomalous white star with a tiny body and rays over 1 *chi* long pointing northwest appeared below Hushi, then retrograded as far as Jing [LM22].
[*Qing shi gao: tianwenzhi*] ch. 39

AD1723 Oct 16 [China]
1st year of the Yongzheng reign period of Emperor Shizong of the Qing Dynasty, 9th month, 18th day; a broom star emerged between Niu [LM 9] and Nü [LM 10], lasting until the 24th day when it was extinguished.
Qing Dynasty Yongzheng period ed. Guangxi [*Cangwu zhi*] ch. 4

AD 1723 Oct 19–Nov 14 [Korea]
(a) 3rd year *guimao* of King Gyoengjong of the Joseon Dynasty, 9th month, day *dingyou* [34]; a broom [star] was seen in Nü [LM 10], seemingly with a tail; it moved and was seen in Niu [LM 9], and by the 17th day of the 10th month (Nov 14) it was no longer seen.
[*Jeungbo munheon bigo*] ch. 6

(b) 3rd year of King Kyŏngjong of the Joseon Dynasty, 9th month, day *dingyou* [34]; during the first watch of the night, a white broom star with a tail was seen within the space of lunar mansion NÜ [LM 10]; after 15 days, it was extinguished.
[*Joseon wangjo sillok: Gyoengjong*] ch. 13

AD 1737 Feb 7–17 [Japan]

(a) 2nd year of the Gembun reign period, 1st month, 8th day (Feb 7); this evening a broom star was seen in the west, its length reaching straight to below azimuth *geng* [W by S]; it was seen for over 20 days.
[*Shishien manpitsu*]

(b) 2nd year of the Gembun reign period, 1st month, 18th day (Feb 17); a fuzzy star was seen in the western sky until the third 10-day week of the month.
[*Ichiwa ichigen*]

AD 1737 Feb 13–Mar 3 [Korea]

(a) 13th year *dingsi* of King Yeongjo of the Joseon Dynasty, 1st month, day *guimao* [40] (Feb 13); a white broom [star] was seen in WEI [LM 12], its tail over 3 *chi* long and its body larger than the big star of KUI [LM 15] (β And); it traversed SHI [LM 13], BI [LM 14], and KUI [LM 15], reaching the west of TIANCANG; through the 2nd month, day *renxu* [59] (Mar 3) it gradually dimmed and was no longer seen.
[*Jeungbo munheon bigo*] ch. 6

(b) 13th year of King Yeongjo of the Joseon Dynasty, 1st month, day *yisi* [42] (Feb 15); a white broom [star] was seen within the space of lunar mansion WEI [LM 12], its tail over 1 *chi* long, and pointing toward azimuth *xun* (SE)…on day *bingwu* [43] (Feb 16), the white broom star moved to appear within the space of lunar mansion SHI [LM 13], its tail 2 *chi* long pointing toward azimuth *xun* (SE)…on day *renzi* [49] (Feb 21), the white broom star moved

and was seen within the space of lunar mansion Bi [LM 14], above Tiezhi, its tail over 3 *chi* long, and pointing at azimuth *xun* (SE)...on day *bingchen* [53] (Feb 25), the white broom star moved and was seen within the space of lunar mansion Kui [LM 15], west of Tiancang, its tail over 2 *chi* long, and pointing at azimuth *xun* (SE)...on day *renxu* [59] (Mar 3), the broom star was seen within the space of lunar mansion Kui [LM 15], to the west of Tiancang, its tail 3 to 4 *cun* long, and pointing at azimuth *xun* (SE), its color white.
[*Joseon wangjo sillok: Yeongjo*] ch. 43

AD 1737 Jul 8 [China] (Ho & Ang 96)
2nd year of the Qianlong reign period of Emperor Gaozong of the Qing Dynasty, 6th month, day *dingmao* [4]; an anomalous white star emerged east of Yougeng, within Lou [LM 16], and proceeded southwestward; on day *bingzi* [13] (Jul 17), it was still visible.
[*Qing shi gao: tianwenzhi*] ch. 39
NB: Ho and Ang make it July 7.

AD 1741 Mar 17–Apr 15 [China]
6th year of the Qianlong reign period of Emperor Gaozong of the Qing Dynasty, 2nd month; a broom star over 1 *zhang* long appeared in the west; in the 7th month, it reappeared and lasted until the 12th month before being extinguished.
Qing Dynasty Guangxu period ed. Shandong [*Dengzhoufu zhi*] ch. 23

AD 1741 May 15–Jun 12 [China]
6th year of the Qianlong reign period of Emperor Gaozong of the Qing Dynasty, 4th month; in the northwest there was a star that spit out a brightness like a broom; in the 9th month, it moved nearer the southwest, lasting until winter before disappearing.
Qing Dynasty Qianlong period ed. Shanghai [*Xu Songnan zhi*] *zaixiang*

AD 1742 Jan 24 [China]

6th year of the Qianlong reign period of Emperor Gaozong of the Qing Dynasty, 12th month, 18th day; there was a star like a broom with horned rays that arose in the northwest corner; it surged upward into the "tail" of SHI [LM 13] and lasted until the beginning of the 2nd month of the following year before ending.

Qing Dynasty Qianlong period ed. Hunan [*Shanhuaxian zhi*] ch. 17

AD 1742 Feb 26–Mar 18 [Japan]

(a) 2nd year of the Kampō reign period, 1st month, day *renwu* [19] (Feb 26); until today (12th day of the 2nd month) (Mar 18), a strange star has been seen at azimuth *gen* [NE], its rayed vapor about 3 *chi* long and pointing toward azimuth *kun* [SW].
[*Taishi kyokusō*]

(b) 2nd year of the Kampō reign period, 1st month (Feb 5–Mar 6); a broom star was seen south of and reaching to HEGU.
[*Shishien manpitsu*]

(c) 2nd year of the Kampō reign period; from the 1st month, 22nd to 23rd days on, a broom star has been seen at midnight in the northeast; by the 2nd month, it faded, but in the second 10-day week of the month, it was seen again.
[*Bunkai zakki*]

AD 1742 Mar 2 [China] (Ho & Ang 97)

7th year of the Qianlong reign period of Emperor Gaozong of the Qing Dynasty, 1st month, day *bingxu* [23]; an anomalous yellowish-white star appeared in the space of lunar mansion DOU [LM 8], above the second star of TIANBIAN, then retrograded northwestward for over 40 days before being hidden from view and setting.

[*Qing chao tong zhi*: *zaixiang*] ch. 122

AD 1742 Mar 3 [China] (Ho & Ang 97)

7th year of the Qianlong reign period of Emperor Gaozong of the Qing Dynasty, 1st month, day *dinghai* [24]; an anomalous star appeared in

the southeast; on day *wuzi* [25], it was at an altitude of over 27 *du*, as big as a crossbow bolt, yellowish, and with a tail over 4 *chi* long pointing southwest, in zodiac sign *chou* (Capricorn) and outside the Xu star of Tianshiyuan; it retrograded for over four 10-day weeks before disappearing.
[*Qing shi gao: tianwenzhi*] ch. 39

AD 1742 Mar 6–Apr 3 [Korea]

(a) 18th year *renxu* of King Yeongjo of the Joseon Dynasty, 1st month, day *gengyin* [27]; a white broom [star] was seen in Dou [LM 8], its tail over 2 to 3 *chi* long and its body as large as the middle star of Hegu; it traversed Niu [LM 9], Nü [LM 10], and reached Ziwei, entering within the wall; by the 2nd month, day *renzi* [49] (Mar 18) it was extinguished.
[*Jeungbo munheon bigo*] ch. 6

(b) 18th year of King Yongjong of the Joseon Dynasty, 1st month, day *gengyin* [27]; a guest star was seen within the space of Dou [LM 8]...in the 2nd month, day *xinmao* [28] (Mar 7), the guest star moved and was seen in Niu [LM 9], its body resembling a broom [star], with a tail 2 *chi* long, and white in color...on day *renchen* [29] (Mar 8), at night, the broom star was seen within the space of lunar mansion Niu [LM 9] in the east, its tail was 2 to 3 *chi* long, its color white, and its body like Hegu...on day *guisi* [30] (Mar 9), the broom star moved and was seen within the space of lunar mansion Nü [LM 10], its body as large as Qian-niu [LM 9] (i.e., β Cap), 2 to 3 *chi* long, and white in color...on day *dingyou* [34] (Mar 13), at night, the broom star was seen at azimuth *gen* (NE), during the third watch of the night, it moved to a position outside the eastern wall of Ziwei, its color white... on day *wushen* [45] (Mar 24), at night, the broom star moved and was seen within Ziweiyuan, its tail 2 *chi* long, and its color white...on day *dingsi* [54] (Apr 3) the broom star was nearly extinguished.
[*Joseon wangjo sillok: Yeongjo*] ch. 55

AD 1742 Dec 27–AD 1743 Jan 25 [China]

(a) 7th year of the Qianlong reign period of Emperor Gaozong of the Qing Dynasty, 12th month; a broom star emerged in lunar mansion KUI [LM 15], its brightness in excess of 1 *chi*, until the 1st month of the next year when it was extinguished.
Qing Dynasty Qianlong period ed. Shandong [*Huiminxian zhi*] ch. 3

(b) 7th year of the Qianlong reign period of Emperor Gaozong of the Qing Dynasty, 12th month; a broom star emerged in lunar mansion KUI [LM 15], its brightness in excess of 1 *chi*, until the 1st month of the next year when it was extinguished. The broom was about 1 *chi* long and emerged in the evening pointing northeast; after some time it emerged at dawn, pointing southwest, in all for over 50 days.
Qing Dynasty Guangxu period ed. Shandong [*Zhanhuaxian zhi*] ch. 4

(c) 7th year of the Qianlong reign period of Emperor Gaozong of the Qing Dynasty, 12th month; a broom star appeared [extending] from BI [LM 14] and SHI [LM 13], its tail reaching KUI [LM 15] and LOU [LM 16], and lasting until the 1st month of spring in the 8th year *guihai*, when it disappeared.
Qing Dynasty Daoguang period ed. Jiangxi [*Yuduxian zhi*] ch. 27

AD 1743 Nov 16–Dec 15 [China]

(a) 8th year of the Qianlong reign period of Emperor Gaozong of the Qing Dynasty, 10th month; a broom star over 1 *chi* long appeared in the northwest, tracking through the spaces of WEI [LM 12] and SHI [LM 13], its bright rays pointing obliquely northeast for over 2 months before being extinguished.
Qing Dynasty Qianlong period ed. Guangdong [*Chaozhoufu zhi*] ch. 11

(b) 8th year of the Qianlong reign period of Emperor Gaozong of the Qing Dynasty, 10th month; a broom star appeared in the west until the 1st month of (year) *jiazi*, when it disappeared.
Qing Dynasty Tongzhi period ed. Jiangxi [*Anrenxian zhi*] ch. 34

AD 1743 autumn [China]

(a) 8th year of the Qianlong reign period of Emperor Gaozong of the Qing Dynasty, autumn; a star appeared in Shi [LM 13] and Bi [LM 14], its brightness surging to a length of 4 to 5 *zhang*, from the west toward the east; in the 3rd month it stopped.
Qing Dynasty Qianlong period ed. Jiangsu [*Wuxian zhi*] ch. 26

(b) 8th year of the Qianlong reign period of Emperor Gaozong of the Qing Dynasty, autumn; a white star over 3 *zhang* long appeared in the west and lasted over a month before disappearing.
Qing Dynasty Qianlong period ed. Shanxi [*Fushanxian zhi*] ch. 34

(c) 8th year of the Qianlong reign period of Emperor Gaozong of the Qing Dynasty, 11th month, day *renxu* [59]; a broom star appeared between Nü [LM 10] and Xu [LM 11] and passed through Wei [LM 12]; it rose up to sweep Shi [LM 13] and Bi [LM 14], lasting 50 days before disappearing.
Qing Dynasty Qianlong period ed. Shandong [*Yexian zhi*] ch. 5

NB: There was no day *renxu* [59] in the 11th month.

AD 1743 Dec 20–AD 1744 Jan 14 [Japan]

(a) 3rd year of the Kampō reign period, 11th month, 5th day *jiashen* [21] (Dec 20); for days a broom star has been seen in the west until the second 10-day week of the 1st month of the following year when it was extinguished.
[*Zoku shigushō*]

(b) 3rd year of the Kampō reign period, 11th month (1743 Dec 16–1744 Jan 14); a broom star was seen in Dongbi [LM 14] its rays pointing obliquely toward Kui [LM 15]; in the 2nd month it diminished slightly; by the 18th day of the 12th month, it was again very bright and nearly twice as long, lasting throughout the evening; by the night of the new moon day of the 1st month of the next year it again increased in brightness, growing larger and longer. After the 15th of the 1st month, it was lost and not seen again. When first seen, it was between the two stars of Dongbi [LM 14]; afterward, it moved slightly northwest toward the lower

star of YINGSHI [LM 13]. Finally, it was close by the lower star of YINGSHI [LM 13], its brightness pointing toward the upper star of DONGBI [LM 14]; in all it was seen for about 50 days before it went out of sight.
[*Shishien manpitsu*]

AD 1744 Jan 1–5 [China] (Ho & Ang 98)

(a) 8th year of the Qianlong reign period of Emperor Gaozong of the Qing Dynasty, 11th month, 17th day (Jan 1); at the beginning of twilight a broom star appeared between lunar mansions KUI [LM 15] and BI [LM 14], until the end of the 1st month of the next year when it appeared at dawn in the east.
Minguo [*Jiangsu sheng tong zhi gao*] ch. 3

(b) 8th year of the Qianlong reign period of Emperor Gaozong of the Qing Dynasty, 11th month, day *jihai* [36] (Jan 4); a yellowish-white broom star as big as a crossbow bolt, and with a tail over 1 *chi* long pointing east, appeared between KUI [LM 15] and BI [LM 14] in zodiacal sign *xu* (Libra); it retrograded until day *xin-mao* [28] in the 1st month of the 9th year, in all for 53 days, during which it traveled over 29 *du*.
[*Qing shi gao: tianwenzhi*] ch. 39

(c) 8th year of the Qianlong reign period of Emperor Gaozong of the Qing Dynasty, 11th month, day *jihai* [36] (Jan 4); a yellowish broom star as big as a crossbow bolt, and with a tail over 1 *chi* long, appeared at a point 2 *du* from the second star of lunar mansion KUI [LM 15]; each night it retrograded while pointing east.
[*Qin ding huang chao wenxian tongkao:*□*xiangwei*] ch. 267

(d) 8th year of the Qianlong reign period of Emperor Gaozong of the Qing Dynasty, 11th month, day *jihai* [36] (Jan 4); a yellowish broom star with a body like a crossbow bolt, and with a tail over 1 *chi* long, appeared between KUI [LM 15] and BI [LM 14] at a point 2 *du* from the second star of lunar mansion KUI [LM 15]; each night it retrograded westward, from zodiac sign *xu* (Libra) to *hai* (Scorpio).
[*Qing chao tong zhi*: *zaixiang*] ch. 122

(e) 8th year of the Qianlong reign period of Emperor Gaozong of the Qing Dynasty, 11th month, 21st day (Jan 5), at night; a broom star appeared in Bi [LM 14].
Qing Dynasty Qianlong period ed. Guangdong [*Jieyangxian zhi*] ch. 7

AD 1744 Jan 6–Mar 6 [Korea]

(a) 19th year *guihai* of King Yeongjo of the Joseon Dynasty, 11th month, day *xinchou* [38]; a broom [star] was seen in Bi [LM 14], moved to Shi [LM 13], and reached Wei [LM 12], its tail was over 2 *zhang* long, and its color pale red, its body was as big as Venus; through the 1st month of the next year, on day *xinmao* [28] (Feb 25), in contrast, when it was seen in the morning twilight in the east but was overcome by the light of dawn and could not be observed further; by day *gengzi* [37] (Mar 6), it was extinguished. [*Jeungbo munheon bigo*] ch. 6

(b) 19th year of King Yongjong of the Joseon Dynasty, 11th month, day *xinchou* [38]; a guest star was seen within the space of lunar mansion Bi [LM 14], in shape it resembled a broom [star]...12th month, day *xinhai* [48] (Jan16), the broom star moved and was seen between lunar mansions Bi [LM 14] and Kui [LM 15], its color pale red, its tail 4 to 5 *chi* long, and pointing toward azimuth *gen* (NE)...on day *guichou* [50] (Jan 18), the broom star moved and was seen within the space of lunar mansion Shi [LM 13]...in the 20th year, 1st month, day *gengchen* [17] (Feb 14), at night, the broom star was seen within the space of lunar mansion Shi [LM 13], to the west of lunar mansion Bi [LM 14]; it was as big as the Tianlang star, its tail over 2 *zhang* long, its color pale red, and pointing at azimuth *gen* (NE)...on day *bingxu* [23] (Feb 20), the broom star moved and was seen west of lunar mansion Shi [LM 13] within the space of lunar mansion Wei [LM 12], its body somewhat larger than the Tianlang star, its tail over 2 *zhang* long, its color pale red, and pointing toward azimuth *gen* (NE)...on day *xinmao* [28] (Feb 25), at morning twilight, the broom star was seen on the horizon in the east, its shape, color and tail all within

range of the light of dawn and unobservable…on day *gengzi* [37] (Mar 6), the broom star was dimly visible at sunrise, then its location was lost.
[*Joseon wangjo sillok: Yeongjo*] ch. 58–59

AD 1744 Feb 6 [China]

(a) 8th year of the Qianlong reign period of Emperor Gaozong of the Qing Dynasty, 12th month, 23rd day; within the space of lunar mansion KUI [LM 15] there was a yellowish-red star, at the root as big as a small wine cup and the tip 6 to 7 *cun* long and as thick as a writing brush tube; from the 24th to the 28th the weather was overcast and rainy every night so it was too dark to see; on the night of the 29th it appeared again, having entered the space of lunar mansion BI [LM 14], its tip sweeping to eastward, as big as a broom and over 1 *zhang* long; in the 1st month of the next year, just before dawn, there were still those who saw it.
Qing Dynasty Qianlong period ed. Jiangxi [*Huichangxian zhi gao*] ch. 34

(b) 8th year of the Qianlong reign period of Emperor Gaozong of the Qing Dynasty, 12th month, 23rd day, at night; there was a star with a yellowish-red gleam within the space of lunar mansion KUI [LM 15], at the root 1 *cun* in diameter and the tip as small as a finger and about 6 to 7 *cun* long and as big as a broom; from the 24th to the 28th, the weather was overcast and rainy every night so it was too dark to see, but on the night of the 29th, it was observed again, now having entered the space of lunar mansion BI [LM 14], its tip sweeping to eastward, as big as a broom and over 1 *zhang* long; in the 1st month of the next year, just before dawn, there were still those who saw it.
Qing Dynasty Daoguang period ed. Jiangxi [*Huichangxian zhi*] ch. 27

AD 1744 Nov 17–18 [Korea]

(a) 20th year of King Yeongjo of the Joseon Dynasty, 10th month, day *dingsi* [54], at night; a guest star was seen within the space

of lunar mansion Jiao [LM 1]...on day *wuwu* [55] (Nov 18), at night, the guest star was seen with a tail within the space of lunar mansion Jiao [LM 1].
[*Joseon wangjo sillok: Yeongjo*] ch. 60

(b) 20th year *jiazi* of King Yeongjo of the Joseon Dynasty, 10th month, day *dingsi* [54]; a star was seen in Jiao [LM 1], seemingly with a tail, 2 days later it was not seen.
[*Jeungbo munheon bigo*] ch. 6

AD 1747 Feb 16 [China]

12th year of the Qianlong reign period of Emperor Gaozong of the Qing Dynasty, 1st month, 8th day; during the hour of *xu* [LT 19–21 h], a broom star appeared with bright rays like the sweep of a broom and over 1 *zhang* long; within a single month, it appeared seven times.
Qing Dynasty Qianlong period ed. Sichuan [*Jiang'anxian zhi*] ch. 3

AD 1747 winter [China]

(a) 12th year of the Qianlong reign period of Emperor Gaozong of the Qing Dynasty, winter; a star appeared in the west with bright rays several tens of *zhang* long and shaped like the sweep of a broom.
Qing Dynasty Qianlong period ed. Guangxi [*Zhaopingxian zhi*] ch. 4

(b) 12th year of the Qianlong reign period of Emperor Gaozong of the Qing Dynasty, winter; a broom star appeared in the southwest.
Qing Dynasty Qianlong period ed. Henan [*Gushixian zhi*] ch. 11

AD 1748 Apr 26 [China]

13th year of the Qianlong reign period of Emperor Gaozong of the Qing Dynasty, 3rd month, day *guichou* [50]; an anomalous star appeared in the east, south of the third star of Ligong; it was as big as a hazelnut, yellowish in color, and with a tail over 2 *chi* long pointing toward the southwest; it proceeded forward 3 *du* until the new moon day *jiayin* [51] of the 4th month, at which point its tail was over 1 *chi*

long, its body small, and its gleam faint; on day *renxu* [59], it reached TENGSHE; on day *yichou* [2], it reached WANGLIANG; on day *bingyin* [3], it was no longer seen.
[*Qing shi gao: tianwenzhi*] ch. 39

AD 1748 May 2–10 [Korea]

(a) 24th year of King Yeongjo of the Joseon Dynasty, 4th month, day *jiwei* [56]; a guest star was seen within the space of lunar mansion SHI [LM 13]...on the night of day *gengshen* [57] (May 3), during the fourth watch, the broom star moved and was seen within the space of lunar mansion BI [LM 14] in the middle of the YINHE (Milky Way), its body was as large as WANGLIANG, its tail over 1 *chi* long, its color pale red, and pointing to azimuth *kun* (SW)...on the night of day *dingmao* [4] (May 10), the broom star moved and was seen west of the WANGLIANG star within the space of lunar mansion KUI [LM 15], its body as big as the WANGLIANG star (β Cas), its tail 1 *chi* long and pale red.
[*Joseon wangjo sillok: Yeongjo*] ch. 67

(b) 24th year *wuchen* of King Yongjong of the Joseon Dynasty, 4th month, day *jiwei* [56]; a broom star as big as WANGLIANG was seen in SHI [LM 13], its tail over 1 *chi* long, its color pale red; having passed BI [LM 14] it was in KUI [LM 15].
[*Jeungbo munheon bigo*] ch. 6

AD 1759 Apr 1–May 16 [Japan]

(a) 9th year of the Hōreki reign period, 3rd month, 4th day (Apr 1); a broom star was seen.
[*Suisei shutsugen nenki*]

(b) 9th year of the Hōreki reign period, 4th month, 1st day xinhai [48] (Apr 27); a broom star 1 *zhang* long was seen in Kaga [i.e., LM 11–12, XU and WEI]; by the second 10-day period of the month it no longer was seen.
[*Zoku shigushō*]

(c) 9th year of the Hōreki reign period, 4th month, 20th day *gengwu* [7] (May 16), clear sky; a broom star...in the early morning sky between

XU [LM 11] and WEI [LM 12], its rays about 5 to 6 *cun* long…it was seen from the second 10-day week of the 3rd month on.
[*Hakkai gyoki*]
NB: This is Comet Halley, return –3.

AD 1759 Apr 1–May 25 [Korea]

(a) 35th year *jimao* of King Yeongjo of the Joseon Dynasty, 3rd month, day *yiyou* [22]; a broom [star] as big as the central star of HEGU was seen in WEI [LM 12], its tail over 2 *chi* long and its color greenish-white; it passed XU [LM 11] and reached NÜ [LM 10]; by day *wushen* [45] (Apr 24), it was extinguished. In the 4th month, day *guichou* [50] (Apr 29), it was seen in YI [LM 27], as big as Venus, its tail over 2 *chi* long; by day *wuyin* [15] (May 24) it was extinguished.
[*Jeungbo munheon bigo*] ch. 6

(b) 35th year of King Yongjong of the Joseon Dynasty, 3rd month, day *bingxu* [23] (Apr 2); a broom star was seen within the space of lunar mansion XU [LM 11], its body like HEGU, its color greenish-white…on day *bingshen* [33] (Apr 12), the broom star was seen within the space of lunar mansion NÜ [LM 10]…in the 4th month, day *jiayin* [51] (Apr 30), at night, the broom star as big as Jupiter was seen below YI [LM 27], its color yellowish white…on the night of day *dingmao* [4] (May 13), the broom star was dimly visible above lunar mansion ZHANG [LM 26]…on the night of day *renshen* [9] (May 18), the broom star was dimly visible at 4 *du* in lunar mansion ZHANG [LM 26]…on day *jimao* [16] (May 25), the broom star was dimly visible.
[*Joseon wangjo sillok: Yeongjo*] ch. 93

NB: This is Comet Halley, return –3.

AD 1759 Apr 8 [China] (Ho & Ang 101)

24th year of the Qianlong reign period of Emperor Gaozong of the Qing Dynasty, 3rd month, day *renchen* [29]; a broom star appeared in the southeast; on day *jiawu* [31], it emerged below the first star of XU [LM 11], as big as a hazelnut, grayish-white in color, with a tail over

1 *chi* long, pointing southwest, and moving forward; on day *guimao* [40], its body was small, its gleam faint, and its tail over 3 to 4 *cun* long; on day *wushen* [45], it vanished completely; on day *wuchen* [5] in the 4th month, the broom star appeared in the southwest, above the second star of ZHANG [LM 26]; on day *jisi* [6], it was 6 *du* from ZHANG [LM 26] and big as a hazelnut, bluish-yellow in color; the brightness of its tail was spread out, but over 2 *chi* long and pointing southeast as it proceeded forward; on day *renshen* [9], the trace of its movement was faint; on day *dingchou* [14], it was even fainter; on day *jimao* [16], it gradually dissipated; by day *renwu* [19], in the 5th month, it completely vanished.
[*Qing shi gao: tianwenzhi*] ch. 39
NB: This is Comet Halley, return –3.

AD 1759 Apr 10 [China]

(a) 24th year of the Qianlong reign period of Emperor Gaozong of the Qing Dynasty, 3rd month, day *jiawu* [31]; a broom star appeared.
[*Qing shi gao: Gaozong ji*] ch. 12

(b) 24th year of the Qianlong reign period of Emperor Gaozong of the Qing Dynasty, 3rd month, day *jiawu* [31]; a grayish-white broom star, with a tail over 1 *chi* long pointing southwest, appeared in the space of lunar mansion XU [LM 11]; it moved forward nightly for over 10 days until it set and was no longer seen; on day *wuchen* [5] in the 4th month, it reappeared in lunar mansion ZHANG [LM 26], its body and influence extremely small and faint as it advanced eastward, until the beginning of the 5th month when it was hidden from view and went out of sight.
[*Qin ding huang chao wenxian tongkao: xiangwei*] ch. 267

(c) 24th year of the Qianlong reign period of Emperor Gaozong of the Qing Dynasty, 3rd month, day *jiawu* [31]; a grayish-white broom star, with a tail over 1 *chi* long pointing southwest, appeared in the space of lunar mansion XU [LM 11]; it advanced nightly for over 10 days until it set and was no longer seen; on day *wuchen*

[5] in the 4th month, it reappeared in the southwest, above the second star of lunar mansion ZHANG [LM 26], its body and radiance extremely small and faint; it advanced eastward until the beginning of the 5th month when it was hidden from view as it set.
[*Qing chao tong zhi: zaixiang*] ch. 122

NB: This is Comet Halley, return –3.

AD 1760 Jan 9 [China]

(a) 24th year of the Qianlong reign period of Emperor Gaozong of the Qing Dynasty, 11th month, day *wuchen* [5]; a anomalous star appeared in the southeast below the fourth star of JING [LM 22]; as big as a hazelnut and grayish-yellow in color, it traveled northwestward; on day *guiyou* [10], it traveled 4 *du* to a point in WEI [LM 17] and had only the faintest tail; on new moon day *dingchou* [14] in the 12th month, it vanished completely.
[*Qing shi gao: tianwenzhi*] ch. 39

(b) 24th year of the Qianlong reign period of Emperor Gaozong of the Qing Dynasty, 11th month, day *wuchen* [5]; a guest star appeared in the southeast below the fourth star of JING [LM 22] in zodiacal sign *wei* (Cancer); from Cancer it moved to the space of WEI [LM 17] in zodiacal sign Virgo; on searching it was hidden from view as it set.
[*Qing chao tong zhi: zaixiang*] ch. 122

AD 1766 Apr 7–11 [Japan]

(a) 3rd year of the Meiwa reign period, 2nd month, 28th day; at dusk a broom star about 8 to 9 *cun* in size was seen on the western horizon south of lunar mansion WEI [LM 17] and east of ZUOGENG, moving westward. 3rd month, 3rd day; overcast, 4th day, it was no longer seen.
[*Tempen chi-i kakinuki*]

(b) 3rd year of the Meiwa reign period, 2nd month, 28th day; dusk, a broom star was seen in lunar mansion WEI [LM 17], its rays about 2 *du* long; its brightness gradually increasing. 3rd month, new moon day; at dusk the sky was overcast; 2nd day (Apr 11), at dusk the broom star's location was as before, its rays about 3 *du* long; through the telescope the broom star was in the middle of lunar mansion WEI [LM 17].

NB: The last sentence of the text is garbled but the meaning is clear.
[*Sansai zusetsu narabi ni Kanpō irai jissoku zusetsu*]

AD 1769 Jul 17–Sep 10 [Japan]

(a) 6th year of the Meiwa reign period, 6th month, 14th day (Jul 17); during the hour of *xu* [LT = 19–21 h], [a broom star] was seen in the shape of a banner from west to east; at the beginning of the 7th month, its brightness was like a white silk 3 *chi* wide and 14 to 15 *chi* long; everyone saw it.
[*Tsugaru henran nikki*]

(b) 6th year of the Meiwa reign period, 7th month, 27th day (Aug 28), at dawn; a broom star was seen in the east in lunar mansion MAO [LM 18], north of DACHUI, and between the first and second stars [of MAO] on the east; after a day it moved eastward, its rays becoming long and full. In the 8th month, 8th day (Sep 7) at dawn; the rays were about 48 *du* long. At dawn on the 10th day of the same month, the rays split in two, each about 65 and 1/2 *du* long. At dawn on the 11th (Sep 10) of the same month, it was outside lunar mansion GUI [LM 23] and south of the first star on the west of WAICHU; one branch was about 72 and 1/2 *du* long and the other about 30 *du* long; from the 16th day to the 24th day, it was overcast. At dawn on the 25th, it was clear and bright, but the broom star was extinguished and no longer seen.
[*Tenpen chi-i kakinuki*]

(c) 6th year of the Meiwa reign period *jichou*; latitude and longitude of the broom star in the 7th month as measured at Bukō (Tokyo) [Aug 24–Oct 30]:

		polar distance
7th month, 23rd day dawn	7 *du* in MAO [LM 18]	80 *du*
8th month, 2nd day	6 *du* in BI [LM 19]	81 *du*

4th day	13 *du* in Bi [LM 19]	82 *du*
6th day	2 *du* in Shen [LM 21]	85 *du*
7th day	7 *du* in Shen [LM 21]	86 *du*
9th day	5 *du* in Jing [LM 22]	88 *du*
10th day	10 *du* in Jing [LM 22]	89 *du*
11th day	16 1/2 *du* in Jing [LM 22]	91 1/2 *du*
12th day	21 *du* in Jing [LM 22]	93 *du*
13th day	26 *du* in Jing [LM 22]	94 1/2 *du*
15th day	36 *du* in Jing [LM 22]	98 *du*

After this, it was rainy and overcast and the broom star was not seen; the broom star gradually moved eastward and approached the Sun's disc; 8th month, third 10-day week of the month, the broom star and the Sun were in conjunction.

10th month, new moon day	3 1/2 *du* in Fang [LM 4]
2nd day	initial *du* of Xin [LM 5]
4th–5th days, overcast	

[*Sansai zusetsu narabi ni Kanpō irai jissoku zusetsu*]

AD 1769 Aug 2–30 [China]

(a) 34th year of the Qianlong reign period of Emperor Gaozong of the Qing Dynasty, 7th month, at night; a broom star about 1 *zhang* long emerged in the northeast, shooting southwestward.
Qing Dynasty Qianlong period ed. Hebei [*Gaoyixian zhi*] ch. 7

(b) 34th year of the Qianlong reign period of Emperor Gaozong of the Qing Dynasty, 7th month; a star whose gleam shot from the northeast to the southwest appeared during the fifth watch every night.
Qing Dynasty Jiaqing period ed. Zhejiang [*Jiashanxian zhi*] ch. 20

(c) 34th year of the Qianlong reign period of Emperor Gaozong of the Qing Dynasty, 7th month; a long star with a gleam over 2 *zhang* long appeared between Shen [LM 21] and Jing [LM 22], until the 9th month was out and it was extinguished.
Qing Dynasty Guangxu period ed. Shanxi [*Pingyaoxian zhi*] ch. 12

AD 1769 Aug 28 [China] (Ho & Ang 104)

(a) 34th year of the Qianlong reign period of Emperor Gaozong of the Qing Dynasty, 7th month, 27th day (*jiachen* [41]); in the southeast a broom star appeared, its tail pointing southwest.
[*Qing Gaozong shilu*] ch. 840

(b) 34th year of the Qianlong reign period of Emperor Gaozong of the Qing Dynasty, 7th month, 27th day (*jiachen* [41]); a broom star appeared in the southeast below MAO [LM 18]; on day *dingwei* [44], it was as big as a crossbow bolt, grayish-white, with a tail 3 *chi* long pointing southwest, and moving forward extremely fast; on day *dingmao* [4] in the 8th month, it was in the same sign as the sun and invisible; on day *xinhai* [48] in the 10th month, it appeared in the west, below the second star of LIESI, its body and influence small and faint, and its tail 1 *chi* long; on day *bingzi* [13], it completely disappeared.
[*Qing shi gao: tianwenzhi*] ch. 39

(c) 34th year of the Qianlong reign period of Emperor Gaozong of the Qing Dynasty, 7th month, 27th day (*jiachen* [41]); a broom star appeared in the space of lunar mansion MAO [LM 18]; it was as big as a crossbow bolt, grayish-white, with a tail over 2 *chi* long pointing just south of due west, and moving forward every night; after the full moon in the 8th month, it was concealed and not seen; in the 10th month, it reappeared, having shifted to the west, then upon searching, it set hidden from view.
[*Qing chao tong: zaixiang*] ch. 122
[*Qing chao wenxian tongkao: xiangwei*] ch. 267

AD 1769 Aug 30–Nov 6 [Korea]

(a) 45th year of King Yeongjo of the Joseon Dynasty, 7th month, day *jiyou* [46]; a star resembling a broom [star] was seen within the space of lunar mansion BI [LM 19]...in the 8th month, new moon day *gengxu* [47] (Aug 31), the broom star was seen within the space of lunar mansion BI [LM 19], its body was as big as BI (α Tau) and its color white...on day *jiayin* [51] (Sep 4), the broom star was seen in lunar mansion SHEN [LM 21], its body and tail

as before...on day *dingsi* [54] (Sep 7), the broom star moved and was seen within the space of lunar mansion JING [LM 22], its shape and color as before...on day *renxu* [59] (Sep 12), the broom star was seen within the space of lunar mansion GUI [LM 23], its color white, its tail 5 to 6 *chi* long...on day *guihai* [60] (Sep 13), the broom star was seen in lunar mansion LIU [LM 24], its shape and color as before...in the 9th month, day *dingwei* [44] (Oct 27), a star was seen in the west above lunar mansion FANG [LM 4], seemingly with a tail...in the 10th month, on day *xinhai* [48] (Oct 31), the broom star was seen within the space of lunar mansion XIN [LM 5], its color pale white...on day *jiayin* [51] (Nov 3), the broom star was seen within the space of lunar mansion XIN [LM 5], sometimes as if with a tail, sometimes not.
[*Joseon wangjo sillok: Yeongjo*] ch. 113

(b) 45th year *jichou* of King Yeongjo of the Joseon Dynasty, 7th month, day *jiyou* [46]; a broom star as big as the large star (α Tau) of BI [LM 19] was seen in BI [LM 19], its tail 3 to 4 *chi* long, its color white; it traversed SHEN [LM 21], JING [LM 22], GUI [LM 23], LIU [LM 24], its tail somewhat long. In the 8th month, day *bingyin* [3] (Sep 16), the body of the star gradually dimmed. In the 9th month, on day *dingwei* [44] (Oct 27), at twilight, in contrast, it was seen in the west above lunar mansion FANG [LM 4], its tail over 1 *chi* long. In the 10th month, day *jiayin* [51] (Nov 3), it moved to lunar mansion XIN [LM 5], on day *dingsi* [54] (Nov 6), it was extinguished.
[*Jeungbo munheon bigo*] ch. 6

AD 1769 Oct 29–Nov 27 [China]

34th year of the Qianlong reign period of Emperor Gaozong of the Qing Dynasty, 10th month; according to a memorial by the directorate of astronomy and the calendar: during the 7th month a broom star appeared, its gleam shining southwestward, for a day, whereupon it was extinguished; now it has appeared again in the southwest, its gleam shining northeastward.
[*Qing Gaozong shilu*] ch. 844

AD 1769 Nov 18 [Japan]

6th year of the Meiwa reign period, 10th month, 21st day *jisi* [6]; yet another broom star several *zhang* long was seen at azimuth *gen* [NE] (dawn observation); after 2 to 3 days it was no longer seen.
[*Zoku shigushō*]

AD 1770 Jun 24–Jul 4 [Japan]

(a) 7th year of the Meiwa reign period, 6th month, 2nd day *dingwei* [44] (Jun 24); there was a fuzzy star about 1 *chi* long emitting bright rays in all directions. It was seen in QIANNIU [LM 9] and reached alongside ZIWEI, but it changed location nightly and after 10 days was no longer seen.
[*Zoku shigushō*]

(b) 7th year *gengyin* of the Meiwa reign period, 6th month, 5th day (Jun 27); there was a fuzzy star in TIANBIAN, but the actual measurement was not correct. On the night of the 6th, using the equatorial armillary the fuzzy star was measured at the initial degree of lunar mansion DOU [LM 8], at a polar distance of 90 *du*; on the night of the 7th, the fuzzy star invaded TIANSHIYUAN; using the equatorial armillary it was measured at 1 *du* in lunar mansion DOU [LM 8], at a polar distance of 83 *du*.
[*Zoku tankai*]

(c) 7th year of the Meiwa reign period, 6th month, 6th day (Jun 28); before midnight a fuzzy star was seen at TIANSHIYUAN east of DONGHAI, moving northward; on the night of the 10th it entered ZIWEIYUAN, north of BEICHEN; on the 11th and 12th, it was overcast; on the 13, it was not seen.
[*Tenpen chi-i kakinuki*]

AD 1770 Jun 26 [China] (Ho & Ang 105)

(a) 35th year of the Qianlong reign period of Emperor Gaozong of the Qing Dynasty, intercalary 5th month, day *jiyou* [46]; an anomalous star, as big as a crossbow bolt and grayish-yellow in color, appeared in the southeast, west of the first star of TIANBIAN;

on day *guichou* [50], it traveled northward 32 *du*; on day *yichou* [2], it was no longer seen.
[*Qing shi gao: tianwenzhi*] ch. 39

(b) 35th year of the Qianlong reign period of Emperor Gaozong of the Qing Dynasty, intercalary 5th month, day *jiyou* [46]; a guest star appeared in the space of lunar mansion Dou [LM 8] and traveled northward nightly; after over 10 days, it went out of sight.
[*Qin ding huang chao wenxian tongkao: xiangwei*] ch. 267

(c) 35th year of the Qianlong reign period of Emperor Gaozong of the Qing Dynasty, intercalary 5th month, day *jiyou* [46]; a grayish-yellow guest star appeared in the space of lunar mansion Dou [LM 8], west of the first star of Tianbian; it traveled northward nightly; after over 10 days, it set hidden from view.
[*Qing chao tong zhi: zaixiang*] ch. 122

AD 1770 Dec 17–AD 1771 Jan 15 [China]

35th year of the Qianlong reign period of Emperor Gaozong of the Qing Dynasty, 11th month; a broom star appeared in the space of lunar mansion Liu [LM 24], below the first star; it was grayish-white, with a tail over 1 *chi* long pointing south, and it advanced nightly over 10 *du* for 7 days, when it set hidden from view.
[*Qin ding huang chao wenxian tongkao: xiangwei*] ch. 267
[*Qing chao tong zhi: zaixiang*] ch. 122

AD 1771 Jan 8 [China] (Ho & Ang 106)

35th year of the Qianlong reign period of Emperor Gaozong of the Qing Dynasty, 11th month, day *yichou* [2]; a broom star over 1 *chi* long appeared in the southeast; on day *bingyin* [3], it was below the second star of Liu [LM 24]; on day *wuchen* [5], its color was grayish-white and its tail pointing southeast as it proceeded northwestward over 10 *du* a day; on day *gengwu* [7], it was faint and dark; on day *xinwei* [8], it completely disappeared.
[*Qing shi gao: tianwenzhi*] ch. 39

AD 1771 Jan 10 [Japan]

(a) 7th year of the Meiwa reign period, 11th month, 25th day *dingmao* [4] [(Jan 10); tonight there was a broom star observed in the space of GUI [LM 23] and LIU [LM 24], its white rays 4 to 5 *du* long and visible in the evening.
[*Zokushi gushō*] ch. 79

(b) 7th year *gengyin* of the Meiwa reign period, 2nd month of winter, 24th day (Jan 10); there was a fuzzy star in LIU [LM 24] with a rays 4 to 5 *du* long starting in the west and pointing east; its radiance very faint; on the 25th, it swung slightly westward to the boundary between GUI [LM 23] and LIU [LM 24]; on the 27th (Jan 13), it continued to swing to the left and invaded the field of JING [LM 22], north of NANHE. At the time, the sun was located at about 10 *du* in lunar mansion DOU [LM 8] and the broom star was at the last *du* of lunar mansion JING [LM 22]. At the start of dusk when the sun set in the west, the broom star was seen in the east at a distance of some 190 *du*…it was observable all night.
[*Sansai zusetsu narabi ni Kanpoo irai jissoku zusetsu*]

AD 1773 Mar 23–Apr 21 [China]

38th year of the Qianlong reign period of Emperor Gaozong of the Qing Dynasty, 3rd month; there was a star bigger than a hen's egg, its head pointing northeast and its tail pointing southwest, with bright rays several *zhang* long flowing outward and shaped like the sweep of a broom; it arose during the third watch and disappeared during the fifth watch, until the 8th month when it desisted.
[*Ming zhai xiao shi*] ch. 1

AD 1776 Aug 14–Sep 12 [China]

41st year of the Qianlong reign period of Emperor Gaozong of the Qing Dynasty, 7th month; a long star shaped like a bolt of silk cloth appeared, with bright rays 5 to 6 *zhang* long that stretched horizontally across the sky in the southeast; it first appeared during the third

watch, and by the fifth watch, it was no longer seen; it was like this for more than 10 nights.
Qing Dynasty Hunan [*Dongtinghu zhi*] ch. 7

AD 1784 Jan 19–Feb 8 [Japan]

(a) 3rd year of the Temmei reign period, late winter [12th month] 27th day to 1st month, third day (Jan 19–Jan 24); a broom star was seen in azimuth *kun* [SW].
[*Gokō nempyō*] ch. 1

(b) 4th year *jiachen* of the Temmei reign period, 1st month (Jan 22–Feb 20); a broom star was seen. At the beginning of dusk on the 6th, it was spotted through a break in the clouds; the body of the star was extremely faint, its rays several *cun* long and pointing eastward. Although located within the space of lunar mansion SHI [LM 13], it was not possible to precisely observe its location. From the 7th day until the full moon the sky was cloudy every night; whenever the night sky happened to clear the Moon's glare made it impossible to observe the rays. On the night of the 18th (Feb 8), it was observed before moonrise; the small broom star was at PILI, and there were some bright rays pointing east, but the body of the star was indistinct and difficult to observe.
[*Sansai zusetsu narabi ni Kanpoo irai jissoku zusetsu*]

AD 1784 Jan 24 [Korea]

(a) 8th year of King Jeongjo of the Joseon Dynasty, 1st month, day *jichou* [26]; there was a star like a broom [star] that was seen within the space of lunar mansion SHI [LM 13]...on day *gengyin* [27] (Jan 25), [the broom star] was seen within the space of lunar mansion SHI [LM 13], east of YULIN[-JUN].
[*Joseon wangjo sillok: Jeongjo*] ch. 17

(b) 8th year of King Jeongjo of the Joseon Dynasty, 1st month, day *jichou* [26]; a broom [star] was seen at the YULIN[-JUN] star.
[*Jeungbo munheon bigo*] ch. 6

NB: Ho and Ang (1970) continued the listing of comets through the end of the Qing Dynasty in 1911. Those records are fairly complete and unremarkable. Since Western parallels exist, they have not been reproduced here.

References

Asher, D. J., Clube, S. V. M., Napier, W. M., & Steel, D. I. (1994). Coherent catastrophism. *Vistas in Astronomy*, *38*(1), 1–27.

Beijing Observatory. (1988). *Zhongguo gudai tianxiang jilu zong ji*. Nanjing: Jiangsu kexue jishu chubanshe.

Biot, E. (1843). *Connaissance des Temps 1846. Additions*, 44–59, 69–84.

Chang, Y. C. (1979). Halley's Comet: Tendencies in its orbital evolution and its ancient history. *Chinese Astronomy*, *3*(1), 120–131.

Chen, M. (Ed.). (1996). *Zhongguo gu xingtu*. Beijing: Liaoning jiaoyu.

Chen, Z. (1984). *Zhongguo tianwenxue shi*. Shanghai: Kexue chubanshe.

Halley, E. (1705). Astronomiae cometicae synopsis. *Philosophical Transactions of the Royal Society*, *24*, 1882–1899.

Hasegawa, I. (1980). Catalogue of ancient and naked-eye comets. *Vistas in Astronomy*, *24*, 59–102.

Hasegawa, I., & Nakano, S. (1995). Periodic comets found in historical records. *Publications of the Astronomical Society of Japan*, *47*, 699–710.

Hind, J. R. (1850). On the past history of the Comet of Halley. *Monthly Notices of the Royal Astronomical Society*, *10*, 51–58.

Ho, P.-y. (1962). Ancient and medieval observations of comets and novae in Chinese sources. *Vistas in Astronomy*, *5*, 127–225.

Ho, P.-y. (1966). *The astronomical chapters of the Chin Shu.* Paris: Mouton & Co.

Ho, P.-y., & Ang, T.-s. (1970). Chinese astronomical records on comets and "guest stars" in the official histories of Ming and Ch'ing and other supplementary sources. *Oriens Extremus*, *17*(1.2), 63–99.

Imoto, S., & Hasegawa, I. (1958). Historical records of meteor showers in China, Korea, and Japan. *Smithsonian Contributions to Astrophysics*, *2*, 131–144.

Kanda, S. (1935). *Japanese historical records of celestial phenomena.* Tokyo: Harashoubou.

Kanda, S. (1936). *General inventory of historical astronomical records*. Tokyo: Harashoubou.

Kiang, T. (1984). Notes on traditional Chinese astronomy. *Observatory*, *104*, 19–23.

Kronk, G. (1999). *Cometography: A catalog of comets.* Cambridge, U.K.: Cambridge University Press.

Lewis, J. S. (1996). *Rain of iron and ice: The very real threat of comet and asteroid bombardment.* New York: Basic Books.

Lewis, J. S. (2000). *Comet and asteroid impact hazards on a populated Earth: Computer modeling*. San Diego, CA: Academic Press.

Needham, J., & Wang, L. (1959). *Science and civilisation in China, vol. 3, mathematics and the sciences of the heavens and the Earth.* Cambridge, U.K.: Cambridge University Press.

Pankenier, D. W. (1981–1982). Astronomical dates in Shang and Western Zhou. *Early China*, *7*, 2–37.

Rufus, W. C. (1936). *Korean astronomy*. Seoul, Korea: The Literary Dept., Chosen Christian College.

Stephenson, F. R. (1994). Chinese and Korean star maps and catalogs. In J. B. Harley & D. Woodward (Eds.), *The History of Cartography* (Vol. 2, book 2, pp. 511–578). Chicago: University of Chicago Press.

Stephenson, F. R., & Green, D. (2002). *Historical supernovae and their remnants*. Oxford, U.K.: Oxford University Press.

Sun, X. C., & Kistemaker, J. (1997). *The Chinese sky during the Han: Constellating stars and society.* Leiden: Brill.

Xu, Z., & Jiang, Y. (1993). Yin Shang huixing jishi kao. *Ziran kexueshi yanjiu, 12*(3), 235–239.

Xu, Z., Pankenier, D. W., & Jiang, Y. (2000). *East Asian archaeoastronomy: Historical records of astronomical observations of China, Japan and Korea.* Amsterdam: Gordon & Breach.

Yeomans, D. K., & Kiang, T. (1981). The long-term motion of Comet Halley. *Monthly Notices of the Royal Astronomical Society*, *197*, 633–646.

Part 2

Meteor Showers

2.1. Ancient East Asian Observations of Meteor Showers

The observation of meteor showers in China has a long history. The *Bamboo Annals* (*Zhushu jinian*) records, "10th year of Emperor *Gui* (*Jie* of the Xia Dynasty); the five planets criss-crossed; in the middle of the night stars fell like rain" (plate 4). This is a late Xia Dynasty record of an astronomical event dating from the 1st quarter of the 16th century BCE. By the pre-Qin period, precise records had already appeared, for example, the 4th century BC narrative history *Zuozhuan*, a commentary on the earlier annalistic chronicle *Spring and Autumn Annals* (*Chunqiu*), records, "7th year of Duke Zhuang of Lu, summer, 4th month, day *xin-mao* [28] (23 Mar 687 BC), at night; the fixed stars did not appear…in the middle of the night stars fell like rain" (figure 2.1). Subsequently, records in the dynastic histories and local gazetteers appear without interruption, providing a wealth of valuable materials.

FIGURE 2.1. Duke Zhuang of Lu's meteor shower.

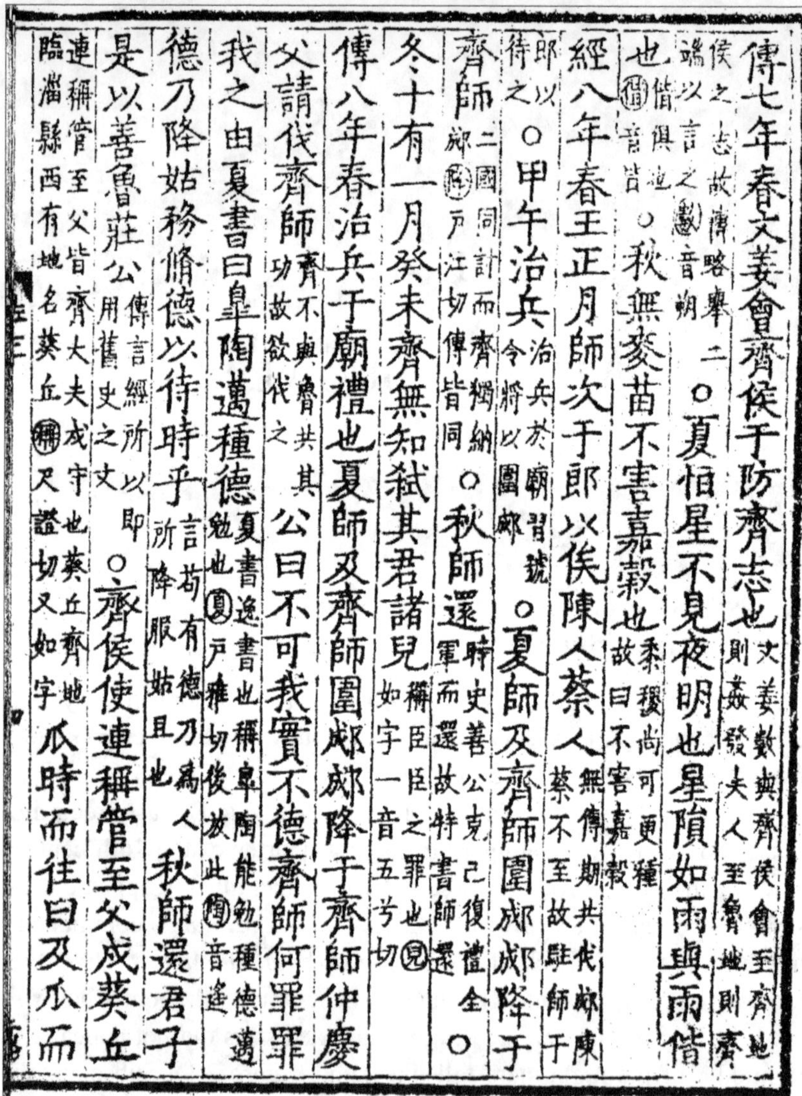

傳七年春文姜會齊侯于防齊志也文姜數與齊侯會至齊地則姦發夫人至魯地則齊
侯之志故傳略舉二端以言之數音朔○夏恒星不見夜明也星隕如雨與雨偕
也偕俱也偕音皆○秋無麥苗不害嘉穀也黍稷尚可更種故曰不害嘉穀
經八年春王正月師次于郎以俟陳人蔡人無傳期共伐郕陳蔡不至故駐師于
郎以待之○甲午治兵治兵於廟習號令將以圍郕○夏師及齊師圍郕郕降于
齊師二國同討而齊獨納郕郕戶江切傳皆同○秋師還時史善公克己復禮全軍而還故特書師還○
冬十有一月癸未齊無知弒其君諸兒稱臣臣之罪也兒如字一音五兮切
傳八年春治兵于廟禮也夏師及齊師圍郕郕降于齊師仲慶
父請伐齊師齊不與魯共其功故欲伐之公曰不可我實不德齊師何罪罪
我之由夏書曰皐陶邁種德夏書逸書也稱皐陶能勉種德邁勉也夏戶雅切後放此陶音遙
德乃降姑務脩德以待時乎言苟有德乃為人所降服姑且也秋師還君子
是以善魯莊公傳言經所以即用舊史之文○齊侯使連稱管至父戍葵丘
連稱管至父皆齊大夫戍守也葵丘齊地臨淄縣西有地名葵丘稱尺證切又如字瓜時而往曰及瓜而

Meteor showers in ancient Chinese sources are generally recorded as "stars fell like rain," "stars streamed as if weaving," or "myriads of stars streamed back and forth," and so forth. Records of meteor showers from the Spring and Autumn period (722–481 BC) down to the Qing Dynasty (AD 1644–AD 1911) number in the many hundreds. Many of these records are detailed, including time of appearance and disappearance, location, number of meteors, color, brightness, duration, and so forth. For example, a detailed record from *Song shu* reads,

> 20th year of the Yuanjia reign period of Emperor Wen of the (Southern Dynasties) Song Dynasty [AD 443], 2nd month, day *yiwei* [32]; meteors as big as peaches emerged from TIANJIN and entered ZIGONG, followed immediately by tiny meteors in fives and threes. Still another large meteor emerged from ZIGONG and entered the bowl of BEIDOU; immediately afterward large meteors emerged from the middle of GUANSUO and passed through TIANSHIYUAN. All the meteors traveled northward in tandem, by daybreak there had been too many to count.

Reading an observation this descriptive it feels as if one had actually witnessed the event.

Ancient sources from Japan and Korea also preserve many observations of meteor showers (figure 2.2). Sometimes the records from Japan are also very concrete and accurate, like the *Sandai jitsuroku* entry from October 28–29, 884:

> 8th year of the *Gangyō* reign period, 8th month, 4th day *renchen* [29]; from the hour of *xu* (LT = 19–21h) to the hour of *zi* (LT = 23–1h) small stars streamed and scattered in all directions, falling like rain. On the 5th day *guisi* [30], from sunset until the hour of repose, meteors too numerous to count moved in and out of ZIWEIGONG, trespassing on many palaces and heavenly offices, some moving in and out of BEIDOU, GUANSUO, [DA]LING, and the lunar mansions outside.

Not only is the exact time of the meteor shower recorded, the precise location in the sky where the meteors appeared and disappeared is also given, making this an exceptionally valuable account.

FIGURE 2.2. Korean record of the Leonid meteor shower of 24 October 1533 from a moveable type edition of the *Veritable Annals of the Joseon Dynasty*.

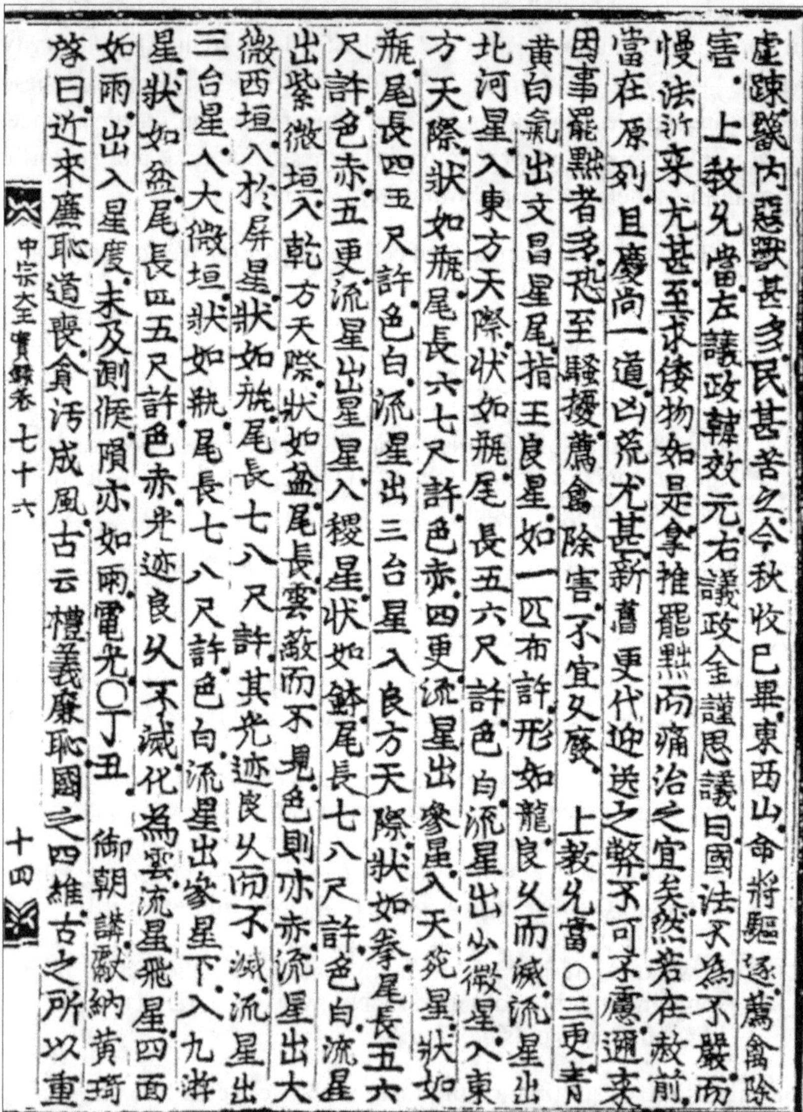
虛踈竊內盜賊甚多民甚苦之今秋收已畢東西山命將驅逐蔫禽除
害。上敎允當左議政韓效元右議政金謹思議曰國法不爲不嚴而
慢法沂來尤甚至求倭物如是拿推罷黜而痛治之宜矣然若在赦前
當在原列且慶尚一道凶荒尤甚新舊吏代迎送之弊不可不慮邇來
因事罷黜者多恐至騷擾蔫禽除害不宜又廢。上敎允當○三更靑
黃白氣出文昌星尾指王良星如一匹布許形如龍良久而滅流星出
北河星入東方天際狀如瓶尾長五六尺許色白流星出少微星入東
方天際狀如瓶尾長六七尺許色赤四更流星出參星入天苑星狀如
瓶尾長四五尺許色白流星出三台星入艮方天際狀如拳尾長五六
尺許色赤五更流星出星星入稷星狀如鉢尾長七八尺許色白流星
出紫微垣入乾方天際狀如盆尾長雲蔽而不見色則亦赤流星出大
微西垣入於屛星狀如瓶尾長七八尺許其光迹良久而不滅流星出
三台星入大微垣狀如瓶尾長七八尺許色白流星出參星下入九斿
星狀如盆尾長四五尺許色赤光迹良久不滅化爲雲流星飛星四面
如雨出入星度未及測候隕亦如雨電光○丁丑 御朝講獻納黃琦
啓曰近來廉恥道喪貪汚成風古云禮義廉恥國之四維古之所以重

中宗大王實錄卷七十六 十四

2.2. Modern Applications of Meteor Shower Records

Modern astronomical research has demonstrated that meteor showers are a consequence of the disintegration of short-period comets. For example, it is known that the August Perseid meteors move in the same orbit as Comet Swift-Tuttle, the Leonids follow that of Comet Tempel-Tuttle, and the May Aquarids follow that of Comet Halley. A great shower of Leonids was reported in AD 931, and Chinese records for that epoch clearly reflect the 33-year period of such storms (Needham, 1959). At the same time, meteor streams and their periodic encounters with the Earth have played an important role in the geological and biological history of our planet (Clube & Napier, 1989; Rasmussen, 1991; Lewis, 1996; Yi, 1998). The Japanese scholar Hasegawa Ichiro has used the ancient East Asian records of meteor showers to identify several short-period comets and determine their orbits, thereby proving the intimate connection between comets and meteor showers. The Korean scholar Yi Tae-jin carried out a statistical study of 460 years' worth of detailed records from the annals of the Joseon Dynasty, including over 3,000 records of individual meteor sightings, to gain insight into climatic effects during the so-called Little Ice Age of the 17th century (Yi, 1998).

Early in the 20th century, Sekiguchi compiled a list of Korean records in secondary sources. In 1933, Zhu Wenxin wrote a short article entitled "*Xuanyuan liuxingyu shilue*" ("A brief history of the Leonid meteor shower") in which he identified seven historical records of Leonid meteor showers. In 1958, Imoto and Hasegawa, after collating and studying the Japanese and Korean records of meteor showers, published their tabulated results demonstrating the important contribution historical meteor records can make to modern astronomical research. In 1966, Zhuang Tianshan, after many years of gathering and studying the records, published a table of 147 Chinese observations of meteor showers, which formed the basis for subsequent compilations. In 1984, in the third volume of *Zhongguo tianwenxue shi* (*A History of Chinese Astronomy*), Chen Zungui took up a further 97 records of meteor showers. *Zhongguo gudai tianxiang jilu zong ji* (*A Comprehensive Catalogue of Ancient Chinese Astronomical Phenomena*), published in 1988, greatly expanded on the number of meteor shower records, reaching a total of over 400. Kevin Yau et al. (1994) compiled and translated Chinese records of observed and recovered meteorites, some involving showers of multiple objects and human casualties, and studied their distribution as a function of time. Sang-Hyeon Ahn (2004) documented and verified some 48 Korean records, in the process correcting and supplementing Imoto & Hasegawa's (1958) catalogue. Most recently, in 2006 Peter Jenniskens published *Meteor Showers and their Parent Comets*, a comprehensive study of historical meteor showers with identifications of the parent bodies that have been made to date (Table 1). A number of the annual Perseid showers have been identified and these are tabulated in Jenniskens (2006, 601–602).

Based on our own research as well as on these prior catalogues, we have examined the original sources and compiled a total of over 600 records, checking in each case the precise language of the original record, the year, and the date. The Korean meteor shower records end in 1696, while Japanese reports continue until 1867. Chinese records of meteor showers after about 1550 vastly outnumber those from Japan and Korea; the last Chinese meteor shower entry is an observation from 1910. There are several reasons for this, mainly having to do with the availability of adequate records. For Korean meteor sightings, after 1696 the available records do not clearly distinguish showers from individual meteors and bolides. The excess of Chinese records of meteor showers after the 16th century is a result of the inclusion of numerous records from the many surviving local gazetteers, records that are not available from Japan and Korea. Since the Chinese gazetteer reports are not available elsewhere we include them here for the sake of completeness.

TABLE 1. Identifications cf historical meteor showers made to date.

Date	Shower (IAU#)	Revolutions Old	Parent Body
BC 687 Mar 23	Lyrids (IAU# 6)	1 rev. dust trail	C/1861 G1 (Thatcher)
BC 15 Mar 27	Lyrids (IAU#6)	1 rev. dust trail	C/1861 G1 (Thatcher)
AD 934 Oct 13.7 (China)	Leonids (IAU#13)	3-7 rev. dust trails	55P/Tempel-Tuttle
AD 967 Oct 14.7 (Japan)	Leonids (IAU#13)	2+4-9 rev. dust trails	55P/Tempel-Tuttle
AD 1002 Oct 14.7 (China)	Leonids (IAU#13)	2 rev. dust trail	55P/Tempel-Tuttle
AD 1037 Oct 14.7 (Japan)	Leonids (IAU#13)	3 rev, dust trail	55P/Tempel-Tuttle
AD 1532 Oct 24/25	Leonids (IAU#13)	15 rev, dust trail	55P/Tempel-Tuttle
AD 1566 Oct 26 14h17m	Leonids (IAU#13)	10 rev, dust trail	55P/Tempel-Tuttle
AD 1602 Nov 06	Leonids (IAU#13)	5 rev, dust trail	55P/Tempel-Tuttle
AD 1625 Nov 5 18:33	Leonids (IAU#13)	8 rev, dust trail	55P/Tempel-Tuttle
AD 1798 Dec 6	Andromedids (IAU#18)	comet	3D/Biela
AD 1825 Nov 10	Dec 9/Dec 3–4 Andromedids (IAU#18)	comet	3D/Biela
AD 1861 Aug 10	Persieds (IAU#7)	Filament of comet	109P/Swift-Tuttle
AD 1862 Aug 10	12: Perseids (IAU#7)	Filament of comet	109P/Swift-Tuttle
AD 1872 Nov 27	Andromedids (IAU#18)	4 rev, dust trail of comet	3D/Biela
AD 1885 Nov 27	Andromedids (IAU#18)	5 rev, dust trail of comet	3D/Biela
AD 1886 Nov 25	Andromedids (IAU#18)	4 rev, dust trail of comet	3D/Biela
AD 1894 Nov 14	Leonids (IAU#13)	comet	55P/Tempel-Tuttle
AD 1899 Aug 12	Perseids (IAU#7)	comet	109P/Swift-Tuttle
AD 1899 Nov 14	Leonids (IAU#13)	comet	55P/ Tempel-Tuttle
AD 1902 Oct 23	Orionids (IAU#8)	comet	1P/Halley
AD 1906 Aug 13	Perseids (IAU#7)	comet	109P/Swift-Tuttle
AD 1906 Oct 24	Orionids (IAU#8)	comet	1P/Halley

Source. After Jenniskens (2006).
Note. The IAU database of meteor showers can be found at the following Web site: http://www.astro.amu.edu.pl/~jopek/.

2.3. EAST ASIAN RECORDS OF METEOR SHOWERS

BC 16th–17th centuries [China]

10th year of King Gui (Jie) of the Xia Dynasty; during the night, stars fell like rain.
[*Zhushu jinian*] ch. *shang*

BC 687 Mar 23 [China]

(a) 7th year of Duke Zhuang of Lu, 4th month, day *xinmao* [28], at night; the fixed stars did not appear; the night was bright. During the night, stars fell like rain, together with the rain.
[*Chunqiu jingzhuan jijie*] ch. 3

(b) 7th year of Duke Zhuang of Lu, 4th month, day *xinmao* [28]; at night the fixed stars did not appear; the night was bright. During the night, stars fell like rain, together with the rain.
[*Chunqiu guliang zhuan*] ch. 5

BC 245 [China]

2nd year of the First Emperor of Qin, 3rd month, day *yiwei* [32]; innumerable large and small meteors traveled westward, and did not stop until late at night.
[*Tong zhi: zaixiang lue*] ch. 74
NB: There was no *yiwei* day during the 3rd month that year. If the record refers to the 2nd month, the date was Feb 15, 245 according to the Xia calendar and Feb 14, 245 according to the Zhuanxu calendar.

BC 74 Apr 7 [China]

(a) 1st year of the Yuanping reign period of Emperor Zhao of Han, 2nd month, day *jiashen* [21]; at dawn there was a meteor as big as the Moon, and a multitude of stars all traveled westward with it.
[*Han shu: Zhao di ji*] ch. 7

(b) 1st year of the Yuanping reign period of Emperor Zhao of Han, 2nd month, day *jiashen* [21]; at dawn there was a star as big as

the Moon, and there were a multitude of stars traveling westward with it.
[*Han shu: tianwen zhi*] ch. 26

BC 15 Mar 27 [China]

(a) 2nd year of the Yongshi reign period of Emperor Cheng of Han, 2nd month, day *guiwei* [20]; stars fell like rain.
[*Han shu: Cheng di ji*] ch. 10

(b) 2nd year of the Yongshi reign period of Emperor Cheng of Han, 2nd month, day *guiwei* [20]; after midnight, stars 1 to 2 *zhang* long and deep scarlet in color fell like rain, but did not reach the earth. This went on till cockcrow.
[*Han shu: wuxing zhi*] ch. 27

BC 15 Sep 23 [Japan]

15th year of Emperor Suinin, 8th month; stars fell like rain.
[Ninjyu kagami]
NB: The record is very doubtful as the first official chronicle Nihon shoki begins 7 centuries later.

BC 12 May 22 [China]

(a) 1st year of the Yuanyan reign period of Emperor Cheng of Han, 4th month, day *dingyou* [34]; it was cloudless but it thundered, the sound and brightness were glaring, and [stars] fell to earth on all sides until twilight.
[*Han shu: Cheng di ji*] ch. 10

(b) 1st year of the Yuanyan reign period of Emperor Cheng of Han, 4th month, day *dingyou* [34]; at the hour of *bu* [LT = 15–17h], the sky was bright and clear and there came a muffled sound of thunder. Dazzling reddish-white meteors over 10 *zhang* long, with big heads like jars came from beneath the Sun and disappeared to the southeast. On all sides, some were as large as bowls and some like hen's eggs, falling brilliantly like rain, not stopping until twilight.
[*Han shu: tianwen zhi*] ch. 26

(c) 1st year of the Yuanyan reign period of Emperor Cheng of Han, 4th month, day *dingyou* [34]; on all sides a multitude of stars streamed and fell in daylight.
[*Han shu: Gu Yong zhuan*] ch. 85

AD 34 [China]
10th year of the Jianwu reign period of Emperor Guangwu of Han, 3rd month, day *guimao* [40]; white meteors like the Moon emerged from TAIWEI and entered the sixth star in the bowl of BEIDOU. Alongside, 10 or more small stars shot off, and when they extinguished, they made a sound like thunder; in a moment it was over.
[*Hou Han shu: tianwen zhi*] ch. 20
NB: There was no day *guimao* in the 3rd month; if it was the 2nd month, day *guimao* was March 4.

AD 36 [China]
12th year of the Jianwu reign period of Emperor Guangwu of Han, 1st month, day *jiwei* [56]; for two nights there have been more than 100 small meteors, some in the northwest, some due north, some in the northeast.
[*Hou Han shu: tianwen zhi*] ch. 20
NB: There was no day *jiwei* in the 1st month; it was day *yiwei* [32], the date was March 16, and if it was 2nd month, day *jiwei*, the date was April 9.

AD 36 Jul 17 [China]
12th year of the Jianwu reign period of Emperor Guangwu of Han, 6th month, day *wuxu* [35]; at dawn more than 100 small meteors flew on all four sides.
[*Hou Han shu: tianwen zhi*] ch. 20

AD 104 Feb 14–Mar 13 [Korea] (Ahn [2004] 1)
25th year of King Pasa of Silla, spring, 1st month; a myriad stars fell like rain, but did not reach the ground.
[*Jeungbo munheon bigo*] ch. 7

AD 268 Jul 27–Aug 24 [China]

(a) 4th year of the Taishi reign period of Emperor Wu of the Jin Dynasty, 7th month; a multitude of stars streamed westward.
[*Jin shu: Wu di ji*] ch. 3

(b) 4th year of the Taishi reign period of Emperor Wu of the Jin Dynasty, 7th month; stars fell like rain, they all streamed westward.
[*Jin shu: tianwen zhi*] ch. 13
[*Song shu: tianwen zhi*] ch. 23

AD 288 Sep 26 [China]

(a) 9th year of the Taikang reign period of Emperor Wu of the Jin Dynasty, 8th month, day *renzi* [49]; stars fell like rain.
[*Jin shu: Wu di ji*] ch. 3
[*Jin shu: tianwen zhi*] ch. 13

(b) 9th year of the Taikang reign period of Emperor Wu of the Jin Dynasty, 8th month, day *renzi* [49]; stars fell like rain in daylight.
[*Tong zhi: zaixiang lue*] ch. 74

AD 308 Jan 20 [China]

1st year of the Yongjia reign period of Emperor Huai of the Jin Dynasty, 12th month, day *dinghai* [24]; stars streamed and scattered thunderously.
[*Jin shu: tianwen zhi*] ch. 13
[*Song shu: tianwen zhi*] ch. 24

AD 401 Apr 8 [China]

(a) 5th year of the Long'an reign period of Emperor An of the Jin Dynasty, 3rd month, day *jiayin* [51]; a multitude of stars streamed westward, passing through TAIWEI.
[*Jin shu: An di ji*] ch. 10

(b) 5th year of the Long'an reign period of Emperor An of the Jin Dynasty, 3rd month, day *jiayin* [51]; a vast number of scarlet

meteors traveled westward through QIANNIU [LM 9], XU [LM 11], WEI [LM 12], TIANJIN, GEDAO, and penetrated TAIWEI and ZIGONG.
[*Jin shu: tianwen zhi*] ch. 13
[*Song shu: tianwen zhi*] ch. 25

AD 443 Apr 9 [China]
20th year of the Yuanjia reign period of Emperor Wen of the (Southern Dynasties) Song Dynasty, 2nd month, 24th day; meteors as big as peaches emerged from TIANJIN and entered ZIGONG, followed immediately by tiny meteors in fives and threes. Still another large meteor emerged from ZIGONG and entered the bowl of BEIDOU. Immediately afterward, large meteors emerged from the middle of GUANSUO and passed through TIANSHIYUAN. All the meteors traveled northward together; by daybreak there had been too many to count.
[*Song shu: tianwen zhi*] ch. 26

AD 447 Feb 1–Mar 2 [China]
24th year of the Yuanjia reign period of Emperor Wen of the (Southern Dynasties) Song Dynasty, 1st month; the stars in heaven streamed westward together, the majority small, the largest not bigger than a hen's egg, with shorter or longer tails. There must have been several hundred, some of which entered BEIDOU and ZIGONG; they stopped only at sunrise when the Sun's light steadied.
[*Song shu: tianwen zhi*] ch. 26

AD 454 Feb 14–AD 455 Feb 2 [Korea] (Ahn [2004] 3)
28th year of King Biyu of Baekje; stars fell like rain.
[*Jeungbo munheon bigo*] ch. 7

AD 458 Mar 31–Apr 28 [China]
4th year of the Taian reign period of Emperor Gaozong of the Northern Wei Dynasty, 3rd month; several tens of thousands of meteors streamed westward.
[*Wei shu: tianxiang zhi*] ch. 105.3

AD 461 Mar 27–Apr 25 [China]

5th year of the Daming reign period of Emperor Wu of the (Southern Dynasties) Song Dynasty, 3rd month; several tens of millions of meteors, some long, some short, some large, some small, traveled westward together and did not stop until daybreak.

[*Song shu: tianwen zhi*] ch. 26

AD 461 Apr 13 [China]

5th year *xinchou* of the Daming reign period of Emperor Wu of the Liu Song Dynasty; in the 3rd month, on day *jiaxu* [11], at night a multitude of stars streamed westward.

Qing Dynasty, Jiaqing period ed. Nanjing [*Jiankang zhi*] ch. 8

AD 461 Apr 20 [China]

2nd year of the Heping reign period of Emperor Gaozong of the Northern Wei Dynasty, 3rd month, day *xinsi* [18]; a hundred large and small long stars more than 1 *pi* (4 *zhang*) long emerged from TIANJIN, extinguished and reemerged again.

[*Wei shu: tianxiang zhi*] ch. 105.3

AD 464 Mar 24–Apr 21 [China]

5th year of the Heping reign period of Emperor Gaozong of the Northern Wei Dynasty, 3rd month;...the Moon entered YUGUI [LM 23]. Innumerable meteors traveled westward (too many to count).

[*Gujin tushu jicheng: Shu zheng dian*] ch. 41

AD 465 Apr 12–May 10 [China]

6th year of the Heping reign period of Emperor Gaozong of the Northern Wei Dynasty, 3rd month; innumerable meteors traveled westward, not stopping till daylight.

[*Wei shu: tianxiang zhi*] ch. 105.3

AD 465 Jul 26 [China]

6th year of the Heping reign period of Emperor Gaozong of the Northern Wei Dynasty, 6th month, day *jimao* [16]; again there were meteors, most traveling southwestward.

[*Wei shu: tianxiang zhi*] ch. 105.3

AD 466 Apr 8 [China]

2nd year of the Taishi reign period of Emperor Ming of the (Southern Dynasties) Song Dynasty, 3rd month, day *yiwei* [32]; innumerable large and small meteors traveled westward, not ceasing until daybreak.
[*Song shu: tianwen zhi*] ch. 26

AD 466 May 22 [China]

2nd year of the Taishi reign period of Emperor Ming of the (Southern Dynasties) Song Dynasty, 4th month, day *jimao* [16]; at the end of the night more than 100 meteors traveled southwestward. One of them emerged from HEGU, was black and as big as a cup, with a tail more than 1 *zhang* long.
[*Song shu: tianwen zhi*] ch. 26

AD 530 Apr 9 [China]

3rd year of the Yong'an reign period of Emperor Zhuang of the Northern Wei Dynasty, 2nd month, day *renshen* [9]; large meteors in their thousands, their trails unbroken, accompanied each other in the northwest.
[*Wei shu: tianxiang zhi*] ch. 105.3

AD 532 Aug 28 [China]

4th year of the Zhong datong reign period of Emperor Wu of the (Southern Dynasties) Liang Dynasty, 7th month, day *jiachen* [41]; stars fell like rain.
[*Liang shu: Wu di ji*] ch. 3
[*Nan shi: Liang Wu di ji*] ch. 7

AD 532 Aug 28 [Korea] (Ahn [2004] 4)

10th year of King Seong of Baekje, 7th month, day *jiachen* [41]; stars fell like rain.
[*Jeungbo munheon bigo*] ch. 7

AD 534 Mar 1–29 [China]

3rd year of the Yongxi reign period of Emperor Xiao Wu of the Northern Wei Dynasty, 2nd month, a multitude of stars streamed northward.
[*Bei shi: Wei Wu di ji*] ch. 5

AD 566 Jun 4–Jul 2 [Korea] (Ahn [2004] 5)
8th year of King Pyeongwon of Goguryeo, 5th month; stars fell like rain.
[*Jeungbo munheon bigo*] ch. 7

AD 581 Mar 20 [Korea] (Ahn [2004] 6)
23rd year of King Pyeongwon of Goguryeo, 2nd month, last day of the month; stars fell like rain.
[*Jeungbo munheon bigo*] ch. 7

AD 585 Sep 23 [China]
5th year of the Kaihuang reign period of Emperor Gaozu of the Sui Dynasty, 8th month, day wushen [45]; several hundred meteors scattered in all directions and came down.
[*Bei shi: Sui Wen di ji*] ch. 11
[*Sui shu: Gaozu ji*] ch. 1

AD 586 Mar 24–Jun 21 [Korea] (Ahn [2004] 7)
(a) 8th year of King Jinpyeong of Silla, summer, 5th month; stars fell like rain.
[*Jeungbo munheon bigo*] ch. 7
(b) 8th year of King Jinpyeong of Silla, summer, 5th month; it thundered and stars fell like rain.
[*Samguk sagi*] ch. 5

AD 643 Nov 1 [Korea] (Ahn [2004] 8)
2nd year of King Bojang of Goguryeo, autumn, 9th month, 15th day; the night was clear but the moon was not seen; a myriad stars streamed westward.
[*Jeungbo munheon bigo*] ch. 7, [*Samguk sagi*] ch. 5

AD 647 Sep 5–Oct 3 [Korea] (Ahn [2004] 9)
(a) 1st year of Queen Jindeok of Silla, autumn, 8th month; a myriad stars streamed northward.
[*Jeungbo munheon bigo*] ch. 7

(b) 1st year of Queen Jindeok of Silla, autumn, 8th month; a broom star emerged in the east, then a myriad stars streamed northward. [*Samguk sagi*] ch. 5

AD 684 Nov 13–Dec 11 [Korea] (Ahn [2004] 10)
4th year of King Sinmun of Silla. winter, 10th month; from dusk till dark meteors [flew] in all directions.
[*Jeungbo munheon bigo*] ch. 7, [*Samguk sagi*] ch. 5

AD 685 Jan 1 [Japan]
12th year of Emperor Temmu, 11th month, day *wuchen* [5]; at dusk, seven stars streamed northeastward then fell there.
[*Nihon shoki*]

AD 685 Jan 3 [Japan]
12th year of Emperor Temmu, 11th month, day *gengwu* [7]; at sunset stars as large as jars fell in the east. By the hour of *xu* (LT = 19–21h) the constellations were all confused, as stars fell like rain.
[*Nihon shoki*]

AD 706 Apr 17–May 16 [Korea] (Ahn [2004] 11)
5th year of King Seongdeok of Silla, spring, 3rd month; a myriad stars streamed westward.
[*Jeungbo munheon bigo*] ch. 7, [*Samguk sagi*]

AD 707 Aug 2–31 [China]
3rd year of the Shenlong reign period of Emperor Zhongzong of the Tang Dynasty, 7th month;...in the clear night innumerable meteors criss-crossed in disorderly profusion. Toward late evening, more than 100 large meteors all with trails, sped and fell in profusion toward the southwest.
[*Kaiyuan zhanjing*] ch. 71

AD 708 Jul 14 [China]
2nd year of the Jinglong reign period of Emperor Zhongzong of the Tang Dynasty, 6th month, day *guiwei* [20], at night; innumerable

meteors sped by and fell in confused profusion, too many to record. The majority emerged from XU [LM 11], WEI [LM 12], HEGU, TIANJIN, GUANSUO, ZHINÜ, YULIN, WANGLIANG, and GEDAO.
[*Kaiyuan zhanjing*] ch. 71

AD 710 Jul 2–30 [China]

(a) 4th year of the Jinglong reign period of Emperor Zhongzong of the Tang Dynasty, 6th month; the stars in heaven scattered and fell like snow.
[*Ri she pian*] ch. 6

(b) 4th year of the Jinglong reign period of Emperor Zhongzong of the Tang Dynasty, 6th month; at the second watch, the stars in heaven scattered and fell like snow.
[*Ri zhi lu*]

AD 713 Jul 15 [China]

2nd year of the Xiantian reign period of Emperor Xuanzong of the Tang Dynasty, 6th month, day *gengxu* [47], at night Innumerable meteors traveled to the north and northwest from YULIN to the ZIWEI Palace, partly from HEGU and ZHINÜ.
[*Kaiyuan zhanjing*] ch. 71

AD 713 Jul 21 [China]

2nd year of the Xiantian reign period of Emperor Xuanzong of the Tang Dynasty, 6th month, day *bingchen* [53], at night; a vast number of small meteors criss-crossed on all sides, too many to record.
[*Kaiyuan zhanjing*] ch. 71

AD 714 Jul 15 [China]

2nd year of the Kaiyuan reign period of Emperor Xuanzong of the Tang Dynasty, 5th month, day *yimao* [52], the last day of the month; stars streamed toward the northwest, some like jars, some like peck measures, piercing BEIJI. The small ones were too numerous to count and the stars in heaven all wavered. This went on till daybreak.
[*Xin Tang shu: tianwen zhi*] ch. 32

AD 718 Oct 28–Nov 26 [Korea] (Ahn [2004] 12)

17th year of King Seongdeok of Silla, winter, 10th month; meteors entered KUI [LM 15] from MAO [LM 18], and a myriad small stars followed them. Heaven Dogs fell in azimuth *gen* (NE).
[*Jeungbo munheon bigo*] ch. 7, [*Samguk sagi*]

AD 735 May 30 [Japan]

7th year of the Tempyō reign period, 5th month, *jiwei* 56]; at night a myriad stars criss-crossed and moved confusedly, with no constant places.
[*Shoku Nihongi*]

AD 764 Dec 30 [China]

(a) 2nd year of the Guangde reign period of Emperor Daizong of the Tang Dynasty, 12th month, day *bingyin* [3]; stars fell like rain from the second watch of night until daylight.
[*Xin Tang shu: tianwen zhi*] ch. 32

(b) 2nd year of the Guangde reign period of Emperor Daizong of the Tang Dynasty, 12th month, day *bingyin* [3]; a multitude of stars fell.
[*Xin Tang shu: Daizong ji*] ch. 6

(c) 2nd year of the Guangde reign period of Emperor Daizong of the Tang Dynasty, 12th month, 3rd day, at night; stars streamed like rain, from the hour of *hai* [LT =21–23h] until daybreak.
[*Jiu Tang shu: tianwen zhi*] ch. 36

AD 764 Dec 31 [China]

2nd year of the Guangde reign period of Emperor Daizong of the Tang Dynasty, 12th month, day *dingmao* [4], at night; stars streamed like rain.
[*Jiu Tang shu: Daizong ji*] ch. 11

AD 765 Jan 7 [Korea] (Ahn [2004] 13)

23rd year of King Gyeongdeok of Silla, 3rd month; a star became fuzzy in the southeast; in winter, the 12th month, 11th day; there were

meteors, some large, some small; observers were unable to count them.
[*Jeungbo munheon bigo*] ch. 7

AD 767 Jul 30–Aug 28 [Korea]
3rd year of King Hyegong of Silla, autumn, 7th month; three stars fell in the King's court. Where touched they glowed like fire, on drawing near they disintegrated.
[*Jeungbo munheon bigo*] ch. 7
NB: Ahn (2004) omitted this record.

AD 773 Jan 10 [Japan]
3rd year of the Hōki reign period, day *jiwei* [56]; stars fell like rain.
[*Shoku Nihongi*]

AD 784 Jul 10 [China]
1st year of the Xingyuan reign period of Emperor Dezong of the Tang Dynasty, 6th month, day *wuwu* [55]; stars fell in fives and tens.
[*Xin Tang shu: tianwen zhi*] ch. 32

AD 784 Dec 24 [Japan]
3rd year of the Enryaku reign period, 11th month, 8th day *yisi* [42]; during the night from the hour of *xu* (LT = 19–21h) to the hour of *yin* (LT = 3–5h), the stars in the sky were all in motion, flying about profusely.
[*Nihon rei yi ki*]

AD 801 Sep 12–Oct 10 [Korea] (Ahn [2004] 14)
(a) 2nd year of King Aejang of Silla, autumn, 8th month; stars fell like rain.
[*Jeungbo munheon bigo*] ch. 7
(b) 2nd year of King Aejang of Silla, autumn, 9th month; Mars entered the moon and stars fell like rain.
[*Samguk sagi*]
NB: *Samguk sagi* date of "9th month" corresponds to Oct 11–Nov 9.

AD 808 Feb 24 [China]

3rd year of the Yuanhe reign period of Emperor Xianzong of the Tang Dynasty, 1st month, 25th day, at night; stars streamed like rain.
[*Kao yi*]
NB: The original was miswritten to read "the fixed stars" fell like rain.

AD 814 Jan 25–Feb 23 [China]

9th year of the Yuanhe reign period of Emperor Xianzong of the Tang Dynasty, 1st month; large stars like half a mat rose up from below, their glow lighting up the ground; multitudes of smaller stars trailed along with them.
[*Xin Tang shu: tianwen zhi*] ch. 32

AD 824 May 3–31 [China]

4th year of the Changqing reign period of Emperor Muzong of the Tang Dynasty, 4th month; within ZIWEI, stars fell in multitudes.
[*Xin Tang shu: tianwen zhi*] ch. 32

AD 830 Jul 22 [China]

(a) 4th year of the Taihe reign period of Emperor Wenzong of the Tang Dynasty, 6th month, day *xinwei* [8], at night; from the first to the fifth watch, large and small meteors [streamed] chaotically, observers were unable to count them all.
[*Jiu Tang shu: Wenzong ji*] ch. 17

(b) 4th year of the Taihe reign period of Emperor Wenzong of the Tang Dynasty, 6th month, day xinwei [8]; from twilight until the fifth watch, there were meteors, some large, some small; observers were unable to count them all.
[*Xin Tang shu: tianwen zhi*] ch. 32

AD 833 [China]

7th year of the Taihe reign period of Emperor Wenzong of the Tang Dynasty, 6th month, day *wuzi* [25]; from twilight until daybreak, there

were more than 100 large and small meteors going up, down, left, and right on all sides.
[*Xin Tang shu: tianwen zhi*] ch. 32
NB: There was no day *wuzi* in the 6th month. If it was 7th month, day *wuzi*, then the date was July 23.

AD 835 Jul 22 [China]

(a) 9th year of the Taihe reign period of Emperor Wenzong of the Tang Dynasty, 6th month, day *dingyou* [34], at night; from the first to the fourth watch, meteors danced chaotically up, down, left, and right in about 20 locations, most near the Milky Way.
[*Jiu Tang shu: tianwen zhi*] ch. 36

(b) 9th year of the Taihe reign period of Emperor Wenzong of the Tang Dynasty, 6th month, day *dingyou* [34], from twilight until the fourth watch, over 20 meteors [shot] up, down, left, and right, appeared and disappeared, most near the Milky Way.
[*Xin Tang shu: tianwen zhi*] ch. 32

AD 837 [China]

2nd year of the Kaicheng reign period of Emperor Wenzong of the Tang Dynasty, 9th month, day *dingyou* [34]; stars like large peck measures 5 *zhang* long streamed northwestward from Shi [LM 13] and Bi [LM 14] and disappeared beneath Dajiao. They moved like the Wangshi star, and the mid-sky thundered; hundreds of small stars trailed along with them.
[*Xin Tang shu: tianwen zhi*] ch. 32
NB: There was no day *dingyou* in that month.

AD 839 Apr 10 [China]

4th year of the Kaicheng reign period of Emperor Wenzong of the Tang Dynasty, 2nd month, day *yihai* (*jihai*) [12]; from the fourth watch to the fifth watch, in mid-sky on all sides there were over 200 meteors both large and small, with trails 2 to 5 *zhang* long, that streamed westward together.

[*Xin Tang shu: tianwen zhi*] ch. 32
NB: An original note reads that the 2nd month had *guichou* [50] as new moon, and hence it lacked day *jihai* [36]; *jihai* is a mistake for *yihai*.

AD 839 Apr 13 [China]
4th year of the Kaicheng reign period of Emperor Wenzong of the Tang Dynasty, 2nd month, 26th day; from the fourth to the fifth watch of the night, on all sides and overhead there were over 200 meteors both large and small, with trails from 2 *zhang* long that streamed westward together.
[*Jiu Tang shu: tianwen zhi*] ch. 36

AD 839 May 10 [China]
4th year of the Kaicheng reign period of Emperor Wenzong of the Tang Dynasty, 3rd month, 23rd day; from the first to the fifth watch, overhead and on all sides more than 100 meteors criss-crossed, appeared and disappeared.
[*Tang hui yao*] ch. 43

AD 841 Jun 23 [China]
1st year of the Huichang reign period of Emperor Wuzong of the Tang Dynasty, 6th month, day *gengzi* [37]; during the fifth watch of the night, over 50 small meteors streamed and scattered chaotically.
[*Jiu Tang shu: Wudi ji*] ch. 18

AD 841 Jul 21 [China]
(a) 1st year of the Huichang reign period of Emperor Wuzong of the Tang Dynasty, 6th month, 29th day; from the first to the fifth watch; over 50 small meteors streamed back and forth and scattered.
[*Jiu Tang shu: Tianwen zhi*] ch. 36
(b) 1st year of the Huichang reign period of Emperor Wuzong of the Tang Dynasty, 6th month, day *wuchen* [5]; from twilight until the

fifth watch; several tens of small stars streamed and scattered up, down, left, right.
[*Xin Tang shu: tianwen zhi*] ch. 32

AD 848 autumn [Korea] (Ahn [2004] 15)

(a) 10th year of King Munseong of Silla, autumn; stars fell like rain.
[*Jeungbo munheon bigo*] ch. 7

(b) 10th year of King Munseong of Silla, winter, 10th month; there was a sound like thunder in the sky.
[*Samguk sagi*]

AD 865 Aug 1 [China]

6th year of the Xiantong reign period of Emperor Yizong of the Tang Dynasty; 7th month, day *yiyou* [22]; during the first watch of night, large meteors several *zhang* long, their glare bright as lightning, flew from south to north, with a multitude of small stars trailing along.
[*Xin Tang shu: tianwen zhi*] ch. 32

AD 878 Jun 13 [Japan]

2nd year of the Gangyō reign period, 5th month, 9th day *jiachen* [41]; at the hour of *hai* (LT = 21–23h) a large meteor, with a reddish gleam and a tail over 2 *zhang* long, emerged from the south of DI [LM 3] and entered between ZHEN [LM 28] and YI [LM 27]. A myriad stars followed along, and where they passed, the leaves of the trees rustled.
[*Sandai jitsu roku*]

AD 881 Sep 10–18 [China]

(a) 1st year of the Zhonghe reign period of Emperor Xizong of the Tang Dynasty, 8th month, day *jichou* [26], at night; stars fell like rain, some like cups and bowls, streaming criss-cross as if weaving; day *gengyin* [27], at night, was also like this and it went on till day *dingyou* [34].
[*Xin Tang shu: tianwen zhi*] ch. 32

(b) 1st year of the Zhonghe reign period of Emperor Xizong of the Tang Dynasty, 8th month, day *jichou* [26]; a multitude of stars fell over Chengdu.
[*Xin Tang shu: Xizong ji*] ch. 9

AD 883 Dec 1–AD 884 Jan 1 [China]
3rd year of the Zhonghe reign period of Emperor Xizong of the Tang Dynasty, 11th month, at night; stars fell in the northwest like rain.
[*Xin Tang shu: tianwen zhi*] ch. 32

AD 884 Oct 28–29 [Japan]
8th year of the Gangyō reign period, 8th month, 4th day *renchen* [29]; from the hour of *xu* (LT = 19–21h) to the hour of *zi* (LT = 23–1h) small stars streamed and scattered in all directions and fell like rain. On the 5th day *guisi* [30], from sunset until the hour of repose (*rending*), meteors too numerous to count moved in and out of ZIWEIGONG, trespassing on many palaces, some moving in and out of BEIDOU, GUANSUO, LING, and outer lunar mansions.
[*Sandai jitsu roku*]

AD 905 Mar 9–Apr 7 [Korea] (Ahn [2004] 16)
9th year of King Hyogong of Silla, spring, 2nd month; stars fell like rain.
[*Jeungbo munheon bigo*] ch. 7

AD 905 Apr 13 [China]
2nd year of the Tianyou reign period of Emperor Aizong of the Tang Dynasty, 3rd month, day *yichou* [2]; at night, large meteors emerged in mid-sky like 5 peck measures and streamed northwestward until they were 10 *zhang* from the ground and then stopped. On top were star-like rays more than 1 and 1/2 *zhang* long that flamed scarlet and yellow like fire, and writhed like snakes. Smaller stars all moved southeastward and fell like rain, disappearing in an instant. Afterward there was a greenish-white vapor like a bamboo grove that rose up to

the zenith, there darkening. The prognostication said: Indeed, it is the WANGSHI star.
[*Xin Tang shu: Tianwen zhi*] ch. 32

AD 924 Jul 21 [China]
2nd year of the Tongguang reign period of Emperor Zhuangzong of the Five Dynasties Tang Dynasty, 6th month, day *jiashen* [21]; a multitude of stars streamed back and forth.
[*Xin Wudai shi: Sitian kao*] ch. 59

AD 924 Jul 23 [China]
2nd year of the Tongguang reign period of Emperor Zhuangzong of the Five Dynasties Tang Dynasty, 6th month, day *bingxu* [23]; a multitude of stars streamed back and forth.
[*Xin Wudai shi: Sitian kao*] ch. 59

AD 925 Jul 22 [China]
(a) 3rd year of the Tongguang reign period of Emperor Zhuangzong of the Five Dynasties Tang Dynasty, 6th month, day *gengyin* [27]; a multitude of stars streamed from the second watch till the third when they stopped.
[*Xin Wudai shi: Sitian kao*] ch. 59
(b) 3rd year of the Tongguang reign period of Emperor Zhuangzong of the Five Dynasties Tang Dynasty, 6th month, day *gengyin* [27]; after the first drum of the night, in the southwest there were over 70 meteors, all with trails, that streamed southwest.
[*Wudai hui yao*] ch. 10

AD 925 Jul 23 [China]
3rd year of the Tongguang reign period of Emperor Zhuangzong of the Five Dynasties Tang Dynasty, 6th month, day *xinmao* [28]; a multitude of small stars streamed in the southwest.
[*Xin Wudai shi: Sitian kao*] ch. 59

AD 926 Jul 22 [China]

1st year of the Tiancheng reign period of Emperor Mingzong of the Five Dynasties Tang Dynasty, 6th month, day *yiwei* [32]; a multitude of small stars streamed back and forth.
[*Xin Wudai shi: Sitian kao*] ch. 59

AD 927 Apr 13 [China]

2nd year of the Tiancheng reign period of Emperor Mingzong of the Five Dynasties Tang Dynasty, 3rd month, day *gengshen* [57]; a multitude of small stars streamed in the northwest.
[*Xin Wudai shi: Sitian kao*] ch. 59

AD 930 Sep 25 [China]

1st year of the Changxing reign period of Emperor Mingzong of the Five Dynasties Tang Dynasty, 9th month, new moon day *xinyou* [58]; a multitude of small stars streamed back and forth and fell.
[*Xin Wudai shi: Sitian kao*] ch. 59

AD 931 Oct 15 [China]

(a) 2nd year of the Changxing reign period of Emperor Mingzong of the Five Dynasties Tang Dynasty, 9th month, day *bingxu* [23]; at the beginning of the second watch of the night, in the northeast small meteors entered the bowl of BEIDOU and extinguished. Coming to the beginning of the fifth watch, north of the stations in the northwest there were meteors shaped like half-pint measures, first small, then bigger, that rapidly streamed into KUI [LM15] and then extinguished. The crooked tracks of their tails condensed in the sky and dissipated like clouds, their brilliance lighting up the ground. Then again in the northeast there were meteors like large peaches that emerged at XIATAI and sped northwestward until they extinguished beside the 3rd star of DOU's handle. After the fifth watch until daylight, there were more than 100 small meteors on all sides and in mid-sky that criss-crossed as they streamed and flowed.
[*Jiu Wudai shi: tianwen zhi*] ch. 139

(b) 2nd year of the Changxing reign period of Emperor Mingzong of the Five Dynasties Tang Dynasty, 9th month, day *bingxu* [23]; from the fifth drum (watch) until daylight, on all sides and in mid-sky more than 100 small meteors criss-crossed as they streamed and flowed.
[*Wudai hui yao*] ch. 10

(c) 2nd year of the Changxing reign period of Emperor Mingzong of the Five Dynasties Tang Dynasty, 9th month, day *bingxu* [23]; a multitude of stars streamed in back and forth.
[*Xin Wudai shi: Sitian kao*] ch. 59

AD 931 Oct 16 [China]

2nd year of the Changxing reign period of Emperor Mingzong of the Five Dynasties Tang Dynasty, 9th month, day *dinghai* [24]; a multitude of stars streamed back and forth and fell.
[*Xin Wudai shi: Sitian kao*] ch. 59

AD 933 Jul 20 [China]

4th year of the Changxing reign period of Emperor Mingzong of the Five Dynasties Tang Dynasty, 6th month, day *gengwu* [7]; a multitude of stars streamed in back and forth.
[*Xin Wudai shi: Sitian kao*] ch. 59

AD 933 Jul 25 [China]

4th year of the Changxing reign period of Emperor Mingzong of the Five Dynasties Tang Dynasty, 7th month, day *yihai* [12]; a multitude of stars streamed back and forth.
[*Xin Wudai shi: Sitian kao*] ch. 59

AD 934 Apr 13 [China]

1st year of the Yingshun reign period of Emperor Min of the Five Dynasties Tang Dynasty, 2nd month, day *dingyou* [34]; a multitude of stars streamed in the northwest.
[*Xin Wudai shi: Sitian kao*] ch. 59

AD 934 Oct 13 [China]

9th year of the Tianxian reign period of Emperor Taizu of the Liao Dynasty, 9th month, day *gengzi* [37]; in the southwest stars fell like rain. [*Liao shi: Taizu ji*] ch. 3

AD 934 Oct 14 [China]

(a) 1st year of the Qingtai reign period of Emperor Mo of the Five Dynasties Tang Dynasty, 9th month, day *xinchou* [38], at night; stars like five peck measures streamed in the southwest, the crooked tracks of their tails several *zhang* long and shaped like dragons. Then a multitude of stars streamed chaotically, too many to count. [*Jiu Wudai shi: Tang Modi ji*] ch. 46

(b) 1st year of the Qingtai reign period of Emperor Mo of the Five Dynasties Tang Dynasty, 9th month, day *xinchou* [38]; a multitude of stars streamed back and forth.
[*Xin Wudai shi: Sitian kao*] ch. 59

AD 967 Oct 14 [Japan]

(a) 4th year of the Kōhō reign period, 9th month, 9th day *jiawu* [31];… at the hour of *hai* (LT = 21–23h) meteors entered the Moon, from azimuth *gen* (NE) to azimuth *kun* (SW), a myriad stars moved confusedly in all directions, streaming and scattering all night.
[*Nihon ki ryaku*]

(b) 4th year of the Kōhō reign period, 9th month, 9th day; at the beginning of night, from the hour of *hai* (LT = 21–23h) to the hour of *yin* (LT = 3–5h), all over the sky a myriad stars streamed from east to west, scattering without interruption, their shapes like knives or swords.
[*Fusō ryaku ki*]

AD 970 [China]

3rd year of the Kaibao reign period of Emperor Taizu of the Song Dynasty, 9th month, day *gengwu* [7]; people in Guangzhou saw a multitude of stars all streaming northward.

[*Song shi: tianwen zhi*] ch. 57
[*Song shi: Nan Han Liu shi*] ch. 481
NB: There was no *gengwu* day in the 9th month. If it was the 10th month, the date was November 3.

AD 989 Jul 24 [Japan]

(a) 1st year of the Eiso reign period, 6th month, 19th day *wuchen* [5]; at the hour of *hai* (LY = 21–23h], in front of the building inside the middle gate of Kamo miyashiro, a large tree toppled over. Several stars emerged from the trees and flew together toward the south. [*Nihon ki ryaku*]

(b) 1st year of the Eiso reign period, 6th month, 19th day; in front of the Kamomiyashiro large trees toppled over. Several stars flew out of the trees and headed south together. [*Hyaku ren shō*]

AD 1002 Oct 12 [China]

5th year of the Xianping reign period of Emperor Zhenzong of the Song Dynasty, 9th month, day *bingshen* [33]; stars as big as peck measures streamed out of the east travelling southwestward and making a sound as of lowing cattle. Several tens of smaller stars trailed along with them and fell.
[*Wenxian tongkao: xiang wei*] ch. 291
NB: *Song shi: tianwen zhi.* chapter 57, has 8th month day *bingshen*, but there was no *bingshen* day in the 8th month.

AD 1002 Oct 14 [China]

(a) 5th year of the Xianping reign period of Emperor Zhenzong of the Song Dynasty, 9th month, day *wuxu* [35]; several tens of stars entered YUGUI [LM 23] and reached ZHONGTAI. All told, several tens of smaller stars trailed along with a large one, among them two were as large as a pint measure. One of these reached the LANG star and the other reached NANDOU [LM 8], where they extinguished. [*Wenxian tongkao: xiang wei*] ch. 291

(b) 5th year of the Xianping reign period of Emperor Zhenzong of the Song Dynasty, 9th month, day *wuxu* [35]; a large star trailed by several tens of smaller stars reached as far as ZHONGTAI, two among them reached the LANG star and NANDOU [LM 8], where they extinguished.
[*Xu Tongzhi: zaixiang*] ch. 171

NB: *Song shi: tianwen zhi*, chapter 57, has 8th month day *wuxu*, but there was no day *wuxu* in the 8th month.

AD 1002 Oct 14–15 [Japan]

(a) 4th year of the Chōhō reign period, 9th month, 6th day *wuxu* [35]; this night…meteors all night. On the 7th day *jihai* [36], on this night, from the hour of *zi* (LT = 23–1h) to the hour of *yin* (LT = 3–5h), there were meteors.
[*Nihon ki ryaku*]

(b) 4th year of the Chōhō reign period, 9th month, 7th day *jihai* [36]; this night there was a heavenly anomaly, from the hour of *zi* (LT = 23–1h) to the marks of the hour of *yin* (LT = 3–5h) there were meteors.
[*Honchō seiki*]

(c) 4th year of the Chōhō reign period, 9th month, 7th day; on the 6th day of this month there was an anomaly involving numerous stars, two gleams were eclipsed together, a myriad stars moved erratically.
[*Teiji kojitsu*]

AD 1007 Jul 20 [Japan]

4th year of the Kankō reign period, 6th month, 4th day *wuxu* [35]; on this night meteors streamed northward, this anomaly lasted several nights.
[*Nihon ki ryaku*]

AD 1008 Mar 26 [China]

1st year of the Dazhong xiangfu reign period of Emperor Zhenzong of the Song Dynasty, 2nd month, day *wushen* [45]; more than 10 orange meteors with trails streamed rapidly into CHONG.
[*Song shi: tianwen zhi*] ch. 57

AD 1012 Sep 11 [China]

5th year of the Dazhong xiangfu reign period of Emperor Zhenzong of the Song Dynasty, 8th month, day *wuwu* [55]; over 20 large and small stars, all with trails, streamed northward. Then one star, its tail more than 1 *zhang* long and 3 *chi* broad and whose brilliance lit up the ground, emerged outside ZIWEIYUAN and streamed northeastward, reaching CHUANSHE, where it vanished.
[*Song shi: tianwen zhi*] ch. 57

AD 1032 Oct 14 [China]

1st year of the Mingdao reign period of Emperor Renzong of the Song Dynasty, 9th month, day *bingzi* [13]; stars emerged from LOU [LM 16] and vanished beside YUNYU, their trails lasting a long time before dissipating. A moment later, other stars whose brightness lit up the ground emerged from TIANDAJIANGJUN, approached KUI [LM 15] and vanished, their trails lasting a long time before dissipating. Following this, stars emerged northwest of BEICHEN and flew rapidly to NEIJIE where they vanished. Another star emerged from TIANYUAN (γ Eri) and disappeared at TIANYUAN (μ1 Eri), its brightness lighting up the ground.
[*Song shi: tianwen zhi*] ch. 57

AD 1035 Oct 14 [Japan]

8th year of the Chōgen reign period 9th month, 11th day; at night during the hour of *yin* [LT 3–5h], there was a meteor [shower] anomaly.
[*Fusō ryaku ki*]

AD 1037 Aug 21 [China]

(a) 4th year of the Jingyou reign period of Emperor Renzong of the Song Dynasty, 7th month, day *wushen* [45]; several hundred stars streamed southwestward reaching DONGBI [LM 14], the brilliance of the largest lit up the ground. Black vapor more than 1 *zhang* long, emerged beneath lunar mansion BI [LM 19].
[*Song shi: Renzong ji*] ch. 10

(b) 4th year of the Jingyou reign period of Emperor Renzong of the Song Dynasty, 7th month, day *wushen* [45]; several hundred stars streamed southwestward together, the largest one reached DONGBI [LM 14] where it disappeared, its brilliance having lit up the ground for a long time before dissipating.
[*Song shi: tianwen zhi*] ch. 57

AD 1037 Oct 14 [Japan]

(a) 1st year of the Chōryaku reign period, 9th month, third day; a myriad stars fell confusedly, flying and scattering in all directions, everyone was amazed.
[*Hyaku ren shō*]

(b) 1st year of the Chōryaku reign period, 9th month, 3rd day; at night, from the hour of *zi* (LT = 23–1h) there was a meteor anomaly.
[*Fusō ryaku ki*]

(c) 1st year of the Chōryaku reign period, 9th month, 4th day; last night there was a heavenly anomaly of meteors.
[*Yukichika ki*]

AD 1038 Feb 7–Mar 8 [China]

6th year of the Chongxi reign period of Emperor Xingzong of the Liao Dynasty (change of reign title from 1st year of the Baoyuan reign period of Emperor Renzong of Song), spring, 1st month; a multitude of stars streamed northwestward.
[*Qidan guo zhi*] ch. 8
NB: *Qidan guo zhi* had Chongxi reign period 6th year, which is a mistake. According to the Western calendar, the conversion is based on the 1st year of the Baoyuan reign period, 1st month. If based on the first month of spring of the 6th year of the Chongxi reign period, the date would be Jan 19, 1037–Feb 17, 1037.

AD 1042 Jul 25 [Korea] (Ahn [2004] 17)

8th year of King Jeongjong of Goryeo, 7th month, day *dingwei* [44]; a myriad stars streamed and swirled.
[*Goryeosa*] ch. 47

AD 1063 Aug 22 [China]

(a) 8th year of the Jiayou reign period of Emperor Yingzong of the Song Dynasty, 7th month, day *yichou* [2]; several hundred large and small stars streamed westward.
[*Song shi: Yingzong ji*] ch. 13

(b) 8th year of the Jiayou reign period of Emperor Yingzong of the Song Dynasty, 7th month, day *yichou* [2]; several hundred stars flew westward, streaming back and forth.
[*Song shi: tianwen zhi*] ch. 57

AD 1091 May 19 [Korea]

8th year of King Sŏnjong of Koryŏ, 4th month, day dingsi [54]; red and green meteor(s) with tail(s) over 10 chi long emerged from FANG [LM 4] and entered XUANYUAN, making a sound like thunder.
[Goryeosa] ch. 47
NB: Ahn (2004), Yang (2005), and Imoto and Hasegawa (1958) did not omit this record.

AD 1095 Jul 25 [Korea] (Ahn [2004] 18)

1st year of King Sukjong of Goryeo, 6th month, day *yiyou* [22]; red meteors as large as papayas, with tails over 9 *chi* long, emerged from the west of SHI [LM 13] and entered the bowl of NANDOU [LM 8]. A myriad of small stars also streamed southward.
[*Goryeosa*] ch. 47

AD 1096 Jul 22–Aug 20 [Japan]

(a) 1st year of the Eichō reign period, 7th month; in azimuth *qian* (NW) there were several tens of stars like strung pearls stretching 4 to 5 *chi*.
[*Kōdai reki*]

(b) 1st year of the Eichō reign period, 7th [month]; in azimuth *qian* (NW) there were several tens of stars like strung pearls.
[*Kō nendai ryaku ki*]

AD 1102 Aug 5 [China]

1st year of the Chongning reign period of Emperor Huizong of the Song Dynasty, intercalary 6th month, day *guiyou* [10]; stars emerged from DOU [LM 8] and streamed slowly toward the southwest, reaching JIAN, where they disappeared. They were greenish-white and their tails left trails behind, while numerous small stars followed them.
[*Song shi: tianwen zhi*] ch. 60

AD 1103 Sep 7 [Korea] (Ahn [2004] 19)

8th year of King Sukjong of Goryeo, 8th month, day *xinhai* [48]; a meteor as large as a torch with a tail over 3 *zhang* long emerged from BEIJI and entered TIANJIN; another as large as a bowl with a tail over 1 *zhang* long emerged from NEIZHUHOU and entered SHANGTAI; another as large as a papaya emerged from WANGLIANG and entered WENCHANG. Then, over a hundred small stars streamed by.
[*Goryeosa*] ch. 47

AD 1106 Jul 27 [Korea] (Ahn [2004] 20)

1st year of King Yejong of Goryeo, 6th month, day *yiyou* [22]; Crooked Arrows flew and witnesses were all amazed. Meteors as large as cups, with tails over 2 *zhang* long, emerged from TIANJIN and entered ZONGREN; two meteors as big as hens' eggs also emerged from XU [LM 11] and entered JIUKAN. Then from dusk till dark a myriad stars streamed in all directions.
[*Goryeosa*] ch. 47

AD 1111 Oct 2 [Korea] (Ahn [2004] 21)

6th year of King Yejong of Goryeo, 8th month, day *wuwu* [55]; a myriad small stars separately streamed in all directions.
[*Goryeosa*] ch. 47

AD 1122 Jun 26–28 [China]

4th year of the Xuanhe reign period of Emperor Huizong of the Song Dynasty, 5th month, days *wuyin* [15], *jimao* [16], *gengchen* [17],

on successive evenings meteors like cups and bowls, their flashing brilliance like a bolt of silk, flew rapidly from south to west from between Ziwei and Wenchang, trespassing against Tianjin and Hegu, and between Milky Way of Wu (kingdom), Dou [LM 8], and Niu [LM 9]. This went on for more than 10 nights without stopping.
Ming Jiajing ed. Hebei [*Xiong cheng*] ch. xia

AD 1134 Jul 23–Aug 21 [China]
4th year jiayin [51] of the Shaoxing reign period of Emperor Gaozong of the Song Dynasty, autumn, 7th month; stars fell until dawn.
Qing Dynasty, Tongzhi period ed.; Jiangxi [*Nancheng xian zhi*] ch. 10

AD 1136 Apr 3 [Korea] (Ahn [2004] 22)
14th year of King Injong of Goryeo, 3rd month, day *wuchen* [5], new moon day; a myriad stars streamed from northeast to southwest.
[*Goryeosa*] ch. 47

AD 1178 Sep 17 [Korea] (Ahn [2004] 23)
8th year of King Myeongjong of Goryeo, 8th month, day *yiwei* [32]; a meteor as large as a papaya emerged from Kui [LM 15] and entered Tusikong; one with a tail over 1 *zhang* long emerged from Neiping and entered Xuanyuan; one as large as a papaya, with a tail over 7 *chi* long, emerged from the seventh star of Qixing [LM 25] and entered Zhang [LM 26]; one with a tail over 5 *chi* long emerged from the 5th star of Que[-qiu] and entered Hu[-shi]; one as large as a papaya, with a tail over 15 *chi* long, emerged from Xiang and entered Changzhen; one as large as a papaya, with a tail over 5 *zhang* long, emerged from Dongjing [LM 22] and entered Wuzhuhou; one with a tail over 1 *zhang* long emerged from Yi [LM 27] and, pointing toward the south, entered the horizon; then, a myriad stars streamed in all directions, too many to count.
[*Goryeosa*] ch. 47
NB: Queqiu only has two stars, so the reference to the "fifth star" here may be a copyist's error deriving from the "5 *chi*" length of the tail; the same may also be true of the reference to the "seventh star" of Qixing.

AD 1179 Apr 17 [Korea] (Ahn [2004] 24)

9th year of King Myeongjong of Goryeo, 3rd month, day *dingmao* [4]; over a hundred small stars streamed from east to west.
[*Goryeosa*] ch. 47

AD 1179 Dec 17 [Japan]

3rd year of the Jishō reign period, 11th month, 17th day *xinwei* [8];… this night in azimuth *kun* (SW) there were large and small meteors that approached and stayed below the Great Star[s], entering and merging with its gleam, then retreating again; now approaching and staying again, now retreating, and so forth.
[*San kai ki*]

AD 1238 Oct 18 [Japan]

1st year of the Ryakunin reign period, 9th month, 7th day *xinsi* [18];… further, from the marks of the hour of *hai* [LT = 21–23h] to the hour of *chou* [LT = 1–3h], there were innumerable reddish white meteors some 7 to 8 *chi* long, some 3 to 4 *chi* long…
[*Azuma kagami*]

AD 1247 Apr 18 [Japan]

1st year of the Hōji reign period, 3rd month, 2nd day *yichou* [2]; it was said that slightly past the marks of the hour of *xu* [LT = 19–21h] there was a bright object, its shape said to be like a large meteor. The glare flew about erratically, not unlike a flame, and after entering the western sky from the east its sound resonated. The eyes and ears of thousands were dazzled; it was unprecedented.
[*Hyaku ren shō*]

AD 1304 Sep 7 [Japan]

2nd year of the Kagen reign period, 8th month, 7th day; at night, just as a great wind and rainstorm reached their peak, five luminous objects flew away from Kasugayama toward azimuth *qian* (NW), their gleam like the disc of the sun.
[*Kōfukuji ryaku nendai ki*]

AD 1321 Jun 17 [Japan]

1st year of the Genkyō reign period, 5th month, 21st day *jiawu* [31]; misty rain fell, on this night everyone said that slightly after the marks of the hour of *hai* [LT = 21–23h] something resembling meteors flew across the sky, their gleam quite like the moon. Some said they were round and greenish-white and as big as peach pits as they flew into the clouds; and again that they flew successively, in great numbers, and so forth. This night there was a sound that resonated, but not the sound of an earthquake or thunder. Some said it was the general's tomb rumbling.

[*Hanazono-in shin ki*]

AD 1325 Dec 9 [Japan]

2nd year of the Shōchū reign period, 11th month, 4th day *gengxu* [47]; it was clear. On this night there was a great meteor and everyone said it burned up…on the 6th day *renzi* [49], Yasuyo privately submitted a prognostication on the heavenly anomaly…"great meteors on successive nights, when anomalies like this recur, can one be anything but frightened."

[*Hanazono-in shin ki*]

AD 1339 Aug 6–Sep 3 [Japan]

4th year of the Engen reign period, 7th month; there were baleful signs of different kinds. Meteors flew repeatedly; some nights there were clouds in the east, their color black with reddish-green rainbows in the center, their glow was like daylight.

[*Nanchō hen nen ki ryaku*]

AD 1363 May 14–Jun 11 [Korea]

12th year of King Gongmin of Goryeo, 4th month; stars battled in mid-sky.

[*Jeungbo munheon bigo: xing bian*]

NB: Ahn (2004) excluded this record as ambiguous (#25).

AD 1368 Mar 19–Apr 17 [China]

28th year of the Zhizheng reign period of Emperor Shun of the Yuan Dynasty, 3rd month; stars streamed in the northeast thundering loudly and trailed by a multitude of small stars.
[*Yuan shi: Shundi ji*] ch. 47

AD 1410 Feb 15 [Korea] (Ahn [2004] 26)

10th year of King Taejong of the Joseon Dynasty, 1st month, day *jimao* [16]; Mars was east of JING [LM 22] and the moon was 1 *chi* to the north of Mars. On this day at midnight, in Godabchon in Hwangju (Hwanghaedo)...presently it began to thunder from the east. At first, the sound was as of cannon fire, then like the sound of a bell, then later like the sound of rifles, and a myriad stars wobbled.
[*Joseon wangjo sillok: Taejong*] ch. 19

AD 1414 Jun 15 [Korea] (Ahn [2004] 27)

14th year of King Taejong of the Joseon Dynasty, 5th month, day *gengzi* [37]; the air was cool like autumn. At dusk, meteors emerged from mid-sky, making noise as they streamed eastward. At first, they were like bottles, by the end not as big, and greenish-red in color, with flames shooting in all directions, their light as bright as the moon. It was like this until midnight.
[*Joseon wangjo sillok: Taejong*] ch. 27

AD 1436 Oct 2 [China]

(a) 1st year of the Zhengtong reign period of Emperor Yingzong of the Ming Dynasty, 8th month, day *yiyou* [22]; from twilight through dawn there were more than 100 large and small meteors.
[*Ming Yingzong shilu*] ch. 21

(b) 1st year of the Zhengtong reign period of Emperor Yingzong of the Ming Dynasty, 8th month, day *yiyou* [22]; from twilight through dawn there were more than 100 large and small meteors.
[*Ming shi: Tianwen zhi*] ch. 27

AD1436 Dec 15 [Japan]

(a) 8th year of the Eikyō reign period, 11th month, 8th day, it was clear; toward the hour of *hai* [LT = 21–23h] at night heavenly anomalies flew. Afterward, above the rooftops objects descended thundering, all faces turned toward heaven and saw it.
[*Kammon gyōki*]

(b) 8th year of the Eikyō reign period, 11th month, 8th day *jihai* [36]; at night, there were meteors streaming from the north to the south. Within their gleam were objects like stars that flew about confusedly; after they disappeared there was a sound.
[*Ya shi*]

AD 1437 Feb 13 [Japan]

(a) 9th year of the Eikyō reign period, 1st month, 8th day; during the marks of the hour of *wei* (LT = 13–15h), bright objects in the middle of LIANCANG flew around for several *du*.
[*Nampō ki ten*]

(b) 9th year of the Eikyō reign period, 1st month, 8th day, *jihai* [36]; on this day, during the hour of *wei* (LT = 13–15h), numerous strange bright objects flew.
[*Zoku shi gushō*]

AD 1439 Oct 5 [China]

(a) 4th year of the Zhengtong reign period of Emperor Yingzong of the Ming Dynasty, 8th month, day *guimao* [40]; from nightfall till dawn there were over 260 large and small meteors.
[*Ming Yingzong shilu*] ch. 58

(b) 4th year of the Zhengtong reign period of Emperor Yingzong of the Ming Dynasty, 8th month, day *guimao* [40]; there were several hundred large and small meteors.
[*Ming shi: tianwen zhi*] ch. 27

AD 1447 Jul 30 [China]

12th year of the Zhengtong reign period of Emperor Yingzong of the Ming Dynasty, 6th month, day *wuyin* [15]; during the fourth watch of the night in the west, there were greenish-white stars like bowls that streamed for more than 1 *zhang*, emitting a brilliance as big as peck measures that lit up the ground. They started from TIANJIN and travelled southeast as far as YULINJUN, their contrails dissipating explosively. Behind, 10 or more smaller stars followed them making a sound like thunder.

[*Ming Yingzong shilu*] ch. 155

AD 1451 Jul 27 [China]

(a) 2nd year of the Jingtai reign period of Emperor Jing of the Ming Dynasty, 6th month, day *bingshen* [33], at night; there were 85 large and small meteors.

[*Ming Yingzong shilu*] ch. 205

(b) 2nd year of the Jingtai reign period of Emperor Jing of the Ming Dynasty, 6th month, day *bingshen* [33], there were over 80 large and small meteors.

[*Ming shi: tianwen zhi*] ch. 27

AD 1451 Sep 11 [China]

(a) 2nd year of the Jingtai reign period of Emperor Jing of the Ming Dynasty, 8th month, day *renwu* [19], at night; there were two meteors, one as big as a hen's egg, one as big as a peck measure. Both were red, and lit up the ground with their brilliance. One emerged from ZIWEI's western boundary and traveled north as far as YINDE while three small stars trailed along. The other emerged from TIANJIN and traveled southeast as far as NANHE, while more than 10 smaller stars trailed along, their trails dispersing explosively and making a sound like thunder.

[*Ming Yingzong shilu*] ch. 207

(b) 2nd year of the Jingtai reign period of Emperor Jing of the Ming Dynasty, 8th month, day *renwu* [19]; there were two scarlet meteors, one like a peach, one like a peck measure, their brilliance lighting up the ground. One emerged from ZIWEI's western boundary and traveled north as far as YINDE while three small stars trailed along. The other emerged from TIANJIN and traveled southeast as far as NANHE, while more than 10 smaller stars trailed along, their contrails dispersing explosively with a sound like thunder.
[*Ming shi: tianwen zhi*] ch. 27

NB: In both cases, NANHE was mistakenly transcribed as Henan.

AD 1457 Aug 20–Sep 18 [China]

Dingchou [14] year of the Tianshun reign period of Emperor Yingzong of the Ming Dynasty, 8th month; stars fell like rain.
Qing Dynasty Kangxi period ed. Hebei [*Ansuxian zhi*] ch. 3

AD 1457 Dec 7 [China]

1st year of the Tianshun reign period of Emperor Yingzong of the Ming Dynasty, 11th month, day *xinsi* [18], at night; greenish-white meteors as large as cups lit up the ground with their brilliance. They emerged from HUAGAI and traveled northeast as far as TIANGOU. More than 10 smaller stars trailed along making a booming sound.
[*Ming Yingzong shilu*] ch. 284

AD 1462 Aug 10 [China]

6th year of the Tianshun reign period of Emperor Yingzong of the Ming Dynasty, 7th month, day *wushen* [45], at night; in the north greenish-white meteors as large as bowls lit up the ground with their brilliance. They emerged from TIANYUAN and traveled northward, entering lunar mansion SHEN [LM 21]. More than 10 smaller stars trailed along.
[*Ming Yingzong shilu*] ch. 342

AD 1466 Jul 28 [China]

(a) 2nd year of the Chenghua reign period of Emperor Xianzong of the Ming Dynasty, 6th month, day *bingchen* [53], at night; brilliant greenish-white meteors as big as oil lamps traveled from due north southeastward into the clouds. Behind more than 10 smaller stars trailed along.
[*Ming Xianzong shilu*] ch. 31

(b) 2nd year of the Chenghua reign period of Emperor Xianzong of the Ming Dynasty, 6th month, day *bingchen* [53], at night; meteors traveled from the north southeastward; several smaller stars trailed along.
[*Guo que*] ch. 34

AD 1467 Jun 2 [China]

(a) 3rd year of the Chenghua reign period of Emperor Xianzong of the Ming Dynasty, 5th month, new moon day *yichou* [2], at night; in the south greenish-white meteors as large as oil lamps lit up the ground with their brilliance. From HEGU they traveled more than 1 *zhang*, the light they emitted as large as a bowl. They traveled northeastward as far as lunar mansion KUI [LM 15], the trails of their tails dissipating explosively. Behind, more than 10 smaller stars trailed along.
[*Ming Xianzong shilu*] ch. 42

(b) 3rd year of the Chenghua reign period of Emperor Xianzong of the Ming Dynasty, 5th month, new moon day *yichou* [2], at night; large stars streamed from HEGU to lunar mansion KUI [LM 15], while several smaller stars trailed along.
[*Guo que*] ch. 35

AD 1468 Apr 7 [Japan]

2nd year of the Ōnin reign period, 3rd month, 16th day *gengzi* [37];… yesterday during light snow, three bright objects flew from the north toward the south.
[*Kyōgaku shi yōshō*]

AD 1477 Apr 18 [Japan]

(a) 9th year of the Bummei reign period, 3rd month, 5th day; in the middle of the capital bright objects were said to fly about, making resonating noises.
[*Nagaoki sukune ki*]

(b) 9th year of the Bummei reign period, 3rd month, 5th day *guiyou* [10]; in the capital there was a strange brightness, flying about to the east, west, south, north.
[*Zoku shi gushō*]

(c) 9th year of the Bummei reign period, 3rd month, 5th day *guiyou* [10]; there were meteors separately flying in all directions.
[*Ya shi*]

AD 1498 Oct 24 [China]

(a) 11th year of the Hongzhi reign period of Emperor Xiaozong of the Ming Dynasty, 10th month, day *renshen* [9]; during the hour of daybreak, scarlet meteors as large as oil lamps streamed in the east, starting in the northeast and traveling for more than 1 *zhang*. The light they emitted was as large as a peck measure, and their brilliance lit up the ground. They traveled southeastward, while several tens of smaller stars trailed along.
[*Ming Xiaozong shilu*] ch. 142

(b) 11th year of the Hongzhi reign period of Emperor Xiaozong of the Ming Dynasty, 10th month, day *renshen* [9]; at daybreak in the east, scarlet stars like bowls traveled southeastward for more than 1 *zhang*, their brilliance lighting up the ground, while several tens of smaller stars trailed along.
[*Ming shi: Tianwen zhi*] ch. 27

AD 1502 Feb 7–AD 1503 Jan 27 [China]

15th year of the Hongzhi reign period of Emperor Xiaozong of the Ming Dynasty; stars fell like rain.
Qing Dynasty Kangxi period ed. Hebei [*Shenzhou zhi*] ch. 7

AD 1504 May 19 [Korea] (Ahn [2004] 28)

10th year of King Yeonsangun, intercalary 4th month, day *bingyin* [3], at night; a meteor emerged from DI [LM 3] and entered ZHENCHE. Another emerged from below the handle of BEIDOU and entered below DAJIAO. Another emerged from ZHEN [LM 28] and entered JIAO [LM 1]. Another emerged from TIANJIANG and entered FU YUE. Another emerged from TUGONG and entered WUJI. Another emerged from below NANDOU and disappeared at the southern horizon. Another emerged from DONGXIAN and entered CONGGUAN. A red one shaped like a pear with a tail 7 to 8 *chi* long emerged below GENGHE and entered ZHEWEI. Another red one shaped like a large pear with a tail 3 to 4 *chi* long emerged below GOUZHEN and entered TIANLI.

[*Joseon wangjo sillok: Yeonsangun*] ch. 52

AD 1505 Jun 13 [Japan]

2nd year of the Eishō reign period, 5th month, 12th day; on this night three stars joined and descended: an evil event.

[*Ni sui ki*]

AD 1506 Mar 25–Apr 22 [China]

(a) 1st year of the Zhengde reign period of Emperor Wuzong of the Ming Dynasty, 3rd month; stars fell like rain.
Qing Dynasty Kangxi period ed. Hebei [*Dachengxian zhi*] ch. 8

(b) 1st year of the Zhengde reign period of Emperor Wuzong of the Ming Dynasty, 3rd month; stars fell like rain, some were as large as the Moon.
[*Zui wei lu*] zhi 1

AD 1510 May 11 [Japan]

(a) 7th year of the Eishō reign period, 4th month, 4th day; during the marks of the hour of *xu* [LT = 19–21h], the mountains shook with sound and numerous bright objects swirled about and changed, and many acolytes 4 to 5 years old assisted at the shrine where the anomaly occurred.
[*Sanetaka kōki*]

(b) 7th year of the Eishō reign period, 4th month, 4th day; at night in front of the shrine numerous bright objects lit up and changed. Many acolytes 5 to 6 years old assisted at the shrine; behind them in the palace the sound resonated, it was outrageous!
[*Go jigen-in dono ki*]

AD 1511 Jun 28 [China]
6th year of the Zhengde reign period of Emperor Wuzong of the Ming Dynasty, 6th month; day *renwu* [19]; in Linjiangfu in Jiangxi Province, during the night, fiery stars streamed back and forth. In a moment, a fire as big as a wagon wheel, with a tail several *zhang* long, fell to the northwest of the city. In Ji'anfu, fiery stars like peck measures streamed from the northwest toward the southeast, their brilliance lighting up the ground and the sound of their falling like thunder.
[*Ming Wuzong shilu*] ch. 76

AD 1517 Feb 2 [Japan]
14th year of the Eishō reign period, 1st month, 12th day; on this night three stars joined together and came down; an evil event.
[*Takayasu shiki*]

AD 1517 Oct 11 [Korea] (Ahn [2004] 29)
12th year of King Jungjong of the Joseon Dynasty, 9th month, day *jihai* [36], at night; a white meteor the size of a fist with a tail about 5 to 6 *chi* long emerged from WENCHANG and entered BEIDOU. A red one the size of a fist and with a tail about 4 to 5 *chi* long emerged from TIANZHU and entered TIANJI. Another red one the size of a fist and about 3 to 4 *chi* long emerged from TIANJIN and entered BAIGUA. Another red meteor the size of a cup, with a tail about 3 to 4 *chi* long emerged from BI [LM 19] and entered SHUIFU. Another red meteor like a fist, with a tail about 5 to 6 *chi* long emerged from TIANZUN and entered CHANGZHEN.
[*Joseon wangjo sillok: Jungjong*] ch. 29

AD 1519 Jul 5 [Korea]

14th year of King Jungjong of the Joseon Dynasty, 6th month, day *xinwei* [8]; on this night in Kyongsangdo of Kyongju there was a heavenly anomaly. At the beginning of dusk, the moon was exceptionally bright, and in the west there was a dim haziness within which there was a gleam like lightning; it was not lightning but resembled flame. At times, it was like streaming arrows all over the sky, and at times, as if meteors streaked past, sometimes like red snakes wriggling, and sometimes like sparks scattering, sometimes bent like a drawn bow, sometimes forked like the tines of a hairpin; displaying a hundred shifting shapes. Now visible, now hidden, criss-crossing in rapid pursuit of one another, alternately appearing and disappearing by turns, constantly without let up, in all like cannon fire. Their light flashed and glittered, lighting up darkened inner rooms, starting from the west and gradually moving northwestward, until the third watch when they were extinguished. The Surveillance Commissioner was informed by Anro Kim of Booyoon.

[*Joseon wangjo sillok: Jungjong*] ch. 36 (Ahn [2004] 30)

AD 1520 Apr 17 [Korea]

15th year of King Jungjong of the Joseon Dynasty, 4th month, day *wuwu* [55]; at night a red meteor the size of a bowl with a tail about 4 to 5 *chi* long emerged from BAGU and entered WUCHE; [another] red one the size of a fist with a tail 3 to 4 *chi* long emerged from GUANSUO and entered FANG [LM 4]; a green one the size of a pear with a tail 3 to 4 *chi* long emerged from the eastern wall of TAIWEI and entered DI [LM 3]; a white one the size of a large pear with a tail 4 to 5 *chi* long emerged from NÜ [LM 10] and entered DOU [LM 8]; a red one the size of a fist with a tail 4 to 5 *chi* long emerged from NÜ [LM 10] and entered DAJIAO.

[*Joseon wangjo sillok: Jungjong*] ch. 39 (Ahn [2004] 31)

AD 1523 Jan 17–1524 Feb 3 [China]

In the guiwei [20] year of the Jiajing reign period of Emperor Shizong of the Ming Dynasty; at midnight stars fell like rain, their scarlet gleam lighting up the ground.

Qing Dynasty Guangxu period ed. Anhui [*Sihongxian zhi*] ch. 19

AD 1528 Feb 11 [China]

7th year of the Jiajing reign period of Emperor Shizong of the Ming Dynasty, 1st month, day *jiawu* [31]; during the night, there suddenly appeared a thousand lamps that scattered and fell on the seashore, as if carried by people as they moved. The local people were greatly agitated, fearing the approach of pirates, but before long they were extinguished. In the 13th year of the Chenghua reign period (1477) it had been seen once before.

Ming Dynasty Jiajing period ed. Zhejiang [*Hejianxian zhi*] ch. 5

AD 1528 Oct 13–Nov 11 [China]

7th year of the Jiajing reign period of Emperor Shizong of the Ming Dynasty, 10th month, in Hejian stars fell like rain.

Qing Dynasty Kangxi period ed. Hebei [*Hejianxian zhi*] ch.11

[*Er shen ye lu*] ch. 4

[*Ming hui yao*] ch. 69

AD 1529 Nov 1–30 [China]

8th year of the Jiajing reign period of Emperor Shizong of the Ming Dynasty, 10th month; stars fell like rain.

Ming Dynasty Wanli period ed. Shandong [*Yizhou zhi*] ch. 1

AD 1531 Jan 18–AD 1532 Feb 5 [China]

(a) 10th year of the Jiajing reign period of Emperor Shizong of the Ming Dynasty; stars fell like rain.

Ming Dynasty Jiajing period ed. Fujian [*Changtaixian zhi*] ch. xia

Qing Dynasty Kangxi period ed. Hebei [*Pingxiangxian zhi*] ch. 3

(b) 10th year of the Jiajing reign period of Emperor Shizong of the Ming Dynasty; large stars scattered and fell like rain.
Qing Dynasty Kangxi period ed. Hebei [*Pingshanxian zhi*] ch. 1
Qing Dynasty Kangxi period ed. Hebei [*Juluxian zhi*] ch. 8

AD 1531 Nov 15 [China]

Xinmao [28] year of the Jiajing reign period of Emperor Shizong of the Ming Dynasty, 10th month, 7th day *bingzi* [13], at night; stars fell like rain.
Qing Dynasty Jiaqing ed. Shandong [*Changlexian zhi*] ch. 1
NB: The 7th day of the 10th month was *dinghai* [24]—not *bingzi* [13].

AD 1532 Feb 6 [China]

11th year of the Jiajing reign period of Emperor Shizong of the Ming Dynasty, spring, 1st month, 1st day of the month, at night; stars fell like peck measures, making a sound like thunder.
[*Jiang ying ke wen ji: Ji zaiyi*]

AD 1532 Oct 24 [Korea]

(a) 27th year *renchen* [29] of King Jungjong of the Joseon Dynasty, 9th month, day *xinwei* [8], at night; stars fell like rain.
[*Jeungbo munheon bigo*] ch. 7

(b) 27th year of King Jungjong of the Joseon Dynasty, 9th month, day *xinwei* [8]; at night, a red meteor with a tail about 4 to 5 *chi* long emerged below XUANYUAN and entered the horizon to the east; a white one with a tail about 1 *zhang* long emerged from BEIDOU and entered the horizon to the north; another white one with a tail 3 to 4 *chi* long emerged from WUCHE and entered BI [LM 19]; another red one with a tail about 4 to 5 *chi* long emerged from SHEN [LM 21] and entered JIUYOU; another white one with a tail about 5 to 6 *chi* long emerged from LIU [LM 24] and entered the horizon to the south; another red one with a tail about 1 *zhang* long emerged from BI [LM 19] and entered TIANCANG, its tail was large, bright and straight, gradually

becoming crooked. After a long while, it turned into a cloud and by the next [double] hour it disappeared. Another red one with a tail about 5 to 6 *chi* long emerged from MAO [LM 18] and entered below TIANQUN, with a bright and straight tail that after a long while turned into a cloud and disappeared. Stars fell like rain; a white broom star with a tail 4 to 5 *chi* long appeared at azimuth *mao* (E).
[*Joseon wangjo sillok: Jungjong*] ch. 73 (Ahn [2004] 32)

AD 1532 Oct 25 [China]
11th year of the Jiajing reign period of Emperor Shizong of the Ming Dynasty, 9th month, 27th day; during the fourth watch of the night, stars fell like rain.
Qing Dynasty Kangxi period ed. Hebei [*Dong anxian zhi*] ch. 1

AD 1532 Oct 28–Nov 26 [China]
(a) 11th year of the Jiajing reign period of Emperor Shizong of the Ming Dynasty, 10th month; stars fell like rain.
Ming Dynasty, Jiajing ed. Guangdong [*Xingningxian zhi*] ch. 1
Qing Dynasty, Qianlong ed. Shanxi [*Fengtaixian zhi*] ch. 12
(b) 11th year of the Jiajing reign period of Emperor Shizong of the Ming Dynasty, 10th month, at night; stars fell like rain.
Qing Dynasty, Shunzhi ed. Shanxi [*Yangchengxian zhi*] ch. 7

AD 1532 Nov 3 [China]
(a) 11th year of the Jiajing reign period of Emperor Shizong of the Ming Dynasty, 10th month, 7th day, at night; stars fell like rain, their brilliance reddening the sky, then they scattered and fell in all directions.
Qing Dynasty, Kangxi period ed. [*Shanxi tongzhi*] ch. 30
Qing Dynasty, Kangxi period ed. Shanxi [*Pingyangfu zhi*] ch. 34
(b) 11th year of the Jiajing reign period of Emperor Shizong of the Ming Dynasty, 10th month, 7th day, at night; stars fell in all directions.
[*Jiang ying ke wen ji: Ji zaiyi*]

AD 1532 Nov 27–Dec 25 [China]

(a) 11th year of the Jiajing reign period of Emperor Shizong of the Ming Dynasty, 11th month, at night; stars fell like rain.
Ming Dynasty, Wanli period ed. Shanxi [*Fenzhoufu zhi*] ch. 16
Qing Dynasty, Kangxi period ed. Shanxi [*Lingshixian zhi*] xiangyi

(b) 11th year of the Jiajing reign period of Emperor Shizong of the Ming Dynasty, 11th month, at night; the stars in the sky showered down like snow, their brilliance lighting up the ground.
Qing Dynasty, Kangxi period ed. Shandong [*Jinanfu zhi*] ch. 10
Qing Dynasty, Kangxi period ed. Shandong [*Binzhou zhi*] ch. 8
Qing Dynasty, Kangxi period ed. Shandong [*Changshanxian zhi*] ch. 7

AD 1532 Dec 6 [China]

11th year of the Jiajing reign period of Emperor Shizong of the Ming Dynasty, 11th month, 10th day, at night; stars fell, their reddish brilliance showering down on all sides, like thrown bricks and stones, without stopping until daybreak.
Qing Dynasty, Kangxi period ed. Henan [*Yanjinxian zhi*] ch. 7

AD 1532 winter [China]

11th year of the Jiajing reign period of Emperor Shizong of the Ming Dynasty, winter, in Zezhou, Yangcheng, Lingshi stars fell like rain.
Qing Dynasty, Yongzheng period ed. [*Shanxi tongzhi*] ch. 163

AD 1533 Jul 22–Aug 19 [China]

12th year of the Jiajing reign period of Emperor Shizong of the Ming Dynasty, 7th month; northwest of Ji'anfu stars fell like rain.
Qing Dynasty, Kangxi period ed. [*Jiangxi tongzhi*] ch. 4
Qing Dynasty, Qianlong period ed. Jiangxi [*Lulingxian zhi*] ch. 1

AD 1533 Aug 20–Sep 17 [China]

(a) 12th year of the Jiajing reign period of Emperor Shizong of the Ming Dynasty, 8th month, at night; stars fell like rain.
Qing Dynasty, Shunzhi period ed. Zhejiang [*Changshanxian zhi*] ch. 1

(b) 12th year of the Jiajing reign period of Emperor Shizong of the Ming Dynasty, 8th month; stars fell like rain.
Qing Dynasty, Kangxi period ed. Zhejiang [*Quzhoufu zhi*] ch. 30

(c) 12th year of the Jiajing reign period of Emperor Shizong of the Ming Dynasty, 8th month; stars fell in even streaks like rain.
Qing Dynasty, Jiajing period ed. Hebei [*Zhendingfu zhi*] ch. 6

(d) 12th year of the Jiajing reign period of Emperor Shizong of the Ming Dynasty, 8th month; stars fell like rain with a brilliance like fire, so that the sky was completely red.
Qing Dynasty, Qianlong ed. Shanxi [*Pingyangfu zhi*] ch. 34

AD 1533 Sep 18–Oct 17 [China]

12th year of the Jiajing reign period of Emperor Shizong of the Ming Dynasty, 9th month; stars fell like rain.
Qing Dynasty, Kangxi period ed. Fujian [*Lianchengxian zhi*] ch. 1
Ming Dynasty, Wanli period ed. Anhui [*Luzhoufu zhi*] ch. 1
Ming Dynasty, Wanli period ed. Anhui [*Hefeixian zhi*] ch. shang

AD 1533 autumn [China]

12th year of the Jiajing reign period of Emperor Shizong of the Ming Dynasty, autumn; in the northwest a multitude of stars fell like rain.
Ming Dynasty, Wanli period ed. Jiangxi [*Yongxinxian zhi*] ch. 5
Qing Dynasty, Shunzhi period ed. Jiangxi [*Ji anxian zhi*] ch. 1
Qing Dynasty, Kangxi period ed. Jiangxi [*Lulingxian zhi*] ch. 2
Qing Dynasty, Kangxi period ed. Jiangxi [*Anfuxian zhi*] ch. 1
Qing Dynasty, Qianlong period ed. Jiangxi [*Longquanxian zhi*] ch. mo
Qing Dynasty, Qianlong period ed. Jiangxi [*Taihexian zhi*] ch. 28

AD 1533 Oct 18–Nov 15 [China]

(a) 12th year of the Jiajing reign period of Emperor Shizong of the Ming Dynasty, 10th month, stars fell like rain.
[*Ming shi: guozonggao zhuan*] ch. 200

(b) 12th year of the Jiajing reign period of Emperor Shizong of the Ming Dynasty, 10th month; in Taiyuan, Yuci, Taigu, Qixian, Yangqu,

Mengxian, Qingyuan, Shouyang, Guoxian, Linfen, Hongdong, Huozhou, Changzhi, and Xiangyuan, stars fell like rain, their brilliance like fire so that the sky was completely red.
Qing Dynasty, Yongzheng period ed. [*Shanxi tongzhi*] ch. 163

(c) 12th year of the Jiajing reign period of Emperor Shizong of the Ming Dynasty, 10th month; in Chaozhou stars fell like rain, in seven counties it was the same.
[*Guangwenlu*] ch. 1

NB: Records in several dozen local gazetteers from Shandong, Hebei, Henan, Gansu, Shanxi, Jiangxi, Anhui, Guangdong, Fujian, and so forth are similar.

AD 1533 Oct 23 [China]

12th year of the Jiajing reign period of Emperor Shizong of the Ming Dynasty, 10th month, day *yihai* [12]; at the beginning of twilight the fixed stars were not seen. By nightfall stars fell like rain, losing their courses and places, streaming back and forth, north and south.
Qing Dynasty, Shunzhi period ed. Shanxi [*Gaopingxian zhi*] ch. 9

AD 1533 Oct 24 [China]

(a) 12th year of the Jiajing reign period of Emperor Shizong of the Ming Dynasty, 10th month, day *bingzi* [13] (7th day); on this night there were scarlet meteors as big as oil lamps whose brilliant light shone on the ground. They began from ZHONGTAI and traveled northeastward to the horizon, the trails of their tails changing into white vapor and then dispersing. From the fourth to the fifth watch, innumerable large and small meteors traveled back and forth, up, down, left, and right, not ceasing until daybreak.
[*Ming Shizong shilu*] ch. 155

(b) 12th year of the Jiajing reign period of Emperor Shizong of the Ming Dynasty, 10th month, day *bingzi* [13]; meteors as big as oil lamps lit up the ground with their brilliance. From ZHONGTAI they traveled northeastward to the horizon, the trails of their tails turning into white vapor. From the fourth to the fifth watch, on all sides innumerable large and small meteors traveled back and forth, up, down, left, and right, not ceasing until daylight.

[*Ming shi: tianwen zhi*] ch. 27

NB: *Ming shi: tianwen zhi* made it the 9th month, day *bingzi*, but there was no *bingzi* day that month, so this is thought to be a mistake for 10th month.

(c) 12th year of the Jiajing reign period of Emperor Shizong of the Ming Dynasty, 10th month, day *bingzi* [13] (7th day); from midnight through daybreak stars fell like rain.
Ming Dynasty, Wanli period ed. Shandong [*Wudingzhou zhi*] ch. 6
Ming Dynasty, Qianlong period ed. Shandong [*Huiminxian zhi*] ch. 3
Ming Dynasty, Jiajing period ed. Shandong [*Qingzhoufu zhi*] ch. 5

(d) 12th year of the Jiajing reign period of Emperor Shizong of the Ming Dynasty, 10th month, day *bingzi* [13] (7th day), at night; brilliant stars fell and the sky was scarlet from top to bottom. By daybreak it still had not stopped.
Ming Dynasty, Jiajing period ed. Hebei [*Qingyuan zhi*] ch. 6

(e) 12th year of the Jiajing reign period of Emperor Shizong of the Ming Dynasty, 10th month, day *bingzi* [13] (7th day), at night; stars fell like rain.
Ming Dynasty, Jiajing period ed. Shanxi [*Huozhouxian zhi*] ch. 6
Ming Dynasty, Longqing period ed. Jiangsu [*Gaoyouzhou zhi*] ch. 2

AD 1533 Oct 24 [Korea]

(a) 28th year of King Jungjong of the Joseon Dynasty, 10th month, day *bingzi* [13]; flying stars and meteors like rain in all directions.
[*Jeungbo munheon bigo*] ch. 7

(b) 28th year of King Jungjong of the Joseon Dynasty, 10th month, day *bingzi* [13], third watch; a greenish, yellowish and whitish vapor, shaped like a dragon and about the size of a bolt of cloth, emerged from WENCHANG, its tail pointing toward WANGLIANG, and after a long while it disappeared. A white meteor the size of a bottle with a tail 5 to 6 *chi* long emerged from BEIHE and entered the horizon

in the east; a red meteor the size of a bottle with a tail about 6 to 7 *chi* long emerged from SHAOWEI and entered the horizon in the east; during the fourth watch, a white meteor the size of a bottle with a tail about 4 to 5 *chi* long emerged from SHEN [LM 21] and entered TIANYUAN; a red meteor the size of a fist with a tail 5 to 6 *chi* long emerged from SANTAI and entered the horizon at azimuth *gen* (NE); during the fifth watch, a white meteor the size of a bowl and with a tail 7 to 8 *chi* long emerged from XING [LM 25] and entered JI; a meteor the size of a platter with a long tail was obscured by a cloud which also reddened; it emerged from ZIWEIYUAN and entered the horizon at azimuth *qian* (NW); a meteor the size of a bottle with a tail 7 to 8 *chi* long emerged from the western wall of TAIWEI and entered PING, its glowing trail lasting a very long time before disappearing. A white meteor the size of a bottle with a tail about 7 to 8 *chi* long emerged from SANTAI and entered TAIWEIYUAN; a red meteor the size of a platter with a tail about 4 to 5 *chi* long emerged below SHEN [LM 21] and entered JIUYOU, its glowing trail lasting a very long time before being extinguished and turning into a cloud. Meteors and flying stars were streaming everywhere like rain, without there being enough time to gauge where they entered and left asterisms, they fell down like rain and lightning flashed.
[*Joseon wangjo sillok: Jungjong*] ch. 76 (Ahn [2004] 33)

AD 1533 Oct 25 [Japan]

(a) 2nd year of the Temmon reign period, 10th month, 8th day; at night, near first light, stars were said to come down like snow in the southwest.
[*Kugyō bunin*]

(b) 2nd year of the Temmon reign period, 10th month, day *dingchou* [14]; stars fell like snow…the sea and the dry land shook.
[*Zoku honchō tsūgan*]

(c) 2nd year of the Temmon reign period, 10th month, 8th [day]; at night, a myriad stars streamed and moved and fell to earth.
[*Zoku tai hei ki*]

(d) 2nd year of the Temmon reign period, 10th month, 8th day *dingchou* [14]; at the end of the night stars fell like snow, generally moving away toward the southwest.
[*Zoku shi gushō*]

AD 1533 Oct 25 [China]

(a) Guisi year (12th year) of the Jiajing reign period of Emperor Shizong of the Ming Dynasty, 10th month, 8th day; during the fourth watch of the night, 10,000 stars streamed and flew up, down, left, right, in an instant falling like rain. This went on till daybreak without stopping.
[*Wu xue bian yu: Zai yi*]

(b) 12th year of the Jiajing reign period of Emperor Shizong of the Ming Dynasty, 10th month, 8th day (*dingchou* [14]); around midnight, stars fell like rain.
Qing Dynasty, Shunzhi period ed. Henan [*Wenxiangxian zhi*] ch. 1
Qing Dynasty, Kangxi period ed. Shanxi [*Tongguanxian zhi*] ch. 2

(c) 12th year of the Jiajing reign period of Emperor Shizong of the Ming Dynasty, 10th month, 8th day (*dingchou* [14]); during the hour of *chou* (LT = 1–3h), stars fell like rain.
Qing Dynasty, Qianlong period ed. Hubei [*Yingshanxian zhi*] ch. 26
Ming Dynasty, Wanli period ed. Anhui [*Liuanzhou zhi*] ch. xia
Ming Dynasty, Wanli period ed. Anhui [*Shuchengxian zhi*] ch. 10

NB: The Qing Dynasty, Kanxi period ed. of Hubei [*Yingshanxian zhi*], made it the hour of *wu* (LT = 11–13h).

AD 1533 Oct 26 [China]

(a) 12th year of the Jiajing reign period of Emperor Shizong of the Ming Dynasty, 10th month, 9th day; during the hour of *chou* (LT = 1–3h), stars fell rain.
Qing Dynasty, Kangxi period ed. Shandong [*Jinanfu zhi*] ch. 10
Qing Dynasty, Kangxi period ed. Shandong [*Haifengxian zhi*] ch. 4
Qing Dynasty, Kangxi period ed. Shandong [*Changshanxian zhi*] ch. 7

(b) 12th year of the Jiajing reign period of Emperor Shizong of the Ming Dynasty, 10th month, 9th day, at night; stars fell like rain.
Ming Dynasty, Jiajing period ed. Zhejiang [*Wenzhoufu zhi*] ch. 6

AD 1533 Oct 27 [China]

(a) 12th year of the Jiajing reign period of Emperor Shizong of the Ming Dynasty, 10th month, 10th day, *jimao* [16]; in the middle of the night stars fell like rain.
Qing Dynasty, Kangxi period ed. Anhui [*Chaoxian zhi*] ch. 4

(b) 12th year of the Jiajing reign period of Emperor Shizong of the Ming Dynasty, 10th month, 10th day, *jimao* [16], at night; stars fell like rain, a fiery gleam shone on doors and windows, blazing like a flame.
Qing Dynasty, Guangxu period ed. Shandong [*Pingduxian zhi*] ch. 12

(c) 12th year of the Jiajing reign period of Emperor Shizong of the Ming Dynasty, 10th month, 10th day, *jimao* [16]; meteors started at ZHONGTAI and were transformed into white vapor. Starting from the fourth watch, innumerable large and small stars criss-crossed, not disappearing until late.
[*Zui wei lu*] zhi 1

AD 1533 Oct 29 [China]

(a) 12th year of the Jiajing reign period of Emperor Shizong of the Ming Dynasty, 10th month, day *xinsi* [18]; at Jingkou stars fell like rain. At Qiongzhou and Chaozhou, it was the same.
[*Ming hui yao*] ch. 69

(b) 12th year of the Jiajing reign period of Emperor Shizong of the Ming Dynasty, 10th month, day *xinsi* [18]; stars fell like rain in daylight. At Jingkou, boatmen did not dare cross. At Chaozhou and Qiongzhou, the star fall was the same.
[*Er shen ye lu*] ch. 4

(c) 12th year of the Jiajing reign period of Emperor Shizong of the Ming Dynasty, 10th month, day *xinsi* [18]; stars fell like rain.
Qing Dynasty, Shunzhi period ed. Hebei [*Yishui zhi*] ch. shang

Qing Dynasty, Qianlong period ed. Jiangxi [*Xinchengxian zhi*] ch. 13
Qing Dynasty, Kangxi period ed. Henan [*Chenzhou zhi*] ch. 4
Qing Dynasty, Qianlong period ed. Henan [*Huainingxian zhi*] ch. 11
Qing Dynasty, Xuantong period ed. Henan [*Xiangchengxian zhi*] ch. 31

AD 1533 Nov 1 [China]

12th year of the Jiajing reign period of Emperor Shizong of the Ming Dynasty, 10th month, full moon, in the middle of the night stars fell like rain.
Qing Dynasty, Qianlong period ed. Gansu [*Zhuanglang zhi lue*] ch. 19

AD 1533 Nov 3 [China]

(a) 12th year of the Jiajing reign period of Emperor Shizong of the Ming Dynasty, 10th month, 17th day, at night; a multitude of stars fell over Renqiu, as if throwing golden stones; by daybreak it still had not stopped.
Qing Dynasty, Qianlong period ed. Hebei [*Hejianfu zhi*] ch. 17

(b) 12th year of the Jiajing reign period of Emperor Shizong of the Ming Dynasty, 10th month, 17th day, at night; stars fell like rain, then the sky reddened.
Ming Dynasty, Chongzhen period ed. [*Shanxi tongzhi*] ch. 36
Qing Dynasty, Shunzhi period ed. Shanxi [*Lu'anfu zhi*] ch. 15
Qing Dynasty, Kangxi period ed. Shanxi [*Changzhixian zhi*] ch. 1

(c) 12th year of the Jiajing reign period of Emperor Shizong of the Ming Dynasty, 10th month, 17th day, at night; stars fell like rain, the sky turned dusky and there were changes in the clouds.
Qing Dynasty, Kangxi period ed. Shanxi [*Xiangyuanxian zhi*] ch. 9

AD 1533 Nov 4 [China]

12th year of the Jiajing reign period of Emperor Shizong of the Ming Dynasty, 10th month, day *dinghai* [24]; boatmen at sea said that after

the third watch stars fell like rain. After this, a scarlet star as large as a peck measure made a humming sound.
[*Wen jian man lu*] ch. shang

AD 1533 Nov 24 [China]

12th year of the Jiajing reign period of Emperor Shizong of the Ming Dynasty, winter, 11th month, 9th day, at night; stars fell like rain.
Ming Dynasty, Wanli period ed.; Zhejiang [*Wenzhoufu zhi*] ch.18

AD 1533 winter [China]

(a) 12th year of the Jiajing reign period of Emperor Shizong of the Ming Dynasty, winter; at Hongyi stars fell like rain.
Qing Dynasty, Qianlong period ed. Anhui [*Sizhou zhi*] ch. 4

(b) 12th year of the Jiajing reign period of Emperor Shizong of the Ming Dynasty, winter; around midnight stars fell like rain, a reddish light lit up the sky.
Qing Dynasty, Kangxi period ed. Anhui [*Hongxian zhi*] ch. 1

(c) 12th year of the Jiajing reign period of Emperor Shizong of the Ming Dynasty, winter; the stars in the sky scattered and fell like rain.
Ming Dynasty, Longqing period ed. Jiangsu [*Yizhenxian zhi*] ch. 13

(d) 12th year of the Jiajing reign period of Emperor Shizong of the Ming Dynasty, winter, at night; stars fell like rain.
Qing Dynasty, Kangxi period ed. Zhejiang [*Yongjiaxian zhi*] ch. 14

AD 1534 Jan 14–AD 1535 Feb 1 [China]

(a) 13th year of the Jiajing reign period of Emperor Shizong of the Ming Dynasty; stars fell like rain.
Ming Dynasty, Jiajing period ed. Henan [*Guangshanxian zhi*] ch. 9
Ming Dynasty, Wanli period ed. Henan [*Xiangchengxian zhi*] ch. 7
Ming Dynasty, Wanli period ed. Henan [*Linxian zhi*] ch. 8

(b) 13th year of the Jiajing reign period of Emperor Shizong of the Ming Dynasty; from east to west stars fell like rain.
Qing Dynasty, Yongzheng period ed. Beijing [*Miyunxian zhi*] ch. 1

(c) 13th year of the Jiajing reign period of Emperor Shizong of the Ming Dynasty; at Suzhou the stars fell like rain.
Qing Dynasty, Qianlong period ed. Jiangsu [*Jiangnan tongzhi*] ch. 197

AD 1534 Sep 8–Oct 6 [China]
13th year of the Jiajing reign period of Emperor Shizong of the Ming Dynasty, 8th month; stars fell like rain.
Ming Dynasty, Jiajing period ed. Hunan [*Hengzhoufu zhi*] ch. 7

AD 1534 Nov 6–Dec 4 [China]
13th year of the Jiajing reign period of Emperor Shizong of the Ming Dynasty, 10th month; stars fell like rain.
Ming Dynasty, Wanli period ed. Shanxi [*Pingyangfu zhi*] ch. 10
Qing Dynasty, Kangxi period ed. Shanxi [*Jiezhou zhi*] ch. 12
Qing Dynasty, Qianlong period ed. Shanxi [*Linfenxian zhi*] ch. 9
Qing Dynasty, Kangxi period ed. Jiangsu [*Jiangnan tongzhi*] ch. 5
Qing Dynasty, Guangxu period ed. Zhejiang [*Leqingxian zhi*] ch. 13

AD 1534 Nov 13 [China]
13th year of the Jiajing reign period of Emperor Shizong of the Ming Dynasty, 10th month, 8th day, at night; stars fell like rain.
Ming Dynasty, Chongzhen period ed. Jiangsu [*Wuxian zhi*] ch. 11
Qing Dynasty, Kangxi period ed. Jiangsu [*Suzhoufu zhi*] ch. 2
Qing Dynasty, Kangxi period ed. Jiangsu [*Changzhouxian zhi*] ch. 2

AD 1534 Nov 20 [China]
13th year of the Jiajing reign period of Emperor Shizong of the Ming Dynasty, 10th month, full Moon, at night; stars fell like rain.
Ming Dynasty, Jiajing period ed. Gansu [*Pingliangfu zhi*] ch. 8

AD 1535 Aug 28–Sep 26 [China]
14th year of the Jiajing reign period of Emperor Shizong of the Ming Dynasty, 8th month; stars fell like rain.
Qing Dynasty, Daoguang period ed. Guangdong [*Changlexian zhi*] ch. 7

AD 1535 Sep 27–Oct 26 [China]

(a) 14th year of the Jiajing reign period of Emperor Shizong of the Ming Dynasty, 9th month; stars as big as peck measures fell over Lincheng, while several hundred stars streamed, their rays entering the southeast. In a moment, the sky rumbled like thunder.
Ming Dynasty, Longqing period ed. Hebei [*Zhaozhou zhi*] ch. 9
Ming Dynasty, Jiajing period ed. Hebei [*Zhendingfu zhi*] ch. 9

(b) 14th year of the Jiajing reign period of Emperor Shizong of the Ming Dynasty, 9th month; stars like peck measures fell, making a sound like thunder.
Qing Dynasty, Kangxi period ed. Hebei [*Linchengxian zhi*] ch. 8
Qing Dynasty, Kangxi period ed. Hebei [*Zhendingfu zhi*] ch. 6

(c) 14th year of the Jiajing reign period of Emperor Shizong of the Ming Dynasty, 9th month; there were stars as big as peck measures, afterward swarms of several hundred stars streamed from northwest to southeast. A moment later, the sky rumbled like thunder.
Ming Dynasty, Jiajing period ed. Hebei [*Shunde zhi*] ch. 17
Ming Dynasty, Longqing period ed. Hebei [*Renxian zhi*] ch. 7

AD 1535 Oct 10 [China]

14th year of the Jiajing reign period of Emperor Shizong of the Ming Dynasty, 9th month, day *renshen* [9], at night; a multitude of stars moved about criss-crossing.
Ming Dynasty, Longqing period ed. Jiangsu [*Gaoyouzhou zhi*] ch. 12

AD 1535 Oct 26 [China]

14th year of the Jiajing reign period of Emperor Shizong of the Ming Dynasty, 9th month, day *wuzi* [25]; in Kaifeng in Henan province, the sky rumbled in broad daylight. A star as big as a bowl streamed out of the southeast, a multitude of smaller stars like pearls following behind.
[*Ming Shizong shilu*] ch. 179
[*Ming shi: tianwen zhi*] ch. 27

AD 1536 Jan 23–1537 Feb 9 [China]

15th year of the Jiajing reign period of Emperor Shizong of the Ming Dynasty; in Shenzhou stars fell like rain.
Ming Dynasty, Jiajing period ed. Hebei [*Zhendingfu zhi*] ch. 6
Qing Dynasty, Kangxi period ed. Hebei [*Shenzhou zhi*] ch. 7

AD 1536 Oct 15–Nov 13 [China]

(a) 15th year of the Jiajing reign period of Emperor Shizong of the Ming Dynasty, 10th month; during the night, the stars in the sky fell chaotically, like rain.
Ming Dynasty, Jiajing period ed. Hebei [*Weixian zhi*] ch. 1
Ming Dynasty, Jiajing period ed. Hebei [*Guangpingfu zhi*] ch. 15

(b) 15th year of the Jiajing reign period of Emperor Shizong of the Ming Dynasty, 10th month; stars fell like rain.
Qing Dynasty, Kangxi period ed. Hebei [*Guangpingfu zhi*] ch. 19
Qing Dynasty, Kangxi period ed. Hebei [*Pingxiangxian zhi*] ch. 3

(c) 15th year of the Jiajing reign period of Emperor Shizong of the Ming Dynasty, 10th month, at night; stars came down like rain.
Qing Dynasty, Shunzhi period ed. Hebei [*Jizexian zhi*] ch. 10

(d) 15th year of the Jiajing reign period of Emperor Shizong of the Ming Dynasty, 10th month, at night; stars came down like rain.
Qing Dynasty, Qianlong period ed. Hebei [*Yongnianxian zhi*] ch. 17

AD 1538 Apr 29–May 27 [China]

(a) 17th year of the Jiajing reign period of Emperor Shizong of the Ming Dynasty, 4th month, at night; in Ninghua stars fell like rain.
Qing Dynasty, Qianlong period ed. Fujian [*Tingzhoufu zhi*] ch. 45

(b) 17th year of the Jiajing reign period of Emperor Shizong of the Ming Dynasty, 4th month; in Tingzhou and Ninghua...at night stars fell like rain.
Qing Dynasty, Daoguang period ed. [*Fujian tongzhi*] ch. 271

AD 1538 Jun 26–Jul 25 [China]

17th year of the Jiajing reign period of Emperor Shizong of the Ming Dynasty, 6th month; stars fell like rain.
Qing Dynasty, Kangxi period ed. Shandong [*Ji nanfu zhi*] ch. 10
Qing Dynasty, Kangxi period ed. Shandong [*Changshanxian zhi*] ch. 7

AD 1538 Oct 23–Nov 21 [China]

17th year of the Jiajing reign period of Emperor Shizong of the Ming Dynasty, 10th month; in Changqing stars fell like rain.
Qing Dynasty, Qianlong period ed. [*Shandong tongzhi*] ch. 33
Qing Dynasty, Kangxi period ed. Shandong [*Changqingxian zhi*] ch. 14
NB: *Changqingxian zhi* did not record the month.

AD 1538 Nov 7 [Korea] (Ahn [2004] 34)

33rd year *wuxu* of King Jungjong of the Joseon Dynasty, 10th month, day *jiachen* [41]; meteors emerged from all directions.
[*Joseon wangjo sillok: Jungjong*] ch. 88

AD 1539 May 30 [China]

18th year of the Jiajing reign period of Emperor Shizong of the Ming Dynasty, 5th month, 13th day, at night; stars fell like rain.
Ming Dynasty, Chongzhen period ed. Fujian [*Tingzhoufu zhi*] ch. 24

AD 1539 Nov 18 [China]

18th year of the Jiajing reign period of Emperor Shizong of the Ming Dynasty, 10th month, 8th day; around midnight, stars fell like rain.
Qing Dynasty, Kangxi period ed. Shanxi [*Puxian xin zhi*] ch. 7

AD 1540 May 6–Jun 4 [China]

19th year of the Jiajing reign period of Emperor Shizong of the Ming Dynasty, 4th month; in Zaoqiangxian, the drums of heaven sounded, at night stars fell like rain.
[*Er shen ye lu*] ch. 4

AD 1546 Jan 14 [China]

24th year of the Jiajing reign period of Emperor Shizong of the Ming Dynasty, 12th month, 12th day; at the hour of *bu* (LT = 15–17h), the ground turned about [?]. Around midnight, the Moon lacked brilliance, and stars fell like rain.
Ming Dynasty, Wanli period ed. Shanxi [*Huayinxian zhi*] ch. 7

AD 1547 Jun 12 [China]

(a) 26th year of the Jiajing reign period of Emperor Shizong of the Ming Dynasty, 5th month, 25th day; stars fell like rain until the hour of *yin* (LT = 3–5h), the drums of heaven sounded, and there was a fiery gleam.
Qing Dynasty, Kangxi period ed. Shandong [*Ji nanfu zhi*] ch. 10
Qing Dynasty, Kangxi period ed. Shandong [*Changshanxian zhi*] ch. 7

(b) 26th year of the Jiajing reign period of Emperor Shizong of the Ming Dynasty, 5th month, 25th day; stars fell like rain and the drums of heaven sounded.
Qing Dynasty, Qianlong period ed. Shandong [*Wudingfu zhi*] ch. 14
Qing Dynasty, Qianlong period ed. Shandong [*Huiminxian zhi*] ch. 3

AD 1548 Aug 24 [Korea]

(a) 3rd year *wushen* of King Myeongjong of the Joseon Dynasty, 7th month, day *jiawu* [31]; large and small meteors criss-crossed in all directions, too many to record.
[*Jeungbo munheon bigo*] ch. 7

(b) 3rd year of King Myeongjong of the Joseon Dynasty, 7th month, day *jiawu* [31]; at night a white meteor the size of a bowl with a tail about 2 *zhang* long emerged from below HEGU and entered the horizon at azimuth *kun* (SW); a white meteor the size of a bowl with a tail 5 to 6 *chi* long emerged from the western wall of TIANSHI and entered the west; a white meteor the size of a fist with a tail about 7 to 8 *chi* long emerged from TIANJIN and entered the horizon at azimuth *kun* (SW); a red meteor the size of a bottle with a tail about

2 to 3 *zhang* long emerged from below KUI [LM 15] and entered the horizon at azimuth *xun* (SE); a white meteor the size of a platter with a tail about 5 to 6 *zhang* long emerged below SHI [LM 13] and entered the horizon in the south, its glow lit up the ground and it took a long time to disappear; a white meteor the size of a bottle with a tail 2 to 3 *zhang* long emerged at the first star of BEIDOU and entered the horizon in the west; a white meteor the size of a bowl with a tail about 2 *zhang* long emerged from BAIGUA and entered NANDOU [LM 8]; a white meteor the size of a bowl with a tail about 5 to 6 *chi* long emerged from TIANJIN and entered below JI [LM 7]; a red meteor the size of a bottle with a tail about 8 to 9 *chi* long emerged from WANGLIANG and entered BI [LM 14]; a white meteor the size of a fist and with a tail about 5 to 6 *chi* long emerged from SIKONG and entered the horizon in the south. From the first watch to the second, in general there were large meteors like this, while small meteors criss-crossed in all directions, too many to count.
[*Joseon wangjo sillok: Myeongjong*] ch. 8 (Ahn [2004] 35)

(c) 3rd year *wushen* of King Myeongjong of the Joseon Dynasty, 7th month, day *gengchen* [17]; the sun had a dim halo and at the hour of *shen* (LT 15–17h) Venus appeared at azimuth *wei* (SSW) and at night the moon had a dim halo. A white meteor the size of a bowl and with a tail about 2 *zhang* long emerged from SHI [LM 13] and entered LOU [LM 16]; a white (meteor) the size of a bottle with a tail about 5 to 6 *chi* long emerged from YULIN and entered the horizon at azimuth *kun* (SW); a red (meteor) the size of a bowl with a tail about 3 to 4 *chi* long emerged from KUI [LM 15] and entered WEI [LM 17]; a white (meteor) the size of a bowl with a tail about 5 to 6 *chi* long emerged from below BEIHE and entered the horizon at azimuth *gen* (NE).
[*Joseon wangjo sillok: Myeongjong*] ch. 8

AD 1553 Sep 8–Oct 6 [China]

(a) 32nd year of the Jiajing reign period of Emperor Shizong of the Ming Dynasty, 8th month, at night; there were meteors as big as peck measures, while several tens of thousands of smaller stars trailed along from east to west, travelling straight as arrows, their blazing brilliance lighting up the sky. After several days, they flew again from west to east.

Qing Dynasty, Kangxi period ed. Henan [*Zhechengxian zhi*] ch. 4

(b) 32nd year of the Jiajing reign period of Emperor Shizong of the Ming Dynasty, 8th month, at night; there were stars as big as peck measures. Innumerable smaller stars trailed along from east to west, traveling straight as arrows. Several days later they flew again from west to east.

Qing Dynasty, Kangxi period ed. Henan [*Luyixian zhi*] ch. 8

AD 1554 Oct 14 [Korea] (Ahn [2004] 36)

9th year of King Myeongjong of the Joseon Dynasty, 9th month, day *bingyin* [3]; there was a letter stating, "Lately the moon has repeatedly had halos and meteors have appeared in succession. Now we have seen the report on the state of affairs by the Intendent of Kyungsang, saying the moon's halo had changed to become still more worrisome. Why has the Director of Astronomy not reported on it for some 14 days? There should be an inquiry into his imprudence." The report of the Intendent of Kyungsang read, "On the night of the 14th of this month, the moon had a halo. At the beginning of night, it was not yet observed, not until the end of the *hai* hour (21–23h) and the beginning of the *zi* hour (23–1h) was it seen. Around the lunar disk there were concentric halos, the innermost comprising four rings, overall about 1 *zhang*, 4 to 5 *chi* in diameter, dazzlingly colorful with greenish-red in between, like a rainbow. The outer ring was round, about 2 *zhang* in diameter, with colors like the inner ones. Next to the moon and where the inner and outer rings touched the colors were slightly paler and seemed to part. The lunar disk was also overwhelmed, its light pale and completely colorless. When the vaporous halos grew slightly dimmer, they rippled and undulated as they dispersed as if blown about by wind or wave. They rejoined two or three times like before and afterward gradually extinguished, the traces of the rings still lingering a long time..."

[*Joseon wangjo sillok: Myeongjong*] ch. 17
NB: Ahn (2004) corrected the date but rejects this report as ambiguous and lacking sufficient detail to confirm a shower.

AD 1556 Jul 26 [China]

(a) 35th year of the Jiajing reign period of Emperor Shizong of the Ming Dynasty, 6th month, 20th day; in the south there was a star that spat out a streak of blazing brilliance more than 1 *zhang* long. During the night, a swarm of more than 30 stars sped southward, with extraordinarily brilliant rays.
Qing Dynasty, Shunzhi period ed. Shandong [*Dengzhoufu zhi*] ch. 1
Qing Dynasty, Kangxi period ed. Shandong [*Fushanxian zhi*] ch. 1

(b) 35th year of the Jiajing reign period of Emperor Shizong of the Ming Dynasty, 6th month, 20th day, at night; in the south a star suddenly spat out a brilliance more than 1 *zhang* long. A swarm of more than 20 stars sped southward.
Qing Dynasty, Kangxi period ed. Shandong [*Laiyangxian zhi*] ch. 9
Qing Dynasty, Qianlong period ed. Shandong [*Haiyangxian zhi*] ch. 3

(c) 35th year of the Jiajing reign period of Emperor Shizong of the Ming Dynasty, 6th month, 20th day, at night; in the south a streaking star emerged with a brilliance more than 1 *zhang* long. During the night, a swarm of more than 30 stars sped southward, their brilliance lighting up the ground.
Qing Dynasty, Kangxi period ed. Shandong [*Ji nanfu zhi*] ch. 10
Qing Dynasty, Kangxi period ed. Shandong [*Changshanxian zhi*] ch. 7
Qing Dynasty, Qianlong period ed. Shandong [*Huiminxian zhi*] ch. 3
Qing Dynasty, Qianlong period ed. Shandong [*Wudingfu zhi*] ch. 14

AD 1557 Aug 11 [China]

36th year of the Jiajing reign period of Emperor Shizong of the Ming Dynasty, 7th month, 17th day, at night; on the southeast horizon large

and small stars fell chaotically into the sea, in a moment, on the northwestern horizon it was the same.
Qing Dynasty, Kangxi period ed. Zhejiang [*Dinghaixian zhi*] ch. 6

AD 1559 Feb 7–AD 1560 Jan 26 [China]
38th year of the Jiajing reign period of Emperor Shizong of the Ming Dynasty; in Dongjing stars fell like rain.
Ming Dynasty, Wanli period ed. Hebei [*Cizhou zhi*] ch. 8

AD 1559 Sep 2–Sep 30 [China]
38th year of the Jiajing reign period of Emperor Shizong of the Ming Dynasty, 8th month; in Liaoyang stars fell like rain.
Ming Dynasty, Jiajing period ed. Liaoning [*Quanliao zhi*] ch. 4

AD 1560 Jun 27 [China]
39th year of the Jiajing reign period of Emperor Shizong of the Ming Dynasty, 6th month, 4th day, at night; stars fell like rain.
Qing Dynasty, Yongzheng period ed. Hebei [*Fuchengxian zhi*] ch. 21

AD 1560 Aug 21–Sep 19 [China]
39th year of the Jiajing reign period of Emperor Shizong of the Ming Dynasty, 8th month; stars fell like rain.
Qing Dynasty, Qianlong period ed. Jiangxi [*Jianchangfu zhi*] ch. 5
Qing Dynasty, Daoguang period ed. Jiangxi [*Nanchengxian zhi*] ch. 27

AD 1560 Aug 24 [China]
(a) 39th year of the Jiajing reign period of Emperor Shizong of the Ming Dynasty, 8th month, 4th day; stars fell like rain.
Qing Dynasty, Kangxi period ed. Beijing [*Miyunxian zhi*] ch. 2
(b) 39th year of the Jiajing reign period of Emperor Shizong of the Ming Dynasty, 8th month, 4th day, at night; stars and constellations scattered and fell like rain.
Qing Dynasty, Kangxi period ed. Shanxi [*Dingxiangxian zhi*] ch. 7

AD 1560 Aug 24 [Korea] (Ahn [2004] 37)

(a) 15th year *gengshen* of King Myeongjong of the Joseon Dynasty, 8th month, *dingyou* [34]; flying stars and meteors like rain in all directions.
[*Jeungbo munheon bigo*] ch. 7

(b) 15th year of King Myeongjong of the Joseon Dynasty, 8th month, day *dingyou* [34]; at night a meteor the size of a bottle with a tail about 3 to 4 *chi* long emerged from SHI [LM 13] and entered YULIN; a meteor the size of a fist with a tail about 4 to 5 *chi* long emerged from XU [LM] and entered TIANLEI; a meteor the size of a fist with a tail 4 to 5 *chi* long emerged from BI [LM 14] and entered the horizon in the south; a meteor the size of a fist with a tail about 3 to 4 *chi* long emerged from SHI [LM 13] and entered WEI [LM 17]; their colors were all red and their glow lasted a long time before disappearing. A meteor the size of a fist with a tail about 2 to 3 *chi* long emerged from BEIHE and entered the horizon in the east; a meteor the size of a pear with a tail about 3 to 4 *chi* long emerged from BEIJI and entered the horizon in the north; a meteor the size of a fist with a tail about 2 to 3 *chi* long emerged from SHEN [LM 21] and entered BEIHE; they were all red. A red flying star the size of a fist with a tail 5 to 6 *chi* long emerged from WUCHE and entered BAGU, lasting a long time before disappearing. From the fourth to the fifth watch, stars flew like rain in all directions without there being enough time to count them or observe where they entered and left.
[*Joseon wangjo sillok: Myeongjong*] ch. 26

AD 1560 Sep 2 [China]

39th year of the Jiajing reign period of Emperor Shizong of the Ming Dynasty, 8th month, 13th day; stars fall like rain.
Ming Dynasty, Chongzhen period ed. Jiangsu [*Taicangzhou zhi*] ch. 15

AD 1563 Jul 3 [China]

42nd year of the Jiajing reign period of Emperor Shizong of the Ming Dynasty, 6th month, 14th day, at night; it rained meteors from south of Guilin to the northwest.

Qing Dynasty, Yongzheng period ed. [*Guangxi tongzhi*] ch. 3

AD 1565 Jul 26 [Japan]

8th year of the Eiroku reign period, 6th month, 29th day, the last day of the month; to the south and north stars flew about, crossing confusedly like fireflies.

[*Shinsen wakan gōzu*]

AD 1566 Oct 13–Nov 11 [China]

45th year of the Jiajing reign period of Emperor Shizong of the Ming Dynasty, 10th month; in the northwest greenish stars fell like rain.

Ming Dynasty, Wanli period ed. Jiangxi [*Yongxinxian zhi*] ch. 5

AD 1566 Oct 25–27 [China]

45th year of the Jiajing reign period of Emperor Shizong of the Ming Dynasty, 10th month, 13th day, at night; stars fell like rain, while making a noise. On the 14th and 15th it was likewise.

[*Xin zhi lu*]

AD 1566 Oct 26 [Korea] (Ahn [2004] 38)

(a) 15th year *gengshen* [57] of King Myŏngjong of the Joseon Dynasty (1560), 8th month, day *dingyou* [34]; flying stars and meteors like rain in all directions. In the 21st year *bingyin* [3], 10th month, day *xinwei* [8], it was the same.
[*Jeungbo munheon bigo*] ch. 7

(b) 21st year *bingyin* of the Joseon Dynasty, 10th month, day *xinwei* [8]; at night a red meteor the size of a bowl with a tail about 5 to 6 *chi* long emerged from XUANYUAN and entered TIANMIAO, disappearing only after a long time; a red meteor the size of a

bowl with a tail 7 to 8 *chi* long emerged from SHEN [LM 21] and entered the horizon in the west, disappearing only after a long time. Small meteors, some like pears, some like peaches, came and went in all directions, too many to record more than a few.
[*Joseon wangjo sillok: Myeongjong*] ch. 33

AD 1566 Nov 25 [China]

45th year of the Jiajing reign period of Emperor Shizong of the Ming Dynasty, intercalary 10th month, day *xinchou* [38], at night; there were meteors that moved as if weaving; two of them were as big as the Moon.
Ming Dynasty, Longqing period ed. Jiangsu [*Gaoyouzhou zhi*] ch. 12

AD 1566 Dec 23–25 [China]

45th year of the Jiajing reign period of Emperor Shizong of the Ming Dynasty, 11th month, 3 days before full Moon, at night; for 3 nights running, large stars fell down and swarms of others followed them, as if raining sounds.
[*Zui wei lu*] zhi 1

AD 1566 Dec 25 [China]

(a) 45th year of the Jiajing reign period of Emperor Shizong of the Ming Dynasty, 11th month, 15th day; during the fourth watch, a large star fell down and a swarm of hundreds of others followed it like rain.
[*Er shen ye lu*] ch. 4
[*Liu qing ri zha*] ch. 4

(b) 45th year of the Jiajing reign period of Emperor Shizong of the Ming Dynasty, 11th month, 15th day; during the fourth watch, a large star fell down and a swarm of hundreds of others followed it.
Qing Dynasty, Guangxu period ed. Zhejiang [*Jiaxingfu zhi*] ch. 35

NB: Qing Dynasty, Jiaqing period ed. [*Jiashanxian zhi*], has 11th month, 11th day.

AD 1578 Feb 12 [China]

(a) 6th year of the Wanli reign period of Emperor Shenzong of the Ming Dynasty, 1st month, 6th day, at night; a multitude of stars streamed in the west, in their midst was one large one.
Ming Dynasty, Wanli period ed. Zhejiang [*Xiushuixian zhi*] ch. 10

(b) 6th year of the Wanli reign period of Emperor Shenzong of the Ming Dynasty, 1st month, 6th day; at night; there was a large star like the Sun that moved westward and a multitude of stars followed it.
Qing Dynasty, Guangxu period ed. Zhejiang [*Jiaxingfu zhi*] ch. 35

(c) 6th year of the Wanli reign period of Emperor Shenzong of the Ming Dynasty, 1st month, 6th day, at night; a large star like the Sun emerged from the west and a multitude of stars surrounded it in the west.
Ming Dynasty, Wanli period ed. Fujian [*Jianningfu zhi*] ch. 47

(d) 6th year of the Wanli reign period of Emperor Shenzong of the Ming Dynasty, 1st month, 6th day, at night; a swarm of stars streamed in the west, in their midst was one extremely large one.
Qing Dynasty, Daoguang period ed. Fujian [*Jianyangxian zhi*] ch. 19

AD 1578 Feb 22 [China]

6th year of the Wanli reign period of Emperor Shenzong of the Ming Dynasty, 1st month, day *wuchen* [5]; a large star like the Sun emerged in the west, while a multitude of stars surrounded it in the west.
[*Ming shi: tianwen zhi*] ch. 27
[*Guo que*] ch. 70
[*Xu tongzhi: zaixiang*] ch. 171

AD 1581 Aug 7 [China]

9th year of the Wanli reign period of Emperor Shenzong of the Ming Dynasty, 7th month, 9th day, at night; in the southwest stars fell like rain.
Qing Dynasty, Kangxi period ed. Fujian [*Fuanxian zhi*] ch. 9

AD 1589 Aug 25 [China]

(a) Jichou [26] year of the Wanli reign period of Emperor Shenzong of the Ming Dynasty, 7th month, 15th day (*gengshen* [57]), at night; in the middle of the Moon white small stars burst out like pearls.
Ming Dynasty, Chongzhen period ed. Shanghai [*Songjiangfu zhi*] ch. 47

(b) Jichou [26] year of the Wanli reign period of Emperor Shenzong of the Ming Dynasty, 7th month, 15th day (*gengshen* [57]), at night; on the Moon above the eastern hills, from the middle of the Moon small white stars were seen to burst outward, like pearls scattering in confusion. First, after a little while, they subsided.
[*Yun jian za shi*] ch. 2

AD 1590 Aug 6 [China]

18th year of the Wanli reign period of Emperor Shenzong of the Ming Dynasty, 7th month, 7th day, at night; stars fell like rain, stopping after an hour.
Qing Dynasty, Qianlong period ed. Gansu [*Didaozhou zhi*] ch. 11

AD 1594 Nov 5 [China]

Jiawu [31] year of the Wanli reign period of Emperor Shenzong of the Ming Dynasty, 9th month, 23rd day, at night; in the northeast there was a greenish-white star as big as a hen's egg. In the southwest there was a star as big as a bowl, also greenish-white, the brilliance of its trail shining on the ground. It traveled southwest, while two smaller stars followed along behind; then again several thousand meteors traveled criss-cross chaotically.
[*Er shen ye lu*] ch. 5

AD 1601 Nov 5 [China]

(a) 29th year of the Wanli reign period of Emperor Shenzong of the Ming Dynasty, 10th month, 11th day; during the fifth watch of the night, the stars changed as if raining.
Ming Dynasty, Wanli period ed. Fujian [*Zhangzhoufu zhi*] ch. 32

(b) 29th year of the Wanli reign period of Emperor Shenzong of the Ming Dynasty, 10th month, 11th day; during the 5th drum of the night, stars fell like rain.
Qing Dynasty, Kangxi period ed. Fujian [*Haichengxian zhi*] ch. 14
Qing Dynasty, Qianlong period ed. Fujian [*Longxixian zhi*] ch. 20

AD 1602 autumn [China]

30th year of the Wanli reign period of Emperor Shenzong of the Ming Dynasty, autumn, at night; there were stars like eggs whose brilliance dispersed and shone on the ground. After them came two smaller stars, and then again two large and small stars that flew to and fro like a weaving shuttle.
Qing Dynasty, Qianlong period ed. Anhui [*Tonglingxian zhi*] ch. 13

AD 1602 Oct 15–Nov 13 [China]

(a) 30th year of the Wanli reign period of Emperor Shenzong of the Ming Dynasty, 9th month; stars fell like rain.
Qing Dynasty, Shunzhi period ed. Guangdong [*Chaozhoufu zhi*] ch. 2

(b) 30th year of the Wanli reign period of Emperor Shenzong of the Ming Dynasty, 9th month; for several days meteors emerged from lunar mansion Shen [LM 21] and entered Tianyuan.
[*Zui wei lu*] *zhi* 1

AD 1602 Nov 5 [China]

30th year of the Wanli reign period of Emperor Shenzong of the Ming Dynasty, 9th month, day *gengchen* [17]; several hundred large and small stars criss-crossed as they traveled.
[*Ming Shenzong shilu*] ch. 376
[*Guo que*] ch. 79

AD 1602 Nov 6 [China]

(a) 30th year (*renyin*) of the Wanli reign period of Emperor Shenzong of the Ming Dynasty, 9th month, day *xinsi* [18] (23rd); several hundred large and small stars criss-crossed as they traveled.

[*Ming shi: tianwen zhi*] ch. 27
[*Wenxian tongkao: xiang wei*] ch. 214
[*Xu tongzhi: zaixiang*] ch. 171

(b) 30th year (*renyin*) of the Wanli reign period of Emperor Shenzong of the Ming Dynasty, 9th month, day *xinsi* [18] (23rd); during the fifth watch of the night, greenish-white stars as big as pheasant eggs with brilliant tails started from XIATAI and traveled northeast almost to the southwest quarter. Another greenish-white star as big as a bowl, with a tail that scattered explosively, lit up the ground with its brilliance. It started from lunar mansion SHEN [LM 21] and proceeded southwest where it entered TIANYUAN. Two smaller stars trailed along behind it, and then several hundred larger and smaller stars criss-crossed chaotically.
[*Er shen ye lu*] ch. 5

(c) 30th year (*renyin*) of the Wanli reign period of Emperor Shenzong of the Ming Dynasty, 9th month, day *xinsi* [18] (23rd), at night; in the north there were greenish-white stars as big as hen's eggs with brilliant tails. One started from XIATAI and traveled northeastward nearly to the southwest quarter. Another star as big as a bowl, with a tail that scattered explosively, lit up the ground with its gleam. It started from lunar mansion SHEN [LM 21] and proceeded northwest where it entered TIANYUAN. Two smaller stars trailed along behind it, and then several tens of hundreds of larger and smaller stars criss-crossed chaotically on all sides.
Qing Dynasty, Kangxi period ed. Fujian [*Anhai zhi*] ch. 8

AD 1602 Nov 7 [China]

30th year of the Wanli reign period of Emperor Shenzong of the Ming Dynasty, 9th month, 24th day; in Yingzhou meteors were seen to writhe like dragons, heading down the sky in the southwest, while over 10,000 smaller stars trailed after them.
Qing Dynasty, Kangxi period ed. Anhui [*Fengyangfu zhi*] ch. 3

AD 1602 Nov 12 [Korea] (Ahn [2004] 39)

35th year of King Sonjo of the Joseon dynasty, 11th month, day *bingyin* [3]; Circuit Intendant Li of Kyungsang composed an urgent letter saying, "During the night of the 28th day of the 9th month past, the fixed stars wobbled, some as if about to change stations. Presently there was a large star that streamed from the northwest to the southeast, its bright rays crossing the sky; it was not extinguished for a long time. In the blink of an eye the whole sky was full of stars falling like hailstones, the tails of the large ones stretching halfway across the sky, the tails of the smaller ones about 1 *chi* long. This went on till daybreak. Many of the local inhabitants saw it; it was an extraordinary anomaly."

[*Joseon wangjo sillok: Sonjo*] ch. 56

NB: Ahn (2004) noted that the Julian date should be Dec 21, but the record gives Nov 12.

AD 1606 Feb 9 [China]

Bingwu [43] year of the Wanli reign period of Emperor Shenzong of the Ming Dynasty, 1st month, 3rd day, at night; stars were seen to fall chaotically in the west.

Qing Dynasty, Kangxi period ed. Hubei [*Zhongxiangxian zhi*] ch. 10

AD 1610 Sep 13 [China]

38th year of the Wanli reign period of Emperor Shenzong of the Ming Dynasty, 7th month, 26th day, at night; in the northern sky stars moved chaotically. After an hour, they settled down.

Qing Dynasty, Kangxi period ed. Anhui [*Tianchangxian zhi*] xing ye

AD 1621 Aug 21 [China]

Xinyou [58] year of the Tianqi reign period of Emperor Xizong of the Ming Dynasty, 7th month, 5th day; during the hour of *hai* (LT = 21–23h), there reddish-scarlet meteors drawn out in a chain, while several hundred smaller stars trailed along behind, having started from the northwest beside lunar mansion Nü [LM 10].

[*Er shen ye lu*] ch. 7
Ming Dynasty, Tianqi period ed. Jiangsu [Huaianfu zhi] ch. 24

AD 1621 autumn [China]

(a) 1st year (xinyou) of the Tianqi reign period of Emperor Xizong of the Ming Dynasty, autumn; stars fell like rain over Lincheng.
Qing Dynasty, Kangxi period ed. Shandong [*Ji nanfu zhi*] ch. 10

(b) 1st year (xinyou) of the Tianqi reign period of Emperor Xizong of the Ming Dynasty, autumn; stars fell like rain over Lincheng. After the third watch, the stars and Moon in the whole sky were all agitated, flashing brilliantly and unsettled. It was like this for several nights.
Qing Dynasty, Kangxi period ed. Shandong [*Changshanxian zhi*] ch. 7

(c) 1st year (xinyou) of the Tianqi reign period of Emperor Xizong of the Ming Dynasty, autumn; stars fell like rain. Each night after the third watch, the stars and Moon in the whole sky were all agitated, flashing brilliantly and unsettled for a long time. It was like this for several nights.
Qing Dynasty, Kangxi period ed. Shandong [*Xinchengxian zhi*] ch. 10

AD 1622 Oct 5–Nov 2 [China]

2nd year of the Tianqi reign period of Emperor Xizong of the Ming Dynasty, 9th month; stars fell like rain.
Qing Dynasty, Kangxi period ed. Gansu [*Jingningzhou zhi*] ch. 11

AD 1622 Oct 25 [China]

2nd year of the Tianqi reign period of Emperor Xizong of the Ming Dynasty, 9th month, day *jiayin* [51]; in Shanxi, Guyuanzhou, stars fell like rain.
[*Ming Xizong shilu*] ch. 26

AD 1622 Dec 3–31 [China]

(a) 2nd year of the Tianqi reign period of Emperor Xizong of the Ming Dynasty, 11th month; a large star as big as a peck measure, trailing several tens of stars in its wake like a string of pearls, flew from the northeast to the southwest and disappeared. The hour was already twilight and its brilliance shone brightly, so that trees and dwellings could be discerned.
Qing Dynasty, Kangxi period ed. Guangxi [*Hepuxian zhi*] ch. 1
Qing Dynasty, Yongzheng period ed. Guangxi [*Qinzhou zhi*] ch. 1
Qing Dynasty, Qianlong period ed. Guangxi [*Lingshanxian zhi*] ch. 3

(b) 2nd year of the Tianqi reign period of Emperor Xizong of the Ming Dynasty, 11th month; a large star as big as a peck measure, trailing several tens of stars in its wake like a string of pearls, flew from the northeast to the southwest and disappeared. The hour was already near nightfall and their brilliance shone bright as daylight, so that dwellings and window frames could be distinguished.
Qing Dynasty, Qianlong period ed. Guangxi [*Lianzhoufu zhi*] ch. 5

AD 1623 Oct 18 [China]

3rd year of the Tianqi reign period of Emperor Xizong of the Ming Dynasty, 9th month, day *renzi* [49] (25th day), at night; in the south, there were scarlet meteors as big as oil lamps trailing brilliant tails. They started from lunar mansion Xu [LM 11] and traveled southeast entering Leibizhen, and 20 other stars trailed along behind them.
[*Ming Xizong shilu*] ch. 38

AD 1623 Oct 20 [China]

3rd year of the Tianqi reign period of Emperor Xizong of the Ming Dynasty, 9th month, *jiayin* [51]; in Guyuanzhou stars fell like rain.
[*Ming shi: tianwen zhi*] ch. 27

[*Xuwenxian tongkao: xiang wei*] ch. 214
[*Xu tongzhi: zaixiang*] ch. 171

AD 1625 Aug 4 [China]

5th year of the Tianqi reign period of Emperor Xizong of the Ming Dynasty, 7th month, 2nd day; in Suichang large stars streamed from west to east, their tails over 20 *zhang* long, with rays as bright as the Moon and making a sound like thunder. From this day, each night meteors flew as if weaving.
Qing Dynasty, Guangxu period ed. Zhejiang [*Chuzhoufu zhi*] ch. 25

AD 1625 Nov 6 [Korea] (Ahn [2004] 40)

(a) 3rd year of King Injo of the Joseon Dynasty, 10th month, day *renwu* [19], at night; more than 10 meteors emerged from the zenith, scattered and entered the horizon in all directions. The largest were fist-sized, or like a cup, each 3 *zhang* long, their light shining on the ground. Again later there were over 30 bright meteors that scattered and entered the horizon.
[*Joseon wangjo sillok: Injo*] ch. 10

(b) 3rd year of King Injo of the Joseon Dynasty, 10th month, day *renwu* [19], first watch of the night; the moon trespassed on the southernmost star of QIANNIU [LM 9]. During the second watch, the moon had a halo. During the third watch, more than 10 meteors emerged from the zenith, scattered and entered the horizon in all directions. The largest were fist-sized, or like a cup, each about 3 *zhang* long, their red lights all shining on the ground. During the fifth watch and the third watch, over 30 bright meteors scattered and entered the horizon.
[*Seungjeongwon ilgi*]

AD 1633 Feb 8 [Korea] (Ahn [2004] 41)

11th year *guiyou* of King Injo of the Joseon Dynasty, [1st month] night of New Year's day; stars appeared as if fighting.
[*Joseon wangjo sillok: Injo*] ch. 28

AD 1636 Jun 30 [China]

9th year of the Chongzhen reign period of Emperor Yizong of the Ming Dynasty, 5th month, day *xinwei* [8]; stars fell like rain over Balihe, outside the prefecture's east gate.

Qing Dynasty, Guangxu period ed. Jiangsu [*Tongzhou Zhilizhou zhi*] ch. mo

AD 1636 Nov 27–Dec 26 [China]

9th year of the Chongzhen reign period of Emperor Yizong of the Ming Dynasty, 11th month; stars fell like rain and the drums of heaven sounded.

Qing Dynasty, Guangxu period ed. Shandong [*Linyixian zhi*] ch. 16

AD 1637 Jan 18 [Korea] (Ahn [2004] 42)

14th year *bingzi* of King Injo of the Joseon Dynasty, 12th month, 23rd day, third watch of the night; a myriad stars surrounded one large star and for a long while scattered toward the west and out of sight.

[*Joseon wangjo sillok: Injo*] ch. 33

AD 1637 Jan 29 [Korea] (Ahn [2004] 43)

15th year of King Injo of the Joseon Dynasty, 1st month, day *jiachen* [41]; stars fought in the southeast, by the next [double] hour they stopped.

[*Joseon wangjo sillok: Injo*] ch. 34

AD 1640 May 21–Jun 18 [China]

(a) 13th year of the Chongzhen reign period of Emperor Yizong of the Ming Dynasty, 4th month; the drums of heaven sounded and [meteors] streamed from the northwest to the southeast; stars fell like rain.

Qing Dynasty, Kangxi period ed. Shanxi [*Lingshixian zhi*] xiang yi

(b) 13th year of the Chongzhen reign period of Emperor Yizong of the Ming Dynasty, 14th month; the drums of heaven sounded in

Lingshi, stars fell with a noise and streamed from the northwest to the northeast; stars fell like rain.
Si ku ben [*Shanxi tongzhi*] ch. 163

AD 1642 May 28–Jun 26 [China]
Renwu [19] year of the Chongzhen reign period of Emperor Yizong of the Ming Dynasty, 5th month; stars streamed as if weaving.
[*Er shen ye lu*] ch. 8

AD 1642 summer [China]
15th year of the Chongzhen reign period of Emperor Yizong of the Ming Dynasty, summer; stars streamed as if weaving.
[*Ming shi: tianwen zhi*] ch. 27
[*Xuwenxian tongkao: xiang wei*] ch. 214
[*Xu tongzhi: zaixiang*] ch. 171

AD 1642 Jul 27–Aug 24 [China]
15th year of the Chongzhen reign period of Emperor Yizong of the Ming Dynasty, 7th month; all the starry images in the sky gathered between DOU [LM 8] and NÜ [LM 10] and streamed away toward the northeast seven times.
Qing Dynasty, Shunzhi period ed. Shandong [*Sishuiixian zhi*] ch. 11

AD 1642 Sep 23 [China]
15th year of the Chongzhen reign period of Emperor Yizong of the Ming Dynasty, 8th month, last day of the month; large stars big as peck measures, trailed by several tens of smaller stars, traveled from northwest to the southeast and fell to earth with a noise. Their brilliant rays were several *zhang* long and the darkness was all lit up.
Qing Dynasty, Kangxi period ed. Zhejiang [*Haiyanxian zhi*]

AD 1643 Jan 18 [China]
15th year of the Chongzhen reign period of Emperor Yizong of the Ming Dynasty, intercalary 11th month, 28th day, at night; stars fell like rain.
Qing Dynasty, Qianlong period ed. Guangdong [*Jieyangxian zhi*] ch. 7

AD 1643 Jul 8 [Korea] (Ahn [2004] 44)

21st year of King Injo of the Joseon Dynasty, 5th month, day *yimao* [52], during the first and second watches of the night, all the stars in the sky shook. During the fifth watch, the moon had a thin halo, red on the inside and green on the outside.
[*Joseon wangjo sillok: Injo*] ch. 44

AD 1644 Mar 9–Apr 6 [China]

(a) 17th year of the Chongzhen reign period of Emperor Yizong of the Ming Dynasty, 2nd month; stars fell like rain.
Qing Dynasty, Daoguang period ed. Hunan [*Yongzhoufu zhi*] ch. 17

(b) 1st year of the Shunzhi reign period of Emperor Shizu of the Qing Dynasty, 2nd month; stars came down like rain.
[*Xiao tian jinian fu kao*] ch. 3

AD 1644 Jun 5–Jul 3 [China]

17th year of the Chongzhen reign period of Emperor Yizong of the Ming Dynasty, 5th month; stars fell like rain. For 24 days in all, in the middle of night stars like peck measures flew criss-cross, now this way, now that, some falling noisily, some disappearing without a sound.
Qing Dynasty, Kangxi period ed. Zhejiang [*Qiantangxian zhi*] ch. 12

AD 1645 Jun 24–Jul 22 [China]

2nd year of the Shunzhi reign period of Emperor Shizu of the Qing Dynasty, 6th month; all night meteors flew back and forth as if weaving.
Qing Dynasty, Qianlong period ed. Zhejiang [*Pinghuxian zhi*] ch. 10

AD 1645 Jul 16 [China]

2nd year of the Shunzhi reign period of Emperor Shizu of the Qing Dynasty, 6th month, 23rd day; from outside the city, stars in the sky were seen to fall like rain inside the city.
Qing Dynasty, Kangxi period ed. Zhejiang [*Jiaxingfu zhi*] ch. 16

AD 1645 Jul 23–Aug 20 [China]

(a) 2nd year of the Shunzhi reign period of Emperor Shizu of the Qing Dynasty, intercalary 6th month; each night stars fell like rain, gradually abating in the 7th month.
Qing Dynasty, Qianlong period ed. Jiangsu [*Wuxian zhi*] ch. 26

(b) 2nd year of the Shunzhi reign period of Emperor Shizu of the Qing Dynasty, intercalary 6th month; at midnight, small stars scattered and fell like snow.
Qing Dynasty, Daoguang period ed. Jiangsu [*Huangjing zhi gao*] ch. 7

AD 1645 Jul 23 [China]

2nd year of the Shunzhi reign period of Emperor Shizu of the Qing Dynasty, intercalary 6th month, new moon day; near midnight in the sky small stars scattered and fell like snow.
Qing Dynasty, Jiaqing period ed. Jiangsu [*Taicangzhou zhi*] ch. 58

AD 1645 Jul 30 [China]

2nd year of the Shunzhi reign period of Emperor Shizu of the Qing Dynasty, intercalary 6th month, 8th day; during the night, there were meteors like the Moon, larger and smaller ones followed each other, their rays brilliant white.
Qing Dynasty, Kangxi period ed. Zhejiang [*Shaoxingfu zhi*] ch. 13

AD 1645 Aug 6 [China]

(a) 2nd year of the Shunzhi reign period of Emperor Shizu of the Qing Dynasty, intercalary 6th month, 15th day, at night; stars fell like rain.
Qing Dynasty, Kangxi period ed. Jiangsu [*Tongzhou zhi*] ch. 1

(b) 2nd year of the Shunzhi reign period of Emperor Shizu of the Qing Dynasty, intercalary 6th month, 15th day, at night; a multitude of stars streamed southward.
Qing Dynasty, Qianlong period ed. Shanghai [*Jiadingxian zhi*] ch. 3

(c) 2nd year of the Shunzhi reign period of Emperor Shizu of the Qing Dynasty, intercalary 6th month, 15th day; stars fell like rain.

Qing Dynasty, Guangxu period ed. Jiangsu [*Tongzhou Zhilizhou zhi*] *juan mo*

AD 1645 Aug 7 [China]

(a) 2nd year of the Shunzhi reign period of Emperor Shizu of the Qing Dynasty, intercalary 6th month, 16th day; a multitude of stars streamed and moved, to and fro, up and down, each with bright rays.
Qing Dynasty, Daoguang period ed. Jiangsu [*Huang jing zhi gao*] ch. 7

(b) 2nd year of the Shunzhi reign period of Emperor Shizu of the Qing Dynasty, intercalary 6th month, day *bingshen* [33]; total lunar eclipse; the stars streamed all night.
[*Xiao tian jinian fukao*] ch. 10

AD 1646 Jul 13–Aug 10 [China]

3rd year of the Shunzhi reign period of Emperor Shizu of the Qing Dynasty, 6th month; large stars fell in the southeast while several hundred smaller stars trailed along.
Qing Dynasty, Kangxi period ed. Zhejiang [*Dongyangxian zhi*] ch. 4

AD 1646 Jul 23 [China]

3rd year of the Shunzhi reign period of Emperor Shizu of the Qing Dynasty, 6th month, 11th day; stars fell like rain.
Qing Dynasty, Kangxi period ed. Zhejiang [*Shaoxingfu zhi*] ch. 13
Qing Dynasty, Kangxi period ed. Zhejiang [*Jiangshanxian zhi*] ch. 10
Qing Dynasty, Kangxi period ed. Zhejiang [*Shanyinxian zhi*] ch. 9
Qing Dynasty, Kangxi period ed. Zhejiang [*Taoyuanxiang zhi*] ch. 8

AD 1646 Aug 11–Sep 8 [China]

3rd year of the Shunzhi reign period of Emperor Shizu of the Qing Dynasty, 7th month; innumerable stars streamed from north to south.
Qing Dynasty, Kangxi period ed. Zhejiang [*Ningpofu zhi*] ch. 30
Qing Dynasty, Kangxi period ed. Zhejiang [*Yinxian zhi*] ch. 24
Qing Dynasty, Qianlong period ed. Zhejiang [*Zhenhaixian zhi*] ch. 4

AD 1647 Feb 5–1648 Jan 24 [China]

4th year of the Shunzhi reign period of Emperor Shizu of the Qing Dynasty; there was a star like an awl and a multitude of smaller stars trailed along behind.

Qing Dynasty, Yongzheng period ed. Yunnan [*Jianshuizhou zhi*] ch. 11

AD 1648 Jan 25–1649 Feb 10 [China]

5th year of the Shunzhi reign period of Emperor Shizu of the Qing Dynasty; stars fell like rain in the southeast.

Qing Dynasty, Qianlong period ed. Guangxi [*Cenxixian zhi*]

AD 1648 Mar 11 [China]

5th year of the Shunzhi reign period of Emperor Shizu of the Qing Dynasty, 2nd month, 18th day, at night; stars fell like rain in the northeast.

Qing Dynasty, Kangxi period ed. Zhejiang [*Changxingxian zhi*] ch. 4
Qing Dynasty, Qianlong period ed. Zhejiang [*Huzhoufu zhi*] ch. 38

AD 1648 spring [China]

5th year of the Shunzhi reign period of Emperor Shizu of the Qing Dynasty, spring; in the southeast stars fell chaotically.

Qing Dynasty, Kangxi period ed. Guangxi [*Pinglexian zhi*] ch. 6
Qing Dynasty, Qianlong period ed. Guangxi [*Zhaopingxian zhi*] ch. 4
Qing Dynasty, Daoguang period ed. Guangxi [*Qixiasi zhi*] ch. shang
NB: Qixiasi is in Guilin.

AD 1649 Jun 10–Jul 9 [China]

6th year of the Shunzhi reign period of Emperor Shizu of the Qing Dynasty, 5th month; swarm after swarm of falling stars flew by, large ones in front and smaller ones trailing behind, in between were some as large as peck measures.

Qing Dynasty, Qianlong period ed. Guangdong [*Guishanxian zhi*] ch. 2

AD 1651 Jun 19 [China]

(a) 8th year of the Shunzhi reign period of Emperor Shizu of the Qing Dynasty, 5th month, day wuyin [15], at night; in the southwest there were white meteors as big as oil lamps with brilliant tails. [*Qing Shizu shilu*] ch. 57

(b) 8th year of the Shunzhi reign period of Emperor Shizu of the Qing Dynasty, 5th month, day *wuyin* [15]; white meteors as big as oil lamps traveled southwestward from lunar mansion KANG [LM 2], while a multitude of smaller stars trailed along behind them into lunar mansion YI [LM 27].
[*Qing shi gao: tianwen zhi*] ch. 39

AD 1651 Sep 15–Oct 13 [China]

(a) 8th year of the Shunzhi reign period of Emperor Shizu of the Qing Dynasty, 8th month; at Hanshan, stars fell like rain. Qing Dynasty, Kangxi period ed. Anhui [*Hezhou zhi*] ch. 4

(b) 8th year of the Shunzhi reign period of Emperor Shizu of the Qing Dynasty, 8th month; a large star flew from the southwest to the northeast making a resonating sound like thunder and with a fiery brilliance shooting in all directions. Trailing along behind it were innumerable smaller stars.
Qing Dynasty, Kangxi period ed. Shandong [*Jiaozhou zhi*] ch. 6

AD 1651 Oct 14 [China]

8th year of the Shunzhi reign period of Emperor Shizu of the Qing Dynasty, 9th month, 1st day *yihai* [12]; at midnight stars fell like rain; near and far there was great fear.
Minguo [*Zhejiang xu tongzhi gao*] ch. 10

AD 1651 Oct 15 [China]

8th year of the Shunzhi reign period of Emperor Shizu of the Qing Dynasty, 9th month, 2nd day *bingzi* [13]; stars fell like rain.
Minguo [*Zhejiang xu tongzhi gao*] ch. 10

AD 1654 Jul 21 [China]

11th year of the Shunzhi reign period of Emperor Shizu of the Qing Dynasty, 6th month, 8th day; during the fourth watch, there was a red star shaped like a broom with brilliant rays more than 1 *zhang* long, while a multitude of smaller stars followed along toward the southeast. On reaching the northwest they fell.
Qing Dynasty, Qianlong period ed. Jiangsu [*Wuxian zhi*] ch. 26

AD 1659 May 21–Jun 19 [China]

16th year of the Shunzhi reign period of Emperor Shizu of the Qing Dynasty, 4th month; stars fell like rain.
Qing Dynasty, Kangxi period ed. Henan [*Neixiangxian zhi*] ch. 11

AD 1659 Jun 6 [China]

16th year of the Shunzhi reign period of Emperor Shizu of the Qing Dynasty, 4th month, 17th day; in the southeastern sky there was a star like a peck measure with several smaller stars trailing along behind. When they reached the southwest, they disappeared. At the time, the scarlet Sun was in the sky and it was cloudless on all sides. There was a muffled booming sound, like the sound of thunder under water.
Qing Dynasty, Daoguang period ed. Jiangsu [*Baoyingxian zhi*] ch. 9

AD 1660 Sep 5–Oct 3 [China]

17th year of the Shunzhi reign period of Emperor Shizu of the Qing Dynasty, 8th month; to the west of the prefectural city boundary a large star streamed northward, with countless other stars trailing after it.
Qing Dynasty, Qianlong period ed. Jiangxi [*Ji'anfu zhi*] ch. 1

AD 1661 Jan 30–1662 Feb 17 [China]

18th year of the Shunzhi reign period of Emperor Shizu of the Qing Dynasty; large stars fell in the west, smaller stars followed like a rain shower.
Qing Dynasty, Kangxi period ed. Yunnan [*Jianshuizhou zhi*] ch. 17
Qing Dynasty, Qianlong period ed. Yunnan [*Zhaozhou zhi*] xiang yi

AD 1661 Sep 23–Oct 22 [China]

18th year of the Shunzhi reign period of Emperor Shizu of the Qing Dynasty, 8th month; at the hour of *you* [LT = 17–19h] in the west large stars streamed in the north, innumerable smaller stars trailed along.

Qing Dynasty, Kangxi period ed. Jiangxi [*Lulingxian zhi*] ch. 2

AD 1661 Oct 1 [China]

(a) 18th year of the Shunzhi reign period of Emperor Shizu of the Qing Dynasty, 8th month, 9th day; at the hour of *shen* [LT = 15–17h], large stars were seen with more than 100 smaller stars following them, making a noise as they flew from the southwest into the north.

Qing Dynasty, Kangxi period ed. Hunan [*Linglingxian zhi*] ch. 14

(b) 18th year of the Shunzhi reign period of Emperor Shizu of the Qing Dynasty, 8th month, 9th day; at the hour of *shen* [LT = 15–17h], there were large stars about 10 *zhang* long with smaller stars following them and a white vapor like a chain. They started in the southwest and traveled rapidly northward, making a booming noise.

Qing Dynasty, Qianlong period ed. Hunan [*Qingquanxian zhi*] ch. 25

Qing Dynasty, Qianlong period ed. Hunan [*Hengzhoufu zhi*] ch. 29

NB: "There were large stars" was miswritten as "there was a man star."

AD 1661 Oct 23–Nov 21 [China]

18th year of the Shunzhi reign period of Emperor Shizu of the Qing Dynasty, 9th month; at Junshan there were flying stars all night.

Qing Dynasty, Qianlong period ed. Jiangsu [*Wushan quanzhi*] ch. 8

NB: Wushan is in Nantong.

AD 1661 Oct 30 [China]

18th year of the Shunzhi reign period of Emperor Shizu of the Qing Dynasty, 9th month, 8th day; toward dark, large stars started in the southwest and traveled northeastward, trailed by several thousand

smaller stars and a swath of white vapor, and making a sound like thunder.
Qing Dynasty, Kangxi period ed. Hunan [*Shaoyangxian zhi*] ch. 6
Qing Dynasty, Kangxi period ed. Hunan [*Baoqingxian zhi*] ch. 22

AD 1661 Dec 11 [China]
18th year of the Shunzhi reign period of Emperor Shizu of the Qing Dynasty, 10th month, 20th day; at midnight stars fell like rain. Starting in the southeast they fell in swarms in the northwest.
Qing Dynasty, Daoguang period ed. Jiangsu [*Huangjing zhi gao*] ch. 7

AD 1665 Nov 17 [China]
4th year (yisi year) of the Kangxi reign period of Emperor Shengzu of the Qing Dynasty, 10th month, 11th day; during the fourth watch, large stars were seen in the southeast with a multitude of smaller stars trailing them, some going up, some down, some going left, some right; when the large stars fell, the smaller ones also followed them and fell.
[*San gang zhi lue*] ch. 5 *buyi*

AD 1665 Nov 18 [China]
4th year of the Kangxi reign period of Emperor Shengzu of the Qing Dynasty, 10th month, 12th day; at the hour of MAO [LT = 5–7h], stars fell like rain.
Qing Dynasty, Kangxi period ed. Zhejiang [*Jiaxingfu zhi*] ch. 2

AD 1666 Jul 23 [China]
(a) 5th year of the Kangxi reign period of Emperor Shengzu of the Qing Dynasty, 6th month, *xinwei* [8] (22nd day), at night; stars in the sky streamed chaotically as if weaving.
Qing Dynasty, Kangxi period ed. Hebei [*Zengbu Lulongxian zhi*] ch. 2
(b) 5th year of the Kangxi reign period of Emperor Shengzu of the Qing Dynasty, 6th month, *xinwei* [8] (22nd day); stars streamed as if weaving.

Qing Dynasty, Kangxi period ed. Hebei [*Shuipingfu zhi*] ch. 3

(c) 5th year of the Kangxi reign period of Emperor Shengzu of the Qing Dynasty, 6th month, *xinwei* [8] (22nd day); stars streamed as if weaving from azimuth *qian* (NW) to *xun* (SE).
Qing Dynasty, Kangxi period ed. Hebei [*Liangzhou Zhilifu zhi*] ch. 31

(d) 5th year of the Kangxi reign period of Emperor Shengzu of the Qing Dynasty, 6th month, *xinwei* [8] (22nd day); stars streamed as if weaving. People in the prefectural city saw it and later from Hangzhou came word of the same.
Qing Dynasty, Jiaqing period ed. Hebei [*Liangzhou zhi*] ch. 1

(e) 5th year of the Kangxi reign period of Emperor Shengzu of the Qing Dynasty, 6th month, *xinwei* [8] (22nd day); stars fell like rain.
Qing Dynasty, Daoguang period ed. Jiangsu Taicang [*Huangjing zhi gao*] ch. 7

AD 1666 Jul 24 [China]

5th year of the Kangxi reign period of Emperor Shengzu of the Qing Dynasty, 6th month, 23rd day; at the hour of *hai* (LT = 21–23h), large stars trailed by more than 1,000 smaller stars streamed from northwest to the southeast where they fell to earth.
Qing Dynasty, Kangxi period ed. Zhejiang [*Haiyanxian zhi buyi*] zai xiang

AD 1666 autumn [China]

5th year of the Kangxi reign period of Emperor Shengzu of the Qing Dynasty, autumn; in the northwest stars fell like rain.
Qing Dynasty, Kangxi period ed. Hunan [*Ningxiangxian zhi*] ch. 2
Qing Dynasty, Qianlong period ed. Hunan [*Yiyangxian zhi*] ch. 1

AD 1666 Nov 7 [China]

5th year of the Kangxi reign period of Emperor Shengzu of the Qing Dynasty, 10th month, 1st day of winter; at night, in the northeast large

stars fell to earth making a noise, while a multitude of smaller stars scattered and fell like rain.
Qing Dynasty, Guangxu period ed. Zhejiang [*Zhongxiu Jiashanxian zhi*] ch. 34

AD 1667 Jan 6 [China]
5th year of the Kangxi reign period of Emperor Shengzu of the Qing Dynasty, 12th month, 12th day; near dawn large stars fell from the southeast to the northeast trailed by a multitude of smaller stars.
Qing Dynasty, Daoguang period ed. Jiangsu [*Huangjing zhi gao*] ch. 7

AD 1668 Apr 21 [China]
7th year of the Kangxi reign period of Emperor Shengzu of the Qing Dynasty, 3rd month, 11th day, at night; there were meteors as big as peck measures, with countless smaller stars surrounding them.
Qing Dynasty, Qianlong period ed. Hubei [*Mianyangzhou zhi*] ch. 13

AD 1671 May 9–Jun 6 [China]
10th year of the Kangxi reign period of Emperor Shengzu of the Qing Dynasty, 4th month; large stars shaped like cartwheels, with rays flaring in all directions, flew from the northwest to the southeast trailing a swath of several tens of smaller stars. They passed straight over Tongjiang; at the time it was already daybreak.
Qing Dynasty, Kangxi period ed. Zhejiang [*Qiantangxian zhi*] ch. 12

AD 1672 May 11 [China]
(a) 11th year of the Kangxi reign period of Emperor Shengzu of the Qing Dynasty, 4th month, 15th day; large stars crossed the sky from the northeast, while several tens of smaller stars trailed along behind. When they reached the southwest, they fell, making a sound like thunder.
Qing Dynasty, Kangxi period ed. Zhejiang [*Jiandexian zhi*] ch. 9
Qing Dynasty, Yongzheng period ed. Zhejiang [*Yanzhoufu zhi*] ch. 22

(b) 11th year of the Kangxi reign period of Emperor Shengzu of the Qing Dynasty, 4th month, 15th day; at the hour of *mao* [LT = 5–7h], in Duchang there was a thing like a wriggly dragon several tens of *zhang* long as big as a bucket and scarlet like fire. A swarm of smaller stars escorted it from the northeast until they disappeared in the southwest, making a sound like thunder.
Qing Dynasty, Kangxi period ed. Jiangxi [*Duchangxian zhi*] ch. 10

AD 1672 May 12 [China]

(a) 11th year of the Kangxi reign period of Emperor Shengzu of the Qing Dynasty, 4th month, 16th day, at night; anomalous stars shaped like dragons moved in the southwest. They were over 1 *zhang* long, their tails a swath of several tens of stars, and where they passed a white vapor trail remained; after an hour they disappeared.
Qing Dynasty, Kangxi period ed. Anhui [*Susongxian zhi*] ch. 3

(b) 11th year of the Kangxi reign period of Emperor Shengzu of the Qing Dynasty, 4th month, 16th day; during the fifth watch, a multitude of stars streamed in swarms from the north to disappear in the south, their fiery gleam disorderly and making a noise as they went.
Qing Dynasty, Kangxi period ed. Jiangxi [*Pengzexian zhi*] ch. 14

AD 1675 Nov 7 [China]

14th year of the Kangxi reign period of Emperor Shengzu of the Qing Dynasty, 9th month, 20th day, at night; stars as big as peck measures fell in the south, while a multitude of smaller stars trailed along and fell with them.
Qing Dynasty, Kangxi period ed. Fujian [*Zhangzhoufu zhi*] ch. 33
Qing Dynasty, Qianlong period ed. Fujian [*Longxixian zhi*] ch. 20

AD 1676 Dec 14 [Korea] (Ahn [2004] 45)

(a) 2nd year of King Sukjong of the Joseon Dynasty, 11th month, day wuzi [25]; a meteor emerged above BEIDOU *and entered the*

eastern wall of Ziwei, after this they emerged continuously. Venus also repeatedly appeared in daylight.
[Joseon wangjo sillok: Sukjong] ch. 5 (Ahn [2004] 45)

(b) 2nd year of King Sukjong of the Joseon Dynasty, 11th month, 10th day wuzi [25]; during the hour of si (9–11h), Venus appeared in the SSW. From the first to the third watches there was a halo around the moon. During the fifth watch, a yellowish-white meteor shaped like a bottle with a tail about 3 to 4 chi long emerged above Beidou and entered Ziwei below the eastern wall. Its light shone on the ground.
[Seungjeongwon ilgi]

NB: Ahn (2004) rejected this report as ambiguous.

AD 1679 Sep 21 [China]

18th year of the Kangxi reign period of Emperor Shengzu of the Qing Dynasty, 8th month, 17th day; during the hour of *shen* [LT = 15–17h], stars as big as peck measures crossed the Milky Way from the northwest and streamed toward the southeast, while countless stars in swarms followed them. Their tails trailed a diaphanous whiteness several hundred *zhang* long; those who saw it were amazed.
Qing Dynasty, Kangxi period ed. Hunan [*Yongzhoufu zhi*] ch. 24

AD 1680 Feb 18 [Korea] (Ahn [2004] 46)

6th year of King Sukjong of the Joseon Dynasty, 1st month, day jiyou [46]; a meteor emerged. After this they emerged daily, or every other day, for over 30 days before stopping.
[Joseon wangjo sillok: Sukjong] ch. 9
NB: Ahn (2004) expressed doubt about this as a record of a shower.

AD 1682 Aug 13–22 [China]

21st year of the Kangxi reign period of Emperor Shengzu of the Qing Dynasty, 7th month, middle, at night; more than 10 large stars each trailing a long pale tail entered the allocated fields of Ji [LM 7] and Wei [LM 6] from the southwest.

Qing Dynasty, Qianlong period ed. [*Fujian tongzhi*] ch. 65
Qing Dynasty, Qianlong period ed. Fujian [*Quanzhoufu zhi*] ch. 73
Qing Dynasty, Qianlong period ed. Fujian [*Jinjiangxian zhi*] ch. 15
Qing Dynasty, Qianlong period ed. Fujian [*Tong anxian zhi*] ch. 13

AD 1682 Sep 11 [China]

21st year of the Kangxi reign period of Emperor Shengzu of the Qing Dynasty, 8th month, 10th day; during the second watch, several hundred large and small stars fell over Jilongshan.
[*Taiwan wai ji*] ch. 26

AD 1683 Oct 26 [Korea]

9th year of King Sukjong of the Joseon Dynasty, 9th month, day *renshen* [9]; a meteor emerged within the western wall of ZIWEI.
9th year of King Sukjong of the Joseon Dynasty, 9th month, day *jiaxu* [11] (Oct 28); a meteor emerged below BEIJI and entered the north; after this they were repeatedly seen.
[*Joseon wangjo sillok: Sukjong*] ch. 14b (Ahn [2004] 47)
NB: After this record, Ahn (2004) reproduced over a dozen records from *Seungjeongwon ilgi* of what appear to be individual meteors or bolides reported during the following 3 weeks.

AD 1696 Oct 29 [Korea] (Ahn [2004] 48)

(a) 22nd year of King Sukjong of the Joseon Dynasty, 10th month, day dinghai [24]; a meteor appeared. They appeared continuously for some nights.
[Joseon wangjo sillok: Sukjong] ch. 30 (Ahn [2004] 48)

(b) 22nd year of King Sukjong of the Joseon Dynasty, 10th month, 4th day dinghai [24], fifth watch of the night; a red meteor the size of a fist with a tail about 2 to 3 chi long emerged below BI [LM 19] and entered the horizon in the south.
[Seungjeongwon ilgi]

NB: Based on the report from Seungjeongwon ilgi, Ahn (2004) concluded this does not qualify as a record of a meteor shower.

AD 1698 Nov 8 [Japan]

11th year of the Genroku reign period, 10th month, day *dingwei* [44]…at night; large meteors flew into lunar mansion JING [LM 22], and hung downward for several *chi*, lasting for several *ke* (quarter hours) before disappearing. On this night stars streamed like weaving, everyone saw it.
[*Jinki roku*] ch. 2

AD 1708 Sep 4 [China]

(a) 47th year of the Kangxi reign period of Emperor Shengzu of the Qing Dynasty, 7th month, 20th day; during the hour of *shen* [LT = 15–17h], several hundred stars blazed forth, arrayed in extremely even ranks like strung gems on a crown, stretching from the east to behind Jinzhongshan.
Qing Dynasty, Kangxi period ed. Fujian [*Luoyuanxian zhi*] ch. 10

(b) 47th year of the Kangxi reign period of Emperor Shengzu of the Qing Dynasty, 7th month, 20th day; at Fuzhoufu during the hour of *shen* [LT = 15–17h], several hundred stars with blazing rays in lines, streamed from the east to behind Jinzhongshan.
Qing Dynasty, Qianlong period ed. [*Fujian tongzhi*] ch. 65

AD 1709 Aug 20 [China]

48th year of the Kangxi reign period of Emperor Shengzu of the Qing Dynasty, 7th month, after the full Moon; during the hour of *you* [LT = 17–19h], stars came out of the south moving extremely slowly, trailing a brilliant white diaphanous vapor behind their tails. Innumerable smaller stars followed along behind, moving away slowly northward and not disappearing for an hour.
Qing Dynasty, Qianlong period ed. [*Hunan tongzhi*] ch. 172
Qing Dynasty, Jiaqing period ed. Hunan [*Ningyuanxian zhi*] ch. 10

AD 1722 Nov 24 [China]

61st year of the Kangxi reign period of Emperor Shengzu of the Qing Dynasty, 10th month, 16th day; during the marks of the hour of *yin*

[LT = 3–5h], a star as big as a peck measure fell from the east to the northwest, carrying along with it as it fell some 10 or so smaller stars, and making a sound like thunder, as its scarlet brilliance traversed the sky.
Qing Dynasty, Yongzheng period ed. Shanxi [*Weinanxian zhi*] ch. 15

AD 1729 Feb 13 [China]
7th year of the Yongzheng reign period of Emperor Shizong of the Qing Dynasty, 1st month, 16th day, at night; a multitude of stars was seen to sway, as if about to fall down. Then some flew and some paced, moving eastward in swarms.
[*Qing Shizong shilu*] ch. 22

AD 1731 Oct 31 [China]
9th year of the Yongzheng reign period of Emperor Shizong of the Qing Dynasty, 10th month, new Moon; stars fell like rain.
Qing Dynasty, Qianlong period ed. Yunnan [*Xinxingzhou zhi*] ch. 1

AD 1732 Jan 27–AD 1733 Feb 13 [China]
10th year of the Yongzheng reign period of Emperor Shizong of the Qing Dynasty; in the southwest stars fell like rain.
Qing Dynasty, Daoguang period ed. Yunnan [*Kunyangzhou zhi*] ch. 2

AD 1732 Dec 15 [China]
(a) 10th year of the Yongzheng reign period of Emperor Shizong of the Qing Dynasty, 10th month, 28th day, at night; stars fell like rain.
Qing Dynasty, Daoguang period ed. Yunnan [*Guangxifu zhi*] ch. 23
(b) 10th year of the Yongzheng reign period of Emperor Shizong of the Qing Dynasty, 10th month, 28th day, at night; stars fell like snow.
Qing Dynasty, Daoguang period ed. [*Yunnan tongzhi gao*] ch. 4

AD 1733 Feb 14–AD 1734 Feb 3 [China]

11th year of the Yongzheng reign period of Emperor Shizong of the Qing Dynasty; stars fell like rain.

Qing Dynasty, Qianlong period ed. Yunnan [*Kaihuafu zhi*] ch. 1

AD 1733 May 14–Jun 11 [China]

11th year of the Yongzheng reign period of Emperor Shizong of the Qing Dynasty, 4th month; stars fell like rain.

Qing Dynasty, Daoguang period ed. [*Yunnan tongzhi gao*] ch. 4

AD 1733 Jun 10 [China]

(a) 11th year of the Yongzheng reign period of Emperor Shizong of the Qing Dynasty, 4th month, 28th day, at night; stars clashed in station SHOUXING [LM 1–2].

Qing Dynasty, Tongzhi period ed. Guangxi [*Cangwuxian zhi*] ch. 18

Qing Dynasty, Guangxu period ed. Guangxi [*Qianjiangxian zhi*] ch. 4

(b) 11th year of the Yongzheng reign period of Emperor Shizong of the Qing Dynasty, 4th month, 28th day, at night; just at the beginning of twilight, stars fell like rain.

Qing Dynasty, Qianlong period ed. Yunnan [*Yongbeifu zhi*] ch. 24

AD 1735 Jan 29 [China]

13th year of the Yongzheng reign period of Emperor Shizong of the Qing Dynasty, 1st month, 6th day; in the southeast corner a star bigger than a bowl rose up from below trailed by five or six smaller stars forming a clashed shape. The large star did not move as the smaller stars fell and then rose up again, repeating this several times until they suddenly fell and disappeared. Then a multitude of smaller stars once again flew into and merged with the large star before disappearing with it.

Qing Dynasty, Qianlong period ed. Guizhou [*Yupingxian zhi*] ch. 1

AD 1735 Apr 23–May 21 [China]

13th year of the Yongzheng reign period of Emperor Shizong of the Qing Dynasty, 4th month; roaming stars flew everywhere.

Qing Dynasty, Qianlong period ed. Yunnan [*Yongbeifu zhi*] ch. 24

AD 1748 Oct 16 [China]

13th year of the Qianlong reign period of Emperor Gaozong of the Qing Dynasty, 8th month, 24th day; during the marks of the hour of *you* [LT = 17–19h], in the east a meteor as big as the Moon, bright as the light of day and more than 1 *zhang* long, gradually reached the zenith. Its shape resembled a washing paddle with innumerable smaller stars gradually trailing behind it that sifted down fiery lights as they fell, making a sound like thunder.

Qing Dynasty, Jiaqing period ed. Anhui [*Jingxian zhi*] ch. 27

AD 1760 May 11 [China]

25th year of the Qianlong reign period of Emperor Gaozong of the Qing Dynasty, 3rd month, 26th day; during the marks of the hour of *xu* [LT = 19–21h], scarlet stars as big as baskets fell from the northwest to the southeast, while behind their tails several tens of smaller stars trailed along, making a sputtering sound and shining a colorful brilliance on people.

Qing Dynasty, Qianlong period ed. Hebei [*Hengshuixian zhi*] ch. 11

AD 1768 Mar 18 [China]

33rd year of the Qianlong reign period of Emperor Gaozong of the Qing Dynasty, 2nd month, day *jiwei* [56]; in the evening stars fell like rain.

[*Qing shi gao lie zhuan*] ch. 528

AD 1769 autumn [China]

34th year of the Qianlong reign period of Emperor Gaozong of the Qing Dynasty, Autumn; stars as big as cart wheels flew several times, some from the north, some from the south.

Qing Dynasty, Daoguang period ed. [*Yunnan tongzhi gao*] ch. 4

AD 1769 Dec 28–AD 1770 Jan 26 [Japan]

6th year of the Meiwa reign period, 12th month; there was a large star which streamed northward while several tens of small stars followed it. [*Zoku shi gushō*]

AD 1770 Mar 27–Apr 25 [China]

35th year of the Qianlong reign period of Emperor Gaozong of the Qing Dynasty, 3rd month; stars fell like rain and the drums of heaven sounded.
Qing Dynasty, Daoguang period ed. Shandong [*Linyixian zhi*] ch. 16
Qing Dynasty, Daoguang period ed. Shandong [*Jinanfu zhi*] ch. 20

AD 1794 Jan 31–Mar 1 [China]

59th year of the Qianlong reign period of Emperor Gaozong of the Qing Dynasty, 1st month; at Jianchuan meteors fell on every side.
Qing Dynasty, Guangxu period ed. [*Yunnan tongzhi*] ch. 4

AD 1794 Dec 11 [China]

59th year of the Qianlong reign period of Emperor Gaozong of the Qing Dynasty, 11th month, 19th day, at night; stars fell like rain all through the 3 hours of *hai* (LT = 21–23h), *zi* (LT = 23–1h), *chou* (LT= 1–3h) before ceasing.
Qing Dynasty, Daoguang period ed. [*Yunnan tongzhi*] ch. 4
Qing Dynasty, Guangxu period ed. Yunnan [*Zhanyizhou zhi*] ch. 4

AD 1794 winter [China]

59th year of the Qianlong reign period of Emperor Gaozong of the Qing Dynasty, winter; in Ya'an a multitude of stars was seen to fall in the west, their aspect like flying fireflies.
Minguo period ed. Sichuan [*Ya'anxian zhi*] ch. 4

AD 1795 Dec 20 [Japan]

7th year of the Kansei reign period, 11th month, 10th day, *dingsi* [54]; the night before the winter solstice flying stars fell like rain.

[*Hōryaku gen rai syū*]
NB: Translated from the Japanese.

AD 1796 Feb 3 [China]
60th year of the Qianlong reign period of Emperor Gaozong of the Qing Dynasty, 12th month, 25th day; during the hour of *xu* [LT = 19–21h], in the south there was a swath of white brilliance. Alongside it was accompanied by several tens of large stars that moved about unsteadily.
Qing Dynasty, Jiaqing period ed. Zhejiang [*Jiashanxian zhi*] ch. 20

AD 1797 Feb 27–Mar 27 [China]
2nd year of the Jiaqing reign period of Emperor Renzong of the Qing Dynasty, 2nd month; in the southwest corner of the sky, stars fell like rain.
Qing Dynasty, Daoguang period ed. Jiangsu [*Huangjing zhi gao*] ch. 7

AD 1797 Oct 20–Nov 17 [China]
2nd year of the Jiaqing reign period of Emperor Renzong of the Qing Dynasty, 9th month, at night; stars flew as if weaving.
Qing Dynasty, Jiaqing period ed. Fujian [*Fudingxian zhi*] ch. 7

AD 1797 Dec 7 [China]
2nd year of the Jiaqing reign period of Emperor Renzong of the Qing Dynasty, 10th month, 20th day; stars fell like rain.
Qing Dynasty, Jiaqing period ed. Anhui [*Jixixian zhi*] ch. 12

AD 1797 Dec 16 [China]
2nd year of the Jiaqing reign period of Emperor Renzong of the Qing Dynasty, 10th month, 29th day; stars fell like rain.
Qing Dynasty, Daoguang period ed. Shanxi [*Taiyuanxian zhi*] ch. 15
Qing Dynasty, Guangxu period ed. Shanxi [*Jiaochengxian zhi*] ch. 1
Qing Dynasty, Guangxu period ed. Shanxi [*Jinci zhi*] ch. 41

AD 1797 Dec 17 [China]

2nd year of the Jiaqing reign period of Emperor Renzong of the Qing Dynasty, 10th month, last day of the month; at midnight, stars fell like rain, all speeding toward the southwest and blazing furiously. It was a long time before they ceased.

Qing Dynasty, Guangxu period ed. Shanxi [*Jinci zhi*] ch. 41

AD 1797 Dec 18–AD 1798 Jan 16 [China]

(a) 2nd year of the Jiaqing reign period of Emperor Renzong of the Qing Dynasty, 11 month, a certain day, at night; stars fell like rain in the north. Then half the sky in the north was also seen to turn greenish-gray, and traces of stars were scarce.

Qing Dynasty, Guangxu period ed. Zhejiang [*Taishun fen jiang lu*] ch. 10

(b) 2nd year of the Jiaqing reign period of Emperor Renzong of the Qing Dynasty, 11th month; stars flew about as if weaving, after a bit they settled down.

Qing Dynasty, Daoguang period ed. Guangdong [*Gaoyaoxian zhi*] ch. 10

AD 1798 Feb 16–Mar 16 [China]

3rd year of the Jiaqing reign period of Emperor Renzong of the Qing Dynasty, 1st month; swarms of stars flew chaotically up, down, left, and right.

Qing Dynasty, Daoguang period ed. Fujian [*Jianyangxian zhi*] ch. 19

AD 1798 May 16–Jun 13 [China]

3rd year of the Jiaqing reign period of Emperor Renzong of the Qing Dynasty, 4th month; stars fell like rain.

Qing Dynasty, Jiaqing period ed. Sichuan [*Danxian zhi*] ch. 41

AD 1798 Jun 14–Jul 12 [China]

3rd year of the Jiaqing reign period of Emperor Renzong of the Qing Dynasty, 5th month; a certain night in Shiwudu stars fell like rain, settling down first after 2 hours.

Qing Dynasty, Guangxu period ed. Hunan [*Yongmingxian zhi*] ch. 43

AD 1798 Oct 9–Nov 7 [China]

3rd year of the Jiaqing reign period of Emperor Renzong of the Qing Dynasty, 9th month; in Shanghai's Qingpu County stars moved about in the sky, as if flying or weaving, and flashing brilliantly in all directions. It was like this for several nights.

[*Ming zhai xiao zhi*] ch. 6

AD 1798 Nov 5–6 [China]

3rd year of the Jiaqing reign period of Emperor Renzong of the Qing Dynasty, 9th month, 28th day, at night; multitudes of stars moved back and forth as if weaving and making a chattering sound, in a moment falling like rain, in all for 2 nights.

Qing Dynasty, Guangxu period ed. Zhejiang [*Zhongxiu Jiashanxian zhi*] ch. 34

AD 1798 Nov 7 [China]

3rd year of the Jiaqing reign period of Emperor Renzong of the Qing Dynasty, 9th month, thirtieth day; stars fell like rain.

Qing Dynasty, Guangxu period ed. Yunnan [*Xu xiu Yimenxian zhi*] ch. 1

AD 1798 Nov 8–Dec 6 [China]

(a) 3rd year of the Jiaqing reign period of Emperor Renzong of the Qing Dynasty, 10th month; stars flew as if weaving.

Qing Dynasty, Jiaqing period ed. Guangdong [*Longshanxiang zhi*] ch. 14

(b) 3rd year of the Jiaqing reign period of Emperor Renzong of the Qing Dynasty, 10th month; stars fell like rain.

Qing Dynasty, Jiaqing period ed. Yunnan [*Jingdongxian zhi gao*] ch. 1

Qing Dynasty, Daoguang period ed. Yunnan [*Jinningzhou zhi*] ch. 11

Qing Dynasty, Daoguang period ed. Yunnan [*Chengjiangfu zhi*] ch. 2

Qing Dynasty, Xuantong period ed. Yunnan [*Xu Mengzixian zhi*] ch. 12

(c) 3rd year of the Jiaqing reign period of Emperor Renzong of the Qing Dynasty, 10th month; stars streamed as if weaving.

Min Guo period ed. Jiangsu [*Rugaoxian zhi*] vol. 2

AD 1798 Nov 20 [China]

3rd year of the Jiaqing reign period of Emperor Renzong of the Qing Dynasty, 10th month, 13th day, at night; stars fell like rain from north to south all night.

Qing Dynasty, Guangxu period ed. Shanghai [*Jiangdong zhi*] ch. 1

AD 1798 Nov 26 [China]

3rd year of the Jiaqing reign period of Emperor Renzong of the Qing Dynasty, 10th month, 19th day, at twilight; stars came down like rain, not stopping till midnight.

Qing Dynasty, Qianlong period ed. Shanxi [*Xunyangxian zhi*] ch. 14
Qing Dynasty, Jiaqing period ed. Shanxi [*Baihexian zhi*] ch. 14

AD 1798 Nov 27 [China]

3rd year of the Jiaqing reign period of Emperor Renzong of the Qing Dynasty, 10th month, 20th day, at night; a multitude of stars streamed back and forth as if weaving.

Minguo period ed. Zhejiang [*Hainingzhou zhi gao*] ch. 40

AD 1798 Nov 29 [China]

3rd year of the Jiaqing reign period of Emperor Renzong of the Qing Dynasty, 10th month, 22nd day; during the hour of *xu* [LT = 19–21h], in the northeast a multitude of stars flew like shuttles, and for a moment stars fell like rain.

Qing Dynasty, Guangxu period ed. Yunnan [*Jiangchuanxian zhi*] ch. 27

AD 1798 Dec 4 [China]

(a) 3rd year of the Jiaqing reign period of Emperor Renzong of the Qing Dynasty, 10th month, 27th day, at night; countless stars moved about.

Qing Dynasty, Daoguang period ed. Henan [*Yanlingxian zhi*] ch. 17
Qing Dynasty, Guangxu period ed. Henan [*Xiangfuxian zhi*] ch. 23

(b) 3rd year of the Jiaqing reign period of Emperor Renzong of the Qing Dynasty, 10th month, 27th day, at night; stars moved about as if weaving.
Minguo period ed. Henan [*Yanlingxian zhi*] ch. 29

AD 1798 Dec 5 [China]

3rd year of the Jiaqing reign period of Emperor Renzong of the Qing Dynasty, 10th month, 28th day, at night; a multitude of stars streamed back and forth as if weaving.
Qing Dynasty, Jiaqing period ed. Zhejiang [*Jiaxingfu zhi*] ch. 35
Qing Dynasty, Daoguang period ed. Zhejiang [*Shimenxian zhi*] ch. 23
Qing Dynasty, Guangxu period ed. Zhejiang [*Haiyanxian zhi*] ch. 13
[*Qing Renzong shilu*] ch. 42

AD 1798 Dec 6 [China]

(a) Wuwu [55] year of the Jiaqing reign period of Emperor Renzong of the Qing Dynasty, 10th month, 29th day, during the night; a multitude of stars streamed back and forth as if weaving.
[*Qing Renzong shilu*] ch. 42
[*Lu yuan conghua*] zaoxiang

(b) Wuwu [55] year of the Jiaqing reign period of Emperor Renzong of the Qing Dynasty, 10th month, 29th (last) day; in the southeast, a multitude of stars streamed as if weaving, stopping only after an hour.
Qing Dynasty, Jiaqing period ed. Zhejiang [*Pinghuxian zhi*] ch. 3

(c) 3rd year of the Jiaqing reign period of Emperor Renzong of the Qing Dynasty, 10th month, 29th (last) day; during the marks of the hour of *hai* [LT = 21–23h], stars streamed as if weaving, lasting an hour before ending.
Qing Dynasty, Jiaqing period ed. Hebei [*Liangzhou zhi*] ch. 1
Qing Dynasty, Daoguang period ed. Hebei [*Jizhou zhi*] ch. 2
Qing Dynasty, Tongzhi period ed. Hebei [*Changlixian zhi*] ch. 1

(d) 3rd year of the Jiaqing reign period of Emperor Renzong of the Qing Dynasty, 10th month, 29th (last) day; stars scattered in the four directions as if weaving.
Qing Dynasty, Jiaqing period ed. Shandong [*Changshanxian zhi*] ch. 4

(e) 3rd year of the Jiaqing reign period of Emperor Renzong of the Qing Dynasty, 10th month, 29th (last) day; during the first watch, stars fell like rain.
Qing Dynasty, Xuantong period ed. Shanghai [*Pengpuli zhi*] ch. 8

(f) 3rd year of the Jiaqing reign period of Emperor Renzong of the Qing Dynasty, 10th month, 29th (last) day, at night; during the hour of *xu* [LT = 19–21h], in the northeast corner, thousands of stars streamed as if weaving, not stopping for about the time it takes to cook a meal.
Qing Dynasty, Jiaqing period ed. Guangdong [*Chaoyangxian zhi*] ch. 12
Qing Dynasty, Tongzhi period ed. Guangdong [*Dapuxian zhi*] ch. 13

(g) 3rd year of the Jiaqing reign period of Emperor Renzong of the Qing Dynasty, 10th month, 29th (last) day; in the northwest, stars fell like rain.
Qing Dynasty, Jiqing period ed. Yunnan [*Lin anfu zhi*] ch. 17

(h) 3rd year of the Jiaqing reign period of Emperor Renzong of the Qing Dynasty, 10th month, 29th (last) day, during the hour of *xu* [LT = 19–21h], stars fell like rain, not stopping until the fourth watch.
Qing Dynasty, Daoguang period ed. Jiangsu [*Kunshan Xinyang liang xian zhi*] ch. 39

AD 1798 Dec 6 [Japan]

(a) 10th year of the Kansei reign period, 10th month, 29th day; it was reported that on this night numerous meteors flew confusedly in all directions.
[*Sanetane kōki*]

(b) 10th year of the Kansei reign period, 10th month, 29th day *wuwu* [55]; the weather was clear; during the hour of *you* (LT = 17–19h), I went out to the garden and saw a myriad large and small stars flying through the air, streaming north to south, east to west, as they fell like rain; it was like this until the hours of *chou* (LT = 1–3h) and *yin* (LT = 3–5h); at dawn, they could not be seen in the glare of the rising sun in the east.
[*Hoku sō zatsu wa*]
NB: Translated from Japanese.

(c) 10th year of the Kansei reign period, 10th month, 29th day; it rained stars all night.
[*Michigawa chūji kiroku*]

AD 1798 Dec 7–1799 Jan 5 [China]

(a) 3rd year of the Jiaqing reign period of Emperor Renzong of the Qing Dynasty, 11th month; swarms of stars flew chaotically from the hour of *you* [LT = 17–19h] until the hour of *hai* [LT = 21–23h] when they first settled down.
Qing Dynasty, Daoguang period ed. Fujian [*Luoyuanxian zhi*] ch. 29

(b) 3rd year of the Jiaqing reign period of Emperor Renzong of the Qing Dynasty, 11th month; chaotic stars flew up, down, left, and right as if weaving for more than one double hour before settling down.
Minguo period ed. Fujian [*Zhaoanxian zhi*] ch. 5

AD 1798 Dec 9 [Japan]

10th year of the Kansei reign period, 11th month, at night; stars flew like before.
[*Mukō nen hyō*]

AD 1798 Dec 12 [China]

3rd year of the Jiaqing reign period of Emperor Renzong of the Qing Dynasty, 11th month, 6th day, at night; stars streamed as if weaving and fell like rain.
Qing Dynasty, Zhejiang ed. [*Haichang xu zai*] ch. 4

AD 1798 Dec 25 [China]

3rd year of the Jiaqing reign period of Emperor Renzong of the Qing Dynasty, 11th month, 19th day, at night; stars fell like rain.
Qing Dynasty, Daoguang period ed. Yunnan [*Zhaozhou zhi*] ch. 3
Qing Dynasty, Daoguang period ed. Yunnan [*Bazhou zhi*] ch. 10

AD 1798 Dec 31 [China]

3rd year of the Jiaqing reign period of Emperor Renzong of the Qing Dynasty, 11th month, 25th day; stars fell like rain.
Qing Dynasty, Daoguang period ed. Fujian [*Shunchangxian zhi*] ch. 9

AD 1799 Jan 6–Feb 4 [China]

3rd year of the Jiaqing reign period of Emperor Renzong of the Qing Dynasty, 12th month; stars fell like rain.
Qing Dynasty, Daoguang period ed. [*Yunnan tongzhi gao*] ch. 4

AD 1799 Feb 2–3 [China]

3rd year of the Jiaqing reign period of Emperor Renzong of the Qing Dynasty, 12th month, 28th and 29th days, at midnight; a multitude of stars streamed back and forth as if weaving.
[*Qingchao xu Wenxian tongkao*] ch. 303

AD 1799 Jun 3–Jul 2 [China]

4th year of the Jiaqing reign period of Emperor Renzong of the Qing Dynasty, 5th month; at Shiwudu stars fell like rain.
Qing Dynasty, Daoguang period ed. Hunan [*Yongmingxian zhi*] ch. 13

AD 1799 Nov 22 [China]

4th year of the Jiaqing reign period of Emperor Renzong of the Qing Dynasty, 10th month, 25th day, at night; stars fell like rain.
Qing Dynasty, Daoguang period ed. Yunnan [*Yongchangfu zhi*] ch. 24

AD 1799 Nov 25 [China]

4th year of the Jiaqing reign period of Emperor Renzong of the Qing Dynasty, 10th month, 28th day, during the night; a multitude of stars streamed back and forth as if weaving.
Qing Dynasty, Guangxu period ed. Shandong [*Ningjinxian zhi*] ch. 11

AD 1799 Nov 26 [China]

(a) 4th year of the Jiaqing reign period of Emperor Renzong of the Qing Dynasty, 10th month, 29th day, beginning of twilight; in the southwest stars fell as if weaving.
Qing Dynasty, Jiaqing period ed. Zhejiang [*Yiwuxian zhi*] ch. 19

(b) 4th year of the Jiaqing reign period of Emperor Renzong of the Qing Dynasty, 10th month, 29th day, during the night; a multitude of stars streamed back and forth as if weaving.
Qing Dynasty, Guangxu period ed. Shandong [*Ningjinxian zhi*] ch. 11

AD 1799 Nov 27 [China]

4th year of the Jiaqing reign period of Emperor Renzong of the Qing Dynasty, 10th month, thirtieth day; during the hour of *you* [LT = 17–19h], stars like crossbow bolts flew back and forth, criss-crossing and streaming.
Qing Dynasty, Jiaqing period ed. Henan [*Xu Jiyuanxian zhi*] ch. 1
NB: The 10th month had only 29 days; therefore, 10th month, 30th day should be 11th month, 1st day.

AD 1800 Nov 17–Dec 15 [China]

(a) 5th year of the Jiaqing reign period of Emperor Renzong of the Qing Dynasty, 10th month; in the east hundreds of stars streamed as if weaving, and in a short while it was over.
Qing Dynasty, Guangxu period ed. Guangdong [*Haiyangxian zhi*] ch. 25

(b) 5th year of the Jiaqing reign period of Emperor Renzong of the Qing Dynasty, 10th month; stars fell like rain.
Minguo period ed. Yunnan [*Yiliangxian zhi*] ch. 1

AD 1801 May 5 [China]

6th year of the Jiaqing reign period of Emperor Renzong of the Qing Dynasty, 3rd month, 23rd day; during the hour of *xu* [LT = 19–21h], there was a star as big as a peck measure and as bright as the Moon, while behind it innumerable tiny stars flew from north to south, making a tinkling sound. Afterward, the drums of heaven thundered and the populations of several provinces simultaneously saw and heard it.
Qing Dynasty, Xianfeng period ed. Guizhou [*Anshunfu zhi*] ch. 12

AD 1801 winter [China]

6th year of the Jiaqing reign period of Emperor Renzong of the Qing Dynasty, winter; at Tongshan stars fell as if weaving.
Qing Dynasty, Tongzhi period ed. Jiangsu [*Xuzhoufu zhi*] ch. 57

AD 1813 Feb 1–AD 1814 Jan 20 [China]

18th year of the Jiaqing reign period of Emperor Renzong of the Qing Dynasty; stars fell like rain.
Qing Dynasty, Guangxu period ed. Shanxi [*Puxian zhi*] ch. 9

AD 1813 Dec 1 [Japan]

guiyou (10th) year of the Bunka reign period, 11th month, 9th day; right at daybreak thin clouds arose in the north when suddenly halfway up the sky there was a noise. Lifting my head to look, some objects emerged from amidst the white clouds, the larger ones the size of handballs, the smaller ones like fists, several tens of them in unison in the space of two 6-foot spaces, their brightness resembling so-called flare meteors. Their flames were also drawn out in white streaks as they flew from the east toward the west, their progress slow like mercury flowing in a basin; finally they entered the clouds and were lost sight of.
Ten yasa ichirō shi syozō bun sho

AD 1814 Aug 14–Sep 13 [China]

19th year of the Jiaqing reign period of Emperor Renzong of the Qing Dynasty, 7th month; during the hour of *hai* [LT = 21–23h], the sky suddenly lit up as if struck open and there was a flash like a long

rainbow more than 1 *zhang* long, oriented south to north. In the middle was a large star and on both sides ranks of numerous smaller stars like strung pearls. The large star flew away eastward and the smaller ones trailed after and disappeared. Everyone saw it.
Qing Dynasty, Daoguang period ed. Guangdong [*Yue xiao ji*] ch. 4

AD 1818 Nov 8 [China]
23rd year of the Jiaqing reign period of Emperor Renzong of the Qing Dynasty, 10th month, day *yihai* [12], at night; stars in the south fell like rain.
Qing Dynasty, Daoguang period ed. Gansu [*Zhenyuanxian zhi*] ch. 7

AD 1822 Feb 3 [China]
Renwu [19] year of the Daoguang reign period of Emperor Xuanzong of the Qing Dynasty, 1st month, 12th day; at Huicheng there was a single large star that streamed from the southwest to the east, while below it were several tens of smaller stars that trailed along, their flashing brilliance lighting up the sky.
Minguo ed. Guangdong [*Guangzhoucheng fang zhi*] ch. 5
NB: *renwu* [19] year of the Daoguang reign period was Daoguang 2nd year.

AD 1822 Sep 9 [China]
2nd year of the Daoguang reign period of Emperor Xuanzong of the Qing Dynasty, 7th month, 24th day; a star as big as a cup went from north to south with several tens of smaller stars trailing along behind it.
Qing Dynasty, Daoguang period ed. Guangdong [*Yue xiao ji*] ch. 4

AD 1824 Dec 29 [China]
4th year jiashen [21] of the Daoguang reign period of Emperor Xuanzong of the Qing Dynasty, 11th month, 10th day, at night; in the northwest quarter stars fell like rain.
[*Lü yuan conghua*] *zaixiang*
NB: *Jiashen* was the 4th year of the Daoguang reign period. The original text had Emperor Renzong, *jiashen* year, which is a mistake.

AD 1825 Mar 4 [China]

5th year of the Daoguang reign period of Emperor Xuanzong of the Qing Dynasty, 1st month, 15th day; large stars streamed in the northeast, while innumerable smaller stars trailed along behind making a sound like thunder.

Qing Dynasty, Guangxu period ed. Shanxi [*Xinzhou Zhilizhou zhi*] ch. 39

AD 1825 Nov 10–Dec 9 [China]

5th year of the Daoguang reign period of Emperor Xuanzong of the Qing Dynasty, 10th month; stars moved about as if weaving.

Qing Dynasty, Daoguang period ed. Henan [*Yanlingxian zhi*] ch. 17

Qing Dynasty, Guangxu period ed. Henan [*Xiangfuxian zhi*] ch. 23

AD 1825 Dec 3–4 [China]

5th year yiyou [22] of the Daoguang reign period of Emperor Xuanzong of the Qing Dynasty, 10th month, on the two nights of the 24th and 25th days, at night; stars moved about as if weaving, all proceeding from the northwest to the southeast.

[*Lü yuan conghua*] zaixiang

NB: *Yiyou* was the 5th year of the Daoguang reign period. The original text had Emperor Renzong, *yiyou* year, which is a mistake.

AD 1832 Apr 16 [China]

12th year of the Daoguang reign period of Emperor Xuanzong of the Qing Dynasty, 3rd month, 16th day, at night; a scarlet star as big as a peck measure started from the northeast and with a clap of thunder fell toward the west, while countless smaller stars trailed along behind it.

Qing Dynasty, Tongzhi period ed. Anhui [*Tongchengxian zhi*] ch. 10

AD 1837 Feb 5–Mar 6 [China]

17th year of the Daoguang reign period of Emperor Xuanzong of the Qing Dynasty, 1st month; in the southwest quarter stars fell like rain.

Minguo period ed. Yunnan [*Kunmingxian zhi*] ch. 7

AD 1837 Jun 17 [China]

17th year of the Daoguang reign period of Emperor Xuanzong of the Qing Dynasty, 5th month, after full Moon; swarms of stars streamed southward like rain.

Qing Dynasty, Guangxu period ed. Hebei [*Dongguangxian zhi*] ch. 11

AD 1837 Dec 27–1838 Jan 25 [China]

17th year of the Daoguang reign period of Emperor Xuanzong of the Qing Dynasty, 12th month; countless stars fell.

Qing Dynasty, Guangxu period ed. Yunnan [*Yunlongzhou zhi*] ch. 2

AD 1838 Jan 26–1839 Feb 13 [China]

18th year of the Daoguang reign period of Emperor Xuanzong of the Qing Dynasty; stars fell like rain.

Qing Dynasty, Daoguang period ed. Yunnan [*Guyuezhou zhi*] ch. 1

Qing Dynasty, Xianfeng period ed. Yunnan [*Nanningxian zhi*] ch. 1

AD 1838 Jul 21–Aug 19 [China]

18th year of the Daoguang reign period of Emperor Xuanzong of the Qing Dynasty, 6th month; stars fell like rain, several nights were all like this.

Minguo period ed. Fujian [*Pingnanxian zhi*] ch. 3

AD 1838 Aug 20–Sep 18 [China]

18th year of the Daoguang reign period of Emperor Xuanzong of the Qing Dynasty, 7th month; stars fell like rain.

Minguo period ed. Yunnan [*Songmingxian zhi*] ch. 4

AD 1838 Dec 6 [China]

18th year of the Daoguang reign period of Emperor Xuanzong of the Qing Dynasty, 10th month, 20th day; at the Prefectural seat the whole sky was filled with flying stars.

Qing Dynasty, Daoguang period ed. Yunnan [*Puerfu zhi*] ch. 2

AD 1839 Jan 15–Feb 13 [China]

18th year of the Daoguang reign period of Emperor Xuanzong of the Qing Dynasty, 12th month; stars fell like rain.
Qing Dynasty, Guangxu period ed. Yunnan [*Luliangzhou zhi*] ch. 8
Minguo period ed. Yunnan [*Luopingxian zhi*] ch. 1

AD 1839 Aug 23 [China]

19th year of the Daoguang reign period of Emperor Xuanzong of the Qing Dynasty, 7th month, 15th day; during the marks of the hour of *hai* [LT = 21–23h], three stars streamed back and forth as if weaving, overtaking each other several times, and disappearing only after several marks on the clock.
Qing Dynasty, Guangxu period ed. Guangdong [*Xiangshanxian zhi*] ch. 22

AD 1841 summer [China]

(a) 21st year of the Daoguang reign period of Emperor Xuanzong of the Qing Dynasty, summer; every night stars of all sizes fell like rain.
Qing Dynasty, Xuantong period ed. Shanghai [*Nansha za zhi*] xiangyi
(b) 21st year of the Daoguang reign period of Emperor Xuanzong of the Qing Dynasty, summer; every night stars fell like rain.
Minguo period ed. Shanghai [*Nanhuixian zhi*] ch. 22

AD 1841 Oct 31 [China]

21st year of the Daoguang reign period of Emperor Xuanzong of the Qing Dynasty, 9th month, 17th day; during the hour of *zi* [LT = 23–1h], stars fell like rain.
Qing Dynasty, Guangxu period ed. Beijing [*Shahexian zhi*] ch. shang

AD 1842 Jan 11 [China]

21st year of the Daoguang reign period of Emperor Xuanzong of the Qing Dynasty, 12th month, new moon day; a large star fell in the southeast, while a multitude of smaller stars trailed along behind.

Qing Dynasty, Guangxu period ed. Zhejiang [*Pinghuxian zhi*] ch. 25
Qing Dynasty, Guangxu period ed. Zhejiang [*Jiashanxian zhi*] ch. 34

AD 1842 Feb 9 [China]
21st year of the Daoguang reign period of Emperor Xuanzong of the Qing Dynasty, last night of the year; stars streamed as if weaving.
Qing Dynasty, Guangxu period ed. Jiangsu [*Dantuxian zhi*] ch. 58

AD 1843 Mar 31–Apr 29 [China]
23rd year of the Daoguang reign period of Emperor Xuanzong of the Qing Dynasty, 3rd month; a giant star fell, and countless smaller stars followed it, their blazing brilliance illuminating the sky.
Qing Dynasty, Guangxu period ed. Jiangsu [*Kun Xin liang xian xuxiu hezhi*] ch. 51

AD 1846 Aug 5 [China]
26th year of the Daoguang reign period of Emperor Xuanzong of the Qing Dynasty, 6th month, 14th day; during the hour of *zi* [LT = 23–1h], stars fell like rain.
Qing Dynasty, Guangxu period ed. Shanghai [*Nanhuixian zhi*] ch. 22

AD 1846 Aug 6 [China]
26th year of the Daoguang reign period of Emperor Xuanzong of the Qing Dynasty, 6th month, full Moon; stars fell like rain.
Minguo period ed. Shanghai [*Nanhuixian erqu Jiuwutuan xiang zhi*] ch. 16

AD 1847 Jul 12–Aug 10 [China]
27th year of the Daoguang reign period of Emperor Xuanzong of the Qing Dynasty, 6th month; there was an earthquake in the capital, and a multitude of stars fell.
Minguo period ed. [*Beijingshi zhi gao*] ch. 4

AD 1847 Jul 24 [China]
27th year of the Daoguang reign period of Emperor Xuanzong of the Qing Dynasty, 6th month, 12th day (*gengshen* [57]); a multitude of stars fell.

Qing Dynasty, Tongzhi period ed. [*Shanghaixian zhi*] ch. 30
Qing Dynasty, Guangxu period ed. Shanghai [*Chuanshaxian zhi*] ch. 14
Qing Dynasty, Guangxu period ed. Shanghai [*Leixian xu zhi*] ch. 12
Qing Dynasty, Guangxu period ed. Shanghai [*Songjiangfu zhi*] ch. 30
Qing Dynasty, Guangxu period ed. Shanghai [*Qingpuxian zhi*] ch. 29
NB: The 12th day was Jul 23 while *gengshen* was Jul 24.

AD 1849 Aug 3 [China]

29th year of the Daoguang reign period of Emperor Xuanzong of the Qing Dynasty, 6th month, 15th day, at night; stars fell like rain.
Minguo period ed. Henan [*Guangshanxian zhi*] ch. 1

AD 1851 Aug 11 [China]

(a) 1st year of the Xianfeng reign period of Emperor Wenzong of the Qing Dynasty, 7th month, 15th day, at night; stars in the south flew like arrows for over 2 double hours before stopping.
Qing Dynasty, Guangxu period ed. Tianjin [*Ninghexian zhi*] ch. 16

(b) 1st year of the Xianfeng reign period of Emperor Wenzong of the Qing Dynasty, 7th month, 15th day, at night; stars fell like rain.
Minguo period ed. Henan [*Guangshanxian zhi*] ch. 1

AD 1852 May 21 [China]

2nd year of the Xianfeng reign period of Emperor Wenzong of the Qing Dynasty, 4th month, 3rd day; at night countless firelights were seen coming from the southeast. They skirted the Commandery seat and scattered only on reaching Xianzhu Peak in the western district.
Qing Dynasty, Guangxu period ed. Jiangxi [*Fuzhoufu zhi*] ch. 84

AD 1853 Feb 8–Mar 9 [China]

3rd year of the Xianfeng reign period of Emperor Wenzong of the Qing Dynasty, 1st month; at Tongzhou, there were fires like stars or phosphorescences numbering in the hundreds of thousands that sped from the southwest to the northeast for 4 to 5 nights before ceasing.
[*Qing shi gao: zaiyi zhi*] ch. 41

AD 1853 Mar 3 [China]

3rd year of the Xianfeng reign period of Emperor Wenzong of the Qing Dynasty, 1st month, day *jisi* [6]; at the end of twilight, there were fires like stars or phosphorescences numbering in the hundreds of thousands that sped from the southwest to the northeast, while a muffled sound as of armored horses was heard for 4 to 5 nights.

Qing Dynasty, Guangxu period ed. Jiangsu [*Tongzhou Zhilizhou zhi*] ch. mo

AD 1853 Aug 19 [China]

3rd year of the Xianfeng reign period of Emperor Wenzong of the Qing Dynasty, 7th month, 15th day; during the hour of *you* [LT = 17–19h], in the southwest quarter stars fell like rain.

Qing Dynasty, Guangxu period ed. Shanxi [*Shouyangxian zhi*] ch. 13

AD 1854 Jan 29–AD 1855 Feb 16 [China]

4th year of the Xianfeng reign period of Emperor Wenzong of the Qing Dynasty; meteors flew like rain in the south for several nights without stopping.

Qing Dynasty, Guangxu period ed. Shandong [*Pingyangxian zhi*] ch. 6

AD 1854 Jul [China]

4th year of the Xianfeng reign period of Emperor Wenzong of the Qing Dynasty, 6th month, 10th day, at noon; heavenly fire fell in profusion in the streets and thoroughfares but on reaching the ground left no trace.

Qing Dynasty, Guangxu period ed. Guangxi [*Guixian zhi*] ch. 6

AD 1855 Aug 16 [China]

5th year of the Xianfeng reign period of Emperor Wenzong of the Qing Dynasty, 7th month, 4th day, at night; stars streamed like copper cash while several tens of smaller stars followed alongside as they flew out of the northwest.

Qing Dynasty, Guangxu period ed. Guangdong [*Haiyangxian zhi*] ch. 25

AD 1856 Oct 31 [China]

6th year of the Xianfeng reign period of Emperor Wenzong of the Qing Dynasty, 10th month, 3rd day of the new moon, at night; swarms of stars fell chaotically.

Qing Dynasty, Guangxu period ed. Yunnan [*Tengyueting zhi*] ch. 1
Minguo period ed. Yunnan [*Longlingxian zhi*] ch. 1

AD 1856 Dec 8 [China]

6th year of the Xianfeng reign period of Emperor Wenzong of the Qing Dynasty, 11th month, 11th day, at night; stars fell like rain.

Minguo period ed. [*Yunnan tongzhi*] ch. 17

AD 1857 Mar 24 [China]

7th year of the Xianfeng reign period of Emperor Wenzong of the Qing Dynasty, 2nd month, 29th day, at night; stars fell like rain.

Qing Dynasty, Guangxu period ed. Fujian [*Guangzexian zhi*] ch. 1

AD 1857 Sep 8 [China]

7th year of the Xianfeng reign period of Emperor Wenzong of the Qing Dynasty, 7th month, 20th day, at night; stars of different sizes fell like rain, from the northeast toward the west. They fell in profusion for over 2 double hours, their speed like arrows and as abundant as the rain.

Qing Dynasty, Guangxu period ed. Shanxi [*Qinyuanxian xu zhi*] ch. 3

AD 1858 Jan 14 [China]

7th year of the Xianfeng reign period of Emperor Wenzong of the Qing Dynasty, 11th month, thirtieth day, at night; stars fell like rain.

Qing Dynasty, Guangxu period ed. Yunnan [*Yongchangfu zhi*] ch. 3

AD 1858 Aug 23 [China]

8th year of the Xianfeng reign period of Emperor Wenzong of the Qing Dynasty, 7th month, 15th day, at night; stars in their thousands streamed like arrows from the northwest to the southeast from the hour of *xu* [LT = 19–21h] until the hour of *chou* (LT = 1–3h) before ceasing.

Qing Dynasty, Guangxu period ed. Shanxi [*Guoxian zhi*] ch. 8

AD 1858 Nov 1 [China]

8th year of the Xianfeng reign period of Emperor Wenzong of the Qing Dynasty, 9th month, 26th day; meteors were seen in daylight moving from the west to the east for over 2 double hours, from the hour of *chen* [LT = 7–9h] to the hour of *wu* (LT = 11–13h), three, five, seven, or eight at a time, all different sizes, but all more than 1 *chi* long.

Qing Dynasty, Guangxu period ed. Jiangsu [*Dantuxian zhi*] ch. 58

AD 1859 Aug 1 [China]

9th year of the Xianfeng reign period of Emperor Wenzong of the Qing Dynasty, 7th month, 3rd day; at the beginning of twilight there were large and small meteors 2 to 3 *zhang* from the ground. Some came from the east and some from the west for about 2 double hours or so before ceasing.

Qing Dynasty, Guangxu period ed. Shanghai [*Chongmingxian zhi*] ch. 5

AD 1859 Aug 13 [China]

(a) 9th year of the Xianfeng reign period of Emperor Wenzong of the Qing Dynasty, 7th month, 15th day, at night; stars streamed as if weaving.

Minguo period ed. Hebei [*Jinxian zhi*] ch. 5

(b) 9th year of the Xianfeng reign period of Emperor Wenzong of the Qing Dynasty, 7th month, 15th day, at night; stars fell like arrows.

Qing Dynasty, Guangxu period ed. Shanxi [*Pingdingzhou zhi*] ch. 5

AD 1860 Aug 17 [China]

10th year of the Xianfeng reign period of Emperor Wenzong of the Qing Dynasty, 7th month, 1st day; stars fell like rain.

Qing Dynasty, Guangxu period ed. Shanxi [*Jiaochengxian zhi*] ch. 1

AD 1860 Aug 23 [China]

10th year of the Xianfeng reign period of Emperor Wenzong of the Qing Dynasty, 7th month, 7th day, at night; in the west stars fell like rain.

Qing Dynasty, Guangxu period ed. Shanxi [*Dingyuanting zhi*] ch. 24

AD 1861 Feb 10–AD 1862 Jan 29 [China]

11th year of the Xianfeng reign period of Emperor Wenzong of the Qing Dynasty; stars fell like rain.

Qing Dynasty, Guangxu period ed. Yunnan [*Yongbeiting zhi*] ch. 1

AD 1861 Jun 15 [China]

11th year of the Xianfeng reign period of Emperor Wenzong of the Qing Dynasty, 5th month, 8th day; during the hour of *you* [LT = 17–19h], the sky was pitch black. In the heart of the sky there was a star as big as a peck measure that was red and dark. Around it to the left and right were several tens of smaller stars that oscillated as if about to set in motion. Suddenly making a tinkling sound, from the heart of the sky the stars scattered as a reddish glow and fell away toward the west.

Minguo period ed. Sichuan [*Wushengxian zhi*] ch. 12

AD 1861 Jun 23 [China]

11th year of the Xianfeng reign period of Emperor Wenzong of the Qing Dynasty, 5th month, 16th day, at night; stars fell like rain, alighting in the northeast.

Minguo period ed. Shanxi [*Baojixian zhi*] ch. 16

AD 1861 Aug 6–Sep 4 [China]

(a) 11th year of the Xianfeng reign period of Emperor Wenzong of the Qing Dynasty, 7th month; meteors flew southward, as dense as a knot of floss.
Qing Dynasty, Tongzhi period ed. Shandong [*Jimoxian zhi*] ch. 11

(b) 11th year of the Xianfeng reign period of Emperor Wenzong of the Qing Dynasty, 7th month; stars fell like rain.
Minguo period ed. Shanxi [*Linfenxian zhi*] ch. 6

(c) 11th year of the Xianfeng reign period of Emperor Wenzong of the Qing Dynasty, 7th month; stars fell like rain from the northeast to the southwest for several marks of the hour.
Minguo period ed. Shanxi [*Liaozhou zhi*] ch. 3

AD 1861 Aug 10 [China]

11th year xinyou [58] of the Xianfeng reign period of Emperor Wenzong of the Qing Dynasty, 7th month, fifth day, at night; stars streamed continuously from north to south.

Minguo period ed. Jiangsu [*Taicangzhou zhi*] ch. 26

AD 1861 Aug 14 [China]

11th year of the Xianfeng reign period of Emperor Wenzong of the Qing Dynasty, 7th month, 9th day, at night; stars fell like rain.

Qing Dynasty, Guangxu period ed. Zhejiang [*Jiashanxian zhi*] ch. 34

Qing Dynasty, Xuantong period ed. Shanghai [*Fengjing xiao zhi*] ch. 10

AD 1861 Aug 19 [China]

11th year of the Xianfeng reign period of Emperor Wenzong of the Qing Dynasty, 7th month, 14th day; a multitude of stars streamed chaotically in the southwest beginning from early twilight and not ending until night.

Qing Dynasty, Guangxu period ed. Shandong [*Xixiaxian zhi*] ch. 8

AD 1861 Aug 20 [China]

(a) 11th year of the Xianfeng reign period of Emperor Wenzong of the Qing Dynasty, 7th month, 15th day (full Moon), at night; meteors sped southwestward without stopping till first light.
Qing Dynasty, Guangxu period ed. Hebei [*Nangongxian zhi*] ch. 8

(b) 11th year of the Xianfeng reign period of Emperor Wenzong of the Qing Dynasty, 7th month, 15th day (full Moon); stars fell like rain.
Qing Dynasty, Guangxu period ed. Hebei [*Huaianxian zhi*] ch. 3

(c) 11th year of the Xianfeng reign period of Emperor Wenzong of the Qing Dynasty, 7th month, 15th day (full Moon), at night; stars fell like rain.
Qing Dynasty, Guangxu period ed. Shanxi [*Taipingxian zhi*] ch. 14
Qing Dynasty, Guangxu period ed. Shanxi [*Jiezhou quanzhi*] ch. 11

Qing Dynasty, Guangxu period ed. Shanxi [*Zhili Jiangzhou zhi*] ch. 20

(d) 11th year of the Xianfeng reign period of Emperor Wenzong of the Qing Dynasty, 7th month, 15th day (full Moon), at night; stars fell like rain in the southwest.
Minguo period ed. Shanxi [*Wuxiangxian zhi*] ch. 10

(e) 11th year of the Xianfeng reign period of Emperor Wenzong of the Qing Dynasty, 7th month, 15th day (full Moon); stars fell like rain toward the south making a muffled sound.
Qing Dynasty, Guangxu period ed. Shanxi [*Qingyuanxiang zhi*] ch. 16

AD 1861 Sep 22 [China]

11th year of the Xianfeng reign period of Emperor Wenzong of the Qing Dynasty, 8th month, day *jiaxu* [11], at night; a large star fell and smaller stars fell like rain.
Qing Dynasty, Guangxu period ed. Shanghai [*Qingpuxian zhi*] ch. 29

AD 1862 Jan 30–AD 1863 Feb 17 [China]

1st year of the Tongzhi reign period of Emperor Muzong of the Qing Dynasty; stars fell like rain.
Minguo period ed. Yunnan [*Jingdongxian zhi*] ch. 1

AD 1862 May 28–Jun 26 [China]

1st year of the Tongzhi reign period of Emperor Muzong of the Qing Dynasty, 5th month; stars fell like rain.
Qing Dynasty, Guangxu period ed. Shanxi [*Suidexian zhi*] ch. 3

AD 1862 Jul 11 [China]

1st year of the Tongzhi reign period of Emperor Muzong of the Qing Dynasty, 6th month, day *bingyin* [3], at night; a multitude of stars streamed southwestward.
Qing Dynasty, Guangxu period ed. Jiangsu [*Wuyang zhi yu*] ch. 5

AD 1862 Jul 27–Aug 24 [China]

(a) 1st year of the Tongzhi reign period of Emperor Muzong of the Qing Dynasty, 7th month; stars fell like rain.
Qing Dynasty, Tongzhi period ed. Shanxi [*Xu Yishixian zhi*] ch. 4
Qing Dynasty, Guangxu period ed. Shanxi [*Linjinxian zhi*] ch. xia
Qing Dynasty, Guangxu period ed. Shanxi [*Xu Qianyangxian zhi*] ch. 14

(b) 1st year of the Tongzhi reign period of Emperor Muzong of the Qing Dynasty, 7th month; meteors filled the sky, starting in the southwest.
Minguo period ed. Shandong [*Qiuxian zhi*] ch. 9

AD 1862 Aug 9 [China]

(a) 1st year of the Tongzhi reign period of Emperor Muzong of the Qing Dynasty, 7th month, 14th day; stars streamed like rain from the northeast toward the southwest.
Qing Dynasty, Tongzhi period ed. Hebei [*Yanshanxian zhi*] ch. 5
Qing Dynasty, Guangxu period ed. Hebei [*Dachengxian zhi*] ch. 10
Qing Dynasty, Guangxu period ed. Hebei [*Nanpixian zhi*] ch. 5

(b) 1st year of the Tongzhi reign period of Emperor Muzong of the Qing Dynasty, 7th month, 14th day; at night stars streamed in profusion from the northeast toward the southwest continuously, without interruption.
Qing Dynasty, Guangxu period ed. Shandong [*Huiminxian zhi buyi*]

(c) 1st year of the Tongzhi reign period of Emperor Muzong of the Qing Dynasty, 7th month, 14th day, at night; meteors continuously crossed the sky in the south all night long.
Minguo period ed. Shandong [*Jiyangxian xu zhi*] ch. 20
Minguo period ed. Shandong [*Xinyangxian zhi*] ch. 2

AD 1862 Aug 10 [China]

(a) 1st year of the Tongzhi reign period of Emperor Muzong of the Qing Dynasty, *zhongyuan* (7th month, 15th day); during the hour of *hai* [LT = 21–23h], a multitude of stars streamed in profusion from the northeast toward the southwest, too many to count. Qing Dynasty, Tongzhi period ed. [*Xu Tianjinxian zhi*] ch. 1

(b) 1st year of the Tongzhi reign period of Emperor Muzong of the Qing Dynasty, *zhongyuan* (7th month, 15th day), at night; stars streamed as if weaving.
Qing Dynasty, Tongzhi period ed. Hebei [*Changlixian zhi*] ch. 1

(c) 1st year of the Tongzhi reign period of Emperor Muzong of the Qing Dynasty, *zhongyuan* (7th month, 15th day); stars fell like rain from the northwest toward the southwest.
Qing Dynasty, Guangxu period ed. Shanxi [*Pingluxian xu zhi*] ch. xia

(d) 1st year of the Tongzhi reign period of Emperor Muzong of the Qing Dynasty, *zhongyuan* (7th month, 15th day); after twilight stars streamed like rain toward the southwest, ending during the second watch.
Qing Dynasty, Guangxu period ed. Shanxi [*Pingyaoxian zhi*] ch. 12

(e) 1st year of the Tongzhi reign period of Emperor Muzong of the Qing Dynasty, *zhongyuan* (7th month, 15th day), at night; stars fell like rain.
Qing Dynasty, Guangxu period ed. Shanxi [*Mizhixian zhi*] ch. 10

(f) 1st year of the Tongzhi reign period of Emperor Muzong of the Qing Dynasty, *zhongyuan* (7th month, 15th day), at night; stars fell like rain.
Minguo period ed. Henan [*Guangshanxian zhi*] ch. 1

(g) 1st year of the Tongzhi reign period of Emperor Muzong of the Qing Dynasty, *zhongyuan* (7th month, 15th day), at night; stars streamed as if weaving.
Qing Dynasty, Guangxu period ed. Shandong [*Lijinxian zhi*] ch. 10

(h) 1st year of the Tongzhi reign period of Emperor Muzong of the Qing Dynasty, *zhongyuan* (7th month, 15th day); a multitude of stars streamed toward the southwest.
Minguo period ed. [*Zhejiang xu tongzhi gao*] ch. 1

(i) 1st year of the Tongzhi reign period of Emperor Muzong of the Qing Dynasty, zhongyuan (7th month, 15th day), at night; a multitude of stars streamed from the southeast toward the northwest.
Minguo period ed. Guangdong [*Shixingxian zhi*] ch. 16

NB: There are several dozen similar records from the gazetteers of a number of provinces.

AD 1862 Aug 10 [Japan]

(a) 2nd year of the Bunkyū reign period, 7th month, 15th day; tonight during the hour of *xu* (LT = 19–21h), stars fell like rain.
[*Kaei Meiji nenkan roku*]

(b) *renxu* year of the Bunkyū reign period, 7th month, 15th day, at night; meteors fell incessantly, from the first watch through the fifth watch there was no let up; they flew fast, all speeding from the northeast to the southwest, following along on the wind. Now, for meteors to fly across the sky is a commonplace event, and though not something to marvel at, for it to be this extreme is most unusual; probably this is what was anciently referred to as "stars fell like rain."
[*Ryūsei suisei kammon*] *zusho ryō zō hon*

(c) 2nd year of the Bunkyū reign period, 7th month, 15th day; on the 15th of this month the meteors were most numerous, numbering over 100, most falling from azimuth *gen* (NE) to *kun* (SW).
[*Ryūsei suisei kambun*]

AD 1862 Aug 11 [China]

1st year of the Tongzhi reign period of Emperor Muzong of the Qing Dynasty, 7th month, 16th day; at the beginning of twilight, a multitude of stars fell criss-cross, most heading southwestward as they moved up, down, left, and right, and continuing until dark.

Qing Dynasty, Guangxu period ed. Shandong [*Dengzhoufu zhi*] ch. 23
Qing Dynasty, Guangxu period ed. Shandong [*Penglaixian xu zhi*] ch. 1
Qing Dynasty, Guangxu period ed. Shandong [*Fushanxian zhi gao*] ch. 8

AD 1862 Aug 12 [China]

(a) 1st year of the Tongzhi reign period of Emperor Muzong of the Qing Dynasty, 7th month, 17th day; a multitude of stars sped rapidly toward the southwest like a swarm of arrows.
Qing Dynasty, Guangxu period ed. Henan [*Nanyangxian zhi*] ch. 12

(b) 1st year of the Tongzhi reign period of Emperor Muzong of the Qing Dynasty, 7th month, 17th day; the whole sky was full of meteors streaming from the northeast toward the southwest, shooting criss-cross through the zenith for more than one watch before stopping.
Minguo period ed. Henan [*Fangchengxian zhi*] ch. 5

AD 1863 Feb 18–1864 Feb 7 [China]

(a) 2nd year of the Tongzhi reign period of Emperor Muzong of the Qing Dynasty; to the west of the city stars fell like rain, making a muffled sound.
Qing Dynasty, Guangxu period ed. Hubei [*Jingshanxian zhi*] ch. 1

(b) 2nd year of the Tongzhi reign period of Emperor Muzong of the Qing Dynasty; stars came falling like rain.
Qing Dynasty, Xuantong period ed. [*Hubei tongzhi*] ch. 76

AD 1863 Aug 28 [China]

(a) 2nd year of the Tongzhi reign period of Emperor Muzong of the Qing Dynasty, 7th month, 15th day, at night; stars streamed southwestward like arrows straight through till daylight.
Qing Dynasty, Guangxu period ed. Hebei [*Xinchengxian zhi*] ch. 10

(b) 2nd year of the Tongzhi reign period of Emperor Muzong of the Qing Dynasty, 7th month, 15th day; meteors covered the sky as if weaving as they streamed southeastward, not stopping until daylight.

Minguo period ed. Hebei [*Xinchengxian zhi*] ch. 22

(c) 2nd year of the Tongzhi reign period of Emperor Muzong of the Qing Dynasty, 7th month, 15th day, at night; stars fell like rain.
Minguo period ed. Shanxi [*Linxian zhi*] ch. 3

(d) 2nd year of the Tongzhi reign period of Emperor Muzong of the Qing Dynasty, 7th month, 15th day; from dusk through the first watch, countless meteors like arrows all streamed southwestward, not stopping until the second watch.
Minguo period ed. Beijing [*Pingguxian zhi*] ch. 3

AD 1863 Nov 11–Dec 10 [China]

2nd year of the Tongzhi reign period of Emperor Muzong of the Qing Dynasty, 10th month; west of Jingshan city stars fell like rain.
Minguo period ed. [*Hubei tongzhi*] ch. 76

AD 1864 Aug 2 [China]

3rd year of the Tongzhi reign period of Emperor Muzong of the Qing Dynasty, 7th month, new moon day, at night; stars fell like rain.
Minguo period ed. Sichuan [*Leshanxian zhi*] ch. 12

AD 1864 Aug 2–Aug 31 [China]

3rd year of the Tongzhi reign period of Emperor Muzong of the Qing Dynasty, 7th month; a multitude of stars streamed out of the southeast.
Qing Dynasty, Tongzhi period ed. Shanxi [*Gaopingxian zhi*] ch. 4

AD 1864 Oct 1–29 [China]

(a) 3rd year of the Tongzhi reign period of Emperor Muzong of the Qing Dynasty, 9th month; there was a white vapor in the sky, for a while stars fell like rain.
Qing Dynasty, Guangxu period ed. Hubei [*Jingshanxian zhi*] ch. 1

NB: The Xuantong period ed. *Hubei tongzhi* has 4th year, 9th month.

(b) 3rd year of the Tongzhi reign period of Emperor Muzong of the Qing Dynasty, 9th month; meteors wove criss-cross.
Qing Dynasty, Guangxu period ed. Yunnan [*Zhanyizhou zhi*] ch. 4

AD 1864 Oct 15 [China]

3rd year of the Tongzhi reign period of Emperor Muzong of the Qing Dynasty, 9th month, 15th day, at night; meteors wove criss-cross, up, down, left, and right.
Minguo period ed. Fujian [*Shanghangxian zhi*] ch. 1

AD 1865 Sep 20–Oct 19 [China]

4th year of the Tongzhi reign period of Emperor Muzong of the Qing Dynasty, 8th month; stars fell like rain.
Qing Dynasty, Tongzhi period ed. [*Shanghaixian zhi*] ch. 30

AD 1866 Aug 24 [China]

5th year of the Tongzhi reign period of Emperor Muzong of the Qing Dynasty, 7th month, 15th day, at night; a multitude of stars streamed southward like arrows.
Qing Dynasty, Guangxu period ed. Henan [*Wenxiangxian zhi*] ch. mo

AD 1866 Nov 3 [China]

5th year of the Tongzhi reign period of Emperor Muzong of the Qing Dynasty, 9th month, 26th day, at midnight; the sky was like daylight, and a golden brilliance came down as if raining millet.
Qing Dynasty, Tongzhi period ed. Jiangxi [*Xinchangxian zhi*] ch. 4
Minguo period ed. Jiangxi [*Yan sheng*] ch. 11

AD 1867 Nov 12 [China]

6th year of the Tongzhi reign period of Emperor Muzong of the Qing Dynasty, 10th month, 17th day, at night; in the northwest, stars fell like rain.
Qing Dynasty, Tongzhi period ed. Anhui [*Tongchengxian zhi*] ch. 10

AD 1867 Nov 13 [Japan]

3rd year of the Keiō reign period, 10th month, 18th day; beginning in the last marks of the hour of *yin* (LT = 3–5h), meteors began to appear uninterruptedly.
[*Shugo kokkai kyū mu*]

AD 1869 Feb 11–Mar 12 [China]

8th year of the Tongzhi reign period of Emperor Muzong of the Qing Dynasty, 1st month, stars fell like rain. For 2 hours, there was no brightness.

Qing Dynasty, Guangxu period ed. Yunnan [*Chenggongxian zhi*] ch. 5

AD 1869 Aug 22 [China]

8th year of the Tongzhi reign period of Emperor Muzong of the Qing Dynasty, 7th month, 15th day; meteors rained toward the southwest with a noise.

Qing Dynasty, Guangxu period ed. Hebei [*Kaizhou zhi*] ch. 1

AD 1869 Dec 15 [China]

8th year of the Tongzhi reign period of Emperor Muzong of the Qing Dynasty, 10th month, day *gengchen* [17], at night; a multitude of stars streamed from azimuth *gen* (NE) to *kun* (SW) and fell like rain.

Qing Dynasty, Guangxu period ed. Jiangsu [*Wuyang zhi yu*] ch. 5

NB: There was no day *gengchen* in 10th month; it should be the 11th month.

AD 1871 Aug 10 [China]

10th year of the Tongzhi reign period of Emperor Muzong of the Qing Dynasty, 6th month, 24th day, at night; in the northeast stars fell like rain.

Qing Dynasty, Xuantong period ed. Jiangsu [*Qiongshanxian zhi*] ch. 28

AD 1872 Aug 4–Sep 2 [China]

11th year of the Tongzhi reign period of Emperor Muzong of the Qing Dynasty, 7th month, at night; stars fell like rain.

Qing Dynasty, Guangxu period ed. Yunnan [*Yunnanxian zhi*] ch. 1

AD 1872 Nov 1 [China]

11th year of the Tongzhi reign period of Emperor Muzong of the Qing Dynasty, 10th month, 1st day, at midnight; stars fell like rain.

Qing Dynasty, Guangxu period ed. Yunnan [*Zhanyizhou zhi*] ch. 4

AD 1872 Nov 27 [China]

11th year of the Tongzhi reign period of Emperor Muzong of the Qing Dynasty, 10th month, 27th day; during the marks of the hour of *zi* [LT = 23–1h], stars fell like rain.

Qing Dynasty, Guangxu period ed. Guangdong [*Raopingxian zhi*] ch. 13

AD 1872 Dec 27 [China]

11th year of the Tongzhi reign period of Emperor Muzong of the Qing Dynasty, 11th month, 27th day, at night; swarms of stars fell chaotically.

Qing Dynasty, Guangxu period ed. Yunnan [*Tengyueting zhi*] ch. 1
Minguo period ed. Yunnan [*Longlingxian zhi*] ch. 1

AD 1874 Jul 31 [China]

Jiaxu [11] year of the Tongzhi reign period of Emperor Muzong of the Qing Dynasty, 6th month, 18th day, at night; swarms of stars changed location, going back and forth as if weaving.

Minguo period ed. Yunnan [*Longlingxian zhi*] ch. 1
Minguo period ed. Hubei [*Yingshanxian zhi*] ch. 14

AD 1874 Aug 4 [China]

13th year of the Tongzhi reign period of Emperor Muzong of the Qing Dynasty, 6th month, 22nd day, at night; a multitude of stars flew chaotically southward.

Qing Dynasty, Guangxu period ed. Guangdong [*Jieyangxian xu zhi*] ch. 4

AD 1875 Jun 9 [China]

1st year of the Guangxu reign period of Emperor Dezong of the Qing Dynasty, 5th month, 6th day, at dusk; the drums of heaven sounded and meteors spilled over on all sides.

Qing Dynasty, Guangxu period ed. Sichuan [*Jingyan zhi*] ch. 32

AD 1875 Nov 17 [China]
1st year of the Guangxu reign period of Emperor Dezong of the Qing Dynasty, 10th month, 20th day, at night; stars fell like rain.
Minguo period ed. Yunnan [*Kunyangxian zhi*] ch. 3

AD 1876 Nov 16–Dec 15 [China]
2nd year of the Guangxu reign period of Emperor Dezong of the Qing Dynasty, 10th month, at night; in Zhuanglang stars fell like rain.
Qing Dynasty, Xuantong period ed. [*Gansu xin tongzhi*] ch. 2

AD 1877 Jun 11–Jul 10 [China]
3rd year of the Guangxu reign period of Emperor Dezong of the Qing Dynasty, 5th month; stars fell like rain.
Minguo period ed. Yunnan [*Luopingxian zhi*] ch. 1

AD 1877 Aug 25–27 [China]
3rd year of the Guangxu reign period of Emperor Dezong of the Qing Dynasty, 7th month, 17th, 18th, 19th days, for 3 nights; the stars in the sky missed their locations, moving back and forth like arrows. All stars in the whole sky were all confused.
Minguo period ed. Guangxi [*Tengxian zhi*] zaiyi

AD 1879 winter [China]
5th year of the Guangxu reign period of Emperor Dezong of the Qing Dynasty, in winter; several tens of meteors as fast as the fastest arrows shot across the sky for a minute, then stopped.
Minguo period ed. Hebei [*Xushui xin zhi*] ch. 10

AD 1881 Mar 30–Apr 27 [China]
7th year of the Guangxu reign period of Emperor Dezong of the Qing Dynasty, 3rd month; stars moved without a fixed direction all over the sky.
Minguo period ed. Yunnan [*Yaoanxian zhi*] ch. 66

AD 1881 Sep 23–Oct 22 [China]

7th year of the Guangxu reign period of Emperor Dezong of the Qing Dynasty, 8th month; stars fell like rain, making a noise.
Minguo period ed. Yunnan [*Guangnanxian zhi*] ch. 3

AD 1882 Sep 2 [China]

8th year of the Guangxu reign period of Emperor Dezong of the Qing Dynasty, 7th month, 20th day; during the hour of *zi* [LT = 23–1h] of the night, stars fell like rain and did not stop for 2 to 3 double hours.
Minguo ed. Yunnan [*Xuanliangxian zhi*] ch. 1
NB: The Minguo ed. of *Luquanxian zhi* had 12th month, 20th day.

AD 1882 Sep 12–Oct 11 [China]

8th year of the Guangxu reign period of Emperor Dezong of the Qing Dynasty, 8th month; the sky was full of large and small stars that all flew southward.
Minguo period ed. Guangxi [*Lingshanxian zhi*] ch. 5

AD 1882 Dec 7 [China]

8th year of the Guangxu reign period of Emperor Dezong of the Qing Dynasty, 10th month, day *gengchen* [17]; a multitude of stars streamed from azimuth *gen* (NE) to *kun* (SW) and fell like rain.
Qing Dynasty, Guangxu period ed. Jiangsu [*Wuyang zhi yu*] ch. 5

AD 1882 Dec 22 [China]

8th year of the Guangxu reign period of Emperor Dezong of the Qing Dynasty, 11th month, 13th day, at night; meteors as numerous as fireworks, their brilliance filled the sky.
Minguo period ed. [*Zhejiang xu tongzhi gao*] ch. 1

AD 1883 Jun 12 [China]

9th year of the Guangxu reign period of Emperor Dezong of the Qing Dynasty, 5th month, 8th day, at night; meteors fell in profusion like rain.
Minguo period, Sichuan [*Bazhongxian zhi*] 4th ed.

AD 1883 Nov 10 [China]

9th year of the Guangxu reign period of Emperor Dezong of the Qing Dynasty, 10th month, day *wuwu* [55], at night; stars fell like rain and not stopping until the fourth watch.
Minguo period ed. Yunnan [*Xu xiu Kunmingxian zhi*] ch. 7

AD 1883 Nov 30–Dec 28 [China]

9th year of the Guangxu reign period of Emperor Dezong of the Qing Dynasty, 11th month, at night; meteors filled the sky and fell on all sides.
Qing Dynasty, Guangxu period ed. Henan [*Nanyangxian zhi*] ch. 12

AD 1884 Jun 28 [China]

10th year of the Guangxu reign period of Emperor Dezong of the Qing Dynasty, intercalary 5th month, 6th day, at night; stars fell like rain.
Minguo period ed. Fujian [*Gutianxian zhi*] ch. 3

AD 1884 Oct 11 [China]

10th year of the Guangxu reign period of Emperor Dezong of the Qing Dynasty, 8th month, 23rd day; swarms of stars fell like rain, from the 7th hour to the 9th, when they stopped.
Minguo period ed. Yunnan [*Zhaotong zhi gao*] ch. 12

AD 1884 Nov 18–Dec 16 [China]

(a) 10th year of the Guangxu reign period of Emperor Dezong of the Qing Dynasty, 10th month, at night; a multitude of stars streamed, their brilliant rays shooting in all directions.
Minguo period ed. Shandong [*Boshanxian zhi*] ch. 1

(b) 10th year of the Guangxu reign period of Emperor Dezong of the Qing Dynasty, 10th month; stars fell like rain.
Minguo period ed. Yunnan [*Kunmingxian zhi*] wuxing

AD 1884 Dec 2 [China]

10th year of the Guangxu reign period of Emperor Dezong of the Qing Dynasty, 10th month, 15th day; stars fell like rain.
Minguo period ed. Gansu [*Zhenyuanxian zhi*] ch. 18

AD 1884 Dec 8 [China]

(a) 10th year of the Guangxu reign period of Emperor Dezong of the Qing Dynasty, 10th month, 21st day, at night; swarms of stars streamed westward, not settling down for about 2 double hours.
Minguo period ed. Henan [*Xuchangxian zhi*] ch. 19
Minguo period ed. Henan [*Linyingxian zhi*] ch. 13

(b) 10th year of the Guangxu reign period of Emperor Dezong of the Qing Dynasty, 10th month, 21st day, at night; stars fell like rain.
Qing Dynasty, Guangxu period ed. Shanxi [*Baihexian zhi*] ch. 13

AD 1884 Dec 9 [China]

(a) 10th year of the Guangxu reign period of Emperor Dezong of the Qing Dynasty, 10th month, 22nd day, at night; stars fell like rain.
Qing Dynasty, Guangxu period ed. Shanxi [*Mizhixian zhi*] ch. 10

(b) 10th year of the Guangxu reign period of Emperor Dezong of the Qing Dynasty, 10th month, 22nd day; stars flew as if weaving, their brilliant rays like white silk, and lasting for a double hour before settling down.
Minguo period ed. Henan [*Xihuaxian xu zhi*] ch. 1

AD 1884 Dec 9–10 [China]

10th year of the Guangxu reign period of Emperor Dezong of the Qing Dynasty, 10th month, 22nd and 23rd days; the stars moved as if weaving, with brilliant rays, some several *chi* long.
Qing Dynasty, Xuantong period ed. Henan [*Xiangchengxian zhi*] ch. 31

AD 1884 Dec 10 [China]

10th year of the Guangxu reign period of Emperor Dezong of the Qing Dynasty, 10th month, 23rd day; in the southwest stars fell like rain.
Minguo period ed. Shanxi [*Ziyangxian zhi*] ch. 5

AD 1884 Dec 11 [China]

10th year of the Guangxu reign period of Emperor Dezong of the Qing Dynasty, 10th month, 24th day, at night; the whole sky was filled with falling stars like rain for 3 nights before ending.
Minguo period ed. Yunnan [*Songmingxian zhi*] ch. 4

AD 1885 Feb 15–AD 1886 Feb 3 [China]

(a) 11th year of the Guangxu reign period of Emperor Dezong of the Qing Dynasty; stars fell like rain.
Minguo period ed. Yunnan [*Yaoanxian zhi*] ch. 66

(b) 11th year of the Guangxu reign period of Emperor Dezong of the Qing Dynasty; stars fell like rain.
Minguo period ed. [*Yunnan tongzhi*] ch. 17

AD 1885 summer [China]

11th year of the Guangxu reign period of Emperor Dezong of the Qing Dynasty, a day in summer; stars streamed like weaving, falling stars came down in pieces like rain.
Minguo period ed. [*Zhejiang xu tongzhi gao*] ch. 4

AD 1885 Aug 10–Sep 8 [China]

(a) 11th year of the Guangxu reign period of Emperor Dezong of the Qing Dynasty, 7th month; stars fell like rain for several nights.
Minguo period ed. Shanxi [*Ziyangxian zhi*] ch. 5

(b) 11th year of the Guangxu reign period of Emperor Dezong of the Qing Dynasty, 7th month; countless stars streamed as if weaving.
Minguo period ed. Shandong [*Danxian zhi*] ch. 14

(c) 11th year of the Guangxu reign period of Emperor Dezong of the Qing Dynasty, 7th month; countless meteors flew back and forth erratically.
Qing Dynasty, Guangxu period ed. Shandong [*Feichengxian zhi*] ch. 10

AD 1885 Aug 30 [China]

11th year of the Guangxu reign period of Emperor Dezong of the Qing Dynasty, 7th month, 21st day, at night; stars fell like rain.
Qing Dynasty, Guangxu period ed. Guangdong [*Chengmaixian zhi*] ch. 12

AD 1885 Sep 9–Oct 7 [China]

(a) 11th year of the Guangxu reign period of Emperor Dezong of the Qing Dynasty, 8th month; all over the sky the stars were chaotic.
Minguo period ed. Guangxi [*Luzhouxian zhi*] ch. 2

(b) 11th year of the Guangxu reign period of Emperor Dezong of the Qing Dynasty, 8th month; at the beginning of twilight, in the middle of Pingyue city people saw the stars in the sky falling in profusion, as if snowing and raining, and not stopping until the third watch.
Qing Dynasty, Guangxu period ed. Guizhou [*Pingyuezhou zhi*] ch. 1

AD 1885 Oct 26 [China]

11th year of the Guangxu reign period of Emperor Dezong of the Qing Dynasty, 9th month, 19th day, at night; stars fell like rain.
Qing Dynasty, Guangxu period ed. Jiangxi [*Nanchangxian zhi*] ch. 55

AD 1885 Oct 28 [China]

11th year of the Guangxu reign period of Emperor Dezong of the Qing Dynasty, 9th month, 21st day, at night; stars fell like rain.
Qing Dynasty, Guangxu period ed. Sichuan [*Fengjiexian zhi*] ch. 11

AD 1885 Nov 2 [China]

11th year of the Guangxu reign period of Emperor Dezong of the Qing Dynasty, 9th month, 26th day; stars streamed as if weaving.
Qing Dynasty, Guangxu period ed. Hubei [*Mianyangxian zhi*] ch. 1

AD 1885 Nov 3 [China]

11th year of the Guangxu reign period of Emperor Dezong of the Qing Dynasty, 9th month, 27th day, at night; stars fell like rain.
Qing Dynasty, Xuantong period ed. Jiangsu [*Suqianxian zhi gao*] ch. 7

AD 1885 Nov 7–Dec 5 [China]

(a) 11th year of the Guangxu reign period of Emperor Dezong of the Qing Dynasty, 10th month; stars fell like rain.
Qing Dynasty, Guangxu period ed. Guangxi [*Xinyixian zhi*] ch. 8
Qing Dynasty, Guangxu period ed. Guangxi [*Songmingzhou zhi*] ch. 2
Minguo period ed. Shanxi [Shenmuxiang tu zhi] ch. 4

(b) 11th year of the Guangxu reign period of Emperor Dezong of the Qing Dynasty, 10th month; stars streamed as if weaving.
Qing Dynasty, Guangxu period ed. Anhui [*Shouzhou zhi*] ch. 35

(c) 11th year of the Guangxu reign period of Emperor Dezong of the Qing Dynasty, 10th month; stars fell like rain.
Qing Dynasty, Guangxu period ed. Jiangxi [*Chongyixian zhi*] ch. 1

(d) 11th year of the Guangxu reign period of Emperor Dezong of the Qing Dynasty, 10th month; all the stars in the sky flew chaotically right, left, up, and down.
Minguo period ed. Guangxi [*Qinzhou zhi*] ch. shou

(e) 11th year of the Guangxu reign period of Emperor Dezong of the Qing Dynasty, 10th month; over several hundred small stars moved about on all sides.
Minguo period ed. [*Yunnan tongzhi*] ch. 17

AD 1885 Nov 14 [China]

11th year of the Guangxu reign period of Emperor Dezong of the Qing Dynasty, winter, 10th month, 8th day; from midnight till cockcrow, in Huatingxian stars fell like rain.
Qing Dynasty, Xuantong period ed. [*Gansu xin tongzhi*] ch. 2

AD 1885 Nov 15 [China]

11th year of the Guangxu reign period of Emperor Dezong of the Qing Dynasty, 10th month, day *jiaxu* [11], at night; the stars in the sky scattered and fell like rain.

Qing Dynasty, Guangxu period ed. Guizhou [*Qianjiangxian zhi*] ch. 5

AD 1885 Nov 19 [China]

11th year of the Guangxu reign period of Emperor Dezong of the Qing Dynasty, 10th month, 13th day; at the juncture of the night hours of *hai* [LT = 21–23h] and *zi* [LT = 23–1h], stars fell like rain while the stars in the sky were dark. For more than one double hour Xian, Chang, Jing, Yuan, and so forth, were among those most severely affected.

Minguo period ed. Shanxi [*Zhouzhixian zhi*] ch. 8

AD 1885 Nov 23–29 [China]

11th year of the Guangxu reign period of Emperor Dezong of the Qing Dynasty, 10th month, 17th to 23rd days; each night large and small stars streamed criss-cross as if weaving, those streaming westward being most numerous.

Qing Dynasty, Guangxu period ed. Hubei [*Deanfu zhi*] ch. 20

AD 1885 Nov 24–25 [China]

11th year of the Guangxu reign period of Emperor Dezong of the Qing Dynasty, 10th month, 18th to 19th days, at night; stars fell like rain.

Minguo period ed. Shanxi [*Huaxian zhi*] ch. 1

AD 1885 Nov 26 [China]

(a) 11th year of the Guangxu reign period of Emperor Dezong of the Qing Dynasty, 10th month, 20th day, at night; in Lingtaixian stars fell like rain. After a brief while, the stars became fewer and were not seen.
Qing Dynasty, Xuantong period ed. [*Gansu xin tongzhi*] ch. 2

(b) 11th year of the Guangxu reign period of Emperor Dezong of the Qing Dynasty, 10th month, 20th day, at night; stars fell like rain.
Minguo period ed. Gansu [*Lingtaixian zhi*] ch. 3

(c) 11th year of the Guangxu reign period of Emperor Dezong of the Qing Dynasty, 10th month, 20th day, the stars moved about as if weaving; the whole night they were unsettled.
Qing Dynasty, Guangxu period ed. Henan [*Xiangfuxian zhi*] ch. 23

(d) 11th year of the Guangxu reign period of Emperor Dezong of the Qing Dynasty,10th month, 20th day, at night; swarms of stars streamed from the northwest to the southeast.
Qing Dynasty, Guangxu period ed. Guangdong [*Linguixian zhi*] ch. 18

AD 1885 Nov 27 [China]

(a) 11th year of the Guangxu reign period of Emperor Dezong of the Qing Dynasty, 10th month, 21st day (*bingxu* [23]); stars fell like rain, from dusk until daybreak.
Qing Dynasty, Guangxu period ed. Shanxi [*Fupingxian zhi*] ch. 10
Qing Dynasty, Guangxu period ed. Shanxi [*Pucheng xin zhi*] ch. 13

(b) 11th year of the Guangxu reign period of Emperor Dezong of the Qing Dynasty, 10th month, 21st day (*bingxu* [23]); stars streamed as if weaving, from the hour of *shen* [LT =15–17h] and not stopping until the marks of the hour of *chen* [LT = 7–9h] the next day.
Minguo period ed. Henan [*Zhengxian zhi*] ch. 1

(c) 11th year of the Guangxu reign period of Emperor Dezong of the Qing Dynasty, 10th month, 21st day (*bingxu* [23]), at night; stars streamed as if weaving.
Qing Dynasty, Guangxu period ed. Shandong [*Zhuchengxian xu zhi*] zongji

(d) 11th year of the Guangxu reign period of Emperor Dezong of the Qing Dynasty, 10th month, 21st day (*bingxu* [23]); after sunset meteors filled the sky, all proceeding from southwest to northeast.
Minguo period ed. Jiangsu [*Gaoyouxian zhi*] buyi

(e) 11th year of the Guangxu reign period of Emperor Dezong of the Qing Dynasty, 10th month, 21st day (*bingxu* [23]); large and small stars numbering in the hundreds and thousands fell, starting from the initial watch of the night and did not stop until the middle of the night.
Qing Dynasty, Guangxu period ed. Shanghai [*Luoxizhen zhi*] ch. 8
Minguo period ed. Shanghai [*Baoshanxian xu zhi*] ch. 17
NB: *Shanghaixian xu zhi* has 11th month, 21st day.

(f) 11th year of the Guangxu reign period of Emperor Dezong of the Qing Dynasty, 10th month, 21st day (*bingxu* [23]), at night; stars fell like rain.
Minguo period ed. Fujian [*Ninghuaxian zhi*] ch. 2

(g) 11th year of the Guangxu reign period of Emperor Dezong of the Qing Dynasty, 10th month, 21st day (*bingxu* [23]); stars fell like rain.
Qing Dynasty, Guangxu period ed. Yunnan [*Lijiangxian zhi*] ch. 2
Minguo period ed. Yunnan [*Luliangxian zhi*] ch. 1
Minguo period ed. Yunnan [*Luopingxian zhi*] ch. 1
NB: Several dozen other local gazetteers also contained similar records.

AD 1885 Nov 27–29 [China]

(a) 11th year of the Guangxu reign period of Emperor Dezong of the Qing Dynasty, 10th month, 21st, 22nd, 23rd days, at night; stars fell like rain.
Qing Dynasty, Guangxu period ed. Hubei [*Wuchangxian zhi*] ch. 10

(b) 11th year of the Guangxu reign period of Emperor Dezong of the Qing Dynasty, 10th month, 21st, 22nd, 23rd days, at night; stars fell like rain. Half the sky was like criss-cross weaving. For 3 nights, it was like this.
Qing Dynasty, Guangxu period ed. Guangdong [*Gaozhoufu zhi*] ch. 51
NB: The Guangxu period ed. of *Dianbaixian zhi* has 11th month.

(c) 11th year of the Guangxu reign period of Emperor Dezong of the Qing Dynasty, 10th month, 21st, 22nd, 23rd days, in the sky above it rained stars, not stopping until the 23rd day.
Qing Dynasty, Guangxu period ed. Yunnan [*Tengyueting zhi*] ch. 1
Minguo period ed. Yunnan [*Longlingxian zhi*] ch. 1

AD 1885 Nov 28 [China]

(a) 11th year of the Guangxu reign period of Emperor Dezong of the Qing Dynasty, 10th month, 22nd day, at night; stars fell like rain.
Qing Dynasty, Guangxu period ed. Shanxi [*Weinanxian zhi*] ch. 11

(b) 11th year of the Guangxu reign period of Emperor Dezong of the Qing Dynasty, 10th month, 22nd day, at night; stars streamed as if weaving.
Qing Dynasty, Guangxu period ed. Hebei [*Dongguangxian zhi*] ch. 11
Minguo period ed. Hebei [*Jiaohexian zhi*] ch. 10
Minguo period ed. Shandong [*Ningjinxian zhi*] ch. 11

(c) 11th year of the Guangxu reign period of Emperor Dezong of the Qing Dynasty, 10th month, 22nd day, at night; stars dashed about on all sides and were agitated and unsettled. People who saw it were terrified.
Qing Dynasty, Guangxu period ed. Henan [*Yongchengxian zhi*] ch. 15

(d) 11th year of the Guangxu reign period of Emperor Dezong of the Qing Dynasty, 10th month, 22nd day; during the hour of *hai* [LT = 21–23h], stars fell like rain, their brilliant rays reflecting and shooting out, and by turns falling to the ground with a noise. It was like this for several nights.
Qing Dynasty, Xuantong period ed. Yunnan [*Xu Mengzixian zhi*] ch. 12

(e) 11th year of the Guangxu reign period of Emperor Dezong of the Qing Dynasty, 10th month, 22nd day, at night; stars fell like rain making a noise.

Minguo period ed. Yunnan [*Menghua zhi gao*] ch. 2
NB: Minguo period ed. *Jianshuixian zhi* had 11th month, 22nd day.

AD 1885 Nov 29 [China]

(a) 11th year of the Guangxu reign period of Emperor Dezong of the Qing Dynasty, 10th month, 23rd day, at night; stars dashed about on all sides and were agitated and unsettled. People who saw it were terrified.
Qing Dynasty, Guangxu period ed. Henan [*Yongchengxian zhi*] ch. 15

(b) 11th year of the Guangxu reign period of Emperor Dezong of the Qing Dynasty, 10th month, 23rd day, at night; stars fell like rain.
Qing Dynasty, Guangxu period ed. Guangdong [*Wuchuanxian zhi*] ch. 10

(c) 11th year of the Guangxu reign period of Emperor Dezong of the Qing Dynasty, 10th month, 23rd day; from the 8th strike of the bell after noon the whole sky was full of stars falling like rain until the 12th strike when fewer were seen.
Minguo period ed. Yunnan [*Weixixian zhi*] ch. 4

AD 1885 Nov 30 [China]

11th year of the Guangxu reign period of Emperor Dezong of the Qing Dynasty, 10th month, 24th day; during the hour of *hai* [LT = 21–23h] of the night, the whole sky was full of stars flying chaotically left, right, up, and down.
Minguo period ed. Guangxi [*Lingshanxian zhi*] ch. 5

AD 1885 Nov 30–Dec 1 [China]

(a) 11th year of the Guangxu reign period of Emperor Dezong of the Qing Dynasty, 10th month, 24th to 25th days; at Liangzhu, all stars flew in streams and fell chaotically like rain.
Qing Dynasty, Guangxu period ed. Shanxi [*Jingbian zhi gao*] ch. 4

(b) 11th year of the Guangxu reign period of Emperor Dezong of the Qing Dynasty, 10th month, 24th to 25th days, at night; at Anhua, extremely many stars fell chaotically.
Minguo period ed. Gansu [*Qingyangxian zhi gao*] ch. 4

AD 1885 Dec 6–15 [China]

11th year of the Guangxu reign period of Emperor Dezong of the Qing Dynasty, 11th month, 1st night; at the 9th striking of the bell, west of the city at the Jiaxinzi Academy, while students studied into the night before resting, in the sky there appeared a multitude of small stars that shuttled back and forth for more than one double hour before ceasing. Everyone saw it.

Minguo period ed. Liaoning [*Liaoyangxian zhi*] ch. 40

AD 1885 Dec 6–AD 1886 Jan 4 [China]

(a) 11th year of the Guangxu reign period of Emperor Dezong of the Qing Dynasty, 11th month; during the night, stars streamed and a vast number of large and small streaking fires filled the sky. Countless numbers dashed chaotically left, right, up, and down the whole night without stopping. The next night some could still be seen.
Minguo period ed. Henan [*Xinyangxian zhi*] ch. 31

(b) 11th year of the Guangxu reign period of Emperor Dezong of the Qing Dynasty, 11th month, at night; stars streamed as if weaving.
Minguo period ed. Henan [*Huaiyangxian zhi*] ch. 20
Minguo period ed. Henan [*Zhongmouxian zhi*] ch. 1

(c) 11th year of the Guangxu reign period of Emperor Dezong of the Qing Dynasty, 11th month; stars streamed like rain.
Qing Dynasty, Guangxu period ed. Jiangsu [*Jiang pu pi sheng*] ch. 39
Minguo period ed. Jiangsu [*Xijin xu zhi xiao lu*] ch. 1
Qing Dynasty, Guangxu period ed. Yunnan [*Puerfu zhi*] ch. 3

AD 1885 Dec 16 [China]

11th year of the Guangxu reign period of Emperor Dezong of the Qing Dynasty, 11th month, 11th day; at the 11th striking of the bell after noon, large and small stars fell chaotically. After a brief while, the stars were fewer.

Minguo period ed. Shanxi [*Qianxian xin zhi*] ch. 8

AD 1885 Dec 26 [China]

(a) 11th year of the Guangxu reign period of Emperor Dezong of the Qing Dynasty, 11th month, 21st day (*yimao* [52]); numerous stars streamed chaotically from the north to the southeast and making noise all night long.
Minguo period ed. Shanghai [*Qingpuxian zhi*] ch. 23

(b) 11th year of the Guangxu reign period of Emperor Dezong of the Qing Dynasty, 11th month, 21st day (*yimao* [52]), at night; stars fell like rain, from the northwest to the southeast.
Qing Dynasty, Guangxu period ed. Zhejiang [*Pinghuxian zhi*] xiangyi.

(c) 11th year of the Guangxu reign period of Emperor Dezong of the Qing Dynasty, 11th month, 21st day (*yimao* [52]); during the marks of the hour of *xu* [LT = 19–21h], stars fell like rain, not settling down until the hour of *chou* [LT = 1–3h].
Qing Dynasty, Guangxu period ed. Fujian [*Puchengxian zhi*] ch. 42

(d) 11th year of the Guangxu reign period of Emperor Dezong of the Qing Dynasty, 11th month, 21st day (*yimao* [52]); during the night, meteors filled the sky from the northwest to the southeast, making a noise like the buzzing of insects all night long without stopping.
Qing Dynasty, Guangxu period ed. Zhejiang [*Jiashanxian zhi*] ch. 34

AD 1885 Dec 28 [China]

11th year of the Guangxu reign period of Emperor Dezong of the Qing Dynasty, 11th month, 23rd day; just over 5 double hours after the hour of *wu* [LT = 11–13h], suddenly a sound came from within the borders to the southeast. On looking in the direction of the noise, falling stars were seen to fly chaotically in the southeastern sky, some as large as eggs, some as big as bowls, others like seed grain, large and small alike falling to the ground in profusion like snowflakes for 30 to 40

minutes. After this was over, some were still seen to suddenly fly up or down with a thundering noise, and after that it stopped.
Minguo period ed. [*Yunnan tongzhi*] ch. 17

AD 1885 winter [China]

(a) 11th year of the Guangxu reign period of Emperor Dezong of the Qing Dynasty, winter; in the zenith a multitude of small stars wove back and forth for more than one double hour.
Minguo period ed. Liaoning [*Liaoyangxian zhi*] ch. shou

(b) 11th year of the Guangxu reign period of Emperor Dezong of the Qing Dynasty, winter; stars fell like rain.
Minguo period ed. Yunnan [*Qiubeixian zhi*] ch. 1

(c) 11th year of the Guangxu reign period of Emperor Dezong of the Qing Dynasty, winter; during the hour of *xu* [LT = 19–21h], in the southeast Corner, stars streamed like rain; it was like this for 3 evenings.
Minguo period ed. [*Yunnan tongzhi*] ch. 17

AD 1886 Jan 2 [China]

(a) 11th year of the Guangxu reign period of Emperor Dezong of the Qing Dynasty, 11th month, 28th day; stars streamed as if weaving.
Qing Dynasty, Guangxu period ed. Henan [*Shanzhou zhi*] ch. 10

(b) 11th year of the Guangxu reign period of Emperor Dezong of the Qing Dynasty, 11th month, 28th day; stars fell like rain from the northwest southward.
Qing Dynasty, Guangxu period ed. Henan [*Wenxiangxian zhi*] ch. mo

AD 1886 Jan 15 [China]

11th year of the Guangxu reign period of Emperor Dezong of the Qing Dynasty, 12th month, 11th day; meteors mostly heading in the *kun* direction (SW).
Minguo period ed. Henan [*Fangchengxian zhi*] ch. 5

AD 1886 Feb 4–AD 1887 Jan 23 [China]

12th year of the Guangxu reign period of Emperor Dezong of the Qing Dynasty, at twilight; in the sky countless meteors spiraled down like a sudden rain shower for 2 hours before abating.
Minguo period ed. [*Yunnan tongzhi*] ch. 17

AD 1886 Jul 31–Aug 28 [China]

12th year of the Guangxu reign period of Emperor Dezong of the Qing Dynasty, a certain day, at night; stars fell like rain.
Qing Dynasty, Guangxu period ed. Hubei [*Lichuanxian zhi*] ch. 9

AD 1886 Sep 4 [China]

12th year of the Guangxu reign period of Emperor Dezong of the Qing Dynasty, 8th month, 7th day, at night; stars streamed like rain.
Jiangsu [*Tangshi zhi bu yi*]

AD 1886 Nov 16 [China]

(a) 12th year of the Guangxu reign period of Emperor Dezong of the Qing Dynasty, 10th month, 21st day, at night; stars fell like rain.
Qing Dynasty, Guangxu period ed. Shanxi [*Yongshouxian zhi*] ch. 10

(b) 12th year of the Guangxu reign period of Emperor Dezong of the Qing Dynasty, 10th month, 21st day, at night; a multitude of stars were seen to fly by streaming from west to east.
Qing Dynasty, Guangxu period ed. Guangxi [*Guixian zhi*] ch. 6

(c) 12th year of the Guangxu reign period of Emperor Dezong of the Qing Dynasty, 10th month, 21st day, at night; stars streamed as if weaving, not stopping till dawn.
Minguo period ed. Henan [*Mixian zhi*] ch. 19

AD 1886 Nov 25 [China]

12th year of the Guangxu reign period of Emperor Dezong of the Qing Dynasty, 10th month, 30th day, at night; stars fell like rain.
Qing Dynasty, Guangxu period ed. Sichuan [*Wushanxian zhi*] ch. 10

AD 1886 Nov 26–Dec 24 [China]

12th year of the Guangxu reign period of Emperor Dezong of the Qing Dynasty, 11th month; stars streamed criss-cross as if weaving.

Minguo period ed. Shandong [*Linzixian zhi*] ch. 8

AD 1886 Dec 15 [China]

12th year of the Guangxu reign period of Emperor Dezong of the Qing Dynasty, 11th month, 20th day, at midnight; in the sky a multitude of stars flashed and flew about in turmoil and streamed criss-cross as if weaving for a double hour before settling down.

Qing Dynasty, Guangxu period ed. Anhui [*Huoshanxian zhi*] ch. 15

AD 1886 Dec 16 [China]

12th year of the Guangxu reign period of Emperor Dezong of the Qing Dynasty, 11th month, 21st day, at night; at the juncture of hours *xu* [LT = 19–21h] and hai [LT = 21–23h], stars streamed like rain for several hours, their profuse bright rays alternately shooting in every direction.

Minguo period ed. Anhui [*Yixian si zhi*] ch. 1

AD 1887 Nov 24 [China]

13th year of the Guangxu reign period of Emperor Dezong of the Qing Dynasty, 10th month, day *guisi* [30], at night; stars shot chaotically about out of position, up, down, left, and right, for more than one quarter of an hour before settling down.

Qing Dynasty, Guangxu period ed. Zhejiang [*Ninghaixian zhi*] ch. 23

AD 1887 Nov 29 [China]

13th year of the Guangxu reign period of Emperor Dezong of the Qing Dynasty, 10th month, 15th day, at night; in the northwest stars fell like rain.

Minguo period ed. Fujian [*Jianningxian zhi*] ch. 27

AD 1888 Mar 16 [China]

14th year of the Guangxu reign period of Emperor Dezong of the Qing Dynasty, 2nd month, day *bingxu* [23]; stars fell like rain for 3 days before returning to normal.

Qing Dynasty, Guangxu period ed. Shandong [*Yixian zhi*] ch. 15

AD 1988 Apr 11–May 10 [China]

(a) 14th year of the Guangxu reign period of Emperor Dezong of the Qing Dynasty, 3rd month, at night; countless meteors battled from north and south, like arrows swiftly shot.
Qing Dynasty, Guangxu period ed. Guangxi [*Yulinzhou zhi*] ch. 4
Minguo period ed. Guangxi [*Luchuanxian zhi*] ch. 2

(b) 14th year of the Guangxu reign period of Emperor Dezong of the Qing Dynasty, 3rd month, at night; countless meteors charged each other from north and south, like arrows being shot back and forth.
Minguo period ed. Guangxi [*Beiliuxian zhi*] bian 12

AD 1888 Aug 22 [China]

14th year of the Guangxu reign period of Emperor Dezong of the Qing Dynasty, 7th month, 15th day; stars fell like rain from the north southward.

Qing Dynasty, Guangxu period ed. Henan [*Wenxiangxian zhi*] ch. mo

AD 1888 Nov 4–Dec 2 [China]

14th year of the Guangxu reign period of Emperor Dezong of the Qing Dynasty, 10th month; a multitude of stars streamed southeastward for several nights before stopping.

Qing Dynasty, Xuantong period ed. Shandong [*Mengyinxian zhi*] ch. xia

AD 1888 winter [China]

14th year of the Guangxu reign period of Emperor Dezong of the Qing Dynasty, winter; stars fell like rain.

Qing Dynasty, Guangxu period ed. Yunnan [*Langqiongxian zhi lue*] ch. 1

AD 1890 Jan 21–AD 1891 Feb 8 [China]

16th year of the Guangxu reign period of Emperor Dezong of the Qing Dynasty, at night; a swarm of stars flew from southwest to northeast, their bright rays several *zhang* long.
Minguo period ed. Shandong [*Boshanxian zhi*] ch. 1

AD 1890 Aug 16–Sep 13 [China]

16th year of the Guangxu reign period of Emperor Dezong of the Qing Dynasty, 7th month, one night; stars fell like rain. Near the horizon stars were sparse and the earth was without illumination.
Minguo period ed. Shanxi [*Qianxian xin zhi*] ch. 8

AD 1891 Apr 9–May 7 [China]

17th year of the Guangxu reign period of Emperor Dezong of the Qing Dynasty, 3rd month; the whole sky was full of falling stars down to the horizon where they suddenly dimmed.
Minguo period ed. Yunnan [*Yaoanxian zhi*] ch. 66

AD 1891 Apr 16 [China]

17th year of the Guangxu reign period of Emperor Dezong of the Qing Dynasty, 3rd month, 8th day; during the hour of hai [LT = 21–23h], stars fell like rain, making noises as they drizzled. For a while the sky was obscured and dark, but after several quarters of an hour returned to its former clarity.
Minguo period ed. Shanxi [*Lantianxian zhi*] ch. 3

AD 1891 Nov 16 [China]

17th year of the Guangxu reign period of Emperor Dezong of the Qing Dynasty, 10th month, full moon day; in Zhuanglang stars fell like rain.
Minguo period ed. [*Gansu tongzhi gao*] bianyi

AD 1891 Dec 1–30 [China]

17th year of the Guangxu reign period of Emperor Dezong of the Qing Dynasty, 11th month; the stars of night were like rain.

Qing Dynasty, Guangxu period ed. Hebei [*Nangongxian zhi*] ch. 8
NB: A winter month means the 11th month.

AD 1891 Dec 27 [China]

17th year of the Guangxu reign period of Emperor Dezong of the Qing Dynasty, 11th month, 27th day, at night; stars fell like rain for several hours before settling down.
Minguo period ed. Henan [*Zhengyangxian zhi*] ch. 3

AD 1892 Mar 4 [China]

18th year of the Guangxu reign period of Emperor Dezong of the Qing Dynasty, 2nd month, 6th day, at night; the whole sky was full of stars flying like shooting arrows.
Minguo period ed. Shandong [*Dingtaoxian zhi*] ch. 9

AD 1892 Aug 22–Sep 20 [China]

18th year of the Guangxu reign period of Emperor Dezong of the Qing Dynasty, 7th month, at night; northeast of Jingning and southwest of Lingzhou, stars fell like rain.
Qing Dynasty, Xuantong period ed. [*Gansu xin tongzhi*] ch. 2

AD 1892 Dec 19–AD 1893 Jan 17 [China]

18th year of the Guangxu reign period of Emperor Dezong of the Qing Dynasty, 11th month, at night; there were stars like rain.
Qing Dynasty, Guangxu period ed. Hebei [*Nangongxian zhi*] ch. 8

AD 1893 Feb 17–AD 1894 Feb 5 [China]

19th year of the Guangxu reign period of Emperor Dezong of the Qing Dynasty; stars fell like rain.
Minguo period ed. Shanxi [*Zhen'anxian zhi*] ch. 10

AD 1893 Oct 10–Nov 7 [China]

(a) 19th year of the Guangxu reign period of Emperor Dezong of the Qing Dynasty, 9th month; stars fell like rain, their reddish gleam filling the sky.

Minguo period ed. Shanxi [Chengcheng xu zhi] ch. 11

(b) 19th year of the Guangxu reign period of Emperor Dezong of the Qing Dynasty, 9th month; in the southern sky stars fell like rain.
Minguo period ed. Yunnan [*Songmingxian zhi*] ch. 4

AD 1893 Nov 10 [China]
19th year of the Guangxu reign period of Emperor Dezong of the Qing Dynasty, 10th month, 3rd day, at night; in Shandanxian stars fell like rain, their brilliant rays illuminating the ground for more than one double hour before extinguishing.
Qing Dynasty, Xuantong period ed. [*Gansu xin tongzhi*] ch. 2

AD 1894 Apr 6–May 4 [China]
20th year of the Guangxu reign period of Emperor Dezong of the Qing Dynasty, 3rd month; stars fell like rain.
Qing Dynasty, Xuantong period ed. Jiangsu [*Suqianxian zhi*] ch. 7

AD 1894 Oct 29–Nov 26 [China]
20th year of the Guangxu reign period of Emperor Dezong of the Qing Dynasty, 10th month; in Zhuanglang stars fell like rain.
Qing Dynasty, Xuantong period ed. [*Gansu xin tongzhi*] ch. 2

AD 1897 Mar 3–Apr 1 [China]
23rd year of the Guangxu reign period of Emperor Dezong of the Qing Dynasty, 2nd month; swarms of stars were seen in daylight, to the south of Gaolan they were especially numerous.
Qing Dynasty, Xuantong period ed. [*Gansu xin tongzhi*] ch. 2

AD 1897 Nov 14 [China]
23rd year of the Guangxu reign period of Emperor Dezong of the Qing Dynasty, 10th month, 20th day, at night; a multitude of stars streamed westward like rain.
Minguo period ed. Anhui [*Taihuxian zhi*] ch. 40

AD 1898 Sep 1 [China]

24th year of the Guangxu reign period of Emperor Dezong of the Qing Dynasty, 7th month, 6th day, at night; Bent Arrow stars (a kind of meteor) flew like weaving shuttles.
Minguo period ed. Anhui [*Mengchengxian zhi*] ch. 12

AD 1899 Aug 12 [China]

25th year of the Guangxu reign period of Emperor Dezong of the Qing Dynasty, 7th month, 7th day, at night; a star as big as a peck measure flew from the northwest to the southeast, its brilliance lighting up the sky, while alongside innumerable smaller star accompanied it.
Qing Dynasty, Guangxu period ed. Zhejiang [*Shouchangxian zhi*] ch. 11

AD 1899 Nov 14 [China]

(a) 25th year of the Guangxu reign period of Emperor Dezong of the Qing Dynasty, 10th month, 12th day, (*bingxu* [23]); during the marks of the hour of *chou* [LT = 1–3h] at night, stars fell like rain.
Qing Dynasty, Guangxu period ed. Shanghai [*Zhengli zhi lue*] ch. 12
Qing Dynasty, Xuantong period ed. Shanghai [*Fengjing xiao zhi*] ch. 10

(b) 25th year of the Guangxu reign period of Emperor Dezong of the Qing Dynasty, 10th month, 12th day, (*bingxu* [23]); stars fell like rain.
Minguo period ed. Shanghai [*Qingpuxian xu zhi*] ch. 23

AD 1900 Jan 31–AD 1901 Feb 18 [China]

26th year of the Guangxu reign period of Emperor Dezong of the Qing Dynasty; during the night, in the west, swarms of stars fell down like rain.
Minguo period ed. [*Yunnan tongzhi*] ch. 17

AD 1901 Apr 19–May 17 [China]
27th year of the Guangxu reign period of Emperor Dezong of the Qing Dynasty, 3rd month; stars fell like rain.
Minguo period ed. Gansu [*Linzexian zhi*] ch. 4

AD 1902 Jul 23 [China]
28th year of the Guangxu reign period of Emperor Dezong of the Qing Dynasty, 6th month, 19th day, at night; stars dispersed and fell toward the east like rain.
Minguo period ed. Sichuan [*Wanyuanxian zhi*] ch. 10

AD 1902 Sep 2–Oct 1 [China]
28th year of the Guangxu reign period of Emperor Dezong of the Qing Dynasty, 8th month; swarms of stars were seen in daylight.
Minguo period ed. Liaoning [*Fengchengxian zhi*] ch. 15

AD 1902 Oct 23 [China]
28th year of the Guangxu reign period of Emperor Dezong of the Qing Dynasty, 9th month, 22nd day; stars fell like rain.
Minguo period ed. Shanxi [*Ziyangxian zhi*] ch. 5

AD 1903 Feb 27–Mar 28 [China]
(a) 29th year of the Guangxu reign period of Emperor Dezong of the Qing Dynasty, 2nd month; stars fell like rain.
Minguo period ed. [*Yunnan tongzhi*] ch. 17
(b) 29th year of the Guangxu reign period of Emperor Dezong of the Qing Dynasty, 2nd month; a multitude of stars streamed southward, like rain they then fell.
Qing Dynasty, Xuantong period ed. [*Gansu xin tongzhi*] ch. 2

AD 1904 autumn [China]
30th year of the Guangxu reign period of Emperor Dezong of the Qing Dynasty, autumn; stars as big as peck measures streamed out of

the north while several tens of smaller stars trailed along with them. On reaching the west, they disappeared.
Minguo period ed. Guangxi [*Lingshanxian zhi*] ch. 5

AD 1906 Aug 13 [China]
32nd year 23rd day, at night; a multitude of stars flew away like rain; from the north to the south they flew.
Qing Dynasty, Guangxu period ed. Guangdong [*Chengmaixian zhi*] ch. 12
Qing Dynasty, Guangxu period ed. Guangxi [*Hepuxian zhi*] ch. 5

AD 1906 Oct 18–Nov 15 [China]
32nd year of the Guangxu reign period of Emperor Dezong of the Qing Dynasty, 9th month; stars fell like rain for several nights in succession, among them one large one more than 3 or 4 *chi* in diameter.
Minguo period ed. [*Yunnan tongzhi*] ch. 17

AD 1906 Oct 24 [China]
32nd year of the Guangxu reign period of Emperor Dezong of the Qing Dynasty, 9th month, 7th day; the night was bright when, from the southwest, there came a noise and in the sky meteors fell chaotically.
Minguo period ed. Shandong [*Linyixian zhi*] ch. 1

AD 1909 Jan 31 [China]
1st year of the Xuantong reign period of Emperor Mo of the Qing Dynasty, 1st month, 10th day; the whole sky was full of stars moving in various directions.
Minguo period ed. Yunnan [*Yaoanxian zhi*] ch. 66

AD 1910 Feb 10–1911 Jan 29 [China]
2nd year of the Xuantong reign period of Emperor Mo of the Qing Dynasty; in the northwest stars fell like rain, but after more than a double hour it was as before.
Minguo period ed. Shanxi [*Qianxian xin zhi*] ch. 8

AD 1910 May 9–Jun 6 [China]

2nd year of the Xuantong reign period of Emperor Mo of the Qing Dynasty, 4th month, at night; stars streamed as if weaving.

Minguo period ed. Jiangsu [*Danyangxian xu zhi*] ch. 41

REFERENCES

Ahn, S. H. (2003). Meteors and showers a millennium ago. *Monthly Notices of the Royal Astronomical Society, 343,* 1095–1100.

Ahn, S. H. (2004). Catalogue of meteor showers and storms in Korean history. *Journal of Astronomy and Space Sciences, 21*(1), 39–72.

Ahn, S. H., Bae, H. J., Cho, H. J., & Jung, S. W. (2002). Meteor showers of 10th to 14th century. *Publications of the Korean Astronomical Society, 17*, 23–40.

Beijing Observatory. (1988). *General compilation of ancient Chinese astronomical records* (*Zhongguo gudai tianxiang jilu zongji*). Nanjing, China: Jiangsu keji jishu chubanshe.

Bone, N. (1993). *Meteors*. Cambridge, MA: Sky Publishing.

Chen, M. (Ed.). (1996). *Zhongguo gu xingtu*. Beijing: Liaoning jiaoyu.

Chen, Z. (1984). *Zhonguo tianwenxue shi*. Shanghai, China: Kexue chubanshe.

Hasegawa, I. (1992). Historical variation in the meteor flux as found in Chinese and Japanese chronicles. Celestial Mechanics and Dynamical Astronomy, *54*, 129–142.

Hasegawa, I. (1993). *Meteoroids and their parent bodies*. Bratislava, Slovakia: Astronomical Institute, Slovak Academy of Sciences.

Hasegawa, I. (1997). Daytime fireballs recorded in China and Japan. In I. S. Nha & F. R. Stephenson (Eds.), *Oriental astronomy from Guo Shoujing to King Sejong* (pp. 127–142). Seoul, Korea: Yonsei University Press.

Hasegawa, I. (1998). Historical variations in the meteor flux as found in the history of the Goryeo Dynasty. In S. Yabushita & J. Henrard (Eds.), *Dynamics of comets and asteroids and their role in the Earth history* (pp. 279–282). Dordrecht, the Netherlands: Kluwer Academic Publishers.

Imoto, S., & Hasegawa, I. (1958). Historical records of meteor showers in China, Korea, and Japan. *Smithsonian Contributions to Astrophysics, 2*, 131–144.

Jenniskens, P. (2006). *Meteor showers and their parent comets*. Cambridge, U.K.: Cambridge University Press.

Kanda, S. (1935). *Japanese historical astronomical records*. Tokyo: Harashoubou.

Kiang, T. (1984). Notes on traditional Chinese astronomy. *Observatory, 104*, 19–23.

Lewis, J. S. (1996). *Rain of fire and ice: The very real threat of comet and asteroid bombardment*. New York: Basic Books.

Liu, X. (1987). Statistical analysis of Chinese historical records of fireballs. *Chinese Astronomy and Astrophysics, 11*(4), 312–319.

Mason, J. W. (1995). The Leonid meteors and Comet 5P/Tempel-Tuttle. *Journal of the British Astronomical Society, 105*, 219–235.

Needham, J., & Wang Ling. (1959). *Science and civilisation in China, vol. 3, mathematics and the sciences of the heavens and the Earth*. Cambridge, U.K.: Cambridge University Press.

Ohsaki, S. (1994). *Japanese historical astronomical records after 1600* (近世時代天文史料). Tokyo.

Rasmussen, K. L. (1990). Historical accretionary events from 700 BC to AD 1850—a 1050-year periodicity? *Quarterly Journal of the Royal Astronomical Society, 31*, 95–108.

Rasmussen, K. L. (1991). Historical accretionary events from 800 BC to AD 1750: Evidence for planetary rings around the Earth. *Quarterly Journal of the Royal Astronomical Society, 32*, 25–34.

Stephenson, F. R. (1994). Chinese and Korean star maps and catalogs. In J. B. Harley & D. Woodward (Eds.), *The history of cartography, Vol. 2: Cartography in the traditional east and southeast Asian societies* (pp. 511–578). Chicago: University of Chicago Press.

Yang, H. J., Park, C. B., & Park, M. G. (2005). Analysis of historical meteor and meteor shower records: Korea, China, and Japan. *Icarus*, *175*, 215–225.

Yau, K., Weissman, P., & Yeomans, D. (1994). Meteorite falls in China and some related human casualty events. *Meteoritics*, *29*, 864–871.

Yi, T. J. (1998). Meteor fallings and other natural phenomena between 1500–1750, as recorded in the annals of the Chosŏn Dynasty (Korea). *Celestial Mechanics and Dynamical Astronomy*, *69*, 199–220.

York, T. J. (Trans.). (2003). *A translation of the observations of meteors recorded in the Goryeosa, Rutherford Appleton Laboratory Technical Report.* Retrieved December 1, 2008, from http://epubs.cclrc.ac.uk/bitstream/313/raltr-2003021.pdf

Zhu, W. (1933). *Tianwen kaogu lu.* Shanghai, China: Commercial Press.

Zhuang, T. (1966). Ancient Chinese records of meteor showers (中国古代流星雨纪录). *Tianwen xuebao*, *14*(1), 37–58; trans., 1977. *Chinese Astronomy*, *1*, 197–220.

Appendix A

Chinese and Western Names of Stars and Asterisms

For identifications of determinative stars, see Sun and Kistemaker (1997) or Stephenson and Green (2002, p. 18). In general, Western equivalents of other asterisms follow Sun and Kistemaker (1997). This list expands and updates the earlier version published in *East Asian Archaeoastronomy: Historical Records of Astronomical Observations of China, Japan, and Korea* (2000). Where appropriate, variant transcriptions in the older Wade-Giles system are given in square brackets.

The 28 Lunar Mansions

Jiao [Chiao]	(LM 1), [Dragon's] Horn, α Vir, 2 stars
Kang [K'ang]	(LM 2), [Dragon's] Neck, κ Vir, 4 stars
Di [Ti]	(LM 3), α^1 Lib, 4 stars
Fang	(LM 4), Room, π Sco, 4 stars
Xin [Hsin]	(LM 5), [Dragon's] Heart, σ Sco, 3 stars
Wei	(LM 6), Dragon's] Tail, μ^1 Sco, 9 stars
Ji [Chi]	(LM 7), Winnowing Basket, γ Sgr, 4 stars
Dou or Nandou [Tou]	(LM 8), Southern Dipper, ϕ Sgr, 6 stars
Niu or Qianniu	(LM 9), Draught Ox, β Cap, 6 stars
Nü or Wunü or Xunü	(LM 10), Serving Girl, ε Aqr, 4 stars
Xu [Hsu]	(LM 11), Void, β Aqr, 2 stars
Wei	(LM 12), Roof, α Aqr, 3 stars
Shi or Yingshi [Shih]	(LM 13), Lay out the Hall, α Peg, 2 stars
Dongbi [Tung-pi]	(LM 14), Eastern Wall, γ Peg, 2 stars
Kui [K'ui]	(LM 15), Swine, ζ And, 16 stars
Lou	(LM 16), Hillock, β Ari, 3 stars
Wei	(LM 17), Belly, 35 Ari, 3 stars

MAO	(LM 18), Pleiades, 17 Tau, 7 stars
BI [Pi]	(LM 19), Hunting net, ε Tau, 8 stars
ZUI [Tsui]	(LM 20), Beak, φ Ori, 3 stars
SHEN	(LM 21), Triaster (Hunter), δ Ori, 10 stars
JING or DONGJING [Ching]	(LM 22), Eastern Well, μ Gem, 8 stars
GUI or YUGUI [Kui]	(LM 23), Spectral Carriage, θ Cnc, 5 stars
LIU	(LM 24), Willow, δ Hya, 8 stars
XING or QI XING [Hsing]	(LM 25), Seven stars, α Hya, 7 stars
ZHANG [Chang]	(LM 26), Spread, ν^1 Hya, 6 stars
YI [I]	(LM 27), Wings, α Crt, 22 stars
ZHEN [Chen]	(LM 28), Carriage Box, γ CrV, 4 stars

OTHER STARS AND ASTERISMS MENTIONED IN THE RECORDS (asterisms made up of several stars are identified by the first in order, according to the Chinese system):

BA [PA] (see BA-SHU)

BAIDIZUO [PAI-TI-TSO], White Emperor's Throne, a star close to β Leo on the west

BAIGUA [PAI-KUA], Broken Melon, ε Del, 5 stars

BAGU [PA-KU], Eight Grains, δ Aur

BANG [PANG], (see TIANBANG)

BA-SHU, Kingdoms of Ba and Shu, ε Ser and α Ser

BEICHEN [PEI-CH'EN], Northern Asterism, α, δ, ε, ζ, η UMi

BEIDOU [PEI-TOU], Northern Dipper, α UMa, 7 stars

BEIJI [PEI-CHI], North-Pole asterism, β UMi, 5 stars

BEILUOSHIMEN [PEI-LUO-SHIH-MEN], Gate of Northern Territories, α PsA

BEIHE [PEI-HO], Northern River, α Gem, 3 stars (see also NANBEIHE)

BIE [PIEH], Soft-shelled Turtle, α Tel, 14 stars in CrA

CANGLONG [TS'ANG-LUNG], Azure Dragon, comprising LM 1–6

CE [TS'E], Riding Crop, γ Cas

CHANGQIU [CH'ANG-CH'IU], Empresses' Palace, α Leo

CHANGZHEN [CH'ANG-CHEN], Permanent Guard of TAIWEI Palace, α_2 CVn, 7 stars

CHANGYUAN [CH'ANG-YÜAN], Long Wall, 53 Leo, 4 stars west of SHAOWEI lining up from S to N

CHESI [CH'E-SSU], Wheelwright, ν Oph, 2 stars

CHEN [CH'EN], 'Asterism', a general reference to the Eastern Dragon constellation, comprising LM 1–6

CHU [CH'U], State of Chu, ε Oph

CHU [CH'U], Pestle, 1 Lac, 3 stars

CHUANSHE [CH'UAN-SHE], Guest Lodgings, GC3947 Cam, 9 stars above HUAGAI [HUA-KAI] along the Celestial River

CHUNSHOU [CH'UN-SHOU], Quail's Head, LM 23 & 24

CIFEI [TZ'U-FEI], Secondary Concubine, ζ Leo

CIJIANG [TZ'U-CHIANG], Subordinate General, of the East = ε Vir, of the West = ι Leo

CONGGUAN [TS'UNG-KUAN], Page, 92 Leo

DACHEN [TA-CH'EN], see CHEN

DACHUI [TA-CH'UI], Heaven's Kitchen TIANCHU [T'IEN-CH'U] in lunar mansion MAO [LM 18]

DAHUO [TA-HUO], Great Fire, α Sco or Antares

DAJIAO [TA-CHIAO], Great Horn, α Boo

DALIANG [TA-LIANG], LM 17–19

DALING [TA-LING], Great Burial Tumulus, 9 Per, 8 stars

DIXING [TI-HSING], Emperor Star, β UMi,

DIZUO [TI-TSO], Emperor's Throne, α_1 Her

DONGCIXIANG [TUNG-TZ'U-HSIANG], Associate Chancellor of the East, δ Vir

DONGHAI [TUNG-HAI], Eastern Sea, η Ser

DONGXIAN [TUNG-HSIEN], East Wall, ψ Oph, 8 stars

DOU [TOU], either BEIDOU = N. Dipper, UMa, 7 stars, or NANDOU (LM 8) = S. Dipper, ϕ Sgr, 6 stars

DOUSHU [TOU-SHU], Dipper Pivot, α UMa

DUANMEN [TUAN-MEN], Main Gate, between β and η Vir

FA, Punitive Attack, 42c Ori, 3 stars

FEI, Concubine, γ Leo

FENMU, Tomb, γ Aqr, 4 stars

FULU, Service Road, ζ Cas

FU YUE [FU YUE], Wise Counsellor, G Sco, 1 star

GANG [KANG], Pole, γ Cam, 9 stars
GANGXING [KANG-HSING] (see GANG)
GEDAO [KE-TAO], Stepped Path, φ Cas, 6 stars
GEN [KEN] of TIANJIN [T'IEN-TSIN], δ Cyg
GENGHE [KENG-HO], Obstructing River, ε Boo
GOU [KOU], (see TIANGOU [T'IEN-KOU])
GOUXING [KOU-HSING], (see TIANGOU)
GOUQIAN [KOU-CH'IEN] (see LING)
GOUZHEN [KOU-CHEN], (-CHEN), Hooked Array, α UMi, 7 stars
GUAN [Kuan] (see TIANGUAN)
GUANSUO [KUAN-SO], Manacles, β CrB, 9 stars
GUANWEI [KUAN-WEI], Beacon, 144 Cnc, 4 stars
HANYU [HAN-YÜ] star, transient phenomenon like a comet or nova, identity unknown
HEGU [HO-KU], River Drum, α Aql, 3 stars
HESHU [HO-SHU], River Garrison, α Gem
HOU, Watcher, α Oph
HOUGONG [HOU-KUNG], Empress's Palace, 4 UMi
HU(-SHI), Bow and Arrow, δ CMa, 9 stars
HUGUA [HU-KUA], Dry Melon, α Del, 5 stars
HUAGAI [HUA-KAI], Resplendent Canopy, ψ Cas, 7 stars
HUANWEI, Surrounding Guards, probably π Cep or α Cam
HUANZHE [HUAN-CHE], Eunuchs, 60 Her, 4 stars
HUANGDI [HUANG-TI], The Yellow Emperor's Throne, β Leo
JISHI [CHI-SHIH], Piled Corpses, π Per
JIAN [CHIEN](SEE JIANBI)
JIANBI [CHIEN-PI], Door-bolt, ν Sco
JIANXING [CHIEN-HSING], Mark Star, π Sgr, 6 stars [alt. ξ Sgr]
JIANGJUN [CHIANG-CHÜN], General, γ_1 And
JISHUI [CHI-SHUI], Bilgewater, λ Per, a star within TIANCHUAN [T'IEN-CHUAN], Celestial Boat
JIXIN [CHI-HSIN], Store-up Kindling, κ Gem
JIXING [CHI-HSING], see BEIJI [PEI-CHI]
JIN [Chin] star, State of Jin, κ Her

Jiu [Chiu], Mortar, χ Peg, 4 stars
Jiukan [Chiu-k'an], Nine Water Wells, ι Sgr, 9 stars
Jiuqi [Chiu-ch'i], Wine Banner, ξ Leo, 3 stars
Jiuyou [Chiu-yu] (or –Liu), Nine Battle Flags, μ Eri, 9 stars
Jun [Chün] (=Junmen), Gate of the Army, ξ Hya, 2 stars
Juanshe [Chüan-she], Curled Tongue, ν Per, 6 stars
Kaiyang [K'ai-yang], Yang Initiator, ζ UMa
Kulou [K'u-lou], Armory tower, δ Cru, 29 stars
Kuxing [K'u-hsing], Weeping Star, μ Cap [alt. δ Χαπ]
Kui [K'ui], Bowl of the Dipper, or specifically, α UMa, Dubhe
Lang (see Tianlang, Langxing)
Langjiang [Lang-chiang], Captain of the Guard, α CVn
Langwei, Board Secretary's Seat, γ Com, 15 stars
Langxing [Lang-hsing], Wolf Star, α CMa
Lei (see Leibizhen)
Leibizhen [Lei-pi-chen], Fortifications and Entrenchments, υ Aqr, a long asterism stretching over 45° along the ecliptic from Aqr to Psc
Leidian [Lei-tien], Thunder and Lightning, ζ Peg, 6 stars
Ligong [Li-kung], Summer Palace, λ Peg, 6 stars
Lian [Lien], Chain, possibly miswritten Jianbi; otherwise unknown
Liang, State of Liang, δ Oph
Liesi [Lieh-ssu], Street of Shops, σ Ser (alt. λ Oph), 2 stars
Ling [Ling] = Gouling [Kou-ling], (var. Gouqin [Kou-chin]鉤鈐), ω_1 Sco, 2 stars
Lingtai [Ling-t'ai], Numinous Platform, 59 Leo, 3 stars west of Mingtang
Liujia [Liu-chia], Six-*Jia* Divination, GC8711 Cam, 6 stars near Ziweigong below Huagai [Hua-kai]
Lu, see Fulu
Mingtang [Ming-t'ang], Hall of Light, η Sex
Nanbeihe [Nan-pei-ho], North and South Rivers (see Nanhe and Beihe)
Nangong [Nan-kung], South Palace, area around 92, 93 Leo in Taiweiyuan
Nanhe [Nan-ho], Southern River, α CMi, (Procyon), 3 stars

NANMEN [NAN-MEN], South Gate, ζ_2 Cen, 2 stars
NEICHU [NEI-CH'U], Interior Kitchen, 7 Dra (or possibly 4 Dra), 2 stars
NEIJIE [NEI-CHIEH], Inner Stairway, o UMa, 6 stars north of WENCHANG
NEIWUZHUHOU [also NEIZHUHOU] [NEI-WU-CHU-HOU], Five Feudal Lords, 39 Com, 5 stars
NIANDAO [NIEN-TAO], Carriage Road, 13 Lyr, 5 stars
NÜSHI [NÜ-SHIH], Lady Astrologer, 34 Dra
NÜYUGONG [NÜ-YÜ-KUNG](SEE YUNÜGONG)
PAOGUA [P'AO-KUA], Calabash, α Del, 5 stars
PILI [P'I-LI], Thunderbolt, β Psc, 5 stars
PINGXING [P'ING-HSING] (PING or NEIPING), Inner Screen, ξ Vir, 4 stars
QIANG [CH'IANG] (see TIANQIANG)
QICHEN JIANGJUN [CH'I-CH'EN CHIANG-CHÜN], Cavalry General, κ Lup, 3 stars
QIGONG [CH'I-KUNG], Seven Eminences, β Boo, 7 stars
QIGUAN [CH'I-KUAN], Cavalry Officers, C_1 Cen, 27 stars
QUAN [CH'ÜAN], Heaven's Beam, δ UMa, Megrez
QUEQIU, [CH'ÜEH-CH'IU], Gate Mounds, 18 Mon, 2 stars
RENXING [JEN-HSING], People Star, κ Peg (or 2 Peg), 4–5 stars
SANGONG [SAN-KUNG], Three Excellencies, 24 CVn, 3 stars south of the handle of BEIDOU
SANNENG [SAN-NENG], (see SANTAI)
SANSHI [SAN-SHIH], Three Magisters, ρ UMa, 3 stars
SANTAI [SAN-T'AI], Three Steps, ι, λ, ν UMa, 6 stars
SHAOFU, Heir Apparent's Tutor, 24 UMa
SHAOZAI [SHAO-TSAI], Lesser Major-domo, η Dra
SHAOWEI, Ranking Officers, 53 Leo, 4 stars
SHANGCHENG [SHANG-CH'ENG], Senior Administrative Officer, GC3947 Cam
SHANGSHU, Imperial Secretary, 27f Dra (alt. 15 Dra), 5 stars
SHANGTAI [SHANG-T'AI], Upper Step, ι UMa, 2 stars
SHANGXIANG [SHANG-HSIANG], Senior Ministers (E & W), γ Vir and σ Leo
SHANGZAI [SHANG-TSAI], Chief Minister, θ Dra
SHENQI [SHEN-CH'I], Banner of Shen, o_1 Ori (alt. π Ori), 9 stars

SHETI [SHE-T'I] (ZUO [TSO] and YOU [YU]) Left and Right Assistant Conductors, o Boo and η Boo, 6 stars on either side of Arcturus
SHILOU [SHIH-LOU], Market Tower, μ Oph, 6 stars
SHUXIN [SHU-HSIN], Heart of the Pivots, i.e., between α and ι Dra, the Left Pivot and Right Pivot
SHUZI [SHU-TZU], Princes, 5 UMi
SHUIWEI, Water Level, 68 Gem, 4 stars
SHUIFU [SHUI-FU], Watery Palace, ν Ori, 4 stars south of JING [LM 22]
SIFU [SSU-FU], Four Advisors, GC14305 Cam, 4 stars
SONG [SUNG], State of Song, η Oph
TAIYANG [-SHOU] [T'AI-YANG-SHOU], Guardian of the Great Yang, χ UMa, 1 star
TAIWEI [-YUAN] [T'AI-WEI-YÜAN], Grand Tenuity Enclosure or Privy Council, β Vir, 10 stars
TAIZI [T'AI-TZU], Crown Prince, γ UMi (alt. 93 Leo in TAIWEI)
TAIZUN [T'AI-TSUN], Venerable Elder, ψ UMa
TENGSHE, Flying Snake, α Lac, 22 stars
TIANBANG [T'IEN-PANG], Celestial Cudgel, ξ Dra, 5 stars
TIANBEI [T'IEN-PEI], Celestial Flail, ι Her, 5 stars
TIANBIAN [T'IEN-PIEN], Celestial Casque, 5 Aql, 9 stars
TIANCANG [T'IEN-TSANG], Celestial Storehouse, ι Cet, 6 stars
TIANCE [T'IEN-TS'E], Celestial Latrine, α Lep, 4 stars
TIANCHU [T'IEN-CH'U], Celestial Kitchen, δ Dra, 6 stars
TIANCHUAN [T'IEN-CH'UAN], Celestial Boat, η Per, 9 stars
TIANCHUANG [T'IEN-CH'UANG], Celestial Couch, GC19189 UMi, 6 stars
TIANDAJIANGJUN [T'IEN-TA-CHIANG-CHUN] (see JIANGJUN)
TIANGOU [T'IEN-KOU], Celestial Hook, β Cep, 9 stars
TIANGUAN [T'IEN-KUAN], Celestial Pass, ζ Tau, 1 star
TIANHUN [T'IEN-HUN], Celestial Pigsty, 21 Cet, 7 stars
TIANHUANG [T'IEN-HUANG], Celestial Pond, 19 Aur, 5 stars within WUCHE
TIANHUANGDADI [T'IEN-HUANG TA-TI], Great Emperor Celestial Theocrat, 32 Dra
TIANJI [T'IEN-CHI], Heaven's Armil, γ UMa, Phecda
TIANJI [T'IEN-CHI], (asterism), Celestial Regulations, ξ CrB, 9 stars

TIANJIU [T'IEN-CHIU], Celestial Stables, θ And, 10 stars

TIANJIANG [T'IEN-CHIANG], Celestial River, θ Oph, 4 stars

TIANJIN [T'IEN-CHIN], Celestial Ford, γ Cyg, 9 stars

TIANLANG [T'IEN-LANG], Wolf of Heaven, α CMa

TIANLAO [T'IEN-LAO], Celestial Dungeon, ω UMa, 6 stars

TIANLEI (= TIANLEICHENG) [T'IEN-LEI-CH'EN], Celestial Ramparts, ξ Aqr, 13 stars

TIANLIN [T'IEN-LIN], Celestial Granary, f Tau, 4 stars

TIANLI [T'IEN-LI], Celestial Judge for the nobility, 4 stars, the bowl of BEIDOU.

TIANMIAO [T'IEN-MIAO], Heavenly Ancestral Temple, α Pyx, 14 stars

TIANMEN [T'IEN-MEN], Gate of Heaven, 53 Vir, 2 stars

TIANPING [T'IEN-P'ING], Celestial Privacy Screen, μ Lep, 2 stars

TIANQIANG [T'IEN-CH'IANG], Celestial Lance, κ Boo, 3 stars

TIANQUN [T'IEN-CH'ÜN], Celestial Granary, α Cet, 13 stars

TIANQUAN [T'IEN-CH'ÜAN], Heaven's Beam, δ UMa, Megrez

TIANSHI or TIANSHIYUAN [T'IEN-SHIH-YÜAN], Celestial Marketplace, ζ Oph, 22 stars

TIANSHU [T'IEN-SHU], Heaven's Pivot, α UMa, Dubhe

TIANTIAN [T'IEN-T'IEN], Celestial Paddy-fields, ω Cap, 10 stars

TIANXIANG [T'IEN-HSIANG], Ministers of Heaven, 17 Sex, 3 stars

TIANYIN [T'IEN-YIN], Heavenly Yin, ζ Ari, 5 stars

TIANYU [T'IEN-YÜ], Celestial Grain in the Field, η For, 3 stars

TIANYUAN [T'IEN-YÜAN], Celestial Abyss, α Sgr, 10 stars

TIANYUAN [T'IEN-YÜAN], Celestial Meadows, γ Eri, 16 stars

TIANYUAN [T'IEN-YÜAN], Celestial Orchard, υ^1 Eri, 13 stars south of Celestial Meadows

TIANYUE [T'IEN-YÜEH], Celestial Gate, 4 Sgr, 7 stars

TIANZHU [T'IEN-CHU], Celestial Pillars, ? Dra, 5 stars inside ZIWEI close to the eastern wall

TIANZUN [T'IEN-TSUN], Celestial Beaker, δ Gem, 3 stars north of JING [LM 22].

TIEZHI [T'IEH-CHIH], Iron Chopping Block, 48 Cet, 5 stars
TUGONG [T'U-KUNG], (K1504) God of the Soil, ω Psc, 2 stars south of DONGBI [LM 14]
TUSI (see TUSIKONG) [T'U-SSU-KUNG]
TUSIKONG [T'U-SSU-KUNG], Butcher Shop, 109 Her, 2 stars
WAICHU [WAI-CH'U], Outer Kitchen, 2 Hya, 6 stars
WAIPING [WAI-P'ING], Outer Fence, δ Psc, 7 stars
WANGLIANG, Charioteer Wang Liang, β Cas, 5 stars
WENCHANG [WEN-CH'ANG], Celestial Secretariat, θ UMa, 6 stars
WUDIZUO or WUDINEIZUO [WU-TI-NEI-TSO], Inner Seats of the Five Sovereigns, γ Cep, 5 stars inside the Purple Tenuity Enclosure below HUAGAI
WUCHE [WU-CH'E], Five Chariots, ι Aur, 14 stars
WUJI [WU-CHI], see TIANJI
WUNÜ [=LM 10], Serving Girl, ε Aqr, 4 stars
WUZHUHOU [WU-CHU-HOU], Five Lords of Heaven, θ Gem, 5 stars
XI or ZHUI [HSI/CHUI] [=LM 20], φ Ori
XIZHONG [HIS-CHUNG], Master of Chariots, κ Cyg, 4 stars
XIANCHI [HSIEN-CH'IH], Pool of Harmony, λ Aur, 3 stars (see also WUCHE).
XIANG [HSIANG], Chancellor, star(s) in the TAIWEI enclosure in Leo
XIATAI [HSIA-T'AI], Lower Step, ν UMa
XINGCHEN [HSING-CH'EN], Favored Official, 2 Com
XUANXIAO [HSÜAN-HSIAO], Somber Emptiness, the winter quadrant or LM 10–12
XUANYUAN [HSÜAN-YÜAN], Yellow Emperor, α Leo, 17 stars
YAOGUANG [YAO-KUANG], Gemmy Light, η UMa
YINDE [YIN-TE], Virtue of Yin, 7, 8 Dra, 7 stars
YOUJIAO [YU-CHIAO], Right Horn of the Dragon, α Vir
YOUSHETI [YU-SHE-T'I], Right Assistant Conductor (see SHETI)
YOUXIA [YU-HSIA], Right Linchpin, α Crv
YOUYEMEN [YU-YE-MEN], Right Side-gate, σ – ι Vir
YOUZHIFA [YU-CHIH-FA], Enforcer of the Right, β Vir
YUE [YUEH], Battle Axe, η Gem

YUHENG [YÜ-HENG], Jade Transverse, ε UMa, Alioth
YUJING [YÜ-CHING], Jade Well, β Eri, 4 stars
YULIN[-JUN] [YÜ-LIN], Imperial Guards, 29 Aqr, 45 stars
YUZUO [YU-TSO], Imperial Throne, β UMi
YUNYU [YÜN-YÜ], Clouds and Rain, κ Psc, 4 stars
YUNÜGONG [YÜ-NÜ-KUNG], Imperial Harem, τ Dra, 4 stars
YUNÜ [YÜ-NÜ], Imperial Lady, 31 Leo
ZAO FU [TSAO-FU], Charioteer Zao Fu, δ Cep, 5 stars
ZHEWEI [CHE-WEI], Deprivers of Dignity, ? Lib, 7 stars south of KANG [LM 2]
ZHENCHE [CHEN-CH'E], σ Lib, Chariot Formation, 3 stars south of DI [LM 3]
ZHIFA [CHIH-FA] (see ZUOZHIFA, YOUZHIFA)
ZHINÜ [CHIH-NÜ], Weaving Maiden, α Lya, 3 stars
ZHONGTAI [CHUNG-T'AI], Middle Step, λ UMa, 2 stars
ZHOUDING [CHOU-TING], Cauldron of Zhou, β Com, 3 stars
ZHOUGONG [CHOU-KUNG], Duke of Zhou star, transient phenomenon, possibly a comet or supernova
ZHU [CHU], variant name for LM 24
ZHU [CHU], Pillar, η Aur, 9 stars
ZIGONG [TZU-KUNG], Purple Tenuity Palace, the circumpolar region, see ZIWEI
ZI [TZU] and SUN, Sons and Grandsons, λ and κ Col, 2 stars each
ZIWEI [TZU-WEI], or ZIWEIYUAN, Purple Tenuity Enclosure, κ Dra, 15 stars
ZIWEIYUAN [TZU-WEI-YÜAN] Palace of Purple Tenuity
ZIYUAN [TZU-YUAN] (see ZIWEIYUAN)
ZONGREN [TSUNG-JEN], Imperial Clansmen, 66 Oph, 4 stars
ZONGZHENG [TSUNG-HENG], Headman of the Imperial Clan, β Oph, 2 stars
ZUOGENG [TSO-KENG], Forester, ν Ari, 5 stars
ZUOXIA [TSO-HSIA], Left Linchpin, η Crv
ZUOYEMEN [TSO-YE-MEN], Left Doorman, η Vir
ZUOZHIFA [TSO-CHIH-FA], Enforcer of the Left, η Vir
ZUIXI [TSUI-HSI], LM 20

Appendix B

Azimuth and Time Indications in Ancient Chinese Astronomical Records

Azimuth Indications

Using 8 trigrams (***bagua***)	**12 branches** (***dizhi***)	**10 stems** (***tiangan***)
坎 *Kan* – N	子 *zi* – N	
艮 *Gen* – NE	丑 *chou* – NNE	甲 *jia* – E by N
震 *Zhen* – E	寅 *yin* – ENE	乙 *yi* – E by S
巽 *Xun* – SE	卯 *mao* – E	丙 *bing* – S by E
离 *Li* – S	辰 *chen* – ESE	丁 *ding* – S by W
坤 *Kun* – SW	巳 *si* – SSE	午 *wu* – S
兑 *Dui* – W	戊 *wu* – S	己 *ji* – S by E
乾 *Qian* – NW	未 *wei* – SSW	庚 *geng* – W by S
	申 *shen* – WSW	辛 *xin* – W by N
	酉 *you* – W	壬 *ren* – N by W
	戌 *xu* – WNW	癸 *gui* – N by E
	亥 *hai* – NNW	

Time Indications

In China from at least the Han Dynasty on—and subsequently in Japan and Korea—the day was divided into 12 double hours, or *shi*, each denoted by 1 of the 12 earthly branches. So, for example, the double hour from 23:00–01:00 is identified as *zi* (LT 23–1h), *chou* is (LT 1–3h), and so on. Once clepsydras (*kelou*) came into general use for keeping time, the day was further subdivided into 100 *ke* "mark, graduation," so that 1 *ke* equals 0.24 h. The hours of darkness from sunset to sunrise were

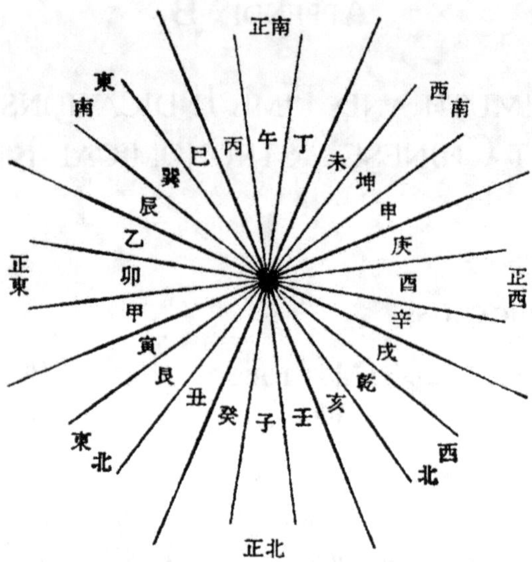

The 24 *shan* "azimuth" notations (south is at the top). In the scheme shown, only four of the trigrams—*Qian*, *Kun*, *Xun*, *Gen*—were traditionally employed for the intercardinal directions, and 2 of the 10 stems—*wu*, *ji*—were omitted to avoid confusion.

divided into five equal *geng*, or "watches," which were also subdivided into five equal parts, denoted *dian*, "points." The length of the watches varied with the seasons from 1.7h to 2.6h. The first watch of the night was sometimes denoted *chugeng* "initial" watch, the others "second," "third," and so forth. For further details on the time intervals corresponding to numbered marks, the halves of the double hours, and so forth, see Xu et al. (2000, p. 7) and Stephenson and Green (2002, p. 15).

Definitions of Certain Observational Terms Commonly Used in the Records

Linear Measure

1 *du* = 0.9856 deg
1 *cun* "inch" (or *fen*) = 0.1 *du*
1 *chi* "foot" = 1 *du* (ca 1.5°; varied slightly over time, see Kiang 1984)
1 *zhang* "rod" = 10 *chi*

Actions and Circumstances (after Ho 1966, 37ff)

守	*shou*	"guard" = remain stationary within 1 to 2 *du* of another body
犯	*fan*	"trespass" = approach within 7 *cun* (0.7 *du*)
侵	*qin*	"invade" = take up station in an inappropriate space and trespass on another body, asterism, and so forth.
合	*he*	"join" = separated by 1 *chi* or less
聚	*ju*	"gather" = within the same lunar mansion (~15 *du*)
斗	*dou*	"battle" = approach then separate
貫	*guan*	"penetrate"

Appendix C

Geographical Coordinates of Historical Capitals of China, Japan, and Korea

Chinese Capitals

Dynasty	City	Latitude	Longitude	Dates
Shang	Yin (Anyang)	36°07’	114° 20E	1576–1047 BC
W. Zhou	Hao	34 10	108 43E	1046–771 BC
E. Zhou	Luoyi	34 45	112 30E	770–256 BC
State of Lu	Qufu	35 08	117 01E	722–476 BC
(Spring & Autumn period)				
Qin	Xianyang	34 23	108 51E	221–206 BC
W. Han	Chang’an	34 21	108 53E	206 BC–AD 25
E. Han	Louyang	34 45	112 28E	AD 25–220
Wei	Louyang	34 45	112 28E	220–265
W. Jin	Louyang	34 45	112 28E	265–317
E. Jin	Jiankang	32 02	118 47E	317–420
Song	Jiankang	32 02	118 47E	420–579
Qi	Jiankang	32 02	118 47E	479–502
Liang	Jiankang	32 02	118 47E	502–557
Chen	Jiankang	32 02	118 47E	557–589
N. Wei	Pingcheng	40 06	113 14E	386–494
	Louyang	34 45	112 28E	494–534
N. Qi	Ye	36 21	114 35E	550–577
N. Zhou	Chang’an	34 16	108 54E	557–581
Sui	Daxing	34 16	108 54E	581–618
Tang	Chang’an	34 16	108 54E	618–907
Liao	Yanjing	39 55	116 25E	907–1125
N. Song	Bian	34 47	114 20E	960–1127
S. Song	Lin’an	30 15	120 10E	1127–1279
Jin	Zhongdu	39 55	116 25E	1115–1234
Yuan	Dadu	39 55	116 25E	1206–1368
Ming	Nanjing	32 02	118 47E	1368–1403
	Beijing	39 55	116 25E	1403–1644
Qing	Beijing	39 55	116 25E	1644–1911

Japanese Capitals

City	Latitude	Longitude	Dates
Nara	34° 41'	135° 49'E	710–784
Nagaoka	37 27	138 50 E	784–794
Heian (Kyoto)	35 02	135 45 E	795–1868
Kamakura	35 19	139 33 E	ca. 1185
Edo (Tokyo)	35 40	139 45 E	1590–1868

Korean Capitals

Dynasty	City	Latitude	Longitude	Dates
Goguryeo	Hwando	41° 10'	126° 12'E	5th c.–AD 700
	Pyongyang	39 01	125 45 E	
Baekje	Weeryesung	37 29	127 09 E	371–475
	Woongjinsung	36 26	127 08 E	476–538
	Sabisung	36 17	126 56 E	539–660
Silla	Kyongju	35 50	129 12 E	5th c.–AD 918
Goryeo	Gaesung	37 58	126 35 E	918–1392
Joseon	Hanyang	37 33	126 58E	1394–

APPENDIX D

CHINESE-WESTERN STAR MAPS

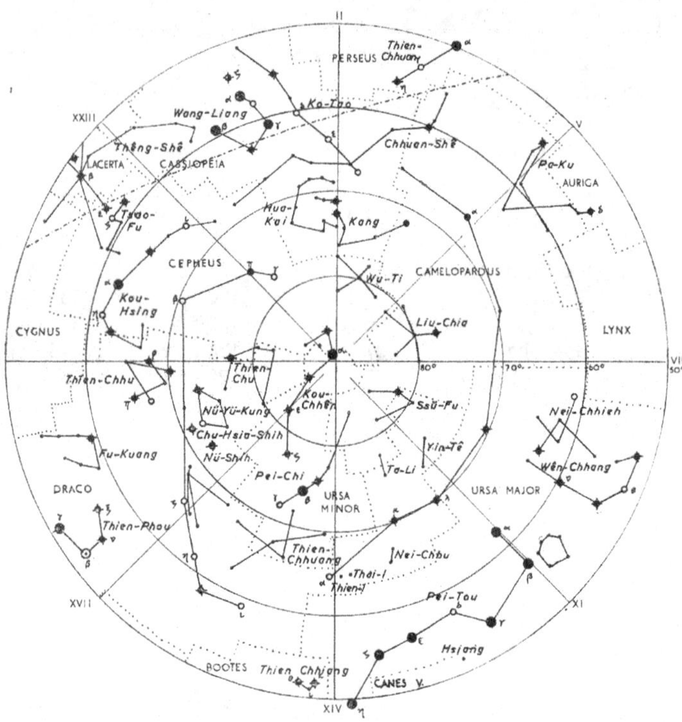

STAR MAP I

STELLAR MAGNITUDES

1 and above

2 and 2½

3 and 3½

4 and 4½

5 and under

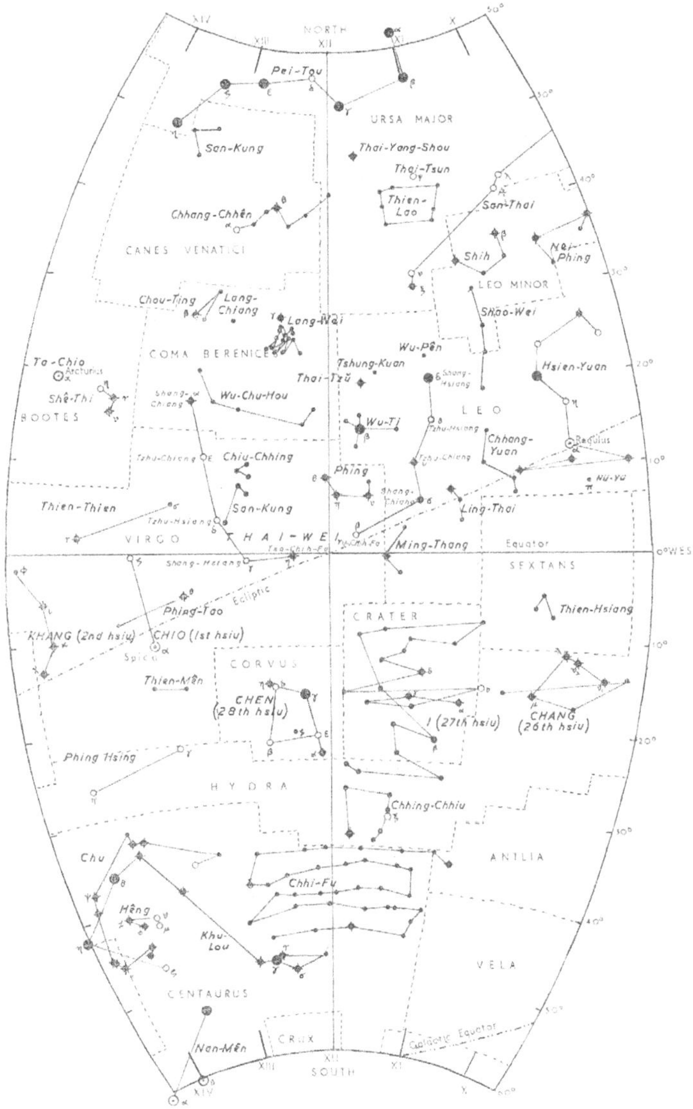
NORTH
XIV
XIII
XII
XI
X
50°
Pei-Tou
URSA MAJOR
San-Kung
Thai-Yang-Shou
Thai-Tsun
Thien-Lao
San-Thai
Chhang-Chhên
CANES VENATICI
Shih
Nei-Phing
LEO MINOR
Chou-Ting
Lang-Chiang
Lang-Wei
Shao-Wei
COMA BERENICES
Wu-Pên
Ta-Chio
Arcturus
Tshung-Kuan
Hsien-Yuan
Shê-Thi
Thai-Tzŭ
Wu-Chu-Hou
BOOTES
LEO
Wu-Ti
Chhang-Yuan
Regulus
Chiu-Chhing
Phing
Nü-Yü
Thien-Thien
San-Kung
Ling-Thai
VIRGO
THAI-WEI
Ming-Thang
Equator
EAST
0° WEST
SEXTANS
Ecliptic
Phing-Tao
Thien-Hsiang
KHANG (2nd hsiu)
CHIO (1st hsiu)
Spica
CRATER
CORVUS
Thien-Mên
CHEN (28th hsiu)
I (27th hsiu)
CHANG (26th hsiu)
Phing Hsing
HYDRA
Chhing-Chhiu
ANTLIA
Chu
Chhi-Fu
Hêng
Khu-Lou
VELA
CENTAURUS
CRUX
Nan-Mên
Galactic Equator
SOUTH
60°

STAR MAP 2

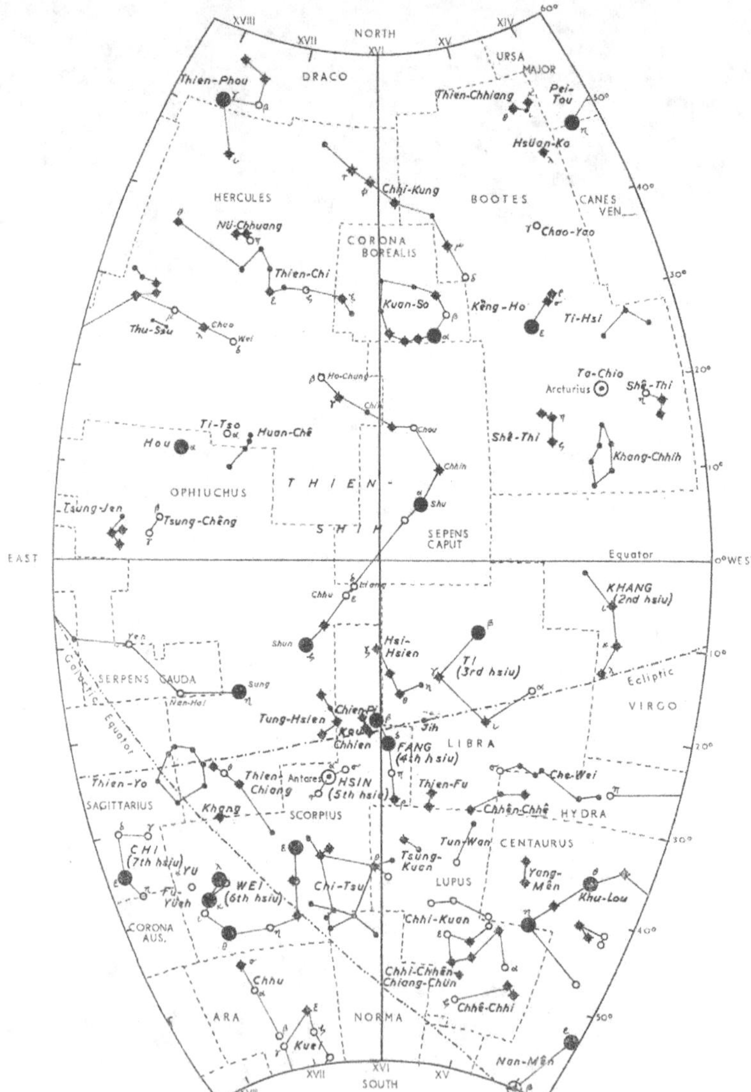

NORTH
XVIII
XVII
XVI
XV
XIV
60°
DRACO
URSA MAJOR
Thien-Phou
Thien-Chhiang
Pei-Tou
50°
Hsüan-Ko
Chhi-Kung
HERCULES
BOOTES
CANES VEN.
40°
Nü-Chhuang
CORONA BOREALIS
Chao-Yao
Thien-Chi
30°
Kuan-So
Kêng-Ho
Ti-Hsi
Thu-Ssu
Chao
Wei
Ho-Chung
Ta-Chio
20°
Arcturius
Shê-Thi
Chou
Chin
Ti-Tso
Huan-Chê
Shê-Thi
Hou
Khang-Chhih
10°
OPHIUCHUS
T H I E N -
Shu
Tsung-Jen
Tsung-Chêng
S H I H
SEPENS CAPUT
EAST
Equator
0° WEST
Liang
Chhu
KHANG
(2nd hsiu)
Yen
Shun
Hsi-Hsien
10°
SERPENS CAUDA
Sung
TI
(3rd hsiu)
Ecliptic
Nan-Hai
VIRGO
Tung-Hsien
Chien-Pi
Jih
Kou-Chhien
FANG
(4th hsiu)
LIBRA
20°
Galactic Equator
Thien-Yo
Thien-Chiang
Antares
HSIN
(5th hsiu)
Thien-Fu
Che-Wei
SAGITTARIUS
Khang
SCORPIUS
Chhên-Chhê
HYDRA
Tun-Wan
CENTAURUS
30°
CHI
(7th hsiu)
YU
WEI
(6th hsiu)
Tsung-Kuan
Chi-Tsu
Yang-Mên
Khu-Lou
Fu-Yüeh
LUPUS
Chhi-Kuan
CORONA AUS.
40°
Chhu
Chhi-Chhên
Chiang-Chün
ARA
Chhê-Chhi
NORMA
50°
Kuei
Nan-Mên
XVIII
XVII
XVI
SOUTH
XV
XIV
60°

STAR MAP 3

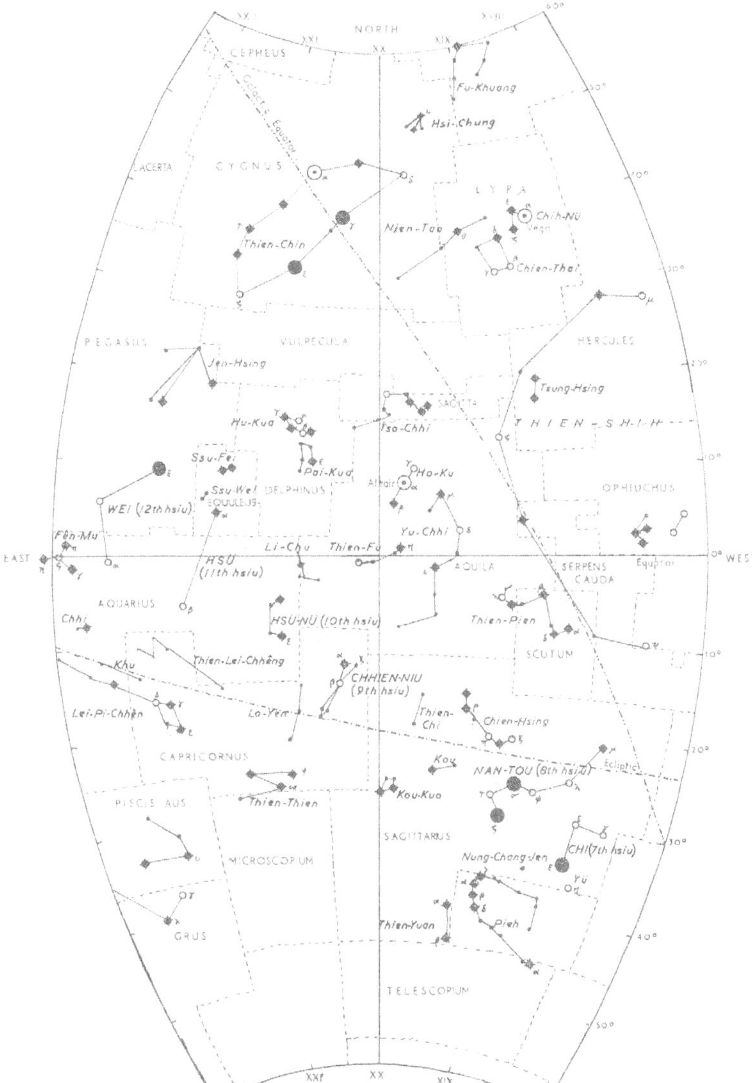
NORTH
SOUTH
EAST
WEST
CEPHEUS
LACERTA
CYGNUS
LYRA
PEGASUS
VULPECULA
HERCULES
SAGITTA
DELPHINUS
EQUULEUS
OPHIUCHUS
AQUILA
SERPENS CAUDA
AQUARIUS
SCUTUM
CAPRICORNUS
PISCIS AUS
MICROSCOPIUM
SAGITTARIUS
GRUS
TELESCOPIUM
Galactic Equator
Equator
Ecliptic
Fu-Khuang
Hsi-Chung
Chih-Nü
Vega
Njen-Tao
Chien-Thai
Thien-Chin
Jen-Hsing
Tsung-Hsing
THIEN-SHIH
Hu-Kua
Tso-Chhi
Ssu-Fei
Pai-Kua
Ho-Ku
Altair
Ssu-Wei
WEI (12th hsiu)
Fên-Mu
Yu-Chhi
Li-Chu
Thien-Fu
HSÜ (11th hsiu)
Chhi
HSÜ-NÜ (10th hsiu)
Thien-Pien
Thien-Lei-Chhêng
Khu
CHHIEN-NIU (9th hsiu)
Lei-Pi-Chhên
Lo-Yen
Thien-Chi
Chien-Hsing
Kou
NAN-TOU (8th hsiu)
Thien-Thien
Kou-Kuo
Nung-Chang-Jen
CHI (7th hsiu)
Yü
Thien-Yuan
Pieh
XXII
XXI
XX
XIX
XVIII
60°
50°
40°
30°
20°
10°
0°

STAR MAP 4

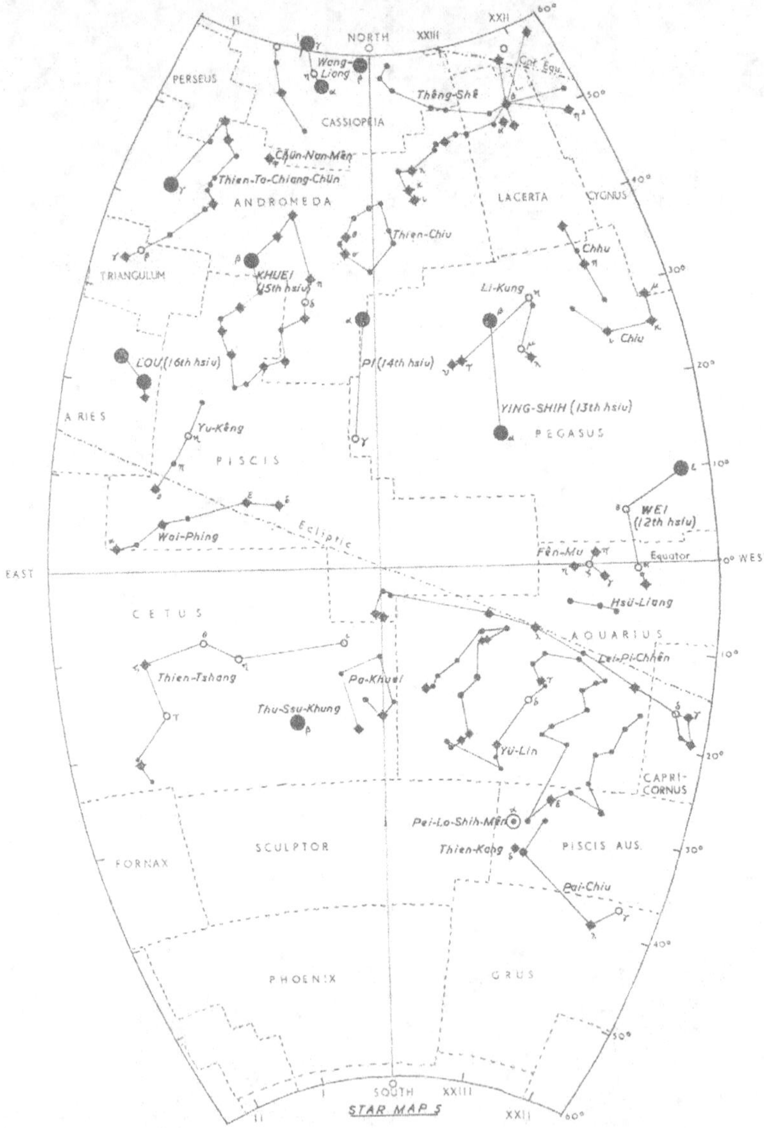
NORTH
PERSEUS
CASSIOPEIA
Wang-Liang
Thêng-Shê
Chün-Nan-Mên
Thien-Ta-Chiang-Chün
ANDROMEDA
LACERTA
CYGNUS
Thien-Chiu
Chhu
TRIANGULUM
KHUEI (15th hsiu)
Li-Kung
Chiu
LOU (16th hsiu)
PI (14th hsiu)
YING-SHIH (13th hsiu)
ARIES
Yu-Kêng
PISCIS
PEGASUS
WEI (12th hsiu)
Wai-Phing
Ecliptic
Fên-Mu
Equator
EAST
WEST
CETUS
Hsü-Liang
AQUARIUS
Lei-Pi-Chhên
Thien-Tshang
Pa-Khuei
Thu-Ssu-Khung
Yü-Lin
CAPRICORNUS
Pei-Lo-Shih-Mên
Thien-Kang
PISCIS AUS.
SCULPTOR
FORNAX
Pai-Chiu
PHOENIX
GRUS
SOUTH

STAR MAP 5

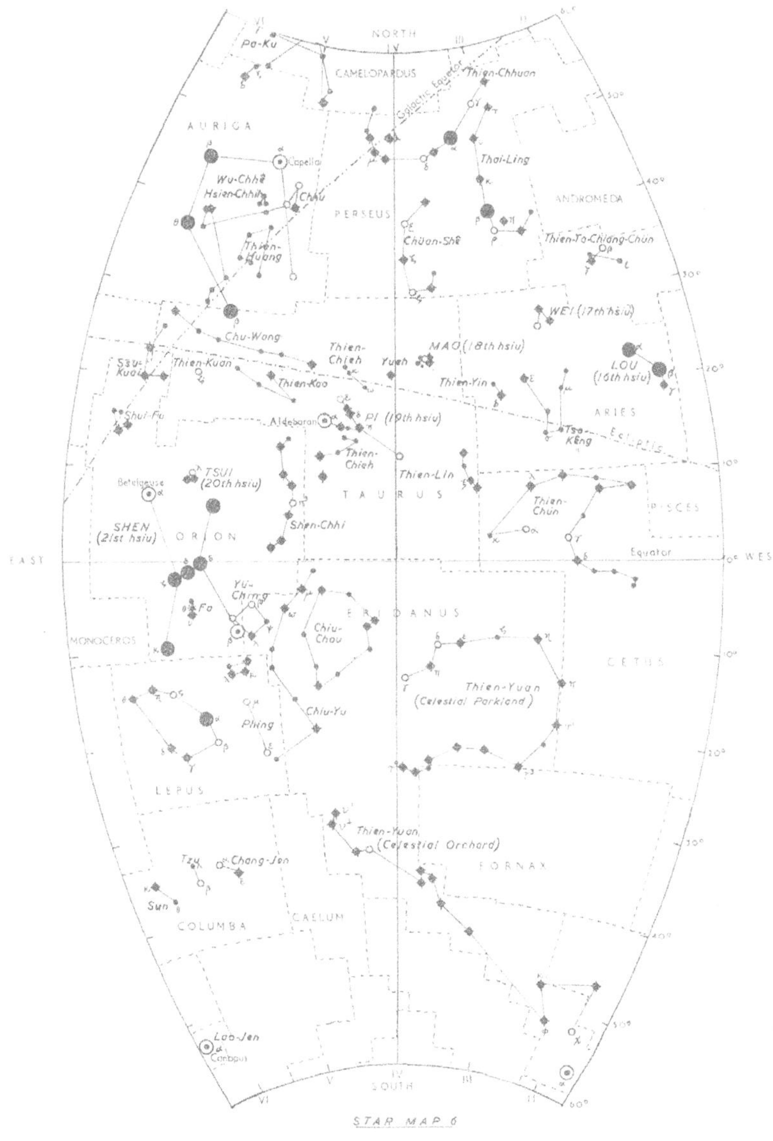

STAR MAP 6

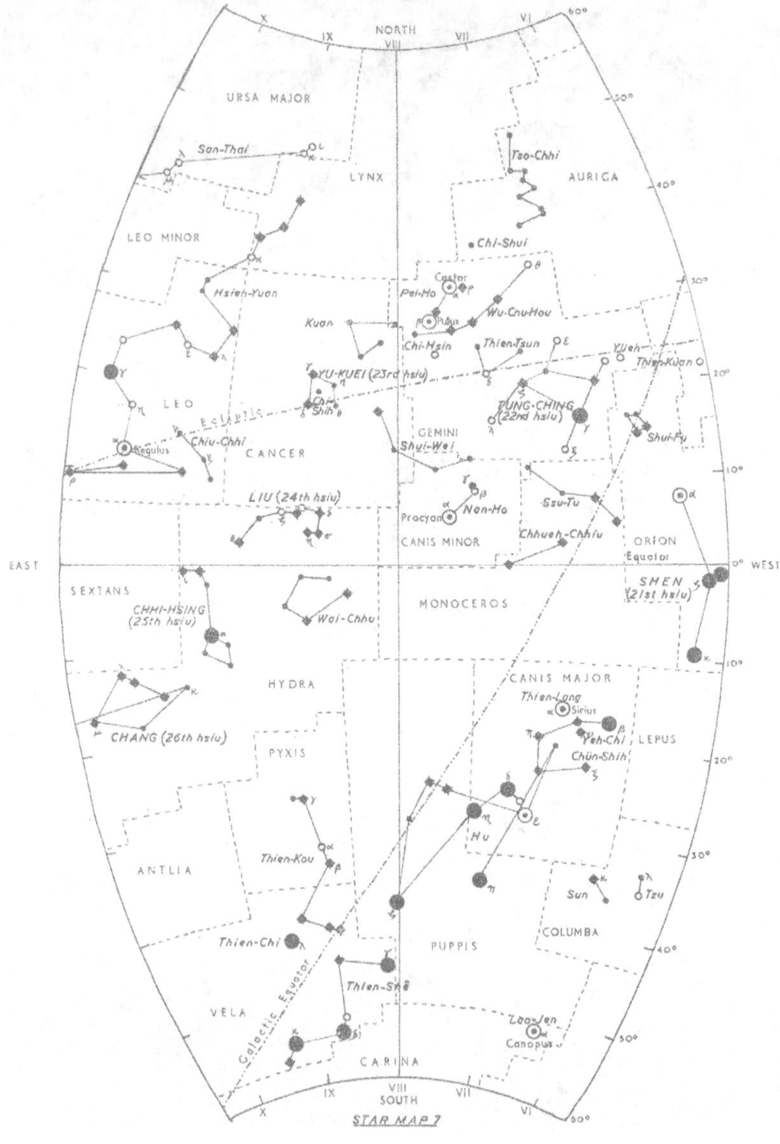
NORTH
SOUTH
EAST
WEST
URSA MAJOR
LYNX
AURIGA
LEO MINOR
LEO
CANCER
GEMINI
CANIS MINOR
ORION
SEXTANS
MONOCEROS
HYDRA
CANIS MAJOR
LEPUS
PYXIS
ANTLIA
COLUMBA
PUPPIS
VELA
CARINA
San-Thai
Tso-Chhi
Chi-Shui
Hsien-Yuan
Castor
Pei-Ho
Pollux
Wu-Chu-Hou
Kuan
Chi-Hsin
Thien-Tsun
Yüeh
Thien-Kuan
YU-KUEI (23rd hsiu)
TUNG-CHING (22nd hsiu)
Shui-Fu
Ecliptic
Regulus
Chiu-Chhi
Shui-Wei
LIU (24th hsiu)
Procyon
Nan-Ho
Ssu-Tu
Chhueh-Chhiu
Equator
SHEN (21st hsiu)
CHHI-HSING (25th hsiu)
Wai-Chhu
Thien-Lang
Sirius
Yeh-Chi
Chün-Shih
CHANG (26th hsiu)
Hu
Thien-Kou
Sun
Tsu
Thien-Chi
Thien-Shê
Galactic Equator
Lao-Jen
Canopus

STAR MAP 7

Appendix E

Original Chinese, Japanese, and Korean Records

Part I. Comets 彗星记录

约公元前1059 ［中国］
西周武王武王伐纣，...彗星出，而授殷人其柄。时有彗星，柄在东方，可以扫西人也！
《淮南子》卷15
注：哈雷彗星纪录

公元前974年2月—4月 ［中国］
西周昭王十九年春，有星孛于紫微。
《竹书纪年》

公元前636—628年 ［中国］
晋文公将与楚成王战于城濮，彗星出楚。楚操其柄，以问咎犯，咎犯对曰："以彗斗，倒之者胜"
《论衡》卷5

公元前632年 ［中国］
晋文公将与楚成王战于城濮，彗星出楚，楚操其柄。
《论衡》卷5

公元613年 ［中国］
(a) 春秋鲁文公十四年(613 BC)彗星入北斗。
《史记·十二诸候年表》卷14
(b) 春秋鲁文公十四年七月(613 BC. 6.7—7.6)，有星孛入于北斗。
《汉书·五行志》卷27

公元前525年　[中国]
(a)　春秋鲁昭公十七年彗星见辰。
《史记·十二诸侯年表》卷14
(b)　春秋鲁昭公十七年冬，有星孛于大辰。
《汉书·五行志》卷27

公元前516年　[中国]
(a)　春秋齐景公三十二年，彗星见。
《史记·十二诸侯年表》卷14
(b)　春秋齐景公三十二年，彗星见。…景公曰："彗星出东北，当齐分野…"
《史记·齐太公世家》卷32

公元前500年　[中国]
秦惠公元年，彗星见。
《史记·十二诸侯年表》卷14

公元前482年10月—12月　[中国]
(a)　春秋鲁哀公十三年，冬十有一月，有星孛于东方。
《春秋左传·哀公十三年》
(b)　春秋鲁哀公十三年十一月，有星孛于东方。
《汉书·五行志》卷27

公元前481年冬　[中国]
春秋鲁哀公十四年冬，有星孛。
《汉书 五行志 》卷27
《春秋左传·哀公十四年》

公元前469年　[中国]
秦厉公七年，彗星见。
《史记·六国年表》卷15

公元前467年　[中国]
战国秦厉共公十年，彗星见。
《史记· 六国年表 》卷15

公元前433年 [中国]
周考王八年，彗星见。
《文献通考》卷286

公元前361年 [中国]
(a) 战国魏惠王十年，彗星见。
《史记·魏世家》卷44
(b) 战国秦孝公元年，彗星见西方。
《史记·六国年表》卷15

公元前305年 [中国]
秦昭公二年，彗星见。
《史记·六国年表》卷15
《史记·秦本记》卷5

公元前303年 [中国]
秦昭公四年，彗星见。
《史记·六国年表》卷15
《史记·秦本记》卷5

公元前296年 [中国]
秦昭公十一年，彗星见。
《史记·六国年表》卷15
《史记·秦本记》卷5

公元前240年 [中国]
(a) 战国秦始皇七年，彗星先出东方，见北方，五月见西方。...彗星复见西方十六日。
《史记·秦始皇本记》卷6
(b) 战国秦始皇七年，彗星见北方西方。
《史记·六国年表》卷15

公元前238年 [中国]
(a) 战国秦始皇九年，彗星见，或竟天。四月...彗星见西方，又见北方，从斗以南八十日。

《史记·秦始皇本纪》卷6

(b) 战国秦始皇九年，彗星见，竟天。嫪毐为乱，迁其舍人于蜀；彗星复见。

《史记·六国年表》卷15

公元前234年2月19日—3月20日 [中国]
战国秦始皇十三年正月，彗星见东方。
《史记·秦始皇本纪》卷6

公元前214年 [中国]
战国秦始皇三十三年，明星出西方。[集解]徐广曰："皇甫谧云彗星见。"
《史记·秦始皇本纪》卷6

公元前204年8月14日—9月12日 [中国]

(a) 汉高祖三年七月，有星孛于大角。

《汉书·高帝纪》卷1

(b) 汉高祖三年七月，有星孛于大角，旬余乃入。

《汉书·五行志》卷27

公元前172年 [中国]
汉文帝前元八年，有长星出于东方。
《汉书·文帝纪》卷4

公元前162年2月6日 [中国]
汉文帝孝文后元二年正月壬寅，天欃夕出西南。
《汉书·天文志》卷26

公元前157年10月2日—31日 [中国]

(a) 汉文帝后元七年九月，有星孛于西方。

《汉书·景帝纪》卷5

(b) 汉文帝后元七年九月，有星孛于西方，其本直尾、箕，末指虚、危，长丈余，及天汉，十六日不见。

《汉书·五行志》卷27

公元前155年9月11日—10月10日 [中国]
汉景帝前元二年八月，彗星出东北。
《史记·景帝纪》卷11

公元前155年11月20日—前154年11月8日 [中国]
(a) 汉景帝前元二年冬十二月（公元前154年1月18日—2月16日），有星孛于西南。
《汉书·景帝纪》卷5
(b) 汉景帝前元二年（公元前155年11月20日-前154年11月8日），是岁彗星出西 南。
《汉书·天文志》卷26

公元前154年2月27日 [中国]
汉景帝前元三年正月乙巳，长星出西方。
《史记·孝景本纪》卷11

公元前148年4月29日—5月27日 [中国]
汉景帝中元二年四月，有星孛于西北。
《汉书·景帝纪》卷5

公元前147年4月18日—5月17日 [中国]
汉景帝中元三年三月，彗星出西北。
《史记·景帝纪》卷11

公元前147年5月13日 [中国]
汉景帝中元三年三月丁酉，彗星夜见西北，色白，长丈，在觜觿，且去益小，十五日不见。
《汉书·天文志》卷26

公元前147年8月6日 [中国]
汉景帝中元三年六月壬戌，蓬星见西南，在房南，去房可二丈，大如二斗器，色白；癸亥，在心东北，可长丈所；甲子，在尾北，可六丈；丁卯，在箕北，近汉，稍小，且去时，大如桃。壬申去，凡十日。

《汉书·天文志》卷26
《西汉会要》卷28

公元前147年10月12日—11月10日 [中国]
汉景帝中元三年九月，有星孛于西北。
《汉书·景帝纪》卷5

公元前138年4月9日—5月8日 [中国]
汉武帝建元三年三月，有星孛于注、张，历太微，干紫宫，至于天汉。
《汉书·天文志》卷26

公元前138年5月9日—6月6日 [中国]
汉武帝建元三年四月，有星孛于天纪，至织女。
《汉书·天文志》卷26

公元前138年8月5日—9月3日 [中国]
汉武帝建元三年七月，有星孛于西北。
《汉书·武帝纪》卷6

公元前137年9月22日—10月20日 [中国]
汉武帝建元四年九月，有星孛于东北。
《汉书·武帝纪》卷6

公元前135年7月3日—8月1日 [中国]
汉武帝建元六年六月，有星孛于北方。
《汉书·五行志》卷27

公元前135年8月31日—9月29日 [中国]
(a) 汉武帝建元六年八月，有星孛于东方，长竟天。
《汉书·武帝纪》卷6
(b) 汉武帝建元六年八月，长星出于东方，长终天，三十日去。
《汉书·五行志》卷27

公元前120年2月—4月 [中国]
汉武帝元狩三年春，有星孛于东方。

《汉书·武帝纪》卷6

公元前119年2月—4月　［中国］
汉武帝元狩四年春，有星孛于东北。
《汉书·武帝纪》卷6

公元前119年5月8日—6月6日　［中国］
汉武帝元狩四年四月，长星又出西北。
《汉书·五行志》卷27

公元前110年5月29日—6月26日　［中国］
汉武帝元封元年五月，有星孛于东井，又孛于三台。
《汉书·五行志》卷27

公元前110年8月—10月　［中国］
(a) 汉武帝元封元年秋，有星茀于东井。后十余日，有星茀于三能。
《史记·封禅书》卷28
(b) 汉武帝元封元年秋，有星孛于东井，又孛于三台。
《汉书·武帝纪》卷6
(c) 汉武帝元封元年秋，有星孛于东井。后十余日，有星孛于三能。
《汉书·郊祀志》卷25

公元前108—107年　［中国］
汉武帝元封中，星孛于河戍，
《汉书·天文志》卷26

公元前102年　［中国］
太初中，星孛于招摇。
《汉书·天文志》卷26

公元前87年 8.10—9.8　［中国］
汉昭帝后元二年七月，有星孛于东方。
《汉书·昭帝纪》卷7

公元前84年3月12日—4月10日 [中国]
汉昭帝始元三年二月，有星孛于西北。
《汉书·昭帝纪》卷7

公元前82年—前83年 [中国]
汉昭帝始元中，汉宦者梁成恢及燕王候星者吴莫如，见蓬星出西方天市东门，行过河鼓，入营室中。
《汉书·天文志》卷26

公元前69年1月27日—2月24日 [中国]
(a) 汉宣帝地节元年正月，有星孛于西方。
《汉书·宣帝纪》卷8
(b) 汉宣帝地节元年正月，有星孛于西方，去太白二丈所。
《汉书·五行志》卷27

公元前69年7月23日 [中国]
汉宣帝地节元年六月戊戌甲夜，客星又居左右角间，东南指，长可二尺，色白。
《汉书·天文志》卷26

公元前69年8月20日 [中国]
汉宣帝地节元年六月丙寅，又有客星见贯索东北，南行，至七月癸酉夜入天市，芒炎东南指，其色白。
《汉书·天文志》卷26

公元前61年7月24日—8月21日 [中国]
汉宣帝神爵元年六月，有星孛于东方。
《汉书·宣帝纪》卷8

公元前49年4月14日—5月12日 [中国]
(a) 汉宣帝黄龙元年三月，有星孛于王良、阁道，入紫宫。
《汉书·宣帝纪》卷8
(b) 汉宣帝黄龙元年三月，客星居王良东北可九尺，长丈余，西指，出阁道间，至紫宫。
《汉书·天文志》卷26

公元前49年4月14日—5月12日 [朝鲜]
新罗始祖九年春三月，有星孛于王良。
《三国史记》卷1
《增补文献备考》卷6

公元前47年6月20日—7月18日 [中国]
汉元帝初元二年五月，客星见昴分，居卷舌东可五尺，青白色，炎长三寸。
《汉书·天文志》卷26

公元前44年3月25日 [朝鲜]
新罗始祖十四年春二月己酉，星孛于河鼓。
《增补文献备考》卷6

公元前44年5月18日—6月16日 [中国]
汉元帝初元五年四月，彗星出西北，赤黄色，长八尺所，后数日长丈余，东北指，在参分。
《汉书·天文志》卷26

公元前44年5月18日—6月16日 [朝鲜]
新罗始祖十四年夏四月，有星孛于参五。
《三国史记》卷1
《增补文献备考》卷6

公元前32年2月6日—3月7日 [中国]
(a) 汉成帝建始元年正月，有星孛于营室。
《汉书·成帝纪》卷10
(b) 汉成帝建始元年正月，有星孛于营室，青白色，长六七丈，广尺余。
《汉书·五行志》卷27

公元前12年8月26日 [中国]
汉成帝元延元年七月辛未，有星孛于东井，践五诸侯，出河戍北率行轩辕、太微，后日六度有余，晨出东方。十三日夕见西方，犯次妃，长秋，斗，填，蠭炎再贯紫宫中。大火当后，达天河，除于妃后之域。南逝度犯大角、摄提，至天市而按节徐行，炎入市，中旬而后西去，五十六日与仓龙俱伏。
《汉书·五行志》卷27

公元前5年1月10日—2月7日 [中国]
汉哀帝建平元年十二月，白气出西南，从地上至天，出参下，贯天厕，广如一疋布，长十余丈，十余日去。
《汉书·天文志》卷26

公元前5年3月9日—4月6日 [中国]
汉哀帝建平二年二月，彗星出牵牛，七十余日。
《汉书·天文志》卷26

公元前4年4月24日 [中国]
汉哀帝建平三年三月己酉，有星孛于河鼓。
《汉书·哀帝纪》卷11

公元22年11月13日—12月12日 [中国]
(a) 新王莽地皇三年十一月，有星孛于张。
《后汉书·光武帝纪》卷1
(b) 新王莽地皇三年十一月，有星孛于张，东南行五日不见。
《后汉书·天文志》卷20
(c) 新王莽地皇三年十一月，有星孛于张，东南行，五日不见。
《汉书·王莽传》卷99

公元39年3月13日 [中国]
(a) 汉光武帝建武十五年正月丁未，有星孛于昴。有星孛于营室。
《后汉书·光武帝纪》卷1
(b) 汉光武帝建武十五年正月丁未，彗星见昴，稍西北行入营室，犯离宫，三月乙未，至东壁灭，见四十九日。
《后汉书·天文志》卷20
注：《东汉会要》卷16作"二月乙未"。

公元46年12月17日—47年1月15日 [朝鲜]
高句丽闵中王三年冬十一月，星孛于南，二十日而灭。
《增补文献备考》卷6
《三国史记》卷14

公元54年2月27日—3月27日 [朝鲜]
新罗儒理王三十一年春二月，星孛于紫宫。
《三国史记》卷1
《增补文献备考》卷6

公元54年4月29日 [中国]
汉光武帝建武三十年闰月癸丑，有星孛于紫宫。
《后汉书·光武帝纪》卷1

公元54年 [中国]
汉光武帝建武三十年闰月甲午，水在东井二十度，生白气，东南指，炎长五尺，为彗，东北行，至紫宫西藩止。五月甲子不见，凡见三十一日。
《后汉书·天文志》卷20
注：是年闰三月，无甲午。

公元55年12月6日 [中国]
汉光武帝建武三十一年十月己亥，七星间有客星，炎二尺所，西南行，至明年二月二十二日，在舆鬼东北六尺所灭，凡见百一十三日。
《后汉书·天文志》卷20

公元59年6月30日—7月28日 [朝鲜]
新罗脱解王三年夏六月，有星孛于天船。
《三国史记》卷1
《增补文献备考》卷6

公元60年8月9日 [中国]
(a) 汉明帝永平三年六月丁卯，有星孛于天船北。
《后汉书·明帝纪》卷2

(b) 汉明帝永平三年六月丁卯，彗星出天船北，长二尺所，稍北行至亢南，（百）[见]三十五日去。

《后汉书·天文志》卷21

注："伏侯古今注曰："彗长三尺所，见三十五日乃去"。《东汉会要》作百三十五日去。

公元61年9月27日 ［中国］

汉明帝永平四年八月辛酉，客星出梗河，西北指贯索，七十日去。

《后汉书·天文志》卷21

公元65年7月29日 ［中国］

汉明帝永平八年六月壬午，长星出柳、张三十七度，犯轩辕，刺天船，陵太微，气至上阶，凡见五十六日去。

《后汉书·天文志》卷21

注："陵"应为"凌"。《东汉会要》作"刺天舡"。

公元前66年2月20日 ［中国］

汉明帝永平九年正月戊申，客星出牵牛，长八尺，历建星至房南（《古今注》曰："历斗、建、箕、房、过角、亢至翼，芒东指。"），灭见至五十日。

《后汉书·天文志》卷21

公元71年3月6日 ［中国］

汉明帝永平十四年正月戊子，客星出昴，六十日，在轩辕右角稍灭。

《后汉书·天文志》卷21

公元75年7月14日 ［中国］

(a) 汉明帝永平十八年六月己未，有星孛于太微。

《后汉书·明帝纪》卷2

(b) 汉明帝永平十八年六月己未，彗星出张，长三尺，转在郎将，南入太微，皆属张。

《后汉书·天文志》卷21

公元76年10月7日 ［中国］

(a) 汉章帝建初元年八月庚寅，有星孛于天市。

《后汉书·章帝纪》卷3

(b) 汉章帝建初元年八月庚寅，彗星出天市，长二尺所，稍行入牵牛三度，积四十日稍灭。

《后汉书·天文志》卷21

公元78年1月18日 [中国]

(a) 汉章帝建初二年十二月戊寅，有星孛于紫宫。

《后汉书·章帝纪》卷3

(b) 汉章帝建初二年十二月戊寅，彗星出娄三度，长八九尺，稍入紫宫中，百六日稍灭。

《后汉书·天文志》卷21

注：《东汉会要》作元年。

公元79年2月22日—4月20日 [朝鲜]

新罗脱解王二十三年春二月，彗星见东方，又见北方，二十日乃灭。

《三国史记》卷1

《增补文献备考》卷6

公元85年5月20日 [中国]

汉章帝元和二年四月丁巳，客星晨出东方，在胃八度，长三尺，历阁道入紫宫，留四十日灭。

《后汉书·天文志》卷21

注：原文"元年四月"，按下文意应为"二年四月"。

《东汉会要》作元年。

公元101年1月12日 [中国]

汉和帝永元十二年十一月癸酉，夜有苍白气，长三丈，起天园，东北指军市，见积十日。

《后汉书·天文志》卷21

公元104年5月30日 [中国]

汉和帝永元十六年四月丁未，紫宫中生白气如粉絮。戊午，客星出紫宫西行至昴，五月壬申灭。

《后汉书·天文志》卷21

公元110年1月27日 [中国]
汉安帝永初三年十二月乙亥，有星孛于天苑。
《后汉书·安帝纪》卷5

公元110年1月9日—2月6日 [中国]
汉安帝永初三年十二月，彗星起天苑南，东北指，长六七尺，色苍白。
《后汉书·天文志》卷21

公元110年7月15日 [中国]
汉安帝永初四年六月甲子，客星大如李，苍白，芒气长二尺，西南指上阶星。
《后汉书·天文志》卷21
注：《东汉会要》作"六月丙子。"

公元117年1月9日 [中国]
汉安帝元初三年十一月甲午，客星见西方，己亥在虚、危，南至胃、昴。
《后汉书·天文志》卷21

公元126年3月23日 [中国]
汉安帝永建元年二月甲午，古今注曰："永建元年二月甲午，客星入太微……"《李氏家书》曰："时天有变气，李郃上书谏曰：……乃月十三日，有客星气象彗孛，历天市、梗河、招摇、枪棓。十六日入紫宫，迫北辰。十七日复过文昌、泰陵，至天船、积水间，稍微不见。"
《后汉书·天文志》卷21

公元128年9月12日—10月11日 [朝鲜]
新罗祇摩王十七年秋八月，长星竟天。
《三国史记》卷1
《增补文献备考》卷6

公元132年1月29日 [中国]
(a) 汉顺帝永建六年十二月壬申，客星出牵牛。

《后汉书·顺帝纪》卷6

(b) 汉顺帝永建六年十二月壬申，客星芒气长二尺余，西南指，色苍白，在牵牛六度。

《后汉书·天文志》卷21

公元132年 [中国]

汉顺帝阳嘉元年十二月戊子，客星出天苑。

《后汉书·顺帝纪》卷6

注：该月无戊子，疑系闰十月之误。

公元133年2月8日 [中国]

汉顺帝阳嘉元年闰月戊子，客星气白，广二尺，长五丈，起天苑西南。

《后汉书·天文志》卷21

公元133年2月9日 [中国]

汉顺帝阳嘉元年闰月十七日己丑，夜，有白气从西方天苑趋入玉井，数日乃灭。

《后汉书·列传》卷30

公元141年3月27日 [中国]

(a) 汉顺帝永和六年二月丁巳(3月27日)，彗星见东方，长六七尺，色青白，西南指营室及坟墓星。丁丑(4月16日)，彗星在奎一度，长六尺。癸未(Apr 22)昏见西北，历昴，毕。甲申(4月23日)，在东井，遂历舆鬼、柳、七星、张、光炎及三台，至轩辕中灭。

《后汉书·天文志》卷21

(b) 汉顺帝永和六年二月丁巳(3月27日)，有星孛于营室。

《后汉书·顺帝纪》卷6

公元149年9月20日—10月19日 [朝鲜]

新罗逸圣王十六年秋八月，有星孛于天市。

《三国史记》卷1

《增补文献备考》卷6

公元149年10月19日 [中国]

(a) 汉桓帝建和三年八月乙丑，有星孛于天市。

《后汉书·桓帝纪》卷7

(b) 汉桓帝建和三年八月乙丑，彗星芒长五尺，见天市中，东南指，色黄白，九月戊辰不见。

《后汉书·天文志》卷22

公元153年11月4日—12月3日 [朝鲜]

新罗逸圣王二十年冬十月，彗星见东方，又见东北方。

《增补文献备考》卷6

《三国史记》卷1

公元161年6月14日 [中国]

(a) 汉桓帝延熹四年五月辛酉，有星孛于心。

《后汉书·桓帝纪》卷7

(b) 汉桓帝延熹四年五月辛酉，客星在营室，稍顺行，生芒长五尺所，至心一度，转为彗。

《后汉书·天文志》卷22

公元178年8月31日—9月29日 [中国]

(a) 汉灵帝光和元年八月，有星孛于天市。

《后汉书·灵帝纪》卷8

(b) 汉灵帝光和元年八月，彗星出亢北，入天市中，长数尺，稍长至五六丈，赤色，经历十余宿，八十余日，乃消于天苑中。

《后汉书·天文志》卷22

公元180年10月—12月 [中国]

(a) 汉灵帝光和三年冬，有星孛于狼、弧。

《后汉书·灵帝纪》卷8

(b) 汉灵帝光和三年冬，彗星出狼、弧，东行至于张乃去。

《后汉书·天文志》卷22

注：《后汉书·灵帝纪》作三年冬闰月，查该年无闰月。

公元182年8月17日—9月15日　［中国］
(a) 汉灵帝光和五年七月，有星孛于太微。
《后汉书·灵帝纪》卷8
(b) 汉灵帝光和五年七月，彗星出三台下，东行入太微，至太子、幸臣，二十余日而消。
《后汉书·天文志》卷22

公元182年8月17日—9月15日　［朝鲜］
高句丽故国川王四年秋七月，星孛于太微。
《增补文献备考》卷6
《三国史记》卷16

公元186年10月31日—11月28日　［朝鲜］
百济肖古王二十一年冬十月，星孛于西北，二十日而灭。
《增补文献备考》卷6
《三国史记》卷23

公元188年3月16日—4月13日　［中国］
(a) 汉灵帝中平五年二月，有星孛于紫宫。
《后汉书·灵帝纪》卷8
(b) 汉灵帝中平五年二月，彗星出奎，逆行入紫宫，后三出，六十余日乃消。
《后汉书·天文志》卷22

公元188年7月28日　［中国］
汉灵帝中平五年六月丁卯，客星如三升碗，出贯索，西南行入天市，至尾而消。
《后汉书·天文志》卷22

公元191年10月6日 - 11月4日　［中国］
汉献帝初平二年九月，蚩尤旗见，长十余丈，色白，出角、亢之南。
《后汉书·天文志》卷22

公元193年12月3日　［中国］
(a) 汉献帝初平四年十月辛丑，有星孛于天市。

《后汉书·献帝纪》卷9

(b) 汉献帝初平四年十月，孛星出两角间，东北行入天市中而灭。

《后汉书·天文志》卷22

公元200年11月6日　［中国］

汉献帝建安五年十月辛亥，有星孛于大梁。

《后汉书·献帝纪》卷9

《后汉书·天文志》卷22

公元204年11月10日—12月9日　［中国］

汉献帝建安九年十月，有星孛于东井。

《后汉书·献帝纪》卷9

公元204年11月10日—12月9日　［朝鲜］

百济肖古王三十九年冬十月，星孛于东井。

《增补文献备考》卷6

《三国史记》卷23

公元204年12月10日—205年1月7日　［中国］

汉献帝建安九年十一月，有星孛于东井舆鬼，入轩辕太微。

《后汉书·天文志》卷22

公元206年1月27日—2月25日　［中国］

(a) 汉献帝建安十一年正月，有星孛于北斗。

《后汉书·献帝纪》卷9

(b) 汉献帝建安十一年正月，星孛于北斗，首在斗中，尾贯紫宫，及北辰。

《后汉书·天文志》卷22

公元207年11月10日　［中国］

汉献帝建安十二年十月辛卯，有星孛于鹑尾。

《后汉书·献帝纪》卷9

《后汉书·天文志》卷22

公元213年1月10日—2月7日　［中国］

汉献帝建安十七年十二月，星孛于五诸侯。

《后汉书·献帝纪》卷9
《后汉书·天文志》卷22

公元217年10月—12月 ［中国］
汉献帝建安二十二年冬，有星孛于东北。
《后汉书·献帝纪》卷9

公元217年11月17日—12月15日 ［朝鲜］
高句丽山上王二十一年冬十月，星孛于东北。
《三国史记》卷16
《增补文献备考》卷6

公元218年4月13日—5月12日 ［中国］
(a) 汉献帝建安二十三年三月，有星孛于东方。
《后汉书·献帝纪》卷9
(b) 汉献帝建安二十三年三月，星孛晨见东方二十余日，夕出西方，犯历五车、东井、五诸候、文昌、轩辕、后妃、太微、锋炎指帝座。
《后汉书·天文志》卷22

公元225年12月9日 ［中国］
魏文帝黄初六年十月乙未，有星孛于少微，历轩辕。
《宋书·天文志》卷23

公元232年12月3日 ［中国］
魏明帝太和六年十一月丙寅，有星孛于翼，近太微上将星。
《三国志·魏书》卷3
《宋书·天文志》卷23

公元236年11月30日 ［中国］
(a) 魏明帝青龙四年十月甲申，有星孛于大辰，乙酉，又孛于东方。
《三国志·魏书·明帝纪》卷3
(b) 魏明帝青龙四年十月甲申，有星孛于大辰，长三尺。乙酉，又孛于东方。

《宋书·天文志》卷23
《晋书·天文志》卷13

公元236年 [中国]
魏明帝青龙四年十一月己亥，彗星见，犯宦者天纪星。
《三国志·魏书·明帝纪》卷3
《宋书·天文志》卷23
注：十一月无己亥。

公元238年7月29日—8月27日 [中国]
魏明帝景初二年八月，彗星见张，长三尺，逆西行，四十一日灭。
《宋书·天文志》卷23

公元238年8月21日 [中国]
魏明帝景初二年八月癸丑，有彗星见张宿。
《三国志·魏书·明帝纪》卷3

公元238年9月30日 [中国]
魏明帝景初二年十月癸巳，客星见危，逆行在离宫北，腾蛇南。甲辰，犯宗星。己酉灭。
《宋书·天文志》卷23

公元240年11月10日 [中国]
魏齐王正始元年十月乙酉，彗星见西方，在尾，长三丈，拂牵牛，犯太白，十一月甲子，进犯羽林。
《宋书·天文志》卷23

公元245年9月18日 [中国]
魏齐王正始六年八月戊午，彗星见七星，长二尺，色白，进至张，积二十三日灭。
《宋书·天文志》卷23

公元246年 [中国]
魏齐王正始七年十一月癸亥，（彗星）又见轸，长一尺，积百五十六日灭。

《宋书·天文志》卷23
注：该月无癸亥。

公元248年4月11日—5月9日 [中国]
魏齐王正始九年三月，（彗星）又见昴，长六尺，色青白，芒西南指。
《宋书·天文志》卷23

公元248年8月7日—9月4日 [中国]
魏齐王正始九年七月，（彗星）又见翼，长二尺，进至轸，积四十二日灭。
《宋书·天文志》卷23

公元251年 [中国]
魏齐王嘉平三年十一月癸亥，有星孛于营室，西行积九十日灭。
《宋书·天文志》卷23
注：《宋书·天文志》作'十一月无癸未'《晋书·天文志》作'十一月癸亥'。

公元252年3月24日 [中国]
魏齐王嘉平四年二月丁酉，彗星见西方，在胃，长五六尺，色白，芒南指贯参，积二十日灭。
《宋书·天文志》卷23

公元253年12月8日—254年1月5日 [中国]
魏齐王嘉平五年十一月，彗星又见轸，长五丈，在太微左执法西，东南指，积百九十日灭。
《宋书·天文志》卷23

公元254年11月27日—12月26日 [中国]
(a) 吴会稽王五凤元年十一月，星茀于斗、牛。
《三国志·吴书》卷48
(b) 高贵乡公正元元年十一月，白气出南斗侧，广数丈，长竟天。
《宋书·天文志》卷23
《晋书·天文志》卷13

公元255年1月25日—2月23日 [中国]

(a) 魏高贵乡公正元二年正月，有彗星数十丈，西北竟天，起于吴楚之分。

《三国志·魏书·毋丘俭传》卷28

(b) 高贵乡公正元元年十一月，白气出南斗侧，广数丈，长竟天。王肃曰："蚩尤之旗也，东南其有乱乎！"二年正月，有彗星见于吴楚分，西北竟天。

《晋书·天文志》卷13

公元257年12月23日—258年1月21日 [中国]

魏高贵乡公甘露二年十一月，彗星见角，色白。

《宋书·天文志》卷23

公元259年11月23日 [中国]

魏高贵乡公甘露四年十月丁丑，客星见太微中，转东南行，历轸宿，积七日灭。

《宋书·天文志》卷23

注：《通志》作"四年"，"七十日灭"。

公元260年7月25日—8月23日 [朝鲜]

新罗沾解王十四年秋七月，星孛于东方，二十五日而灭。

《增补文献备考》卷6

《三国史记》卷2

公元262年12月2日 [中国]

魏陈留王景元三年十一月壬寅，彗星见亢，色白，长五寸，转北行，积四十五日灭。

《宋书·天文志》卷23

注：《三国会要》作"长五丈"。

公元265年6月1日—30日 [中国]

魏陈留王咸熙二年五月，彗星见王良，长丈余，色白，东南指，积十二日灭。

《宋书·天文志》卷23

公元268年2月18日 [中国]

(a) 晋武帝泰始四年正月丙戌，有星孛于轸。

《晋书·武帝纪》卷3

(b) 晋武帝泰始四年正月丙戌，彗星见轸，青白色，西北行，又转东行。

《宋书·天文志》卷23

《晋书·天文志》卷13

公元269年10月13日—11月10日 [中国]

晋武帝泰始五年九月，有星孛于紫宫。

《宋书·天文志》卷23

《晋书·武帝纪》卷3

《晋书·天文志》卷13

公元269年10月13日—11月10日 [朝鲜]

百济古尔王三十六年秋九月，星孛于紫宫。

《三国史记》卷24

《增补文献备考》卷6

公元275年1月14日—2月12日 [中国]

晋武帝泰始十年十二月，有星孛于轸。

《宋书·天文志》卷23

《晋书·武帝纪》卷3

《晋书·天文志》卷13

公元276年6月29日—7月27日 [中国]

(a) 晋武帝咸宁二年六月，星孛于氐……七月，星孛大角……八月，星孛太微，至翼，北 斗、三台。

《宋书·天文志》卷23

(b) 晋武帝咸宁二年六月甲戌，星孛于氐……七月，星孛大角……八月，星孛太微，至翼，北斗、三台。

《晋书·天文志》卷13

(c) 晋武帝咸宁二年六月甲戌，星孛于氐……七月，星孛大角……八月，星孛太微。九月 ，又孛于翼。

《晋书·武帝纪》卷3
注：六月无甲戌。

公元277年2月20日-6月17日 [中国]
(a) 晋武帝咸宁三年正月（2月20—3月21），星孛于西方。三月，星孛于胃……四月，星孛女御……五月，又孛于东方。七月，星孛紫宫。
《晋书·天文志》卷13
(b) 晋武帝咸宁三年四月（5月20—6月17），星孛女御……五月，又孛于东方。七月，星孛紫宫。
《宋书·天文志》卷23

公元278年5月9日—6月7日 [中国]
晋武帝咸宁四年四月，蚩尤旗见东井。
《晋书·天文志》卷13
《宋书·天文志》卷23

公元279年3月30日-4月27日 [中国]
(a) 晋武帝咸宁五年三月，星孛于柳。
《宋书·天文志》卷23
(b) 晋武帝咸宁五年三月，星孛于柳，四月，又孛于女御。
《晋书·武帝纪》卷3
《晋书·天文志》卷13

公元279年7月26日—8月24日 [中国]
晋武帝咸宁五年七月，星孛于紫宫。
《晋书·武帝纪》卷3
《晋书·天文志》卷13

公元281年9月1日—29日 [中国]
晋武帝太康二年八月，有星孛于张。
《宋书·天文志》卷23
《晋书·武帝纪》卷3
《晋书·天文志》卷13

公元281年11月28日—12月27日 [中国]
晋武帝太康二年十一月，有星孛于轩辕。
《宋书·天文志》卷23
《晋书·武帝纪》卷3
《晋书·天文志》卷13

公元283年4月22日 [中国]
晋武帝太康四年三月戊申，星孛于西南。
《宋书·天文志》卷23
《晋书·天文志》卷13

公元287年10月24日—11月22日 [中国]
晋武帝太康八年九月，有星孛于南斗，长数十丈，十余日灭。
《宋书·天文志》卷23
《晋书·天文志》卷13

公元295年5月1—30日 [中国]
晋惠帝元康五年四月，有星孛于奎，至轩辕、太微、经三台、大陵。
《宋书·天文志》卷24；《晋书·惠帝纪》卷4
《晋书·天文志》卷13

公元300年12月28日—301年1月25日 [中国]
(a) 晋惠帝永康元年十二月，彗出牵牛之西，指天市。
《晋书·天文志》卷13
《宋书·天文志》卷24
(b) 晋惠帝永康元年十二月，彗星见于东方。
《晋书·惠帝纪》卷4

公元301年5月24日—6月22日 [中国]
晋惠帝永康二年四月，彗星见齐分。
《晋书·天文志》卷13
《宋书·天文志》卷24

公元302年5月14日—6月11日　[中国]
晋惠帝太安元年四月，彗星昼见。
《晋书·惠帝纪》卷4
《晋书·天文志》卷13

公元303年4月3日—5月2日　[中国]
晋惠帝太安二年三月，彗星见东方，指三台。
《宋书·天文志》卷24
《晋书·天文志》卷13

公元302年5月14日—6月11日　[朝鲜]
百济汾西王五年夏四月，彗星昼见。
《三国史记》卷24
《增补文献备考》卷6

公元305年9月5日—10月4日　[中国]
晋惠帝永兴二年八月，有星孛于昴、毕。
《宋书·天文志》卷24
《晋书·天文志》卷13

公元305年11月21日　[中国]
晋惠帝永兴二年十月丁丑，有星孛于北斗。
《宋书·天文志》卷24
《晋书·惠帝纪》卷4
《晋书·天文志》卷13

公元329年8月11日—9月9日　[中国]
(a)　晋成帝咸和四年七月，有星孛于西北，二十三日灭。
《宋书·天文志》卷24
(b)　晋成帝咸和四年七月，有星孛于西北。
《晋书·成帝纪》卷7
(c)　晋成帝咸和四年七月，有星孛于西北，犯斗，二十三日灭。
《晋书·天文志》卷13

公元336年2月16日 [中国]
(a) 晋成帝咸康二年正月辛巳，彗星夕见西方，在奎。
《宋书·天文志》卷24
《晋书·天文志》卷13
(b) 晋成帝咸康二年正月辛巳，彗星见于奎。
《晋书·成帝纪》卷7

公元336年2月16日 [朝鲜]
百济比流王三十三年春正月辛巳，彗星见于奎。
《增补文献备考》卷6

公元340年3月25日 [中国]
晋成帝咸康六年二月庚辰，有星孛于太微。
《晋书·成帝纪》卷7
《晋书·天文志》卷13

公元343年12月8日
(a) 晋康帝建元元年十一月六日，彗星见亢，长七尺，尾白色。
《宋书·天文志》卷24
(b) 晋康帝建元元年十一月六日，彗星见亢，长七尺，白色。
《晋书·天文志》卷13

公元349年12月2日—350年3月30日 [中国]
(a) 晋穆帝永和五年十一月乙卯，彗星见于亢。芒西向，色白，长一丈。六年正月丁丑，彗星又见于亢。
《晋书·天文志》卷13
(b) 晋穆帝永和五年十一月乙卯，彗星见于亢。芒西向，色白，长一丈。
《宋书·天文志》卷24
(c) 晋穆帝永和六年闰月丁丑（3月30日），彗星见于亢。
《晋书·穆帝纪》卷8
注：(1) 十一月无乙卯，十一月己卯为12月2日。
(2) 六年闰二月。

公元358年6月26日 [中国]

(a) 晋穆帝升平二年五月丁亥，彗出天船，在胃度中。

《宋书·天文志》卷24

(b) 晋穆帝升平二年五月丁亥，彗星出天船，在胃。

《晋书·天文志》卷13

公元363年8月26日—9月23日 [中国]

(a) 晋哀帝兴宁元年八月，星孛大角、亢，入天市。

《宋书·天文志》卷24

(b) 晋哀帝兴宁元年八月，有星孛于角、亢，入天市。

《晋书·哀帝纪》卷8

《晋书·天文志》卷13

公元374年3月4日

(a) 晋孝武帝宁康二年二月丁巳(374年3月4日)，有星孛于女虚。三月丙戌(4月2日)，彗星见于氐。

《晋书·孝武帝纪》卷9

(b) 晋孝武帝宁康二年正月丁巳（注：正月无丁巳，系二月之误。）

(374年3月4日)，有星孛于女虚，经氐、亢、角、轸、翼、张。

《宋书·天文志》卷25

(c) 晋孝武帝宁康二年正月（注：系二月之误）丁巳(374年3月4日)，有星孛于女虚，经氐、亢、角、轸、翼、张。至三月丙戌(4月2日)，彗星见于氐…九月丁丑(11月19日)，有星孛于天市。

《晋书·天文志》卷13

公元390年7月28日—8月26日 [朝鲜]

百济辰斯王六年秋七月，星孛于北河。

《增补文献备考》卷6

公元390年8月7日 [中国]

晋孝武帝太元十五年七月丁巳，有星孛于北河。

《晋书·孝武帝纪》卷9

公元390年8月22日 ［中国］
(a) 晋孝武帝太元十五年七月壬申，有星孛于北河戒，经太微、三台、文昌，入北斗，长十余丈。八月戊戌，入紫微，乃灭。
《宋书·天文志》卷25
(b) 晋孝武帝太元十五年七月壬申，有星孛于北河戍，经太微、三台、文昌，入北斗，色白,长十余丈。八月戊戌，入紫宫乃灭。
《晋书·天文志》卷13

公元395年3月8日—4月5日 ［朝鲜］
百济阿莘王四年春二月，星孛于西北，二十日而灭。
《增补文献备考》卷6
《三国史记》卷25

公元395年9月30日—10月29日 ［中国］
晋孝武帝太元二十年九月，有蓬星如粉絮，东南行，历女虚及哭星。
《宋书·天文志》卷25
《晋书·天文志》卷13
注：《晋书·天文志》作“至哭星”。

公元396年7月22日—8月19日 ［中国］
北魏太祖皇始元年六月，有星彗于髦头。
《魏书·天象志》卷105

公元400年3月12日—4月9日 ［朝鲜］
百济阿莘王九年春二月，星孛于奎娄。
《增补文献备考》卷6
《三国史记》卷25

公元400年3月19日 ［中国］
(a) 晋安帝隆安四年二月己丑，有星孛于奎，长三丈，上至阁道紫宫西蕃，入斗魁，至三台、太微、帝座、端门。
《宋书·天文志》卷25

(b) 晋安帝隆安四年二月己丑，有星孛于奎、娄，进至紫微。
《晋书·安帝纪》卷10
(c) 晋安帝隆安四年二月己丑，有星孛于奎，长三丈，上至阁道，紫宫西蕃，入北斗魁，至三台。三月，遂经于太微帝座端门。
《晋书·天文志》卷13

公元400年4月10日—5月9日 ［中国］
北魏太祖天兴三年三月，有星孛于奎，历阁道，至紫微西蕃，入北斗魁，犯太阳守，循下台，
轥南宫，履帝座，遂由端门以出。
《魏书·天象志》卷105

公元400年4月10日—5月9日 ［中国］
晋安帝隆安四年三月，彗星见于太微。
《晋书·安帝纪》卷10

公元401年1月2日 ［中国］
(a) 晋安帝隆安四年十二月戊寅，有星孛于贯索、天市、天津。
《宋书·天文志》卷25
《晋书·天文志》卷13
(b) 晋安帝隆安四年十二月戊寅，有星孛于天市。
《晋书·安帝纪》卷10

公元402年11月11日—12月10日 ［中国］
晋安帝元兴元年十月，有客星色白如粉絮，在太微西，至十二月入太微。
《宋书·天文志》卷25
《晋书·天文志》卷13

公元415年6月21日—24日 ［中国］
(a) 北魏太宗神瑞二年四月辛巳（6月21日），有星孛于天市。五月甲申，彗星出天市，扫帝座，在房心北。
《魏书·天象志》卷105

(b) 晋安帝义熙十一年五月甲申（6月24日），彗星出天市，扫帝座，在房、心。

《宋书·天文志》卷25

(c) 晋安帝义熙十一年五月甲申（6月24日），彗星二出天市，扫帝座，在房心北。

《晋书·天文志》卷13

公元415年8月8日 ［中国］

北魏太宗神瑞二年六月己巳，有星孛于昴南。

《魏书·天象志》卷105

公元418年6月24日—7月6日 ［中国］

(a) 晋安帝义熙十四年五月庚子（6月24日），有星孛于北斗魁中。七月癸亥，彗星出太微西，柄起上相星下，芒渐长至十余丈，进扫北斗、紫微、中台。

《晋书·天文志》卷13

(b) 晋安帝义熙十四年五月壬子（7月6日），有星孛于北斗魁中……七月…癸亥，彗星出太微西，柄起上相星下，芒渐长至十余丈，进扫北斗、紫微、中台。

《宋书·天文志》卷25

公元418年10月16日—11月13日 ［中国］

北魏太宗泰常三年九月，长彗星孛于北斗，轹紫微。辛酉，入南宫，凡八十余日。

《魏书·天象志》卷105

公元419年1月12日—2月10日 ［中国］

北魏太宗泰常三年十二月，彗星出自天津，入太微，迳北斗，干紫宫，犯天棓，八十余日，及天汉乃灭。

《魏书·天象志》卷105

《魏书·崔浩传》卷35

公元419年2月17日 ［中国］

晋恭帝元熙元年正月戊戌，有星孛于太微西藩。

《晋书·恭帝纪》卷10
《晋书·天文志》卷13

公元419年2月17日 ［朝鲜］
百济腆支王十五年春正月戊戌，星孛于太微。
《增补文献备考》卷6

公元422年3月16日—3月26日 ［中国］
(a) 北魏太宗泰常七年二月辛巳（3月16日），有星孛于虚、危，向河津。
《魏书·天象志》卷105
(b) 南朝宋武帝永初三年二月丙戌（3月21日），有星孛于虚危。
《南史·宋本纪》卷1
(c) 南朝宋武帝永初三年二月辛卯（3月26日），有星孛于虚危，向河津，扫河鼓。
《宋书·天文志》卷26

公元422年12月14日 ［中国］
北魏太宗泰常七年十一月甲寅，彗星出室，扫北斗，及于口门。
《魏书·天象志》卷105
注：按位置推断，“口门”应是“垣（紫微）门”。

公元422年12月18日 ［中国］
(a) 南朝宋武帝永初三年十一月戊午，有星孛于营室。
《宋书·少帝纪》卷4
《南史·宋本纪》卷1
(b) 南朝宋武帝永初三年十一月戊午，有星孛于室壁。
《宋书·天文志》卷26

公元423年1月28日—2月25日 ［中国］
北魏太宗泰常八年正月，彗星出奎南，长三丈，东南扫河。
《魏书·天象志》卷105

公元423年2月13日 ［中国］
(a) 南朝宋少帝景平元年正月乙卯，有星孛于东壁。

《宋书·少帝纪》卷4
《南史·宋本纪》卷1
(b) 南朝宋少帝景平元年正月乙卯，有星孛于东壁南，白色，长二丈余，拂天苑，二十日灭。
《宋书·天文志》卷26

公元423年12月13日—14日 [中国]
(a) 南朝宋少帝景平元年十月戊午（12月13日），有星孛于氐北，尾长四丈，西北指，贯摄提，向大角，东行，日长六七尺，十余日灭。
《宋书·天文志》卷26
(b) 南朝宋少帝景平元年十月己未（12月14日），有星孛于氐，指尾，贯摄提，向大角，仲月在危，季月扫天仓而后灭。
《宋书·少帝纪》卷4

公元423年12月19日—424年1月16日 [中国]
北魏太宗泰常八年十一月，彗星孛于土司空。
《魏书·天象志》卷105

公元435年6月12日—7月11日 [中国]
北魏太武帝太延元年五月，彗出轩辕。
《魏书·天象志》卷105

公元436年6月21日 [中国]
北魏太武帝太延二年五月壬申，有星孛于房。
《魏书·天象志》卷105

公元442年10月20日—11月18日 [中国]
(a) 南朝宋文帝元嘉十九年九月丙辰（11月1日），有客星在北斗，因为彗，入文昌，贯五车扫毕，拂天节，经天苑，冬季乃灭。
《南史·宋本纪》卷2
(b) 北魏太武帝太平真君三年九月乙丑（11月10日），有星孛于天牢，入文昌、五车，经昴、毕之间，至天苑，百余日与宿俱入西方。

《魏书·天象志》卷105

(c) 南朝宋文帝元嘉十九年九月（10月20日—11月18日），客星见北斗，渐为彗星，至天苑末灭。

《宋书·天文志》卷26

公元449年6月7日—7月5日 [中国]

北魏太武帝太平真君十年五月，彗星出于昴北。

《魏书·天象志》卷105

公元449年11月1日—30日 [中国]

(a) 南朝宋文帝元嘉二十六年十月癸卯（11月11日），彗星见于太微。

《南史·宋本纪》卷2

(b) 南朝宋文帝元嘉二十六年十月（11月1日-30日），彗星入太微。

《宋书·天文志》卷26

(c) 北魏太武帝太平真君十年十月辛巳，彗星见于太微。

《魏书·天象志》卷105

注：十月无辛巳。

公元451年 [中国]

(a) 南朝宋文帝元嘉二十八年四月己卯(6月10日)，彗星见于昴。五月壬子(7.13)，彗星见太微中，对帝座。

《南史·宋本纪》卷2

(b) 北魏太武帝正平元年五月(6月15日—7月14日)，彗星见卷舌，入太微。六月辛酉(7.22)，彗星进逼帝座；七月乙酉(8月15日)，犯上相，拂屏，出端门，灭于翼轸。

《魏书·天象志》卷105

(c) 南朝宋文帝元嘉二十八年五月(6月15日—7月14日)，彗星见卷舌，入太微，逼帝座，犯上相，拂屏，出端门，灭翼、轸。

《宋书·天文志》卷26

公元453年2月24日—3月25日 [中国]

北魏文成帝高宗兴安二年二月，有星孛于西方。

《魏书·天象志》卷105

公元454年2月14日—455年2月2日 [朝鲜]
百济毗有王二十八年，星孛于西北，长二丈许。
《增补文献备考》卷6

公元460年10月31日—11月28日 [中国]
北魏文成帝和平元年十月，有长星出于天仓，长丈余。
《魏书·天象志》卷105

公元461年4月20日 [中国]
二年三月辛巳，有长星出天津，色赤，长匹余，灭而复出，大小百数。
《魏书·天象志》卷105

公元464年12月15日—465年1月12日 [中国]
北魏文成帝和平五年十一月，长星出织女，色正白，彗之象也。
《魏书·天象志》卷105

公元579年11月5日 —12月3日 [朝鲜]
百济威德王二十六年冬十月，长星竟天，二十日而灭。
《增补文献备考》卷6
《三国史记》卷27

公元498年11月29日—12月28日 [中国]
北魏孝文帝太和二十二年十一月，彗星起轩辕，历鬼南，及天汉。
《魏书·天象志》卷105

公元501年4月14日 [中国]
南齐和帝中兴元年三月乙巳，是夜彗星竟天。
《南史·齐本纪》卷5

公元507年8月15日 [中国]
北魏宣武帝正始四年七月己卯，有星孛于东北。
《魏书·天象志》卷105

公元520年10月1日 [中国]
梁武帝普通元年九月乙亥，有星晨见东方，光烂如火。
《梁书·武帝纪》卷3

《南史·梁本纪》卷7
《隋书·天文志》卷21

公元520年10月7日 [中国]
北魏孝明帝正光元年九月辛巳，有彗星光焰如火，出于东方。
《魏书·天象志》卷105

公元530年8月29日 [中国]
北魏孝庄帝永安三年七月甲午(8月29日)，有彗星晨见东北方，在中台东一丈，长六尺，色正白，东北行，西南指；丁酉(9月1日)，距下台上星西北一尺而晨伏；庚子(9月4日)，夕见西北方，长尺，东南指，渐移入氐。至八月己未(9月23日)，渐见；癸亥(9月27日)灭。
《魏书·天象志》卷105

公元533年3月1日 [中国]
梁武帝中通五年正月己酉，长星见。
《梁书·武帝纪》卷3
《隋书·天文志》卷21

公元535年2月18日—536年2月7日 [中国]
东魏孝静帝天平二年，有星孛于太微，历下台，及室壁而灭。
《魏书·天象志》卷105

公元539年11月17日 [中国]
(a) 东魏孝静帝兴和元年十月辛丑，有彗星出于南斗，长丈余，至十一月丙戌，距太白三尺，长丈余，东南指；二月乙卯，至娄始灭。
《魏书·天象志》卷105
(b) 梁武帝大同五年十月辛丑，彗出南斗，长一尺余，东南指，渐长一丈余。十一月乙卯，至娄灭。
《隋书·天文志》卷21

公元560年10月9日 [中国]
(a) 陈文帝天嘉元年九月癸丑，彗星见。

《陈书·世祖纪》卷3
《南史·陈本纪》卷9
(b) 陈文帝天嘉元年九月癸丑，彗星长四尺见,芒指西南。
《隋书·天文志》卷21

公元565年7月22日—24日 [中国]
(a) 北周武帝保定五年六月庚申（7月22日），彗星出三台，入文昌，犯上将，后经紫宫西垣 ，入危，渐长一丈余，指室、壁。后百余日，稍短，长二尺五寸，在虚、危灭。
《周书·武帝纪》卷5
《隋书·天文志》卷21
(b) 北周武帝保定五年六月庚申（7月22日），彗星出三台，入文昌，犯上将，经紫宫入危，渐长丈余，百余日乃灭。
《北史·周本纪》卷10
(c) 陈文帝天嘉六年六月辛酉（7月23日），彗星见于台北。
《南史·陈本纪》卷9
(d) 北齐后主天统元年六月壬戌（7月24日），彗星出文昌东北，其大如手，后稍长，乃至丈余，百日乃灭。
《北齐书·帝纪》卷8
《北史·齐本纪》卷8
(e) 北齐后主天统元年六月壬戌（7月24日），彗星见于文昌，长数寸，入文昌，犯上将，然后经紫微宫西垣，入危，渐长一丈余，指室壁。后百余日，在虚危灭。
《隋书·天文志》卷21

公元568年7月20日 [中国]
(a) 北周武帝天和三年六月甲戌，有星孛于东井，北行一月，至舆鬼，乃灭。
《周书·武帝纪》卷5
(b) 北周武帝天和三年六月甲戌，有星孛于东井。
《北史·周本纪》卷10
(c) 北周武帝天和三年六月甲戌，彗见东井，长一丈，上白下赤而锐，渐东行，至七月癸卯，在鬼北八寸所乃灭。
《隋书·天文志》卷21

公元568年8月9日—9月6日 [中国]

(a) 北周武帝天和三年七月己未（9月3日），客星见房，渐东行，入天市，犯营室，至奎，四十余日乃灭。

《周书·武帝纪》卷5

(b) 北周武帝天和三年七月己未（9月3日），客星见房心，白如粉絮，大如斗，渐大，东行；八月，入天市，长如匹所，复东行，犯河鼓右将；癸未，犯瓠瓜，又入室，犯离宫；九月壬寅，入奎，稍小；壬戌，至娄北一尺所灭，凡六十九日。

《隋书·天文志》卷21

(c) 北齐后主天统四年七月（8月9日—9月6日），孛星见房心，白如粉絮，大如斗，东行。八月，入天市，渐长四丈，犯瓠瓜，历虚危，入室，犯离宫；九月，入奎，至娄而灭。

《隋书·天文志》卷21

公元574年4月4日 [中国]

北周武帝建德三年二月戊午，客星大如桃，青白色，出五车东南三尺所，渐东行，稍长二尺所；至四月壬辰，入文昌，丁未，入北斗魁中，后出魁，渐小。凡见九十三日。

《隋书·天文志》卷21

公元574年5月31日—6月2日 [中国]

(a) 北周武帝建德三年四月乙卯（5月31日），星孛于紫宫垣外，大如拳，赤白，指五帝座，渐东南行，稍长一丈五尺；五月甲子，至上台北灭。

《隋书·天文志》卷21

(b) 北周武帝建德三年四月丁巳（6月2日），有星孛于东北紫宫垣，长七尺。

《周书·武帝纪》卷5

(c) 北周武帝建德三年四月丁巳（6月2日），有星孛于东井。

《北史·周本纪》卷10

公元575年4月27日 [中国]

陈宣帝太建七年四月丙戌，有星孛于大角。

《陈书·宣帝纪》卷5

《南史·陈本纪》卷10

《隋书·天文志》卷21

公元579年11月5日 —12月3日
百济威德王二十六年冬十月，长星竟天，二十日而灭。
《增补文献备考》 卷6
《三国史记》 卷27

公元588年11月22日 ［中国］
隋文帝开皇八年十月甲子，有星孛于牵牛。
《北史·隋本纪》卷11
《隋书·高祖纪》卷2

公元595年1月9日 ［中国］
(a) 隋文帝开皇十四年十一月癸未，有星孛于角、亢。
《北史·隋本纪》卷11
《隋书·高祖纪》卷2
(b) 隋文帝开皇十四年十一月癸未，有彗星孛于虚危及奎娄。
《隋书·天文志》卷21

公元595年1月9日 ［朝鲜］
百济威德王四十一年冬十一月癸未，星孛于角亢。
《增补文献备考》卷6
《三国史记》卷27

公元607年 ［中国］
(a) 隋炀帝大业三年春正月丙子(2月28日)，长星竟天，出于东壁，二旬而止。...二月己丑(3月13日)，彗星见于东井，文昌，历大陵、五车、北河、入太微，扫帝座，前后百余日而止。
《北史·隋本纪》卷12
(b) 隋炀帝大业三年春正月丙子(2月28日)，长星竟天，出于东壁，二旬而止。...二月己丑(3月13日)，彗星见于奎，扫文昌，历大陵、五车、北河，入太微，扫帝座，前后百余日而止。
《隋书·炀帝纪》卷3
(c) 隋炀帝大业三年三月辛亥(4月4日)，长星见西方，竟天。干历奎娄、角亢而没；至九月辛未(10月21日)，转见南方，亦竟天，又干角亢，频扫太微帝座，干犯列宿，唯不及参、井。经岁乃灭。

《隋书·天文志》卷21

公元608年10月22日 [中国]

隋炀帝大业四年九月戊寅，彗星出于五车，扫文昌，至房而灭。

《北史·隋本纪》卷12

《隋书·炀帝纪》卷3

公元615年7月2日—30日 [中国]

隋炀帝大业十一年六月，有星孛于文昌东南，长五六寸，色黑而锐，夜动摇，西北行，数日至文昌，去宫四五寸，不入，却行而灭。

《隋书·天文志》卷21

公元617年7月9日—8月6日 [中国]

隋炀帝大业十三年六月，有星孛于太微五帝座，色黄赤，长三四尺所，数日而灭。

《隋书·天文志》卷21

公元617年10月6日—11月3日 [中国]

隋炀帝大业十三年九月，彗星见于营室。

《隋书·炀帝纪》卷4

《北史·隋本纪》卷12

公元626年3月26日 [中国]

(a) 唐高祖武德九年二月二十三日，夜，星孛于胃、昴间，凡二十八日，又孛于卷舌。

《旧唐书·天文志》卷36

注：校勘记谓此处“凡”字当作“至”。

(b) 唐高祖武德九年二月壬午，有星孛于胃、昴；丁亥，孛于卷舌。

《新唐书·高祖纪》卷1

(c) 唐高祖武德九年二月壬午，有星孛于胃、昴间；丁亥，孛于卷舌。

《新唐书·天文志》卷32

公元634年8月29日—9月27日　　［日本］
日本舒明六年秋八月，长星见南方，时人曰彗星。
《日本书纪》23；《一代要记》甲

公元634年9月20日—22日　　［中国］
(a) 唐太宗贞观八年八月二十三日（9月20日），星孛于虚、危，历于玄枵，凡十一日而灭。
《旧唐书·天文志》卷36
(b) 唐太宗贞观八年八月甲子（9月22日），有星孛于虚、危，历于氐，十一月上旬乃灭。
《旧唐书·太宗纪》卷3
(c) 唐太宗贞观八年八月甲子（9月22日），有星孛于虚、危。
《新唐书·太宗纪》卷2
(d) 唐太宗贞观八年八月甲子（9月22日），有星孛于虚、危，历玄枵，乙亥不见。
《新唐书·天文志》卷32

公元635年1月24日—2月22日　　［日本］
日本舒明七年春正月，彗星廻见于东。
《日本书纪》23；《一代要记》甲
《日本纪略》前7
注：《日本纪略》前7和《一代要记》甲 记为三月。

公元639年3月5日　　［日本］
(a) 日本舒明十一年正月己巳，长星见西北，时旻师曰，彗星也。
《日本书纪》23
(b) 日本舒明十一年正月，长星见西北。
《扶桑略记》一本四

公元639年4月30日　　［中国］
(a) 唐太宗贞观十三年三月乙丑，有星孛于毕、昴。
《旧唐书·太宗纪》卷3
《新唐书·太宗纪》卷2

《新唐书·天文志》卷32
(b) 唐太宗贞观十三年三月二十二日，夜，星孛于毕、昴。
《旧唐书·天文志》卷36

公元641年8月1日 [中国]
(a) 唐太宗贞观十五年六月己酉，有星孛于太微，犯郎位……秋七月甲戌，孛星灭。
《旧唐书·太宗纪》卷3
(b) 唐太宗贞观十五年六月己酉，有星孛于太微。
《新唐书·太宗纪》卷2
(c) 唐太宗贞观十五年六月己酉，有星孛于太微，犯郎位，七月甲戌不见。
《新唐书·天文志》卷32
(d) 唐太宗贞观十五年六月十九日，星孛于太微，犯郎位，七月甲戌灭。
《旧唐书·天文志》卷36

公元663年9月29日 [中国]
(a) 唐高宗龙朔三年八月癸卯，彗星见于左摄提。
《旧唐书·高宗纪》卷4
《新唐书·高宗纪》卷3
(b) 唐高宗龙朔三年八月癸卯，有彗星于左摄提，长二尺余，乙巳不见。
《新唐书·天文志》卷32

公元667年5月24日 [中国]
唐高祖乾封二年四月丙辰，有彗星于东北，在五车、毕、昴间，乙亥不见。
《新唐书·天文志》卷32
注：二年四月无乙亥。

公元668年5月17日—6月14日 [中国]
(a) 唐高宗总章元年四月丙辰（5月18日），有彗星见于毕、昴之间…乙亥，彗星灭。

《旧唐书·高宗纪》卷5

(b) 唐高宗总章元年四月丙辰（5月18日），有彗星出于五车。

《新唐书·高宗纪》卷3

(c) 唐高宗总章元年四月（5月17日—6月14日），彗见五车……星虽孛而光芒小…星孛于东北…二十二日星灭。

《旧唐书·天文志》卷36

公元668年5月17日—6月14日 [朝鲜]

(a) 新罗文武王八年四月，彗守天船。

《三国史记》卷6

《增补文献备考》卷6

(b) 高句丽宝藏王二十七年夏四月，彗星见於毕昴之间。

《增补文献备考》卷6

公元672年9月27日—10月25日 [朝鲜]

新罗文武王十二年秋九月，彗星七出北方。

《三国史记》卷7

《增补文献备考》卷 6

公元675年10月24日—11月22日 [中国]

(a) 唐高宗上元二年十月壬午（11月4日），星孛于角、亢之南，长五尺。

《旧唐书·高宗纪》卷5

(b) 唐高宗上元二年十月壬午（11月4日），有彗星出于角、亢。

《新唐书·高宗纪》卷3

(c) 唐高宗上元二年十月（10月24日—11月22日），彗见于角、亢南，长五尺。

《旧唐书·天文志》卷36

公元676年8月15日—9月12日 [日本]

(a) 日本天武四年七月，是月，有星出于东，长七八尺，至九月竟天。

《日本书纪》 29

(b) 日本天武四年秋七月，有星出东，其长八尺。
《扶桑略记》5

公元676年8月15日—9月12日 [朝鲜]
新罗文武王十六年秋七月，彗星出北河积水之间，长六七许步。
《增补文献备考》卷6
《三国史记》卷7

公元676年9月4日 [中国]
(a) 唐高宗上元三年七月二十一日，彗见东井，指南河、积薪，长三尺余，渐向东北，光芒益衰，长三丈，扫中台，指文昌，经五十八日而灭。
《旧唐书·天文志》卷36
(b) 唐高宗上元三年七月丁亥，有彗星于东井，指北河，长三尺余；东北行，光芒益盛，长三丈，扫中台，指文昌。九月乙酉，不见。
《新唐书·天文志》卷32

公元681年10月17日 [中国]
(a) 唐高宗永隆二年（开耀元年）九月丙申（一日），彗星见于天市，长五尺。
《旧唐书·高宗纪》卷5
(b) 唐高宗永隆二年（开耀元年）九月丙申（一日），夜，彗见西方天市中，长五尺，渐小，向东行，出天市，至河鼓右旗，十七日灭。
《旧唐书·天文志》卷36
(c) 唐高宗永隆二年（开耀元年）九月丙申（一日），有彗星出于天市。
《新唐书·高宗纪》卷3
(d) 唐高宗永隆二年（开耀元年）九月丙申（一日），有彗星于天市中，长五丈，渐小，东行至河鼓，癸丑不见。
《新唐书·天文志》卷32

公元683年4月20日 [中国]
(a) 唐高宗永淳二年（弘道元年）三月丙午（十八日），彗见五车北，二十五日而灭。

《旧唐书·高宗纪》卷5

(b) 唐高宗永淳二年（弘道元年）三月丙午（十八日），彗见五车之北，凡二十五日而灭。

《旧唐书·天文志》卷36

(c) 唐高宗永淳二年（弘道元年）三月丙午（十八日），有彗星于五车北，四月辛未不见。

《新唐书·天文志》卷32

(d) 唐高宗永淳二年（弘道元年）三月丙午（十八日），有彗星出于五车。

《新唐书·高宗纪》卷3

公元683 年10月25日—11月23日 ［朝鲜］

新罗神文王三年十月，彗出五车。

《三国史记》卷8

《增补文献备考》卷 6

公元684年9月6日—685年1月10日 ［中国］

(a) 唐则天皇后文明元年七月二十二日（辛未）(9月6日)，夕，有彗星于西方，长丈余，八月甲辰(10月9日)不见。是谓天搀。光宅元年九月丁丑(11月11日)…有星如半月，见于西方。

《新唐书·天文志》卷32

(b) 唐则天皇后文明元年七月二十二日（辛未）(9月6日)，西方有彗，长丈余，凡四十九日灭。

《旧唐书·天文志》卷36

(c) 唐则天皇后文明元年七月二十二日（辛未）(9月6日)，西北方有彗，长丈余，经四十二日灭。

《唐会要》卷43

(d) 唐则天皇后文明元年七月(8月16日—9月14日)，彗星见西北方，长二丈余，经三十三日乃灭。

《旧唐书·则天皇后》卷6

公元684年9月6日—685年1月10日 ［日本］

(a) 天武十二年七月二十三日壬申(9月7日)，彗星出于西北，长丈余。十一月(684年12月12日—685年1月10日)，是月有星孛于中央，与昴星双而行之，及月尽失焉。

《日本书纪》29

(b) 天武十二年七月二十三日壬申(684年9月7日)，彗星出西，长一丈余。

《一代要记》甲

公元699年3月7日—4月5日 [朝鲜]

新罗孝昭王八年春二月，白气竟天，星孛于东。

《增补文献备考》卷6

《三国史记》卷8

公元701年3月14日—4月12日 [朝鲜]

新罗孝昭王十年春二月，彗星入月。

《三国史记》卷8

《增补文献备考》卷6

公元707年11月16日 [中国]

(a) 唐中宗景龙元年十月壬午（十八日），彗见于西，月余而灭。

《旧唐书·中宗纪》卷7

(b) 唐中宗景龙元年十月壬午（十八日），彗见西方，凡四十三日而灭。

《旧唐书·天文志》卷36

(c) 唐中宗景龙元年十月壬午（十八日），有彗星出于西方。

《新唐书·中宗纪》卷4

(d) 唐中宗景龙元年十月壬午（十八日），有彗星于西方，十一月甲寅不见。

《新唐书·天文志》卷32

公元708年7月28日 [中国]

(a) 唐中宗景龙二年七月七日（丁酉），星孛胃、昴之间。

《旧唐书·天文志》卷36

(b) 唐中宗景龙二年七月七日（丁酉），有星孛胃、昴。

《新唐书·中宗纪》卷4

(c) 唐中宗景龙二年七月七日（丁酉），有星孛于胃、昴间。

《新唐书·天文志》卷32

公元709年9月16日　　[中国]

(a)　唐中宗景龙三年八月壬辰（八日），有星孛于紫宫。

《旧唐书·中宗纪》卷7

《旧唐书·天文志》卷36

《新唐书·中宗纪》卷4

《新唐书·天文志》卷32

(b)　唐中宗景龙三年八月壬辰（八日），有星孛于紫微垣。

《唐会要》卷43

公元712年7月9日—8月6日　　[中国]

(a)　唐睿宗太极元年七月四日（8月10日），彗入太微。

《旧唐书·天文志》卷36

(b)　唐睿宗先天元年七月辛未（8月10日），有彗星入于太微。

《新唐书·睿宗纪》卷5

(c)　唐睿宗延和元年六月（7月9日—8月6日），有彗星自轩辕入太微，至大角灭。

《新唐书·天文志》卷32

公元718年12月8日　　[日本]

日本养老二年十一月壬寅，彗星守月。

《续日本纪》8

公元730年6月30日　　[中国]

(a)　唐玄宗开元十八年六月甲子（十一日），彗星见于五车。

《旧唐书·玄宗纪》卷8

(b)　唐玄宗开元十八年六月甲子（十一日），彗见五车。

《旧唐书·天文志》卷36

(c)　唐玄宗开元十八年六月甲子（十一日），有彗星出于五车。

《新唐书·玄宗纪》卷5

(d)　唐玄宗开元十八年六月甲子（十一日），有彗星于五车。

《新唐书·天文志》卷32

公元730年7月9日—19日　　[中国]

(a)　唐玄宗开元十八年六月癸酉（7月9日），有星孛于毕、昴。

《旧唐书·玄宗纪》卷8
《新唐书·玄宗纪》卷5
(b) 唐玄宗开元十八年六月癸酉（7月9日），有彗孛于毕、昴。
《新唐书·天文志》卷32
(c) 唐玄宗开元十八年六月三十日（7月19日），有星孛于毕、昴。
《旧唐书·天文志》卷36

公元759年4月2日—5月1日
新罗景德王十八年春三月，彗星见，至秋乃灭。
《增补文献备考》 卷6
《三国史记》 卷9

公元738年4月1日 ［中国］
(a) 唐玄宗开元二十六年三月丙子（八日），有星孛于紫微垣中，历斗魁，十余日，因阴云不见。
《旧唐书·玄宗纪》卷9
《旧唐书·天文志》卷36
(b) 唐玄宗开元二十六年三月丙子（八日），有星孛于紫微。
《新唐书·玄宗纪》卷5
(c) 唐玄宗开元二十六年三月丙子（八日），有星孛于紫宫垣，历北斗魁，旬余，因云阴不见。
《新唐书·天文志》卷32

公元759年4月2日—5月1日 ［朝鲜］
新罗景德王十八年春三月，彗星见，至秋乃灭。
《增补文献备考》卷6
《三国史记》卷9

公元760年5月16日 ［中国］
(a) 唐肃宗乾元三年（上元元年）四月丁巳（二十七日）(760年5月16日)，彗见于东方，在娄胃间，色白，长四尺，疾行向东北，历昴毕觜参井鬼柳轩辕宿，至太微西，右执法西七尺许灭，凡经五十余日。
《唐会要》卷43

(b) 唐肃宗乾元三年（上元元年）四月丁巳（二十七日）(760年5月16日) ，夜五更，彗出东方，色白，长四尺，在娄、胃间，疾行向东北角，历昴、毕、觜、参、井、鬼、柳、轩辕、至太微右执法七寸所，凡五十余日方灭。

《旧唐书·天文志》卷36

(c) 唐肃宗乾元三年（上元元年）四月丁巳（二十七日）(760年5月16日)，有彗星于东方，在娄、胃间，色白，长四尺，东方疾行，历昴、毕、觜、觿、参、东井、舆鬼、柳、轩辕至右执法西，凡五旬余不见。

《新唐书·天文志》卷32

公元767年1月22日 ［中国］

(a) 唐代宗大历元年十二月己亥，彗起匏瓜，其长尺余，犯宦者星。

《旧唐书·代宗纪》卷11

(b) 唐代宗大历元年十二月己亥，彗星出匏瓜，长尺余，犯宦者星。

《旧唐书·天文志》卷36

(c) 唐代宗大历元年十二月己亥，有彗星出于匏瓜。

《新唐书·代宗纪》卷6

(d) 唐代宗大历元年十二月己亥，有彗星于匏瓜，长尺余，经二旬不见，犯宦者星。

《新唐书·天文志》卷32

公元770年5月26日 ［中国］

(a) 唐代宗大历五年四月己未（二十七日），夜，彗起五车，长三丈……五月己卯夜，彗起北方，其色白……六月己未，彗星始灭。

《旧唐书·代宗纪》卷11

(b) 唐代宗大历五年四月己未（二十七日），夜，彗出五车，蓬孛，光芒长三丈。五月己卯夜，彗出北方，其色白。癸未夜，彗随天东行，近八谷……癸卯，彗去三公二尺，己未，彗星 灭。

《旧唐书·天文志》卷36

(c) 唐代宗大历五年四月己未（二十七日），有彗星出于五车。
《新唐书·代宗纪》卷6
(d) 唐代宗大历五年四月己未（二十七日），有彗星于五车，光芒蓬勃，长三丈。五月己卯，彗星见于北方，色白，癸未东行近八谷中星；六月癸卯近三公，己未不见。
《新唐书·天文志》卷32
(e) 唐代宗大历五年四月己未（二十七日），彗出于五车，长五丈余。六月二十八日灭。
《唐会要》卷43

公元770年6月9日 [朝鲜]
(a) 新罗惠恭王六年夏五月十一日，彗星出五车北，至六月十二日灭。
《增补文献备考》卷6
(b) 新罗惠恭王六年五月十一日，彗星出五车北。
《三国史记》卷9

公元770年6月28日—8月25日 [日本]
(a) 日本宝龟元年九月壬午，是年六七月，彗星入於北斗。
《续日本纪》 30
(b) 日本宝龟元年七月癸未，自六七月，彗星入北斗焉。
《日本纪略》 前11

公元773年1月17日 [中国]
(a) 唐代宗大历七年十二月丙寅，长星出于参。
《旧唐书·代宗纪》卷11
《新唐书·代宗纪》卷6
(b) 唐代宗大历七年十二月丙寅，有长星于参下，其长亘天。长星。彗属。
《新唐书·天文志》卷32
(c) 唐代宗大历七年十二月丙寅，夜，长星出于参。
《旧唐书·天文志》卷36

公元773年1月20日 [日本]
日本宝龟三年十二月己巳，彗星见南方。

《续日本纪》32
《日本纪略》 前12

公元815年4月13日—5月12日　［中国］
唐宪宗元和十年三月，有长星于太微，尾至轩辕。
《新唐书·天文志》卷32

公元815年9月7日—10月6日　［朝鲜］
新罗宪德王七年秋八月，大星出翼轸间，指庚，芒长六许尺，广二许寸。
《增补文献备考》卷6
《三国史记》卷10

公元817年2月17日　［中国］
(a) 唐宪宗元和十二年正月戊子，夜，彗出毕南，长丈余，指西南，凡三日，南近参旗而没。
《旧唐书·宪宗纪》卷15
(b) 唐宪宗元和十二年正月戊子，彗出毕南，长二尺余，指西南，凡三日，近参旗没。
《旧唐书·天文志》卷36
(c) 唐宪宗元和十二年正月戊子，有彗星出于毕。
《新唐书·宪宗纪》卷7
(d) 唐宪宗元和十二年正月戊子，有彗星于毕。
《新唐书·天文志》卷32

公元821年2月26日—27日　［中国］
(a) 唐穆宗长庆元年正月戊午（2月26日），夜，星孛于翼。
《旧唐书·穆宗纪》卷16
(b) 唐穆宗长庆元年正月己未（2月27日），有星孛于翼。
《新唐书·穆宗纪》卷8
《新唐书·天文志》卷32
(c) 唐穆宗长庆元年正月己未（2月27日），夜，星孛于翼。
《旧唐书·天文志》卷36

公元821年3月7日 [中国]

(a) 唐穆宗长庆元年正月丁卯，星孛于辰，近太微西垣南第一星。
《旧唐书·穆宗纪》卷16

(b) 唐穆宗长庆元年正月丁卯，有星孛于太微。
《新唐书·穆宗纪》卷8

(c) 唐穆宗长庆元年正月丁卯，星孛于太微西上将。
《新唐书·天文志》卷32

(d) 唐穆宗长庆元年正月丁卯，夜，星孛于辰上，去太微西垣南第一星七寸所。
《旧唐书·天文志》卷36

公元821年7月3日—8月1日 [中国]

(a) 唐穆宗长庆元年六月，有彗星出于昴。
《新唐书·穆宗纪》卷8

(b) 唐穆宗长庆元年六月，有彗星于昴，长一丈，凡十日不见。
《新唐书·天文志》卷32

公元823年2月19日 [日本]

日本弘仁十四年正月辛酉，有星孛于西南，三日而不见。
《日本纪略》 前14

公元828年9月3日 [中国]

(a) 唐文宗大和二年七月甲辰，有彗星于右摄提。
《新唐书·文宗纪》卷8

(b) 唐文宗大和二年七月甲辰，有彗星于右摄提南，长二尺。
《新唐书·天文志》卷32

公元834年10月9日 [中国]

(a) 唐文宗大和八年九月辛亥，夜，彗起太微，近郎位，西指，长丈余，西北行，凡九夜，越郎位西北五尺灭。
《旧唐书·文宗纪》卷17

(b) 唐文宗大和八年九月辛亥，夜五更，太微宫近郎位有彗星，长丈余，西指，西北行，凡九夜，越郎位星西北五尺灭。
《旧唐书·天文志》卷36

(c) 唐文宗大和八年九月辛亥，有彗星出于太微。

《新唐书·文宗纪》卷8

(d) 唐文宗大和八年九月辛亥，有彗星于太微，长丈余，西北行，越郎位，庚申不见。

《新唐书·天文志》卷32

(e) 唐文宗大和八年九月辛亥，彗起太微，越帝位西北，五日乃灭。

《唐会要》卷34

公元834年10月18日 [中国]

(a) 唐文宗大和八年九月庚申，夜，彗出东方，长三尺，辉耀甚伟。

《旧唐书·文宗纪》卷17

(b) 唐文宗大和八年九月庚申，夜，彗星出东方，长三尺，芒耀甚猛。

《旧唐书·天文志》卷36

公元837年6月26日 [中国]

唐文宗开成二年五月壬午，客星如孛，在南斗天籥旁。

《新唐书·天文志》卷32

公元837年3月22日 [中国]

(a) 唐文宗开成二年二月丙午(837年3月22日)，夜，彗出东方，长七尺余，在危初度，西指。戊申(3月24日)夜，危之西南，彗长七尺，芒耀俞猛，亦西指。癸丑(3月29日)夜，彗在危八度。庚申(4月5日)夜，在虚三度半。辛酉(4月6日)夜，彗长丈余，直西行，稍南指，在虚一度半。壬戌(4月7日)夜，彗长二丈，其广三尺，在女九度。癸亥(4月8日)夜，彗愈长广，在女四度。三月甲子(4月9日)朔，其夜，彗长五丈，歧分两尾，其一指氐，其一掩房，在斗十度。丙寅(4月11日)夜，彗长六丈，尾无歧，北指，在亢七度。文宗召司天监朱子容问星变之由，...其夜彗长五丈，阔五尺，却西北行，东指。戊辰(4月13日)夜,彗长八丈有余，西北行，东指，在张十四度。...癸未(4.28)夜，彗长三尺，出轩辕之右，东指，在张七度。 ...八月丁酉(9月9日)，彗出虚、危之间。

《旧唐书·天文志》卷36

(b) 唐文宗开成二年二月丙午(837年3月22日)，有彗星于危，长七尺余，西指南斗；戊申(3月24日)在危西南，芒耀愈盛；癸丑(3月29日)在虚；辛酉(4月6日)，长丈余，西行稍南指；壬戌(4月7日)，在婺女，长二丈余，广三尺；癸亥(4月8日)，愈长且阔；三月甲子(4月9日)，在南斗，乙丑(4月10日)，长五丈，其末两岐，一指氐，一掩房，丙寅(4月11日)，长六丈，无歧，北指，在亢七度；丁卯(4月12日)，西北行，东指；己巳(4月14日)，长八丈余，在张；癸未(4月28日)，长三尺，在轩辕右不见。凡彗星晨出则西指，夕出则东指，乃常也，未有遍指四方，凌犯如此之甚者。 甲申(4月29日)，客星出于东井下。戊子(5月3日)，客星别出于端门内，近屏星。四月丙午(5月21日)，东井下客星没。五月癸酉(6月17日)，端门内客星没。壬午(6月26日)，客星如孛，在南斗天籥旁。八月丁酉(9月9日)，有彗星于虚、危。

《新唐书·天文志》卷32

(c) 唐文宗开成二年八月丁酉，彗出虚、危之间。

《旧唐书·文宗纪》卷17

公元837年4月12日［日本］承和四年三月四日丁卯(837年4月12日)，彗星见于东南，其光芒东至天涯。壬申(4月17日)，彗星犹见，但为月光所夺，其光芒微少耳。

《续日本后纪》 6

公元838年11月9日 ［中国］

(a) 唐文宗开成三年十月十九日，彗见，长二丈余；二十日夜，长二丈五尺；二十一日夜，长三丈；二十二日夜，长三丈五尺，并在辰上，西指轸、魁。

《旧唐书·天文志》卷36

(b) 唐文宗开成三年十月十九日，有彗出于辰上，长二丈余；二十日夜，见于辰上，长三丈五尺；二十一日夜，见于辰，长三丈余，西指轸东南星；二十二日夜，见于辰，长三丈五尺

余，南指轸魁。

《唐会要》卷43

公元838年11月11日 [中国]

(a) 唐文宗开成三年十月乙巳，有彗星出于轸。

《新唐书·文宗纪》卷8

(b) 唐文宗开成三年十月乙巳，有彗星于轸魁，长二丈余，渐长，西指。

《新唐书·天文志》卷32

(c) 唐文宗开成三年十月乙巳，夜，彗起于轸，其长三丈，东西指。

《旧唐书·文宗纪》卷17

公元838年11月12日 [日本]

(a) 日本承和五年十月丙午，夜彗星见东南，其气赤白，竟天数许里，须臾而不见。十一月辛未，彗星见东方。是星起十月廿二日，至今月十七日，每夜寅刻见东方，其长七尺许。

《续日本后纪》 7

(b) 日本承和五年十月丙午，是夜彗星见东南，其气赤白，竟天数许里，须臾而不见。庚戌，又见。十一月辛未，彗星见东方，是星起十月廿二日，至今月十七日，每夜寅刻见东方，其长七八许丈。承和六年正月丙子，彗星见兑方，长一许丈。

《诸道勘文》 45

公元838年11月21日 [中国]

(a) 唐文宗开成三年十一月乙卯，有彗星见于东方，在尾、箕，东西亘天；十二月壬辰不见。

《新唐书·天文志》卷32

(b) 唐文宗开成三年十一月乙卯，夜，彗孛东西竟天。

《旧唐书·文宗纪》卷17

(c) 唐文宗开成三年十一月乙卯，夜，彗出东方，东西竟天。

《旧唐书·天文志》卷36

公元839年2月7日—17日 [中国]

(a) 唐文宗开成四年正月癸酉（2月7日），有彗星出于羽林。闰月丙午，出于卷舌。

《新唐书·文宗纪》卷8

(b) 唐文宗开成四年正月癸酉（2月7日），有彗星于羽林。闰月丙午，有彗星于卷舌西北；二月己卯不见。

《新唐书·天文志》卷32

(c) 唐文宗开成四年正月癸酉（2月7日），夜，彗出于西方，在室十四度。闰月二十三日，又见于卷舌北，凡三十三日，至二十六日夜灭。

《旧唐书·天文志》卷36

注：校勘记曰：此处二十六日前当脱"二月"二字。查是年闰正月，自正月癸酉至闰月二十三日丙午，计三十三日，无误。

(d) 唐文宗开成四年正月三十日（2月17日），有彗见于室南，历壁、奎、娄、胃等宿，至闰月二十三日，又见于卷舌北。凡三十有三日。后二十四日，二十五日阴，二十六日晴，已灭。

《唐会要》卷43

公元840年3月20日 ［中国］

(a) 唐文宗开成五年二月庚申，有彗星出于室、壁。

《新唐书·武宗纪》卷8

(b) 唐文宗开成五年二月庚申，彗出于营室、东壁间，二十日灭。

《新唐书·天文志》卷32

公元840年12月3日 ［中国］

(a) 唐文宗开成五年十一月戊寅，有彗星出于东方。

《新唐书·武宗纪》卷8

(b) 唐文宗开成五年十一月戊寅，有彗星于东方。

《新唐书·天文志》卷32

公元841年7月22日—8月19日 ［中国］

(a) 唐武宗会昌元年七月，彗复出室壁之间。

《旧唐书·武宗纪》卷18

(b) 唐武宗会昌元年七月，有彗星出于羽林。

《新唐书·武宗纪》卷8

(c) 唐武宗会昌元年七月，有彗星于羽林，营室、东壁间也。

《新唐书·天文志》卷32

公元841年12月22日　［中国］

(a) 唐武宗会昌元年十一月壬寅（六日），夜，彗起于室，凡五十六日而灭。

《旧唐书·武宗纪》卷18

(b) 唐武宗会昌元年十一月壬寅（六日），彗见西南，在室初度，凡五十六日而灭。

《旧唐书·天文志》卷36

(c) 唐武宗会昌元年十一月壬寅（六日），有彗星出于营室。

《新唐书·武宗纪》卷8

(d) 唐武宗会昌元年十一月壬寅（六日），有彗星于北落师门，在营室，入紫宫，十二月辛卯不见。

《新唐书·天文志》卷32

(e) 唐武宗会昌元年十一月壬寅（六日），有彗见于西南室宿之分，凡五十六日而灭。

《唐会要》卷43

公元841年12月22日　［日本］

日本承和八年十一月壬寅，彗星见西方，丁巳，彗星犹见。

《续日本后纪》10

公元852年3月14日　［日本］

日本仁寿二年二月丁巳，是夕彗星出于西方，长可五丈。

《文德实录》4

《日本纪略》前16

公元852年3月25日—4月22日　［中国］

唐宣宗大中六年三月，有彗星出于觜参。

《新唐书·宣宗纪》卷8

《新唐书·天文志》卷32

日本855年2月23日　［日本］

日本齐衡二年二月癸丑，有长星出于东北。

《文德实录》7

《日本纪略》前16

公元857年9月22日 [中国]

(a) 唐宣宗大中十一年九月乙未，彗出于房初度，长三尺。

《旧唐书·宣宗纪》卷18

(b) 唐宣宗大中十一年九月乙未，有彗星出于房。

《新唐书·宣宗纪》卷8

(c) 唐宣宗大中十一年九月乙未，有彗星于房，长三尺。

《新唐书·天文志》卷32

公元864年4月10日—5月9日 [中国]

唐懿宗咸通五年三月，彗星出于娄，司天监以为瑞星。

《世史正纲》卷20

公元864年6月21日 [中国]

(a) 唐懿宗咸通五年五月己亥，有彗星出于娄。

《新唐书·懿宗纪》卷9

(b) 唐懿宗咸通五年五月己亥，夜漏未尽一刻，有彗星出于东北，色黄白，长三尺，在娄。

《新唐书·天文志》卷32

公元864年4月23日 [日本]

日本贞观六年三月十四日庚子，彗星见东，在营室宿，长四许尺。

《三代实录》8

《日本纪略》 前17

公元868年1月21日 [日本]

(a) 日本贞观九年十二月廿三日戊子，彗星见紫微宫西，贯内阶，长可五尺。

《三代实录》 14

(b) 日本贞观九年十二月廿三日戊子，时戌刻彗星见紫微宫西蕃许，贯内阶，星长五尺。

《诸道勘文》45

注：《三代实录》原文为“十一月廿三日戊午”，系“十二月廿三日戊子”之误。

公元868年1月29日—2月26日 [中国]

唐懿宗咸通九年正月，有彗星出于娄、胃。

《新唐书·懿宗纪》卷9
《新唐书·天文志》卷32

公元869年9月10日—10月8日 ［中国］
(a) 唐懿宗咸通十年八月，有彗星出于大陵。
《新唐书·懿宗纪》卷9
(b) 唐懿宗咸通十年八月，有彗星于大陵，东北指。
《新唐书·天文志》卷32

公元875年3月12日—4月9日 ［朝鲜］
新罗景文王十五年春二月，星孛于东，二十日乃灭。
《增补文献备考》卷6
《三国史记》卷11

公元875年6月5日 ［日本］
日本贞观十七年四月廿八日庚辰，卯时白彗见东北，其色赤，以成芒角，至五月二日，其体长可丈余，始出五车，稍扫八榖星，其气虽耗减，而未灭芒。五月十六日丁酉，夜有云气竟天，形如幡，头挂西山，尾挂东山，十八日己亥，夜有星孛东北。
《三代实录》 27

公元885年1月20日—886年2月7日 ［中国］
唐僖宗光启元年，有彗于积水、积薪之间。
《新唐书·天文志》卷32

公元886年6月6日—7月5日 ［中国］
(a) 唐僖宗光启二年五月丙戌（6月13日），有星孛于箕、尾。
《新唐书·僖宗纪》卷9
(b) 唐僖宗光启二年五月丙戌（6月13日），有星孛于尾、箕，历北斗、摄提。
《新唐书·天文志》卷32
(c) 唐僖宗光启二年五月（6月6日—7月5日），星孛于箕尾，历北斗摄提。
《旧唐书·僖宗纪》卷19

公元891年5月12日 ［中国］
(a) 唐昭宗大顺二年四月庚辰，有彗星入于太微。

《新唐书·昭宗纪》卷10

(b) 唐昭宗大顺二年四月庚辰，有彗星于三台，东行入太微、扫大角，天市，长十丈余，五月甲戌不见。

《新唐书·天文志》卷32

公元892年11月23日—12月22日 [中国]

唐昭宗景福元年十一月，有星孛于斗、牛。

《新唐书·昭宗纪》卷10

《新唐书·天文志》卷32

公元892年12月28日 [中国]

唐昭宗景福元年十二月丙子，天搀出于西南；己卯，化为云而没。

《新唐书·天文志》卷32

公元893年5月6日 [中国]

(a) 唐昭宗景福二年四月乙酉，有彗星入于太微。

《新唐书·昭宗纪》卷10

(b) 唐昭宗景福二年四月乙酉，夜，有彗星于上台，长十余丈，东行入太微，扫大角，入天市，经三旬有七日，益长，至二十余丈，因云阴不见。

《新唐书·天文志》卷32

公元894年2月10日—3月10日 [中国]

唐昭宗乾宁元年正月，有星孛于鹑首。

《新唐书·昭宗纪》卷10

《新唐书·天文志》卷32

公元896年11月9日—12月7日 [中国]

唐昭宗乾宁三年十月，有客星三，一大二小，在虚、危间，乍合乍离，相随东行，状如鬥，经三日而二小星没，其大星后没。

《新唐书·天文志》卷32

公元902年2月11日—3月12日 [中国]
唐昭宗天复二年正月，客星如桃，在紫宫华盖下，渐行至御女。丁卯，有流星起文昌，抵客星，客星不动；己巳，客星在杠，守之，至明年犹不去。
《新唐书·天文志》卷32

公元905年5月18日 [中国]
(a) 唐哀帝天祐二年四月庚子，有彗星出于西北；甲辰，出于北河。
《新唐书·哀帝纪》卷10
(b) 唐哀帝天祐二年四月庚子，夕，西北隅有星类太白，上有光似彗，长三四丈，色如赭，辛丑夕，色如缟。咸曰五车之水星也，一曰昭明星也。
《新唐书·天文志》卷32

公元905年5月21日 [日本]
日本延喜五年四月十五日癸卯……又乾方见彗星，十六、十八、十九日，同见彗星。廿四日壬子，乾方见彗星，长卅余丈，光芒指巽方。廿五日癸丑，乾方见彗星，长竟於天。廿六七八九日，同见。五月一日己未，彗星今夜渐以细薄。三日辛酉，乾方见彗星，自今夜不见。
《扶桑略记》 23
注：《诸道勘文》45 有相似记录。

公元905年5月22日 [中国]
(a) 唐哀帝天祐二年四月甲辰，夜，彗起北河，贯文昌，其长三丈，在西北方……五月乙酉夜，西北彗星长六七十丈，自轩辕大角及天市西垣，光辉猛怒，其长竟天。
《旧唐书·哀帝纪》卷20
(b) 唐哀帝天祐二年四月甲辰，有彗星于北河，贯文昌，长三丈余，陵中台，下台；五月乙丑夜，自轩辕左角及天市西垣，光芒猛怒，其长亘天；丙寅云阴，至辛未少霁，不见。

《新唐书·天文志》卷32
注："陵"应为"凌"。

公元905年6月6日—7月4日 [中国]
唐哀帝天祐二年五月，西北长星竟天，扫太微，文昌、帝座诸宿。
《旧唐书·列传》卷179

公元906年4月26日—5月25日 [中国]
唐哀帝天祐三年四月，彗出西北，扫文昌、轩辕、天市。
《新五代史·唐六臣传》卷35
注：疑系二年之误。

公元907年4月7日 [日本]
日本延喜七年二月二十二日。二月二十四日辛未，彗星食太白，长三长许。二月二十二日己巳，戌刻彗星出酉戌方，廿九日，彗星出同方。卅日丁丑，不见。
《扶桑略记》 23
注：原文“二月二十五日辛未”应为“二月二十四日辛未”之误。

公元912年5月13 - 15日 [中国]

(a) 后梁太祖乾化二年四月甲戌(AD 912.5.15)，夜，彗见于灵台之西。

《旧五代史·天文志》卷139
《五代会要》卷10

(b) 后梁太祖乾化二年四月壬申(AD 912.5.13)，彗出于张，甲戌(5.15)彗出灵台。

《新五代史·司天考》卷59

公元912年7月19 - 28日 [日本]

(a) 延喜十二年六月三日己卯，始自今夜，戌亥彗星见。五、六日同。八日甲申，见辰巳。九日乙酉，见乾方。十二日戊子，见西方。

《扶桑略记》 23

(b) 延喜十二年六月三日，彗星始见戌亥方，七个夜。

《一代要记》丙

公元918年11月7日 ［日本］
日本延喜十八年十月一日，彗星始见西南方，经三箇夜。
《一代要记》丙；《诸道勘文》 45

公元923年11月11日—12月10日 ［中国］
辽太祖天赞三年（同光元年）十月，彗星见，出舆鬼，长丈余。
《契丹国志》卷1
注：天赞三年非同光元年，公元纪年按同光元年查出。

公元923年12月8日
后唐庄宗同光元年十月戊戌，彗星见舆鬼，长丈余。
《永乐大典》卷7159

公元928年12月13日 ［中国］
唐明宗天成三年，十月庚午夜，西南有孛，长丈余，东南指，在牛五度。
《旧五代史·天文志》卷139

公元930年5月31日—7月27日 ［日本］
(a) 日本延长八年五月以后七月以前，客星入羽林中。
《明月记》野宫本 23
(b) 日本延长八年五月以后七月以前，客星入羽林，始西行於羽林。
《一代要记》 丙

公元934年12月19日 ［中国］
后唐末帝清泰元年十一月丁未，彗出虚、危，扫天垒及哭星。
《新五代史·司天考》卷59

公元936年9月21日 ［中国］
后唐末帝清泰三年九月己丑，彗出虚危，长尺余，形细微，经天垒、哭星。
《旧五代史·天文志》卷139

公元936年10月19日 ［中国］
后唐末帝清泰三年十月丁巳，夜，彗星出虚危，长尺余。
《旧五代史·唐书》卷48

公元941年3月31日—4月28日 [日本]
日本天庆四年春三月，相当西方有星，其光如白虹，本细末渐广，程十里许，经三箇月。
《扶桑略记》 25

公元941年8月25日—9月23日 [中国]
南唐烈祖昇元五年八月，有星孛于天市，长数尺，七十日没。
《南唐书·烈祖纪》卷1

公元941年10月18日 [中国]
后晋高祖天福六年九月壬午，夜，有彗星出于西方，长二丈余，在房一度，尾迹穿天市垣东行，逾月而灭。
《旧五代史·晋书》卷80

公元943年11月5日 [中国]
(a) 后晋高祖天福八年十月庚戌，夜，有彗见于东方，西指，尾长一丈，在角九度。
《旧五代史·天文志》卷139
(b) 后晋高祖天福八年十月庚戌，是夕五更，有彗见于东方，在角，旬日而灭。
《旧五代史·晋书》卷82

公元947年9月12日 [中国]
(a) 后汉高祖天福十二年闰七月丁丑，有彗出于张，旬日而灭。
《旧五代史·汉书》卷100
(b) 南唐元宗保大五年闰七月丁丑，夜，有彗出东方，近浊，其尾迹近侧扫少微及长垣，至八月壬辰乃没。
《南唐书·元宗纪》卷2

公元956年3月13日 [中国]
(a) 周太祖显德三年正月壬戌，夜，有彗孛于参角，其芒指于东南。
《旧五代史·天文志》卷139
(b) 周太祖显德三年正月壬戌，有星孛于参。
《新五代史·司天考》卷59

公元962年1月28日 ［中国］
宋太祖建隆二年十二月己酉，客星出天市垣宗人星东，微有芒彗。三年正月辛未，西南行入氐宿，二月癸丑至七星没。
《宋史·天文志》卷56

公元967年1月8日 ［日本］
(a) 康保三年十一月廿五日乙卯，夜亥时彗星见南方。
《诸道勘文》45
(b) 康保三年十一月二十五日乙卯，亥时彗星见西方。
《一代要记》丙

公元975年 ［中国］
南唐后主乙亥岁三月丁巳，彗出五车，色白，长五尺，夏六月转见西方，犯太微，六十日灭。
《南唐书·后主纪》卷3
注：该月无丁巳。

公元975年8月2日 ［日本］
(a) 日本天延三年六月廿二日癸亥，晓彗星见艮方，其形如团扇，长五六尺，连夜久见。廿三日甲子，晓月犯毕，又客星如白虹。
《日本纪略》后6
(b) 日本天延三年六月廿二日，彗星见，经五箇日，如打羽，长五六丈，色白。
《诸道勘文》45

公元975年8月3日 ［中国］
(a) 宋太祖开宝八年六月甲子，彗出柳，长四丈，辰见东方。
《宋史·太祖纪》卷3
(b) 宋太祖开宝八年六月甲子，彗出柳，长四丈，辰见东方，西南指，历舆鬼至东壁，凡十一舍，八十三日而灭。
《宋史·天文志》卷56

公元983年4月3日 ［中国］
宋太宗太平兴国八年二月甲辰，客星出太微垣端门东，近屏星北行。
《宋史·天文志》卷56

公元988年7月18日 [中国]
宋太宗端拱元年六月二日，彗星见于井鬼，不数日彗灭。
《宋会要辑稿》 第52册

公元989年7月23日 [中国]
宋太宗端拱二年六月十八日，彗见积水西，光芒长五尺，行拂右摄提星，至今月十九日隐大西方，计见四十日。
《宋会要辑稿》 第52册
注："今月"指"七月"。

公元989年 [中国]

(a) 宋太宗端拱二年七月丁亥(989年8月12日)，客星出北河星西北，稍暗，微有芒彗，指西南。

《宋史·天文志》卷56

(b) 宋太宗端拱二年七月戊子(989年8月13日)，彗星又出东井积水西，青白色，光芒渐长，辰见东北，旬日夕见西北，历右摄提，凡三十日至亢没。

《宋史·天文志》卷56

(c) 宋太宗端拱二年七月戊子(989年8月13日)，有彗出东井，...八月丙辰(9月10日)，是夕彗不见。

《宋史·太宗二》卷5

公元989年7月6日 [日本]

(a) 永祚元年六月一日庚戌(989年7月6日)，其日彗星见东西天。七月中旬(8月14日～8月23日)，连夜彗星见东西天。

《日本纪略》 后9

(b) 永延三年七月十三日(989年8月16日)，彗星见东方，经数夜，长五尺许。

《诸道勘文》 45

公元989年10月18日 [朝鲜]
成宗八年九月甲午(989年10月18日)，彗星见。
《高丽史》卷47

公元990年2月2日 [中国]
宋太宗淳化元年正月辛巳，客星出轸宿，逆至张，七十日，经四十度乃不见。
《宋史·天文志》卷56

公元998年2月12日—23日 [中国]
(a) 宋真宗咸平元年正月十三日（2月12日），彗出营室北，光芒尺余…二十六日彗星灭。
《宋会要辑稿》 第52册
(b) 宋真宗咸平元年正月甲申（2月23日），彗出营室北。
《宋史·真宗纪》卷6
(c) 宋真宗咸平元年正月甲申（2月23日），彗星又出营室北，光芒尺余，至丁酉，凡十四日灭。
《宋史·天文志》卷56

公元998年2月25日 [日本]
(a) 日本长德四年正月廿六日丙戌，寅时彗星见东方，长四尺许。
《诸道勘文》 45
(b) 日本长德四年正月廿六日，寅刻彗星见东方，长四尺许。
《一代要记》 丁

公元1003年12月24日 [中国]
(a) 宋真宗咸平六年十一月甲寅，有星孛于井、鬼。
《宋史·真宗纪》卷7
(b) 宋真宗咸平六年十一月甲寅，有彗孛于井、鬼，大如杯，色青白，光芒四尺余，历五诸侯及五车入参，凡三十余日没。
《宋史·天文志》卷56

公元1005年10月1日—4日 [中国]
(a) 宋真宗景德二年八月辛丑（10月1日），有星孛于紫微。
《宋史·真宗纪》卷7

(b) 宋真宗景德二年八月甲辰（10月4日），客星出紫微天棓侧，孛孛然如粉絮，稍入垣内，历御女、华盖，凡十一日没。

《宋史·天文志》卷56

公元1014年2月12日 [中国]

辽圣宗开泰三年正月丁酉，是夕，彗星见西方。

《辽史·圣宗纪》卷15

公元1014年2月24日 [中国]

宋真宗大中祥符七年正月二十二日，夜，有星出东方，光芒二尺余，司天言；含誉星也。

《宋会要辑稿》 第52册

公元1014年2月27日 [朝鲜]

高丽显宗五年正月壬子，彗见于五车。二月庚申，彗入大陵。

《高丽史》卷47

公元1014年3月1日 [日本]

日本长和三年正月二十七日，自去正月廿七日戌刻，彗星见天船北卷舌南，长二尺。

《小右记》

注：“天船北卷舌”误写为“天船此卷舌”。

公元1014年9月29日 [中国]

宋真宗大中祥符七年九月丙戌，含誉星又见，似彗，有尾而不长。

《宋史·天文志》卷56

公元1018年8月2日—3日 [日本]

(a) 日本宽仁二年六月十八日（8月2日）。六月廿三日甲寅…始自去十八日，戌亥角彗星出…如此异星，先例参勘文之后不经几灭藏云云。

《左经记》

(b) 日本宽仁二年六月十九日（8月3日），彗星见，经数日，长二丈余。

《诸道勘文》 45

公元1018年8月3日 ［朝鲜］
高丽显宗九年六月庚戌，彗出北斗第二星，光射文昌、天牢，长四丈余。
《高丽史》卷47

公元1018年8月4日 ［中国］
(a) 宋真宗天禧二年六月辛亥（二十日），彗出北斗魁…七月丁亥，彗没。
《宋史·真宗纪》卷8
(b) 宋真宗天禧二年六月辛亥（二十日），彗出北斗魁第二星东北，长三尺许，与北斗第一星齐，北行经天牢，拂文昌，长三丈余，历紫微、三台、轩辕速行而西，至七星，凡三十七日没。
《宋史·天文志》卷56
(c) 宋真宗天禧二年六月辛亥（二十日），有彗出北斗魁，北行经天…二十七日夕，彗始灭，凡三十七日。
《宋会要辑稿》 第52册
注：《古今图书集成》和《宋会要辑稿》均作三年六月辛亥。《宋会要辑稿》"二十七日夕"当漏"七月"二字。

公元1019年2月6日 ［朝鲜］
高丽显宗九年十二月丁巳，彗见于天市垣、宗正、宗人、市楼间，指西。
《高丽史》卷47

公元1020年1月26日 ［朝鲜］
高丽显宗十年十二月辛亥，彗见宗正、宗人、市楼间。
《高丽史》卷47

公元1021年5月25日 ［中国］
(a) 宋真宗天禧五年四月丙辰，客星出轩辕。
《宋史·真宗纪》卷8
(b) 宋真宗天禧五年四月丙辰，客星出轩辕前星西北，大如桃，速行，经轩辕大星入太微垣，掩右执法，犯次将，历屏星西北，凡七十五日入浊没。

《宋史·天文志》卷56

(c) 宋真宗天禧五年四月丙辰，有星出轩辕前星，大如桃，状若粉絮，犯次将，入太微垣，历屏星，凡七十五日，入浊没。

《宋史·天文志》卷57

公元1021年5月31日 [朝鲜]

高丽显宗十二年四月壬戌，妖星见于太微。

《高丽史》卷47

公元1032年7月15日 [中国]

宋仁宗明道元年六月乙巳，客星出东北方，近浊，有芒彗，至丁巳，凡十三日没。

《宋史·天文志》卷56

公元1033年3月5日 [中国]

明道二年二月戊戌，含誉星见东北方，其色黄白，光芒长二尺许。

公元1033年3月5日 [日本]

(a) 日本长元六年二月二日，彗星见，长六七尺，经十箇夜。同六年二月三日己亥，晓寅刻彗星见东方，长六尺许，色白。

《诸道勘文》 45

(b) 日本长元六年二月二日，彗星见，长六七尺，经十二箇夜。长元六年二月三日己亥，晓寅时彗星见东方，长六尺许，色白。

《一代要记》 丁

注：原文均误写为“长元二年”，系“六年”之误。

公元1034年9月20日 [中国]

(a) 宋仁宗景祐元年八月壬戌，有星孛于张、翼。

《宋史·仁宗纪》卷10

(b) 宋仁宗景祐元年八月壬戌，夜，有星孛于张、翼，长七尺，阔五寸，十二日而没。

《宋史·天文志》卷56

(c) 宋仁宗景祐元年八月壬戌，青黄白气如彗，长七尺余，出张、翼之上，凡三十三日不见。
《宋史·天文志》卷60

公元1037年3月19日 [朝鲜]
高丽靖宗三年二月癸酉，彗星五出，长各五六尺。
《高丽史》卷47

公元1041年8月30日—9月27日 [朝鲜]
高丽靖宗七年八月，彗星见东方，长三十尺许，二十余日乃灭。
《高丽史》卷47

公元1041年9月1日 [中国]
宋仁宗庆历元年八月庚辰，夜，东方有白气，长十尺许，在星宿度中，至十日，长丈余，冲天相，居星宿大星南，九十余日没。
《宋史·天文志》卷60

公元1041年10月28日—11月26日 [朝鲜]
高丽靖宗七年十月，彗星长三十尺许，出东方十余日 。
《高丽史》卷47

公元1049年3月10日 [中国]
宋仁宗皇祐元年二月丁卯，彗出虚，晨见东方，西南指，历紫微至娄，凡一百一十四日而没。
《宋史·天文志》卷56
注：《契丹国志》作"凡一百二十四日而没。"

公元1053年2月25日 [朝鲜]
高丽文宗七年二月乙亥，彗出库楼入翼，长丈余。
《高丽史》卷47

公元1056年8月14日—9月11日 [中国]
(a) 宋仁宗嘉祐元年七月，彗出紫微垣，长丈余。八月癸亥，是夕彗灭。

《宋史·仁宗纪》卷12

(b) 宋仁宗嘉祐元年七月，彗出紫微，历七星，其色白，长丈余。至八月癸亥灭。

《宋史·天文志》卷56

公元1056年9月10日 [日本]

日本天喜四年七月廿八日，晓彗星始见东方，及八月七八日，长二丈许。

《诸道勘文》 45

公元1056年12月10日—1057年1月7日

高丽文宗十年十一月，彗见于轸。

《增补文献备考》 卷6

公元1060年12月22日 [日本]

(a) 日本康平三年十一月二十七日，彗星见南方，长五尺许，经七箇夜。

《诸道勘文》 45

(b) 日本康平三年十一月二十七日，彗星见东方，长五尺许，七个夜。

《一代要记》 丁

公元1063年6月8日 [朝鲜]

高丽文宗十七年五月庚戌，有星出大角，孛于氐。

《高丽史》卷47

公元1063年8月22日 [朝鲜]

高丽文宗十七年七月乙丑，有星尾长数尺，出坎，孛于氐，大如斗。

《高丽史》卷47

公元1066年 [中国]

(a) 宋英宗治平三年三月己未(1066年4月2日)，彗出营室，晨见东方，长七尺许，西南指危泊坟墓，渐东速行近日而伏；

至辛巳(4月24日)，夕见西南，北有星无芒彗，益东方，别有白气一，阔三尺许，贯紫微极星并房宿，首尾入浊，益东行，历文昌、北斗贯尾；至壬午(4月25日)，星复有芒彗，长丈余，阔三尺余，东北指，历五车，白气为歧横天，贯北河、五诸候，轩辕、太微五帝座内五诸候及角、亢、氐、房宿；癸未(4月26日)，彗长丈五尺，星有彗气如一升器，历营宿至张，凡一十四舍，积六十七日，星气孛皆灭。

《宋史·天文志》卷56

(b) 宋英宗治平三年春三月庚申(1066年4月3日)，彗星晨见于室。…辛巳(4月24日)，彗星见于昴，如太白，长丈有五尺。壬午(4月25日)，孛于毕，如月。 …五月乙丑(6月7日)，彗至张而没。

《宋史·英宗》卷13

(c) 辽道宗咸雍二年三月壬午(1066年4月25日)，彗星见于西方。

《辽史·道宗二》卷22

公元1066年4月3日 [日本]

(a) 治历二年三月六日(1066年4月3日)，晓，彗星东方见。四月一日(4月27日)，酉刻又见西方。

《十三代要略》 2

(b) 治历三年（注：二年之误）三月(1066年3月29日 - 4月26日)，见东方，经二十日不见，其后又见西方。

《山槐记》

(c) 治历二年三月六日(1066年4月3日)晓，彗星东方见；二十八日壬午（4月5日），亥刻，天有光景；四月一日(4月27日)酉刻，彗星西方见，司天之所奏，灾孽可慎。

《扶桑略记》29

公元1066年4月20日 [朝鲜]

文宗二十年三月丁丑(1066年4月20日)，有星出乾方，大如月，俄变为彗孛。

《高丽史》卷47

公元1075年11月17日　[中国]

(a) 宋神宗熙宁八年十月乙未，彗出轸。丁未，彗不见。

《宋史·神宗纪》卷15

(b) 宋神宗熙宁八年十月乙未，星出轸度中，如填，青白。丙申，西北生光芒，长三尺，斜指轸，若彗。丁酉，光芒长五尺，戊戌，长七尺，斜指左辖，至丁未入浊不见。

《宋史·天文志》卷56

公元1075年11月18日　[朝鲜]

高丽文宗二十九年十月丙申，彗见于轸星，长七尺余。

《高丽史》卷47

公元1075年11月19日　[日本]

(a) 日本承保二年十月九日，晓彗星见，五箇夜。

《诸道勘文》 45

(b) 日本承保二年十月九日，彗星见，五个夜。

《一代要记》 戊

公元1080年1月6日　[中国]

辽道宗大康五年十二月丙午，彗星犯尾。

《辽史·道宗纪》卷24

公元1080年8月10日　[中国]

(a) 宋神宗元丰三年七月癸未，彗出西北太微垣郎位南，白气长一丈，斜指东南，在轸度中，丙戌，向西北行，在翼度中，戊子，长三尺，斜穿郎位，癸卯，犯轩辕，至丁酉入浊不见，庚子晨，复出于张度中，至戊子，凡三十有六日，没不见。

《宋史·天文志》卷56

(b) 宋神宗元丰三年七月癸未，彗出太微垣…八月戊午，彗不见。

《宋史·神宗纪》卷16

注：两条比对，癸卯应为癸巳，戊子应为戊午。

公元1092年1月8日 [中国]
宋哲宗元祐六年十一月辛亥，客星出参度中，犯掩厕星，壬子，犯九游星，十二月癸酉入奎，至七年三月辛亥乃散。
《宋史·天文志》卷56

公元1097年9月9日—10月7日 [中国]
(a) 宋哲宗绍圣四年八月己酉（10月6日），彗出西方...九月戊辰，彗灭。
《宋史·哲宗纪》卷18
(b) 宋哲宗绍圣四年八月己酉（10月6日），彗出氐度中，如填，有光，色白，气长三丈，斜指天市左星，九月壬子，光芒长五尺，入天市垣，己未，犯天市垣宦者，庚申，犯天市垣
帝座，戊辰，没不见。
《宋史·天文志》卷56
(c) 辽道宗寿昌三年八月（9月9日—10月7日），彗出氐，斜指天市垣，光芒三尺余，越三夕，长丈余，扫巴星。
《契丹国志》卷9

公元1097年10月4日 [朝鲜]
高丽肃宗二年八月丁未，彗见氐房间，光芒射天市垣。
《高丽史》卷47

公元1097年10月8日 [日本]
(a) 日本承德元年九月一日，今月一日辛亥，昏酉刻彗星出于西方，芒气及天市垣，长一许丈，色白，其后东行。同二三四日见天市垣中，指震方。同五日，天阴不见。六日北行见天市垣北，长七八丈，光渐微。其后连夜天阴不见。十八日夜天晴，消而不见。
《诸道勘文》 45
(b) 日本承德元年九月一日，彗星见西方。同十八日以后不见。
《一代要记》 戊

公元1106年2月9日 [日本]

(a) 日本嘉承元年正月四日，今月四日丁酉，昏酉刻彗星出于坤方，芒指震方，触天仓星及天苑星间，长十许丈，色白，同六日夜，光芒已微，长一许丈。同七日夜，颇东行，长三四尺许。同八九十日夜，天阴不见。同十一日夜，又东行。同十二三四日夜，天阴不见。同十五日夜，东行，长二丈许。

《诸道勘文》 45

(b) 日本嘉承元年正月四日，彗星见于坤方，长十许丈，卅余日而灭。

《百錬抄》 5

公元1106年2月9日 [朝鲜]

高宗睿宗元年正月丁酉，彗见西南，长十尺许，月余乃灭。

《高丽史》卷47

公元1106年2月10日 [中国]

(a) 宋徽宗崇宁五年正月戊戌，彗出西方，其长竟天…三月丙申，诏星变已消，罢求直言。

《宋史·徽宗纪》卷20

(b) 宋徽宗崇宁五年正月戊戌，彗出西方，如杯口大，光芒散出如碎星，长六丈，阔三尺，斜指东北，自奎宿贯娄、胃、昴、毕，后入浊不见。

《宋史。天文志》卷56

公元1110年1月22日—1111年2月9日 [中国]

宋徽宗大观四年，有星孛于中宫，长数丈，始出王良、造父间，遂历阁道，逆行入紫宫，几遍扫垣内外座，已退，俄又进扫帝座者再，前后三十余日乃灭。

元 江苏《金陵新志》卷3

公元1110年5月29日 [中国]

(a) 宋徽宗大观四年五月丁未，彗出奎、娄。

《宋史·徽宗纪》卷20

(b) 宋徽宗大观四年五月丁未，彗出奎、娄，光芒长六尺，北行入紫微垣，至西北入浊不见。

《宋史·天文志》卷56

公元1110年5月31日 ［朝鲜］

高丽睿宗五年五月己酉，彗星入紫微。乙卯，夜二更，彗星发天将军阁道星间，至晓乃灭。丁巳夜，彗星发路策星。戊午，发于王良星西北。己未，发华盖传舍间。庚申夜，发华盖中。辛酉，发华盖下六甲星北。癸亥夜，行女御宫星内。

《高丽史》卷47

公元1110年6月1日 ［日本］

(a) 日本天永元年五月十二日，彗星现东方，其光五六尺许。

《东寺长者补任》2

(b) 日本天永元年五月十二日...去五月十二日出东方，其光六尺云云...（六月）九日，远晴，彗星灭光。十日，星光失了。

《孔雀经御修法记》

公元1111年8月12日 ［中国］

宋徽宗政和元年七月六日，有彗星出紫微垣，历七星，其色白，长丈余，自是至八月十四日夜灭。

《宋会要辑稿》第52册

公元1113年8月15日 ［朝鲜］

高丽睿宗八年七月辛巳，有星孛于营室。

《高丽史》卷47

公元1123年2月2日 ［中国］

宋徽宗宣和五年正月五日，彗出西方，由奎贯娄、胃、昴、毕...至二十五日彗没。

《宋会要辑稿》第52册

公元1123年8月11日　[朝鲜]
高丽仁宗元年七月己巳，有星孛于北斗。
《高丽史》卷47

公元1126年7月19日　[中国]
宋钦宗靖康元年六月壬戌，彗出紫微垣。
《宋史·钦宗纪》卷23

公元1126年7月22日　[日本]
(a) 日本大治元年七月一日乙丑以后，彗星见北，长或三尺，或丈馀。
《诸道勘文》45
(b) 日本大治元年七月朔日，彗星见北方，长三尺许。
《一代要记》戊

公元1127年1月15日　[中国]
宋钦宗靖康元年十二月壬戌，彗出紫微垣东北，长数丈，拂帝座，扫文昌。彗见，有白气出太微垣。
元 江苏《金陵新志》卷3

公元1132年1月5日　[中国]
(a) 宋高宗绍兴元年十二月戊寅，彗星见于胃。
《宋史·天文志》卷56
(b) 宋高宗绍兴元年十二月戊寅，以彗出，求直言。
《宋史·高宗纪》卷26

公元1132年10月5日　[日本]
(a) 日本长承元年八月二十五日。八月廿八日…去月廿五日壬子夜，晓寅刻彗星见于北方，长三尺，其色白，尾指西，在觜度，近八穀星。同廿六日夜，天阴不见。廿七日，夜亥刻见寅方，与娄第三星相近，长三丈余，光芒殊盛，尾指戌亥。廿八日夜，南行，其光衰微，长一丈余，芒气相及奎宿。廿九日夜，在同度，犹南行，与土司空星相並，芒角减少，长二三尺所。卅日夜，今月（九月）一日夜，共天阴不见。二日夜以后，消不见。

《中右记》
注：“八穀星”误写为“入敷星”；“土司空星”误写 为“下司空星”。
(b) 日本长承元年八月二十五日，今晓彗星见北方。二十七日，行巽方，长三丈余。
《十三代要略》 2

公元1132年10月5日 [朝鲜]
高丽仁宗十年八月壬子，彗星见八穀，指东南。甲寅，指西北，长三尺。
《高丽史》卷47

公元1132年10月6日—7日 [中国]
(a) 宋高宗绍兴二年八月二十六日（10月6日），太史言："是夜四更，彗出位（胃）宿度内，如火木星，" ...夜来初更奏；又犯土司空星...至九月十七日彗星消。
《宋会要辑稿》 第52册
(b) 宋高宗绍兴二年八月甲寅（10月7日），彗出胃...九月甲戌，彗没。
《宋史·高宗纪》卷27
(c) 宋高宗绍兴二年八月甲寅（10月7日），彗星见于胃，丙辰，行犯土司空，至九月甲戌始灭。
《宋史·天文志》卷56

公元1138年8月27日 [日本]
日本保延四年七月二十日，彗星见戌亥方，长五六尺，色白，两三日以后渐微少不见。
《一代要记》 戊
注：原文“四年”误写为“五年”。

公元1138年8月27日 [日本]
日本保延四年七月二十日，彗星见戌亥方，长五六尺，色白，两三日以后渐微少不见。
《一代要记》 戊
注：原文“四年”误写为“五年”。

公元1138年9月3日 [中国]
宋高宗绍兴八年七月辛亥，彗出东方...八月丁丑，彗灭。
《宋史·高宗纪》卷29

公元1142年8月8日 [中国]
宋高宗绍兴十二年七月丙午，彗见东北井宿内。
清 康熙《陕西通志》卷30

公元1145年4月26日 [中国]
(a) 宋高宗绍兴十五年四月戊寅(1145年4月26日)，彗星见东方，丙申(5月14日)，复见于参度，五月丁巳(6月4日)，化为客星，其色青白，壬戌(6月9日)，留守张，至六月丁亥(7月4日)乃消。
《宋史·天文志》卷56
(b) 宋高宗绍兴十五年四月戊寅（三日）(1145年4月26日)，彗星出东方。...丁亥(5月5日)，以彗出大赦。癸巳(5月11日)，彗没。...五月丙辰(6.3)，客星见。...六月丁亥(7月4日)，客星没。
《宋史·高宗七》卷30
(c) 金熙宗皇统五年四月丙申(1145年5月14日)，彗星见于西北，长丈余，至五月壬戌(6月9日)始灭。
《金史·天文》卷20

公元1145年5月3日 [日本]
(a) 久安元年四月十日乙酉(1145年5月3日)天文权博士晴道未云，有星孛于奎。十六日辛卯(5月9日)，晓，彗星甲重，其光指西，二许丈。廿日乙未(5月13日)，师长未云，自四日(4月27日)夜，始见彗星。闾巷说，去月晦(4月23日)顷始出云云。廿一日丙申(5月14日)彗星奉币即问师安，申云，天禄元年依彗星奉币用复日，何览定乎申刻许师安未云，自今晓彗星没不见。廿三日戊戌(5月16日)，彗星见辛方，光长五许尺，指东，但,其残光云掩不见，长短难定， 余出见之。廿四日己亥(5月17日)，天晴，彗星见，光长二丈许。廿五日庚子(5月18日)，天晴，午以后阴不雨光房未，使头仰云，东方彗星没。复自二十三日(5月16日)，又见西方。五月七日壬子(5月30日)，彗星光渐微细，但若被消月光欤。十二日丁巳(6月4日)，彗星犹见，但无光芒。（十三日(6月5日)又同）去八日其光薄细，九、十、十一、三个日天阴，今夕唯有星而已，无芒气，师长云，先例芒气尽时，星又不见，未知如此之例。十六日(6月8日)夜见之，光芒三许尺，以之思之，十二日依月近其光不见乎，十六日(6月8日)依与月远，其光见乎，十四、五日(6月7日，6月8日)天阴不见。十六日辛酉(6月8日)，彗星又见光芒。廿日乙丑(6月12日)，彗星犹不消，孔雀经法，又无其验六月七日辛巳(6月28日)，师安耒云，去夜彗星不见，去月廿六日(6月18日)夜，天晴彗星见，其后每夜阴，不知有无，去夜属晴不见。

《台记》 5

(b) 久安元年四月五日庚辰(1145年4月28日)，寅时彗星见东方，(长一丈，色白，自今日至某日)。廿三日戊戌(5月16日)，自今夕彗星见西方。廿五日庚子(5月18日)，依彗星御祈，被立廿二社奉币使，上卿内大臣，宣命云，今月五日晓，彗星见东方，贯犯奎宿，既经数夜，至廿日(5月13日)芒气不见，天文道奏，然间廿三日(5月16日)夕见西方。七月廿二日丙寅，有改元事，依彗星也。

《本朝世纪》 29

公元1145年5月14日 [朝鲜]

仁宗二十三年四月丙申(1145年5月14日)，彗星见乾方十五日，长丈余。

《高丽史》卷47

公元1146年12月29日 [中国]

宋高宗绍兴十六年十一月丙寅，彗星见西南危宿。

《宋史·天文志》卷56

公元1147年1月4日—2月13日 [日本]

(a) 日本久安二年十二月三日戊戌（1月6日），彗星在危室之间，光长一丈许也。十九日甲寅，今夜彗星微薄，如不见。

《本朝世纪》 31

(b) 日本久安三年正月十二日丙子（2月13日），今晓寅刻彗星见东方，光长一丈，在女虚间，其光近匏瓜星也。十三日丁丑，今晓彗星度分同昨日，但颇倚北。廿四日戊子，今夜彗星不见。

《本朝世纪》 32

注：原文"匏瓜星"误写为"匏攸星"。

(c) 日本久安二年十二月一日（1月4日），彗星见坤方。三年正月十二日，彗星见东方。

《百錬抄》 7

公元1147年1月6日 [中国]

(a) 宋高宗绍兴十六年十二月戊戌（三日），彗见西南方，乙巳，灭。

《宋史·高宗纪》卷30

(b) 宋高宗绍兴十六年十二月戊戌（三日），是夜彗见西南方...至是月九日夜消。

《宋会要辑稿》 第52册

公元1147年2月8日 [中国]

金熙宗皇统七年正月辛未，彗星出东方，长丈余，凡十五日灭。

《金史·天文志》卷20

公元1147年2月12日 [中国]

宋高宗绍兴十七年正月乙亥，妖星出东北方女宿内，小如岁星，光芒长五丈，二月丙寅始消。

《宋史·天文志》卷60

公元1156年7月25日—26日 [中国]

(a) 宋高宗绍兴二十六年七月丙午（7月25日），彗星见东井，约长一丈，光芒二尺，癸丑，又犯五诸侯。

《宋史·天文志》卷56

(b) 宋高宗绍兴二十六年七月丙午（7月25日），彗见东北方井宿内，丁未其星如木星，光芒长二尺。

《文献通考》卷286

(c) 宋高宗绍兴二十六年七月丁未（7月26日），彗出井，丙辰，彗灭。

《宋史·高宗纪》卷31

(d) 宋高宗绍兴二十六年七月八日（7月26日），彗出井宿…至是月十七日消伏。

《宋会要辑稿》第52册

公元1156年7月29日　［日本］

日本保元元年七月十一日，晓彗星见寅方，长六尺许，色白。同十二三日间，东北行，至十五日晓，犯五诸侯三公星，相去三寸许。

《一代要记》己

公元1161年7月22日　［中国］

宋高宗绍兴三十一年六月己巳，彗星见北斗天权星东北，太史妄称为含誉。

《宋史·天文志》卷56

公元1166年4月23日　［日本］

(a) 日本仁安元年三月廿二日乙丑，夜亥时客彗见于太微宫中，在屏星口，其光芒指右执法星，长三尺许，色白。同廿三日以后，彗星北。廿九日，近五帝座，相去一尺许，芒气指黄帝并白帝座。卅日，天阴不见。其后连阴。四月十一日，天晴，彗星销不见。

《泰亲朝臣记》

(b) 日本仁安元年三月二十二日，彗星见太微宫中，在右掖门中，出屏西，色白，长三尺，贯执法星。四月十一日，消不见，现二十箇日。

《一代要记》己

公元1166年5月1日　［中国］

乾道二年三月癸酉，出太微垣内五帝坐大星西，微小，色青白。

《宋史·天文志》卷56

公元1178年1月14日　［日本］

(a) 日本治承元年十二月二十四日。治承二年正月七日壬寅…今晓巽方彗星出……去年十二月廿四日出，其后不出，今晓又出也。

《山槐记》

(b) 日本治承元年十二月二十四日。治承二年正月七日，寅刻彗星见巽方之由，泰亲朝臣奏闻，又去年十二月廿四日出现云云。

《百錬抄》 8

公元1189年3月16日 [日本]

日本文治五年二月二十八日，彗星见东方，在太微北郎位傍，长丈余，色赤白。

《一代要记》 庚

注：原文将"太微北郎位"误写为"太微北即位"。

公元1210年10月19日 [日本]

(a) 日本承元四年九月卅日乙卯，晴，戌刻西方天市垣第三星傍见奇星，光指东方，三尺余，芒气殊盛，长一许丈，此星如本文者为彗星之由，有申之辈云云。

《吾妻镜》 19

(b) 日本承元四年九月卅日，戌时彗星见坤方。

《一代要记》庚

公元1220年1月25日 [日本]

日本承久元年十二月十八日，戌时彗星见乾方，在腾蛇星东，长一尺，色赤。

《一代要记》 庚

公元1220年2月6日 [朝鲜]

高丽高宗七年正月壬辰，彗出钩星，尾指西北，长三尺许。

《高丽史》卷48

公元1220年3月21日 [朝鲜]

高丽高宗七年二月丙子，有星孛于轩辕。

《高丽史》卷48

公元1220年12月27日—1221年1月24日 [朝鲜]

高丽高宗七年十二月，有星孛于北斗。

《高丽史》卷48

公元1222年9月10日 [中国]

(a) 金宣宗兴定六年（元光元年）八月己卯(1222年9月10日)，彗星出于亢宿、右摄提、周鼎之间，指大角。九月丁未(10月8日)灭。

《金史·天文》卷20

(b) 宋宁宗嘉定十五年八月甲午(1222年9月25日)，有彗星出于氐。...九月壬戌(10月23日)，彗星没。

《宋史·宁宗四》卷40

(c) 宋宁宗嘉定十五年八月甲午(1222年9月25日)，彗星见右摄提，光芒三尺余，体类岁星，凡两月，历氐、房、心乃没。

《宋史·天文志》卷56

公元1222年9月8日 [日本]

(a) 贞应元年八月二日丁丑(1222年9月8日)，霁，戌刻，彗星见戌方，轴星大如半月，色白光芒赤，长一丈七尺余。十三日戊子(9月19日)，晴，自二日至去夜（十二日），彗星连夜出见十五日庚寅(9月21日)，霁今夜彗星光才一尺余，自二日每夜出现云云。...廿二日丁酉(9月28日)，彗星犹见坤方，长五尺。廿九日甲辰(10月5日)，晴，彗星变不见之由，司天等申之。

《吾妻镜》 26

(b) 贞应元年八月一日(1222年9月7日)，自今夜彗星见乾方。

《百炼抄》13

公元1222年9月3日 [朝鲜]

高宗九年七月壬申(1222年9月3日)，彗星出三台中，尾指西，长三尺许。甲戌(9月5日)流星出虚入羽林，大如缶，长三尺许。乙亥(9月6日)彗星见西北，长三尺许。八月丁丑(9月8日)见乾方，长二十尺许，戊寅(9月9日)昼见，又太白昼见，经天。

《高丽史》卷48

(g) （越南）建嘉十二年秋八月(1222年9月7日~10月6日)，彗星见于西南方。

《越史略》

公元1230年12月4日　　[日本]
日本宽喜二年十月二十八日西方客星出…三日庚寅…四日辛卯，天晴…夜天晴见奇星，此星胧胧光薄，其势非小。去二日泰俊朝臣示送，自一昨日夜晦日，奇星现辛方，在织女东天津艮奚仲傍，其体客彗之疑候，未及光芒…十一月二十五日，客星晓迴南方。
《明月记》 野宫本23

公元1230年12月15日　　[中国]
宋理宗绍定三年十一月丁酉，有星孛于天市垣屠肆星之下，明年二月壬午乃消。
《宋史·天文志》卷56

公元1231年　　[中国]
金哀宗正大七年十二月庚寅，有星出天津下，大如镇星而色不明，初犯辇道，二日见于东北，在织女南，乙未，入天市垣，戊申方出；癸丑，历房北，复东南行，入积薪，凡二十五日而灭。
《金史·天文志》卷20
注：该月无庚寅。

公元1232年10月16日—11月13日　　[中国]
宋理宗绍定五年闰九月，彗星见东方，十月己未始消。
《宋史·天文志》卷56

公元1232年10月17日—18日　　[中国]

(a) 金哀宗天兴元年闰九月己酉（10月17日），彗星见东方，色白，长丈余，弯曲如象牙，出角、轸南行，至十二日长二丈，十六日月烛不见。二十七日五更复出东南，约长四丈余，至十月一日始灭，凡四十有八日。

《金史·天文志》卷20

(b) 宋理宗绍定五年闰月庚戌（10月18日），彗星出于角。

《宋史·理宗纪》卷41

公元1232年10月19日 ［日本］
日本贞永元年闰九月四日辛亥，霁，寅刻彗气见乙方，指庚方，长二尺，广八寸，色白赤…八日乙卯，此两三日，或阴或雨降，今晓适见青天，彗星犹出现增光，芒气长二丈，广一尺余，指同方，逆行一许丈，南行四尺，去东山五尺许也…十五日壬戌，晴，彗微薄，芒气不见…十月五日辛巳，晴，自去月廿八日至于今晓，彗星连夜出现，光芒虽微薄，指乾方见南，长四丈。
《吾妻镜》 吉川本27

公元1240年1月27日—29日 ［日本］
(a) 日本仁治元年正月二日丁卯（1月27日），天晴…戌刻彗星出现申方，芒气三尺，指巽方，色白赤…七日壬申，天晴，戌刻彗星现岁星傍，相去三尺余所，芒气指艮方，光芒五尺，轴星大如太白。八日癸酉，戌刻彗星近岁星，相去二尺余所…十一日丙子，戌刻彗星犯壁一星，相去一尺余所。同时月入东井星。十八日癸未，彗星近奎。十九日甲申，彗星入奎中…自去十七日至今夜，光芒次第盛也。廿六日辛卯，戌刻彗星出现，犯王良第五星，光芒微薄。二月六日辛丑…入夜彗星出现，自正月四日至今日不消没。
《吾妻镜》 33
(b) 日本仁治元年正月四日己巳（1月29日），戌刻彗星见坤方，光芒三丈许，万人驚目。五六日天阴之间，彗星不见。八日又出见，芒气著东方，二丈余，或时一丈余也。
《百錬抄》 14

公元1240年1月31日 ［中国］
(a) 宋理宗嘉熙四年正月辛未，彗星出营室。甲午，彗星犯王良第二星。三月辛未，彗星消伏。
《宋史·理宗纪》卷42
(b) 宋理宗嘉熙四年正月辛未，彗星见于室，至三月辛未乃消。
《宋史·天文志》卷56

公元1245年2月24日—25日 [日本]

(a) 日本宽元三年正月廿六日壬戌（2月24日），天晴，今晓客星出现云云…三月一日丙申，天晴，今晓彗星见东方…六日辛丑，自今晓彗星不见。

《百鍊抄》 15

(b) 日本宽元三年正月廿七日癸亥（2月25日），天晴，丑刻客星出现于天市垣巽斗度云云。廿八日甲子，天晴，寅刻客星犹牛宿南出现…二月一日丙寅，天晴，客星见牵牛度，行度二夜三丈口也云云…三月一日丙申，天晴，寅刻彗星见室壁之间，长二尺云云。

《吾妻镜》 36

公元1264年7月21日 [日本]

(a) 日本文永元年六月二十六日。七月四日丙子，今晓彗星见于寅方，长四、五尺，色白。去月廿六日，始见于戌方云云。五日丁丑，卯刻彗星见于东方，光芒弥大也…十一日癸未，彗星犹出现，六七丈许云云…十七日己丑，彗星犹出现，光芒渐微…廿一日癸巳，彗星犹出现，光芒又增长…九月二日癸酉，彗星出现，光芒十余丈。

《外记日记》

(b) 日本文永元年六月廿六、七日两日，彗星见艮。七月四日，晓彗星见东方，长三尺许。七日，晓光芒大焉。八月二十二日，彗星不见。二十六日，彗星又见。

《一代要记》 壬

公元1264年7月26日 [中国]

(a) 宋理宗景定五年七月甲戌，彗星出柳。己卯，彗星退于鬼。辛巳，彗星退于井。戊戌，彗星退于参。八月戊午，彗星消伏。甲子，彗星复见于参。辛未，彗星化为霞气。

《宋史·理宗纪》卷45

(b) 宋理宗景定五年七月甲戌，彗星见于柳。芒角烛天，长十余丈，日高方敛，凡月余；己卯，退行见于舆鬼，辛巳，在井；丙申，见于参；戊戌，在参宿度内，八月末，光芒稍减，凡四月乃灭。

《宋史·天文志》卷56

公元1264年7月26日 ［中国］

(a) 元世祖至元元年七月甲戌，彗星出舆鬼，昏见西北，贯上台，扫紫微、文昌及北斗，且见东北，凡四十余日。

《元史·世祖纪》卷5

《元史·天文志》卷48

(b) 元世祖至元元年七月甲戌，彗星出鬼柳，昏见西北，光烛天，长数十丈，贯上台，扫紫微、文昌、北斗，且见东北，凡四十余日。

《元史类编》卷2

公元1264年7月26日 ［朝鲜］

高丽元宗五年七月甲戌，彗星始见于艮方，尾长七、八尺，渐分为五，向西北方…八月壬寅，光芒复合为一而渐长。壬戌，彗星光芒益炽，竟天…九月乙酉，彗星灭。自七月至是，凡七十二日。

《高丽史》卷48

公元1266年1月18日—21日 ［日本］

(a) 日本文永二年十二月十一日乙亥（1月18日），今晓彗星见出申方，光芒一尺余…三年正月一日乙未，今夕彗星犹出现西方，光芒一丈许…九日癸卯，今日彗星出现。

《外记日记》

(b) 日本文永二年十二月十四日戊寅（1月21日），霁，今晓彗星见东方…廿七日辛卯，霁，今日彗星见西方，在室宿，芒气二尺余，色白。三年正月一日乙未，天晴风静…昏黑彗星见西，壁八度。

《吾妻镜》 52

注：原文将“在室宿”误写为“有室宿”；将“壁八度”误写为“辟八度”。

公元1273年2月5日—6日 ［日本］

(a) 日本文永十年正月十六日（2月5日），戌时许彗星出现西方，长丈余，其色白，此后夜夜出现。

《师守记》 5

(b) 日本文永十年正月十七日（2月6日），酉刻彗星见，一尺余，及数日。

《关东评定传》

公元1273年4月9日 ［中国］
元世祖至元十年三月癸酉，客星青白如粉絮，起毕，度五车北，复自文昌贯斗杓，历梗河，至左摄提，凡二十一日。
《元史·世祖纪》卷8
《元史·天文志》卷48

公元1277年3月9日 ［日本］
日本建治三年二月四日，今夜彗星出现，光芒三尺许云云。昨日夜同出现，此后度度出现。
《师守记》 5

公元1293年11月7日 ［中国］
元世祖至元三十年十月庚寅，彗星入紫微垣，抵斗魁，光芒尺许，凡一月乃灭。
《元史·世祖纪》卷17
《元史·天文志》卷48
注：《续文献通考》作"抵北斗"。

公元1293年11月13日 ［日本］
日本永仁元年十月十四日，卯刻客现出东方，经旬日，芒气现，为彗星。
《北條九代记》下

公元1293年11月21日 ［朝鲜］
高丽忠烈王十九年十月甲辰，彗见东方。戊申，彗星见于太微左掖门，长尺五寸许…十一月丙子，彗星犯紫微，又犯北斗。
《高丽史》卷49

公元1297年3月12日 ［朝鲜］
高丽忠烈王二十三年二月辛亥，彗见东井，六日乃灭。三月甲子，彗见东井。
《高丽史》卷49

公元1297年3月13日 ［日本］
(a) 日本永仁五年二月十九日，今夜彗星见东方。

《师守记》 5

(b) 日本永仁五年二月十九日，戌刻彗星出现，芒气五六尺。

《北條九代记》 下

公元1297年9月14日—21日 ［日本］

(a) 日本永仁五年八月廿六日（9月14日），丑刻客星见东方，其体如粉絮。

《北條九代记》 下

(b) 日本永仁五年九月三日（9月21日），戌刻彗星出现东方乙方，光芒一尺余，其色白，此后夜夜出现。

《师守记》 5

公元1299年1月6日 ［日本］

(a) 日本永仁六年十二月三日，今晓彗星出现南方，当角宿度，光芒指乾方，长一尺七寸许，其色白，此后夜夜出现。七年正月一日，彗星出现，此后度度出现。

《师守记》 5

(b) 日本永仁六年十二月三日，晓彗星现南方，光芒一尺余。廿日，夜彗星出现南方，光芒七尺。

《北條九代记》 下

公元1299年1月24日 ［中国］

元成宗大德二年十二月甲戌，彗出子孙星下。

《元史·成宗纪》卷19

《元史·天文志》卷48

公元1299年10月23日 ［日本］

日本正安元年九月廿八日，寅刻彗星出现辰方，光芒二丈。

《北條九代记》 下

公元1301年 ［中国］

(a) 元成宗大德五年八月庚辰(1301年9月16日)，彗出井二十四度四十分，如南河大星，色白，长五尺，直西北，后经文昌斗魁，南扫太阳，又扫北斗、天玑、紫微垣、三公、贯索，星长丈余，至天市垣，巴蜀之东，梁楚之南，宋星上，长盈尺，凡四十六日而灭。

《元史·天文志》卷48

公元1301年9月15日 [日本]

(a) 正安三年八月十三日(1301年9月15日)，寅刻彗星出现东方，光芒三尺余，数日增光一丈。

《北条九代记》下

(b) 正安三年八月十四日(1301年9月16日)，今晓彗星出现东方。此后夜夜出现云云。同廿一日，今夜戌刻彗星又出现戌亥方，光芒一丈余。

《师守记》5

公元1301年9月14日 [朝鲜]

忠烈王二十七年八月戊寅(1301年9月14日)，彗星见于北斗、紫微。壬午(9月18日)见于北斗，乙未(10月1日)见于上台，入天市垣。九月壬寅(10月8日)，月入南斗，癸卯(10月9日)彗星见于天市垣。

《高丽史》卷49

公元1303年7月27日 [日本]

日本嘉元元年六月十三日，彗星见艮方，芒气一尺余，其色白，有轴星。去十日以后每晓出现云云。

《师守记》5

公元1304年2月3日 [朝鲜]

高丽忠烈王二十九年十二月庚戌，彗见西方。三十年正月甲寅，彗见于奎。

《高丽史》卷49

公元1304年2月4日 [日本]

日本嘉元元年十二月廿八日，彗星出现。二年正月一日，戌刻彗星出现未申方，光芒二尺余，此后夜夜出现。

《师守记》5

公元1304年4月18日 [中国]

(a) 元成宗大德八年三月乙丑，自去岁十二月庚戌，彗星见，约盈尺，在室十一度，入紫微垣，至是灭，凡七十四日。

《元史·成宗纪》卷21

(b) 元成宗大德八年三月乙丑，自去岁十二月庚戌，彗星见，约盈尺，指东南，色白，测在室十一度，渐长尺余，复指西北，扫腾蛇，入紫微垣，至是灭，凡七十四日。

《元史·天文志》卷48

公元1304年12月24日 [朝鲜]

高丽忠烈王三十年十一月乙亥，彗见虚、危间。十二月戊寅，彗星犯虚。庚辰，贯虚。辛巳，入危。

《高丽史》卷49

公元1304年12月26日 [日本]

日本嘉元二年十一月廿九日，今夜彗星见西方庚方，在虚宿南星之南，芒气一尺余，指东，色白，此后夜夜出现。

《师守记》 5

公元1305年6月19日 [日本]

日本嘉元三年五月廿七日，寅刻彗星出现艮方，六九已后不见云云。

《武家年代记》 下

注：“六九”指“六月九日”。

公元1307年8月24日 [朝鲜]

高丽忠烈王三十三年七月戊子，彗星见于尾。

《高丽史》卷49

公元1313年4月13日 [中国]

元仁宗皇庆二年三月丁未，彗出东井。

《元史·仁宗纪》卷24

《元史·天文志》卷48

公元1313年4月13日　　[朝鲜]
高丽忠宣王五年三月丁未，有星彗于东井。
《高丽史》卷49

公元1313年 4月14日　　[日本]
(a) 日本正和二年三月十八日，今夜戌时彗星见西方，芒气指东，色白。十九日同。
《师守记》5
(b) 日本正和二年三月十八日。三月十九日己酉，晴…去夜彗星见西方云云…彗星今夜又出现。廿九日己未，晴，今夜彗星见。
《花園院宸记》

公元1315年10月29日　　[中国]
元仁宗延祐二年十月丙子朔，客星见太微垣。十一月丙午，客星变为彗，犯紫微垣，历轸至壁十五宿，明年二月庚寅乃灭。
《元史·仁宗纪》卷25
《元史·天文志》卷48

公元1337年5月4日　　[中国]
元顺帝至元三年四月甲戌，有星孛于王良，至七月壬寅没于贯索。
《元史·顺帝纪》卷39

公元1337年6月25日　　[朝鲜]
高丽忠肃王后六年五月丙寅，彗见，　长丈余，自天船北至王良阁道。六月庚午朔，彗见艮方。癸酉，又见紫微、西藩、华盖、勾陈、北极。七月癸卯，又见紫微东藩。丁未，又犯贯索。八月庚辰，彗见天市垣，四十日乃灭。
《高丽史》卷49
《增补文献备考》卷6
注：《高丽史》将“七月癸卯”　误写为“七月癸巳”；将“丁未，又犯贯索” 误写为“丁酉，又犯贯星”；“八月庚辰”误写为“八月庚子”。

公元1337年6月26日　　[中国]
(a) 元顺帝至元三年五月丁卯，彗星见于东北，大如天船星，色白，约长尺余，彗指西南，至八月庚午始灭…八月庚午彗星不见，自五月丁卯始见，至是凡见六十三日，自昴至房，凡历一十五宿而灭。

《元史·顺帝纪》卷39

(b) 元顺帝至元三年五月丁卯，彗星见于东北，如天船星大，色白，约长尺余，彗指西南，测在昴五度…八月庚午，彗星不见。其星自五月丁卯始见，戊辰往西南行，日益渐速，至六月辛未，芒彗愈长，约二尺余，丁丑扫上丞，己卯光芒愈甚，约长三尺余，入圜卫。壬午扫华盖、杠星，乙酉扫勾陈大星及天皇大帝，丙戌贯四辅，经枢心，甲午出圜卫，丁酉出紫微垣，戊戌犯贯索，扫天纪，七月庚子扫河间，癸卯经郑晋，入天市垣，丙午扫列肆，己酉太阴光盛，微辨芒彗，出天市垣，扫梁星，至辛酉，光芒微小，瞻在房宿键闭之上、罚星中星正西，难测，日渐南行，至是凡见六十有三日，自昴至房，凡历一十五宿而灭。

《元史·天文志》卷49

公元1337年6月27日—7月19日 [日本]

(a) 日本建武四年五月廿八日(6月27日)，今夜寅刻彗星见东方。

《师守记》 5

(b) 日本建武四年六月廿一日(7月19日)，彗星侵北极。

《皇年代略记》；《皇年代私记》

公元1340年3月24日 [中国]

(a) 元顺帝至元六年二月己酉，彗星如房星大，色白，壮如粉絮，尾迹约长五寸余。彗指西南，渐向西北行。三月辛巳彗星不见，自二月己酉至三月庚辰，凡见三十二日。

《元史·顺帝纪》卷40

(b) 元顺帝至元六年二月己酉，彗星如房大，色白，壮如粉絮，尾迹约长五寸余。彗指西南，测在房七度，渐往西北行。三月辛巳，是夜彗星不见。自二月己酉至三月庚辰，凡见三十二日。

《元史·天文志》卷49

(c) 元顺帝至元六年二月己酉，有星孛于房，星大色白，壮如粉絮，尾迹五寸余。彗指东南，渐向西北行，阅三十二日始灭。

《元史类编》卷10

公元1340年3月25日 [朝鲜]

高丽忠惠王后元年二月庚戌，彗见东方，入太微。甲寅，又见东方。

《高丽史》卷49

公元1340年4月1日 [日本]

日本历应三年三月四日…降庭拜彗星…今夜己犯太微宫东蕃次相星。

《玉英记抄》神事

公元1345年7月31日 [朝鲜]

高丽忠穆王元年七月甲申，彗见紫微垣。丁亥，彗见北河北。

《高丽史》卷49

公元1345年8月2日 [日本]

日本兴国六年七月四日…今晓寅刻妖星出现云云…五日…东北方光芒已四尺许也，其色白殊重云云。

《園太历》 5

公元1351年11月24日—26日 [中国]

(a) 元顺帝至正十一年十一月辛亥（11月24日），孛星见于奎宿。癸丑，孛星见于娄宿。甲寅，孛星见于胃宿。乙卯亦如之。丙辰，孛星见于昴宿。丁巳，孛星微见于毕宿。

《元史·天文志》卷49

(b) 元顺帝至正十一年十一月癸丑（11月26日），有星孛于娄宿。甲寅，孛星见于胃宿。乙卯、丙辰，也如之。丁巳，孛星微见于毕宿。

《元史·顺帝纪》卷42

公元1356年9月21日 [中国]

(a) 元顺帝至正十六年八月甲戌，彗星见张宿，色青白，彗指西南，长尺余，至十月戊午始灭。

《元史·顺帝纪》卷44

(b) 元顺帝至正十六年八月甲戌，彗星见于正东，如轩辕左角大，色青白，彗指西南，约长尺余，测在张宿十七度一十分，至十月戊午灭迹，西北行四十余日。

《元史·天文志》卷49

注：《元史类编》 作"至十二月戊午朔始灭"。

公元1362年3月5日 [中国]

(a) 元顺帝至正二十二年二月乙酉，彗星见于危宿，光芒长丈余，色青白。丁酉，彗星犯离宫西星，至二月终，光芒长二丈余。三月戊申，彗星不见星形，惟有白气，形曲竞天，西指扫大角。壬子，彗星行过太阳前，惟有星形，无芒，在昴宿，至戊午始灭。

《元史·顺帝纪》卷46

(b) 元顺帝至正二十二年二月乙酉，彗星见，光芒约长尺余，色青白。测在危七度二十分。丁酉，彗星犯离宫西星，至二月终，光芒约长二丈余。三月戊申，彗星不见星形，惟有白气，形曲竞天，西指，扫大角。壬子，彗星行过太阳前，惟有星形，无芒，如酒杯大，昏朦，色白，测在昴宿六度，至戊午始灭迹焉。

《元史·天文志》卷49

公元1362年4月25日 [中国]

元顺帝至正二十二年四月丙子朔，长星见，其形如练，长数十丈，在虚、危之间，后四十余日乃灭。

《元史·顺帝纪》卷46

公元1362年6月29日 [中国]

(a) 元顺帝至正二十二年六月辛巳，彗星见紫微垣，光芒长尺余，东南指，西南行。戊子，彗星光芒扫上宰。七月乙卯，彗星灭迹。

《元史·顺帝纪》卷46

(b) 元顺帝至正二十二年六月辛巳，彗星见于紫微垣，测在牛二度九十分，色白，光芒约长尺余，东南指，西南行。戊子，彗星光芒扫上宰。七月乙卯，彗星灭迹。

《元史·天文志》卷49

公元1362年6月29日 [朝鲜]

高丽恭愍王十一年六月辛巳，彗见紫微垣华盖下，长尺许，凡三日。

《高丽史》卷49

公元1363年3月16日 [中国]

元顺帝至正二十三年三月辛丑朔，彗星见东方，经月乃灭。

《元史·顺帝纪》卷46

公元1364年3月30日 [朝鲜]

高丽恭愍王十三年二月辛酉，彗见一在太微南，一在大角边，一在北斗东北，一在氐北，色赤，长尺余。

《高丽史》卷49

公元1366年10月25日 [中国]

元顺帝至正二十六年九月庚子，孛星见于紫微垣北斗权星之侧，色如粉絮，约斗大，往东南行，过犯天棓星。辛丑，孛星测在尾十八度五十分。壬寅，孛星测在女二度五十分。癸卯，孛星测在女九度九十分。甲辰，孛星测在虚初度八十分。乙巳，孛星出紫微垣北斗权星、玉衡之间，在于轸宿，东南行，过犯天棓，经渐台、辇道，去虚宿、垒壁阵西方星，始消灭焉。

《元史·天文志》卷49

公元1366年10月26日 [朝鲜]

高丽恭愍王十五年九月辛丑，有星孛于房虚上星。壬寅，又见于东南方。

《高丽史》卷49

公元1368年2月7日 [中国]

(a) 元顺帝至正二十八年正月庚寅，彗星见于昴、毕之间。

《元史·顺帝纪》卷47

(b) 明太祖洪武元年正月庚寅，彗星见于昴、毕。

《明史·天文志》卷27

公元1368年3月7日 [朝鲜]

高丽恭愍王十七年二月己未，彗见于西方，长丈余。三月甲申，见西方。庚寅，出大陵积尸间。辛卯，出大陵、卷舌间，射天船九星。丙申，出卷舌上。己亥，出大陵上。四月辛丑，又见。

《高丽史》卷49

公元1368年3月8日 [日本]

日本正平二十三年二月十九日。三月廿二日壬辰，晴，去夜戌刻，乾方有星，光芒甚长...去月十九日，戌刻，出现西方……四月……十二日壬子，晴，彗星此间不见云云。

《愚管记》 12

公元1368年4月8日 [中国]

(a) 明太祖洪武元年三月辛卯，夜，彗星出……四月己酉，夜，彗星没。先是，三月辛卯，彗出昴北大陆（应为大凌）、天船间，芒长约八尺余。辛丑，指文昌，近五车。是夜，没于五车北。

《明太祖实录》卷31

(b) 明太祖洪武元年三月辛卯，彗星出昴北大陵、天船间，长八尺余。指文昌，近五车，四月己酉，没于五车北。

《明史·天文志》卷27

(c) 明太祖洪武元年三月辛卯，夜，彗见昴北。四月辛丑夜，彗指文昌，近五车。己酉夜，彗灭。

《国榷》卷3

公元1373年4月23日—5月22日 [中国]

明太祖洪武六年四月，彗星三入紫微垣。

《明史·天文志》卷27

公元1374年3月11日—20日 [日本]

(a) 日本文中三年正月廿八日甲午（3月11日），今夜孛星，光芒四出，见西…三十日丙申，自今夜有彗星，见南。

《续史愚抄》 27

(b) 日本文中三年二月七日癸卯（3月20日）…去月卅日戌刻彗星见南方。一日、二日依阴气不现。三日以来每夜出现云云…十七日癸丑…彗星事相寻亲宣朝臣之处，犹未消云云。

《愚管记》 18

公元1374年3月15日 [朝鲜]

高丽恭愍王二十三年二月戊戌，彗见东方，长丈余。凡四十五日乃灭。

《高丽史》卷49

公元1376年6月22日—8月8日 [中国]

(a) 明太祖洪武九年六月戊子（6月22日），有星大如弹丸，白色。止天仓，经外屏、卷舌，入紫微垣，扫文昌，指内厨，入于张。七月乙亥灭。

《明史·天文志》卷27

(b) 明太祖洪武九年七月乙亥（8月8日），客星没。先是六月戊子，客星大如弹丸，白色，止天仓。止癸巳，益有光。甲午，经外屏。壬寅，经卷舌，入紫微垣。庚戌，扫文昌，指内厨。壬子，扫文昌。是月癸亥，入于张，至是夕始灭。

《明太祖实录》卷107

公元1376年7月9日—10日 [日本]

(a) 日本天授二年六月二十三日（7月10日），寅刻彗星出现丑方云云。今夜亥初彗星出乾方，晓更又如夜夜见丑方云云。七月…六日戊午…彗星犹出现，在乾，行度颇西行，几入紫微宫事，天安以来五六个度也，然而犯天皇大帝星之条，今度之外无例云云。

《愚管记》 20

(b) 日本天授二年六月廿二日（7月9日），寅刻彗星丑方见，光三尺余，又其色白云云。同廿七日，戌刻亥方现光一丈余云云，其后又晓如本丑方见。

《鸠嵛杂事记》

公元1376年7月12日 [朝鲜]

高丽辛禑二年六月戊申，彗见文昌，光芒射紫微垣…七月甲寅，彗星现文昌西隔四五尺，光射斗魁。

《高丽史》卷49

公元1376年7月21日 [中国]

明太祖洪武丙辰七月五日，有星孛于北斗。

《二申野录》卷1

公元1378年9月22日—10月21日 [中国]

明太祖洪武戊午九月，有星孛于天矢。

《二申野录》卷1

公元1378年9月26日 - 11月15日 [中国]

明太祖洪武十一年九月己丑(1378年10月11日)，客星扫天弁，西南行入尾。先是，甲戌(9月26日)，见于五车东北。丁丑(9月29日)，入紫微，戊寅(9月30日)，发芒长丈余，扫内阶。庚辰(10月2日)，入紫微宫，扫北极五星。壬午(10月4日)，犯东垣少宰。甲申(10月6日)，入天帝垣。丁亥(10月9日)，入天市。自是至十月己未(10月11日)，云阴不见。

《明太祖实录》卷119

公元1378年10月1日 [日本]

天授四年九月九日己卯(1378年10月1日)，晴，被下赦书，去夜彗星出现之由，有世朝臣申之。十月十五日甲寅(11月5日)，晴，自今夜公家被行五坛法，是彗星御祈云云。廿五日甲子(11月15日)，晴，…是彗星御祈云云。

《愚管记》 22

公元1378年10月1日 [朝鲜]
辛禑四年九月己卯(1378年10月1日)，有星孛于紫微西藩，犯四辅、北极，出东藩，犯天棓。
《高丽史》卷49

公元1381年11月7日 [朝鲜]
高丽辛禑七年十月壬申，彗见于氐，长丈余，十五日乃灭。
《高丽史》卷49

公元1381年11月8日 [日本]
日本弘和元年十月廿二日晓，彗星出东方，光芒长一丈五六尺许云云。廿七日戊寅，晴，今晓见彗星，光芒诚长…十一月十五日，入夜为世朝臣持来彗星勘文，此两三夜消没云云。今日密奏云云，今度彗星无轴星，是又有例云云。
《愚管记》 22

公元1382年9月19日 [朝鲜]
高丽辛禑八年八月戊子，彗星见太微东藩，长丈余。
《高丽史》卷49

公元1385年5月10日—6月7日 [中国]
明太祖洪武乙丑四月，彗星见，扫翼，复扫天庙。
《二申野录》卷1

公元1385年10月23日 [中国]

(a) 明太祖洪武十八年九月戊寅，客星入太微垣。辛巳夜，客星犯右执法，出端门。乙酉夜，客星入翼，彗长丈余。丁亥夜，客星在轸宿西南，彗扫翼宿。戊子，客星又入太微。十月庚寅夜，客星犯军门，彗星扫天庙。

《明太祖实录》卷175

(b) 明太祖洪武十八年九月戊寅，有星见太微垣，犯右执法，出端门。乙酉入翼，彗长丈余。至十月庚寅，犯军门，彗扫天庙。

《明史·天文志》卷27

公元1385年11月10日 [中国]
明太祖洪武十八年十月丙申，有星孛于斗。
《国榷》卷8

公元1389年8月6日 [中国]
(a) 明太祖洪武二十二年七月辛巳，彗见紫微旁，乙卯灭。
《国榷》卷9
(b) 明太祖洪武二十二年七月辛巳，彗星见紫微侧，在牛度九十分，色白，光约长丈余，东南指，西北行。戊子，彗光扫上宰。七月乙卯灭。
明万历 浙江《秀水县志》卷10
注：原文六月，六月无辛巳，是七月之误。七月无乙卯。

公元1391年5月23日 [中国]
(a) 明太祖洪武二十四年四月丙子，夜，彗星二：一入紫微垣阊阖门，犯天床；一犯六甲，扫五帝座。
《明太祖实录》卷208
(b) 明太祖洪武二十四年四月丙子，彗星二：一入紫微垣阊阖门，犯天床；一犯六甲，扫五帝内座。
《明史·天文志》卷27

公元1402年2月20日 [朝鲜]
李朝太宗二年壬午正月壬寅，彗见于奎东，光芒指东，长六尺许…甲辰，有星见于东方，长丈余，光芒四射…二月己巳，彗星没。自前日壬寅始见，至此日乃灭。
《朝鲜王朝实录·太宗》卷3

公元1402年2月21日—4月1日 [日本]
(a) 日本应永九年正月十九日（2月21日），昏戌时彗星出西方，在奎度，光芒三尺所，色白…（二月）十六日己巳，晴，彗星至今夜出现云云，自十七日不见云云。
《吉田家日次记》
(b) 日本应永九年二月（3月3日—4月1日），彗星见鎌仓。
《续本朝通鑑》百五十四

公元1410年10月28日—11月25日 [中国]
明太宗永乐八年十月，有星孛于南斗。
《明太宗实录》卷167

公元1421年1月9日 [日本]
日本应永二十七年十二月六日，戌时彗星见乾方，光芒五尺许。
《迎阳记》
注：原文将“五尺许”误写为“五见许”。

公元1430年11月14日 [中国]
(a) 明宣宗宣德五年十月丙申，夜，有蓬星色白如粉絮，见外屏南，渐东南行，经天仓、天庾，凡八日始灭。
《明宣宗实录》卷71
(b) 明宣宗宣德五年十月丙申，蓬星见外屏南，东南行，经天仓、天庾，八日而灭。
《明史·天文志》卷27
(c) 明宣宗宣德五年十月丙申，蓬星色白如粉絮，见外屏。南行经天仓、天庾，八日灭。
《国榷》卷21

公元1431年5月12日—6月9日 [中国]
(a) 明宣宗宣德六年四月戊戌（5月15日），昏刻，彗出东井，芒角蓬勃，长五尺余。
《明宣宗实录》卷78
(b) 明宣宗宣德六年四月戊戌（5月15日），有星孛于东井，长五尺余。
《明史·天文志》卷27
(c) 明宣宗宣德六年四月（5月12日—6月9日），有星孛于东井。
明 万历《四川总志》卷27

公元1432年2月3日 [中国]
(a) 明宣宗宣德七年正月壬戌，夜，有彗出东方，光芒长丈余，尾扫天津，渐东南行，十日始灭…戊子，夜，复有彗出西方，十有七日始灭。
《明宣宗实录》卷86

(b) 明宣宗宣德七年正月壬戌，彗星出东方，长丈余，尾扫天津，东南行，十月始灭…是月戊子，又出西方，十有七日而灭。

《明史·天文志》卷27

注：{b} 原文"十月始灭"是"十日始灭"之误.

公元1433年9月15日 [中国]

(a) 明宣宗宣德八年闰八月壬子，昏刻，有彗出天仓旁，光芒长丈，己巳，入贯索，扫七公。己卯，入天市垣，扫晋星，又二十有四日灭。

《明宣宗实录》卷105

(b) 明宣宗宣德八年闰八月壬子，彗星出天仓旁，长丈许。己巳，入贯索，扫七公。己卯，入天市垣，扫晋星，二十有四日而灭。

《明史·天文志》卷27

公元1433年10月2日 [朝鲜]

李朝世宗十五年癸丑闰八月己巳，彗星始见贯索度内；甲戌，慧见天市西垣。九月戊子，彗星微；庚子，彗星在西方天际稀微；壬寅，在西方不见。

《朝鲜王朝实录·世宗》卷61

公元1433年10月3日 [中国]

(a) 明宣宗宣德八年闰八月庚午，昏刻，有彗出贯索内，尾长五尺余，渐西南行，凡三十有四日始灭。

《明宣宗实录》卷105

(b) 明宣宗宣德八年闰八月庚午，昏刻，彗出贯索，三十四日灭。

《国榷》卷22

(c) 明宣宗宣德八年闰八月壬子，彗星出天仓旁，长丈许。己巳，入贯索，扫七公。己卯，复入天市垣，扫晋星，二十有四日而灭。

《明史·天文志》卷27

公元1433年10月8日 [日本]

(a) 日本永享五年八月廿五日，昏戌时彗星见酉与戌之间，在尾度近贯索，其色白。

《看闻御记》

(b) 日本永享五年八月廿五日，晴，戌初当西方星光指东…九月…今月十九日光芒微薄，分明不见候。同廿日，廿一日不见候之间，消灭候哉云云。

《满济准后日记》

公元1439年3月15日—4月13日 [日本]

(a) 日本永享十一年二月十五日（3月29日），彗星见之。

《大乘院日记目录》2

(b) 日本永享十一年二月（3月15日—4月13日），彗星出北方，廿三日见定由，司天申之…今度彗星是也，自十八九日比粗见及，出现於鬼宿。

《建内记》 9

(c) 日本永享十一年二月廿三日辛丑（4月6日），今夜彗星见紫微者，廿五日安三位奏之。

《续史愚抄》 36

公元1439年3月25日 [中国]

(a) 明英宗正统四年闰二月己丑，夜，东南张宿旁生彗星，大如弹，色白，西行。丁酉夜，彗星芒长五尺余，西北行，扫酒旗庚子夜，彗星西北行，长七尺余，犯鬼宿。

《明英宗实录》卷52

(b) 明英宗正统四年闰二月己丑，彗星见张宿旁，大如弹，色白，西行。丁酉，长五丈（应为尺）余，西行，扫酒旗，迤北，犯鬼宿。

《明史·天文志》卷27

公元1439年4月1日—12日 [朝鲜]

(a) 李朝世宗二十一年闰二月丙申（4月1日），彗见；丁酉，彗星见；丁未，慧见井；戊申，彗星见。三月辛亥，彗星见，凡十二日而灭。

《朝鲜王朝实录·世宗》卷84

(b) 李朝世宗二十一年己未闰二月丙申（4月1日），彗见于东井，凡二十二日。

《增补文献备考》卷6

公元1439年7月12日 ［中国］

(a) 明英宗正统四年六月戊寅，彗星见毕宿，光芒长一丈余，指西南，至七月二十五日乃没。

《明英宗实录》卷56

(b) 明英宗正统四年六月戊寅，彗星见毕宿旁，长丈余，指西南，计五十五日乃灭。

《明史·天文志》卷27

公元1444年8月6日 ［中国］

(a) 明英宗正统九年七月庚午，昏刻，太微东垣彗星光芒长一丈，渐长。至闰七月初二日，入角宿没。

《明英宗实录》卷118

(b) 明英宗正统九年七月庚午，彗星见太微东垣，长丈许，累日渐长，至闰七月己卯，入角没。

《明史·天文志》卷27

公元1444年8月7日 ［日本］

(a) 日本文安元年闰六月廿三日，后闻，今夜彗星出现云云。廿六日，彗星出现事，今日闻之间，入夜见之，西方有之，光芒二尺许也，小星也。

《师乡记》 11

(b) 日本文安元年闰六月廿三日辛未，有彗星见戌方，至翌月上旬不见。

《续史愚抄》 37

公元1449年12月20日 ［中国］

(a) 明英宗正统十四年十二月壬子，晓刻，彗星见于天市垣市楼星旁。癸丑晓刻，彗星见于尾十二度，色苍白，芒长尺余，扫正西。乙卯晓刻，彗星见于尾十一度，长二尺余。丙辰晓刻，彗星见尾宿十一度，西行，至乙亥夜不见。

《明英宗实录》卷186

(b) 明英宗正统十四年十二月壬子，彗星见天市垣市楼旁，历尾度，长二尺余，至乙亥没。

《明史·天文志》卷27

公元1449年12月26日 ［朝鲜］

李朝世宗三十一年己巳十二月十二日戊午，彗见于东方天市垣内，其后密云不得测。至二十一日丁卯，彗又见于天市，尾长五六尺许...三十二年正月癸巳，彗星始不见。

《朝鲜王朝实录·世宗》卷126-127

公元1450年1月19日 ［中国］

(a) 明英宗景泰元年正月壬午，彗星出天市垣，扫天纪星。

《明英宗实录》卷187

《明史·天文志》卷27

(b) 明英宗景泰元年正月壬午，夜，彗见天市垣外，扫天纪。

《国榷》卷29

公元1452年3月21日 ［中国］

明英宗景泰三年三月甲午朔，夜，孛星见于毕。

《明英宗实录》卷214

公元1456年5月27日 ［中国］

(a) 明英宗景泰七年四月壬戌，夜，彗星见东北方，在胃宿，光芒长二丈，尾指西南...五月癸酉夜，彗星光芒渐长丈余，指西南...戊子昏刻，彗星见西北方，在柳宿，尾长九尺余，扫犯轩辕星...甲午昏刻，彗星见张宿，光芒七尺余，尾扫太微西垣 ，往西南行...六月壬寅夜，彗星入太微西垣内上将星旁，芒长五尺余。

《明英宗实录》卷265—卷267

(b) 明英宗景泰七年四月壬戌，彗星东北见于胃，长二尺，指西南。五月癸酉，渐长丈余。戊子，西北见于柳，长九尺余，扫犯轩辕星。甲午，见于张，长七尺余，扫太微北，西南行。六月壬寅，入太微垣，长尺余。

《明史·天文志》卷27

公元1456年5月31日 [日本]
日本康正二年四月二十七日。五月七日乙亥，晚头以后雨下，后闻，今晓彗星见艮方，光芒三尺许云云。或人云，去月廿七日见之云云…六月一日己亥，彗星今夜犹见也。
《师乡记》 22

公元1456年6月6日 - 29日 [朝鲜]
世祖二年5月壬申，彗星见；癸酉(6月7)、甲戌(6月8)、乙亥(6月9)、戊寅(6月12)、己卯(6月13)、乙酉(6月19)、戊子(6月22)、辛卯(6月25)、甲午(6月28)、乙未(6月29)，彗星见。
《朝鲜王朝实录·世祖》卷4

公元1456年8月1日—30日 [中国]
(a) 明英宗景泰七年七月间，昼漏当申刻之末，彗星如洗帚状，微见于西方，至酉刻以后，渐长如扫帚，人呼曰扫帚星，日既没，其长竟半天，如此两月而灭。
《闲中古今录》
(b) 明英宗景泰七年七月间，彗星昼见西方，自申刻至日没，其长竟半天，凡两月而灭。
《二申野录》卷2

公元1457年1月14日 [中国]
(a) 明英宗景泰七年十二月甲寅，夜，彗星复见于毕宿，光芒长五寸，徐徐东南行，光芒渐长，自是日至于癸亥。
《明英宗实录》卷273
(b) 明英宗景泰七年十二月甲寅，彗星复见于毕，长五寸，东南行，渐长，至癸亥而没。
《明史·天文志》卷27

公元1457年6月15日 [中国]
(a) 明英宗天顺元年五月丙戌，夜，彗星见于危宿，状如粉絮，色青白，拂拂摇动，至丁亥夜东行一度，微芒长五寸，指西南…六月癸巳夜，彗星见于室宿，芒角长丈余…己亥夜，彗星犯壁宿，尾指东壁二星…乙巳夜，彗星犯天大将军…壬子夜，彗星经犯卷舌第三星…七月戊寅夜，彗星犯井宿…辛卯夜，彗星犯水位南第二星…八月甲午，上躬祷于昊天上帝，词曰：兹者彗星出现于今两月，光芒尚在，垂变不消。

《明英宗实录》卷278 - 卷281

(b) 明英宗天顺元年五月丙戌，彗星见于危，若动摇者，东行一度，芒长五寸，指西南。六月癸巳朔，见室，长丈余，由尾至东壁，犯天大将军、卷舌第三星，井宿水位南第二星。

《明史·天文志》卷27

公元1457年6月23日—7月16日 [朝鲜]

世祖四年6月甲午，彗星见；乙未(6月24)、丙申(6月25)、丁酉(6月26)、戊戌(6月27)、壬寅(7月1)、丁未(7月6)、己酉(7月8)、更戌(7月9)、辛亥(7月10)、壬子(7月11)、甲寅(7月13)、丙辰(7月15)、丁巳(7月16)，彗星见。

《朝鲜王朝实录·世祖》卷84—85

公元1457年7月30日—8月19日 [日本]

(a) 日本长禄元年七月九日辛未（7月30日），今日立秋也，晓天起见彗星，卯刻当寅方出了，其色薄，尾长五尺许在之。

《经觉私要钞》 34之6

(b) 日本长禄元年七月廿九日（8月19日），彗星未消，在东方，今晓予见之，芒光七八尺斗也。

《宗贤卿记》柳原家记录 157

公元1457年10月26日 [中国]

(a) 明英宗天顺元年十月己亥，夜，彗星复见于角宿，芒长五寸余，尾指北…己酉夜晓刻，彗星犯角宿北星…壬子夜，彗星犯平道东星。

《明英宗实录》卷283

(b) 明英宗天顺元年十月己亥，彗星见于角，长五寸余，指北，犯角北星及平道东星。

《明史·天文志》卷27

公元1458年12月24日 [中国]

(a) 明英宗天顺二年十一月癸卯，夜，有客星见于星宿，色白，西行，至丙午夜，其体微大，状如粉絮，在轩辕之旁…庚戌夜，客星生芒，可五寸，经犯 位西北星…十二月壬戌夜，客星没于井宿。

《明英宗实录》卷297-卷298

(b) 明英宗天顺二年十一月癸卯，有星见于星宿，白色，西行，至丙午，其体微，状如粉絮，在轩辕旁。庚戌，生芒五寸，犯 位西北星，至十二月壬戌，没于东井。

《明史·天文志》卷27

公元1458年12月29日 [朝鲜]

世祖四年十一月戊申，至十二月乙卯彗星见；丙辰至癸亥彗星见，其光微；壬午至五年正月丙戌、戊子彗星见；甲午彗星见在娄北，光微；乙未至二月甲寅朔彗星见，其光微；乙卯至三月朔癸未彗星见；戊子彗星见，其光微。

《朝鲜王朝实录·世祖》卷84—85

公元1459年2月5日 [朝鲜]

李朝世祖五年己卯正月丙戌，彗见于娄，光芒微，至于三月戊子。

《增补文献备考》卷6

公元1461年7月30日 [中国]

明英宗天顺五年六月壬辰，夜，天市垣宗正星旁有异星见，色粉白，至乙未夜，化为白气而消。

《明英宗实录》卷329

公元1461年8月5日 [中国]

(a) 明英宗天顺五年六月戊戌，夜，彗星见东方，光芒长三尺余，尾指西南…七月庚子夜晓刻，彗星在井四度，光芒长一尺余，尾指西南…丙午夜晓刻，彗星仍见井宿，至丙寅夜灭。

《明英宗实录》卷329 - 卷330

(b) 明英宗天顺五年六月戊戌，彗见东方，指西南，入井度。七月丙寅始灭。

《明史·天文志》卷27

(c) 明英宗天顺五年六月戊戌，夜，彗星见东方，芒二尺余；七月庚子夜，彗见井宿；丙午夜，彗见井宿；丙寅夜，灭。

《国榷》卷33

公元1462年6月29日 [中国]

(a) 明英宗天顺六年六月丙寅，夜，客星见于策星旁，色苍白…己巳夜，客星入紫微垣内…乙亥夜，客星入犯天牢星…癸未夜，客星居中台星下，形渐微。

《明英宗实录》卷341

(b) 明英宗天顺六年六月丙寅，有星见策星旁，色苍白，入紫微垣，犯天牢，至癸未，居中台下，形渐微。

《明史·天文志》卷27

公元1465年3月27日—4月24日 [中国]

明宪宗成化元年三月，彗星又见西北，长三丈余，三阅月而没。

《明史·天文志》卷27

明万历 湖北《襄阳府志》卷33

注：山西[康熙]临汾县志作二月，文同。

公元1468年7月19日—8月17日 [中国]

明宪宗成化四年七月，有星孛于台、斗。

明嘉靖 湖北《随志》卷上

公元1468年9月18日 [中国]

(a) 明宪宗成化四年九月己未，夜，客星见星五度，东北行…癸亥夜，客星色苍白，光芒长三丈余，尾指西南，变为彗星…戊辰，彗星晨见东北方…己巳，彗星昏见西南方…庚午昏刻，彗星犯三公星…辛未昏刻，彗星犯北斗摇光星…丁丑昏刻，彗星犯七公西第四星…壬午昏刻，彗星入天市垣…十月乙巳，彗星出天市垣，其体渐小…甲寅，彗星犯天屏西第一星…十一月庚申夜，彗星灭。

《明宪宗实录》卷58 - 卷60

(b) 明宪宗成化四年九月己未，有星见星五度，东北行。越五日，芒长三丈余，尾指西南，变为彗星。其后晨见东方，昏见室，南犯三公、北斗、瑶光、七公，转入天市垣。出垣渐小，犯天屏西第一星。十一月庚辰，始灭。

《明史·天文志》卷27

公元1468年9月22日—10月6日 ［日本］

(a) 日本应仁二年九月六日（9月22日），彗星现，见三十日。

《大乘院日记目录》 3

(b) 日本应仁二年九月二十日丁丑（10月6日），彗星出於西，光芒丈余。佥曰，自上旬某夕而见。

《碧山日录》 5

公元1468年9月22日 ［朝鲜］

(a) 李朝世祖十四年戊子九月癸亥，彗见天市，尾长三丈许。

《增补文献备考》卷6

(b) 李朝世祖十四年九月己未，彗星见。庚申（9月14日）、壬戌（9月16日），彗星见。李朝成宗元年九月，甲子（9月18日）、丙寅（9月20日）、戊辰（9月22日）、己巳（9月23日）、庚戌（9月24日）、辛未（9月25日）、壬申（9月26日）、癸酉（9月27日）、甲戌（9月28日）、乙亥（9月29日）、丁丑（10月1日）、戊寅（10月2日）、己卯（10月3日）、庚辰（10月4日）、辛巳（10月5日）、壬戌（10月6日）、癸未（10月7日）、甲申（10月8日）、乙酉（10月9日）、丙戌（10月10日），彗星见。

《朝鲜王朝实录·世祖》卷47

《朝鲜王朝实录·成宗》卷1

公元1468年9月27日 ［中国］

明宪宗成化四年九月戊辰，[按明昭代典]彗星晨见东北方。己巳，彗星昏见西南方。九月丁丑昏刻，彗星犯七宫西等四星。壬午昏刻，彗星入天市垣。十月甲寅，彗星犯天屏西第一星。十一月戊午夜，彗星灭。

《古今图书集成·庶征典·星变部》卷55

公元1468年10月2日 [中国]
明宪宗成化四年九月癸酉，[按大政纪]夜，客星色苍白，光芒长三丈余，尾指西南，变为彗星。丁丑昏刻，彗星犯七公等四星。壬午昏刻，彗星入天市垣。十月甲寅，彗星犯天屏西第二星。十一月戊午，彗星灭。
《古今图书集成·庶征典·星变部》卷55

公元1471年12月12日—1472年1月9日 [中国]
(a) 明宪宗成化七年十一月，彗星见，出天田入太微垣。
明万历 山西《洪洞县志》卷8
(b) 明宪宗成化七年十一月，彗出轩辕。十二月丁丑夜，彗星北行，横扫太微垣郎位星己(应为乙)卯度，彗星光芒东西竞天，自十一日北行二十八度余，犯天枪，尾扫北斗、三公太阳。丙戌立春昏刻，彗星犯天河星。
《二申野录》卷2
(c) 明宪宗成化七年十二月甲戌，彗星见天田，西指，寻北行，犯右摄提，扫太微垣上将及幸臣、太子、从官，尾指正西，横扫太微垣郎位。己卯，光芒长大，东西竟天。北行二十八度余，犯天枪，扫北斗、三公、太阳，入紫微垣内，正昼犹见。自帝星、北斗、魁、庶子、后宫、勾陈、天枢、三师、天牢、中台、天皇大帝、上卫、阁道、文昌、上台，无所不犯。乙酉，南行犯娄、天河、天阴、外屏、天囷。
《明史·天文志》卷27

公元1472年1月7日 [朝鲜]
(a) 李朝成宗二年十一月乙丑夜，有星见角，度天田星上，微有光芒……十二月乙亥，彗星见角，度周鼎星下，尾长二丈许…己卯，彗星见紫微垣天仓星，尾长二丈许…庚辰，彗星见华盖杠星，尾长二丈许…辛巳，彗星见王良北…甲申，彗星见奎南，尾长一丈许…三年正月己亥，彗星见外屏星南，尾长五尺许…丙午，彗星微见。
《朝鲜王朝实录·成宗》卷13-14

(b) 李朝成宗二年辛卯十一月乙丑，彗见于角，过周鼎抵华盖，尾长二丈许。十二月庚 寅历王良移奎南。翌年正月丙午乃灭。

《增补文献备考》卷6

公元1472年1月13日—2月9日 ［日本］

(a) 日本文明三年十二月三日（1月13日），今月三日，子时彗星见东方，贯太微第三星，光芒三尺所，其色白。

《亲长卿记》2

注：原文"贯大微"系"贯太微"之误。

(b) 日本文明四年正月一日（2月9日），天晴，今晓见西天彗星，未消，出初之时有东方，次廻北，次廻西讫。

《亲长卿记》3

(c) 日本文明三年十二月廿六日（2月5日），自当月初彗星出现，光长十丈许，初晓现东，近日初夜时分西方出现。

《大乘院日记目录》 3

公元1472年1月16日 ［中国］

(a) 明宪宗成化七年十二月甲戌，彗见天田，光芒西指…乙亥夜，彗星北行，光芒渐著，犯右摄提，扫太微垣上将及幸臣、太子、从官…丁丑夜，彗星北行五度余，尾指正西，其光益著，横扫太微垣郎位星…己卯夜，彗星光明长大，东西竟天，自十一日北行二十八度余，犯天枪，尾扫北斗、三公、太阳…庚辰，彗星北行入紫微垣内，正昼犹见，历犯帝星、北斗魁第二星，庶子、后宫、勾陈下星、北斗魁第一星、勾陈第三星、天枢、三师、天牢、中台、天皇大帝上卫星…辛巳昏刻，彗星入紫微垣，历犯阁道、文昌、上台星…乙酉昏刻，彗星南行犯娄宿；丙戌昏刻，彗星犯天河星…己丑昏刻，彗星犯天阴星…癸巳昏刻，彗星犯外屏星…乙未昏刻，彗星犯天囷星……八年正月丙午夜，彗星行奎宿外屏星下，形渐消小。

《明宪宗实录》 卷99 - 卷100

(b) 明宪宗成化七年十二月甲戌，彗星见天田，西指，寻北行，犯右摄提，扫太微垣上将及幸臣、太子、从官，尾指正西，横扫太微垣郎位。己卯，光芒长大，东西竟天，北行二十八度余，犯天枪，扫北斗、三公、太阳，入紫微垣内，正昼犹见。自帝星、北斗、魁，庶子、后宫勾陈、天枢、三师、天牢、中台、天皇大帝、上卫、阁道、文昌、上台，无所不犯。乙酉，南行犯娄、天河、天阴、外屏、天囷。八年正月丙午，行奎宿外屏，渐微，久之始灭。

《明史·天文志》卷27

公元1472年1月24日 [中国]

明宪宗成化七年十二月壬午，彗星入紫微垣。

《明史·宪宗纪》卷13

公元1490年12月31日 [中国]

(a) 明孝宗弘治三年十一月戊戌，昏刻，彗星见于天津南，芒长尺余，尾指东北…辛丑昏刻，彗星犯人星…十二月戊申，彗星入室宿…庚申昏刻，彗星犯天仓。

《明孝宗实录》卷45 - 卷46

(b) 明孝宗弘治三年十一月戊戌，彗星见天津南，尾指东北。犯人星，历杵臼。十二月戊申朔，入营室。庚申，犯天仓。

《明史·天文志》卷27

(c) 明孝宗弘治三年十一月戊戌，昏刻，彗见天津，芒尺余。辛丑昏刻，彗犯人星。己未昏刻，彗犯天仓。

《国榷》卷42

公元1491年1月3日 [朝鲜]

(a) 李朝成宗二十一年十一月辛丑，前夜一更，虚星度有微光，长三、四尺…癸卯，去夜微光星移入危星度…乙巳，去夜彗星移在危六，去极六十五度…丁未，去夜彗星移在危十一度，去极七十六度，光长丈余…近日彗星出虚危。十二月己酉，去夜彗星移入室星二度，去极八十一度，观象监书启曰，去夜一更彗星移入室南大星东，光长一丈许…庚戌，去夜彗星移在室星四度半，去极八十四度…观象监书启，去夜一更，彗星在室星度雷电星东，长一丈许…辛亥，今夜密云彗星初不见，四更在太微西垣市二三星间…甲寅，去夜彗星移在室星度内云雨星东第一星南，三四尺许…乙卯，去夜彗星移在壁星南…戊午，去夜彗星移在垒壁星天溷星间…己未，去夜彗星移在天仓西第一星上…庚申，去夜彗星移在天仓西第二星西…乙亥，去夜一更彗星移在天仓东第一星东，无光芒，其体甚微。

《朝鲜王朝实录·成宗》卷247-248

(b) 李朝成宗二十一年庚戌十一月辛丑，彗见于虚，渐东在危，尾长丈余，过东壁仓，翌年正月乃灭。

《增补文献备考》卷6

公元1491年1月4日—8日 ［日本］

(a) 日本延德二年十一月廿四日（1月4日），晴，后闻，今夜彗星出西方，光长五尺计云云。廿七日，晴，予见彗星，有西方，光长三尺许。

《亲长卿记》

(b) 日本延德二年十一月廿八日（1月8日），戌时彗星见西方，在室宿，其长五尺许，其色白。

《后法兴院政家记》 15

公元1491年1月19日 ［中国］

(a) 明孝宗弘治三年十二月丁巳，夜，客星见天市垣内，色苍白，徐徐东南行。戊辰昏刻，客星见天仓下，色苍白，渐向壁宿。

《明孝宗实录》卷46

(b) 明孝宗弘治三年十二月丁巳，有星见天市垣，东南行。戊辰，见天仓下，渐向壁。

《明史·天文志》卷27

公元1495年1月7日 ［中国］

(a) 明孝宗弘治七年十二月丙寅，晓刻，尾宿天江旁客星见，自庚午至庚辰，徐徐近斗宿。八年正月庚戌晓刻，客星入危宿，犯上星。

《明孝宗实录》卷95 - 卷96

(b) 明孝宗弘治七年十二月丙寅，有星见天江旁，徐行近斗，到八年正月庚戌，入危。

《明史·天文志》卷27

公元1499年8月16日 [中国]

(a) 明孝宗弘治十二年七月戊辰，昏刻，客星见天市垣宗星旁…丙子昏刻，客星光指东南，行至紫微垣东蕃少宰星旁…丁丑昏刻，客星入紫微垣东蕃内尚书星旁…戊寅夜，客星行至紫微垣内太子星旁，渐微…己卯夜，客星在紫微垣内后星旁…甲申昏刻，客星行至紫微垣西蕃内少辅旁星，渐消…丁亥昏刻，客星行出紫微垣西蕃外…八月己丑夜，客星不见。

《明孝宗实录》卷152 - 卷153

(b) 明孝宗弘治十二年七月戊辰，有星见天市垣宗星旁，入紫微垣东蕃，经少宰、尚书、抵太子后宫，出西蕃少辅旁，至八月己丑灭。

《明史·天文志》卷27

公元1500年5月8日 [中国]

(a) 明孝宗弘治十三年四月甲午，彗星见室宿垒壁阵上…丙申夜，彗星见室宿西北…己亥夜，彗星见室、壁之间，芒长尺余…乙巳，彗星芒长三尺余，尾指离宫…五月甲寅夜，彗星扫造父星…丁卯夜，彗星行过紫微垣，渐微…辛巳夜，彗星入紫微垣，近女史…六月丁亥，彗星连犯尚书星…戊戌夜，彗星不见。

《明孝宗实录》卷161 - 卷163

(b) 明孝宗弘治十三年四月甲午，彗星见垒壁阵上，入室壁间，渐长三尺余。指离宫，扫造父，过太微垣，渐微。入紫微垣，近女史，犯尚书，六月丁酉没。

《明史·天文志》卷27

(c) 明孝宗弘治十三年四月甲午，彗见室宿。己亥，彗见室、壁间，芒尺余。乙巳夜，彗长三尺余。五月甲寅夜，彗扫文昌。丁卯夜，彗历紫微星，渐微。辛巳夜，彗入紫微垣，近女史，六月丁酉灭。六月丁亥，彗连犯尚书星。戊戌夜，彗灭。

《国榷》卷44

公元1500年5月18日 [朝鲜]

(a) 李朝燕山君六年庚申四月甲辰，彗星见，尾长五六尺，至于六月戊子。

《增补文献备考》卷6

(b) 李朝燕山君六年庚申四月甲辰，彗星见，长二三尺，尾指西南间。

《朝鲜王朝实录·燕山君》卷37

公元1500 年5月24日 [日本]

(a) 日本明应九年四月二十六日（5月24日）庚戌，晴，近日彗星见天。轴星在艮，彗尾向坤也。

《和长卿记》一

(b) 日本明应九年五月十四日（6月11日）丁卯，晴，此间彗星出现云云。今月四日，子时彗星见丑方。在内杵，星长二尺所，未方指，光芒色白。

《后法兴院政家记》 25

公元1502年11月28日 [中国]

(a) 明孝宗弘治十五年十月戊辰，客星见张宿天庙星旁......己巳，客星行至翼宿......十一月癸酉，客星复在张宿，戊寅夜，客星灭。

《明孝宗实录》卷192 - 卷193

(b) 明孝宗弘治十五年十月戊辰，有星见天庙旁，自张抵翼，复退至张，戊寅灭。

《明史·天文志》卷27

公元1506年7月31日 [中国]

(a) 明武宗正德元年七月己丑，夜，有星见北方紫微西蕃外，如弹丸，色苍白。壬辰夜，又见其星，有微芒，连三夕，数见参、井之间。至戊戌夜，芒渐长二尺，偏扫如帚，徐徐西北至文昌...壬寅夜，彗星光芒长五尺许，扫犯下台上星...癸卯昏刻，彗星入太微垣。

《明武宗实录》卷15

(b) 明武宗正德元年七月己丑，有星见紫微西蕃外，如弹丸，色苍白。越数日，有微芒见参、井间，渐长二尺，如帚，西北至文昌。七月庚子，彗星见，有光，流东南，长三尺。越三日，长五尺许，扫下台上星，入太微垣。

《明史·天文志》卷27

(c) 明武宗正德元年七月己丑，彗星见参井，扫太微垣，至戊戌乃灭。

明万历《陕西通志》卷4

公元1506年8月7日—10日 [日本]

(a) 日本永正三年七月十八日（8月7日）。七月廿一日，晴，彗星出现。去十八日夜诸人见之，今日宣卿进勘文云云。

《宣胤卿记》

(b) 日本永正三年七月廿一日（8月10日），戌时彗星亥方，在内阶，星长一丈所，东方指，光芒色白。

《后法成寺尚通公记》 1

(c) 日本永正三年七月廿一日己亥（8月10日），晴，彗星出现，今夜人人见之云云...去月十九日，戌时彗星出乾方，而光芒长丈余。

《实隆公记》

公元1518年9月5日—10月4日 [中国]

明武宗正德十三年八月，有星孛于南斗。

清道光 广东《恩平县志》卷1

公元1523年7月23日—8月10日 [中国]

明世宗嘉靖二年六月，有星孛于天市。

《明史·天文志》卷27

公元1529年9月1日 [朝鲜]

李朝中宗二十四年己丑七月壬戌，彗见于西方，长四、五尺，色白。己卯，移见庚方。

《增补文献备考》卷6

公元1530年7月11日　　［中国］
明世宗嘉靖九年六月乙亥，彗星出东井。
清光绪 陕西《凤县志》卷9

公元1530年秋　　［中国］
明世宗嘉靖九年秋，彗星见，次于毕尾，经月乃灭。
明万历　山东《青城县志》卷2

公元1531年8月5日　　［中国］

(a) 明世宗嘉靖十年闰六月乙巳，彗星见于东井，光芒长尺余，指西南...庚戌夜，彗星行扫轩辕北第一星，光芒渐长...七月乙卯，彗星行翼度，光芒七尺余，尾东北指，扫大尊星...戊午，入太微垣，扫郎位星...丙子，彗星行角度，尾东南指，扫亢北第二星，芒渐敛，越三日而没。

《明世宗实录》卷127 — 卷128

(b) 明世宗嘉靖十年闰六月乙巳，彗星见于东井，长尺余，扫轩辕第一星。芒渐长，至翼，长七尺余。东北扫天樽，入太微垣，扫郎位，行角度，东南扫亢北第二星，渐敛，积三十四日而没。

《明史·天文志》卷27

公元1531年8月8日　　［日本］
日本享禄四年六月廿六日戊申，今夜寅刻彗星见于艮，长五尺，色青白，数日见云...七月十日壬戌，自今日七箇日於诸社有彗星御祈。
《续史愚抄》 46

公元1531年8月10日　　［朝鲜］

(a) 李朝中宗二十六年辛卯闰六月庚戌，彗见太微东垣外，尾长十余尺，色白。

《增补文献备考》卷6 ［朝鲜］

(b) 李朝中宗二十六年闰六月庚戌，艮方有彗，尾十余尺。

《朝鲜王朝实录·中宗》卷71

公元1532年2月6日—1533年1月24日 [中国]

(a) 明世宗嘉靖十一年，彗星见东方，其指西南丈余，至冬尽乃没。

明嘉靖 福建《龙溪县志》卷5

(b) 明世宗嘉靖十一年，明嘉靖初，何栋上书：今岁八月彗出井度，晨见东方，历鬼、柳、星、张，入轸度，过太微垣，犯上将，位可知矣。始出形长数尺，渐东行乃几数丈，形可见矣。徘徊井度，始行颇迟，既历列宿，行之渐速，自八月初五至九月十八共四十三日行几八十余度，其行又可考矣。始则苍白，后乃变赤，其色有可见矣。八月初五，其日庚辰，由黄道顺行，至九月十五日，其日庚申，犯太微上将最急，其日辰又可考矣。

清嘉庆 陕西《长安县志》卷27

注：据"八月初五，其日庚辰"和"九月十五日，其日庚申"查出该年为嘉靖十一年。

公元1532年3月9日 [中国]

(a) 明世宗嘉靖十一年二月壬午，夜五更，有星见东南方，苍白色，微芒，至十九日始灭。

《明世宗实录》卷135

(b) 明世宗嘉靖十一年二月壬午，有星见东南，色苍白，有芒，积十九日灭。

《明史·天文志》卷27

公元1532年6月7日 [中国]

明世宗嘉靖壬辰五月五日，彗星见东方，芒长尺余，复东北行，历天津，望井宿，芒渐至丈余，扫太微垣及角宿天门。八月己卯，彗星见东井。

《二申野录》卷4

公元1532年6月21日 [日本]

日本天文元年五月十八日…彗星艮方出，增光。七月十日迄有之，其后者隐毕。

《快元僧都记》

公元1532年9月2日 [中国]

(a) 明世宗嘉靖十一年八月己卯，夜，彗星见于东井，芒长尺余，后东北行，历天津星宿，芒渐长至丈余，扫太微垣诸星及角宿天门，至十二月甲戌，凡一百十有五日而灭。

《明世宗实录》卷141

(b) 明世宗嘉靖十一年八月己卯，彗星见东井，长尺许。后东北行，历天津，渐至丈余。扫太微垣诸星及角宿、天门，至十二月甲戌，凡一百十五日而灭。

《明史·天文志》卷27

注：《明会要》误作"一百五十有五日乃灭"。

公元1532年9月14日 [朝鲜]

李朝中宗二十七年壬辰八月辛卯，彗见于柳。九月己酉，移见于卯地，尾长一丈许，色白。至十二月戊寅，其光微。

《增补文献备考》卷6

公元1532年9月25日 [日本]

(a) 日本天文元年八月廿六日，彗星出现东方，至十月十余日。

《公卿补任》

(b) 日本天文元年八月壬寅，彗星见东，至十月不消。

《续本朝通鑑》 183

公元1533年6月27日 [朝鲜]

李朝中宗二十八年癸巳六月丁丑，彗见于五车北，尾长七八尺，色白，入天船，过王良，历螣蛇，抵天津，至八月丁丑，其形渐微。

《增补文献备考》卷6

公元1533年7月1日 [中国]

(a) 明世宗嘉靖十二年六月辛巳，彗星见于五车，芒长五尺余，尾指西南...壬午，彗星复出毕、昴之间，光芒稍见...己亥夜，彗星扫大陵及天大将军，芒长丈余...七月甲寅，彗星扫阁道，行螣蛇，至八月二十八日而灭。

《明世宗实录》卷151 - 卷152

(b) 明世宗嘉靖十二年六月辛巳，彗星见于五车，长五尺余，扫大陵及天大将军，渐长丈余，扫阁道，犯滕蛇，至八月戊戌而灭。

《明史·天文志》卷27

公元1533年7月15日 [日本]

日本天文二年六月廿三日，彗星在北方，孛丈余…七月三日，彗星二更现，北方，孛颇薄。

《实隆公记》

公元1533年8月20日—9月17日 [中国]

明世宗嘉靖十二年八月，彗星复出东井，扫太微垣，至于十二月。

《二申野录》卷4

公元1534年6月12日 [中国]

(a) 明世宗嘉靖十三年五月丁卯朔，夜四更，客星见于螣蛇，历天厩，入阁道，至二十四日而灭。

《明世宗实录》卷163

(b) 明世宗嘉靖十三年五月丁卯朔，有星见螣蛇，历天厩，入阁道，二十四日灭。

《明史·天文志》卷27

公元1536年3月24日—4月27日 [中国]

(a) 明世宗嘉靖十五年三月戊午（3月24日），夜，客星见于天禧(棓)星旁。

《明世宗实录》卷185

(b) 明世宗嘉靖十五年三月戊午（3月24日），有星见天棓旁，东行历天厨，西入天汉，至四月壬辰没。

《明史·天文志》卷27

(c) 明世宗嘉靖十五年三月戊午（3月24日），夜，客星见紫微垣。四月壬辰灭。

《国榷》卷56

(d) 明世宗嘉靖十五年四月壬辰（4月27日），夜，客星见于天棓旁，渐东行，历天厨西，入天汉，至是日（原文做"始日"）始没。

《明世宗实录》卷186

公元1537年9月4日—10月3日 [中国]

明世宗嘉靖十六年八月，彗星见于西南，月余始灭。

清顺治 河南《阌乡县志》卷1

公元1538年1月21日 [朝鲜]

李朝中宗三十二年丁酉十二月丙寅，彗见于酉地，长三丈许，色白，至于甲戌，微有气。

《增补文献备考》卷6

公元1539年4月20日 [朝鲜]

(a) 李朝中宗三十四年四月庚子，有星见于三台星上台下，形体如彗，尾长一尺许…戊申夜，彗星见于轩辕尾第二星…辛酉夜，彗星见於轩辕第十星西，尾长五六尺许，色白…五月己巳夜，彗星见于轩辕星西角，尾长四五尺许，色白…甲戌夜，彗星微见。

《朝鲜王朝实录·中宗》卷90

(b) 李朝中宗三十四年己亥四月庚子，彗见于轩辕，尾长五六尺，色白，五月过酒旗，抵星南。

《增补文献备考》卷6

公元1539年4月30日 [中国]

(a) 明世宗嘉靖十八年四月庚戌，夜，彗星见，芒长三尺许，光指东南，扫轩辕北第八星，旬日始灭。

《明世宗实录》卷223

(b) 明世宗嘉靖十八年四月庚戌，彗星见，长三尺许，光指东南。扫轩辕北第八星，旬日始灭。

《明史·天文志》卷27

公元1540年4月30日 [中国]

明世宗嘉靖十九年三月丙辰，有星如彗，长丈余，历胃、室、壁，闰月丙辰朔入娄。

清同治 山东《即墨县志》卷11
注：康熙志，乾隆志均无此记录。

公元1545年12月26日 [中国]
(a) 明世宗嘉靖二十四年十一月壬午，夜，客星出天棓 ，入箕宿。越三日，转东北方，逾月始没。
《明世宗实录》卷305
(b) 明世宗嘉靖二十四年十一月壬午，有星出天棓，入箕，转东北行，逾月没。
《明史·天文志》卷27

公元1546年3月2日—31日 [中国]
明世宗嘉靖二十五年二月，彗星见北方，长丈余，凡三十余日乃灭。
清乾隆 福建《龙溪县志》卷20

公元1549年3月7日 [朝鲜]
李朝明宗四年己酉二月己酉，彗见于艮方。三月移见卯方。
《增补文献备考》卷6

公元1554年5月2日—30日 [中国]
明世宗嘉靖三十三年四月，彗星见，出北斗口，尾长亘天，五日而灭。
明隆庆 江苏《仪真县志》卷13

公元1554年6月23日 [中国]
(a) 明世宗嘉靖三十三年五月癸亥，夜，彗星见北斗天权星旁…六月辛未夜，彗星渐西北行，犯文昌…己丑夜，彗星行入近浊，始没。
《明世宗实录》卷410 — 卷411
(b) 明世宗嘉靖三十三年五月癸亥，彗星见天权旁，犯文昌，行入近浊，积二十七日没。
《明史·天文志》卷27

公元1554年6月26日 ［朝鲜］

(a) 李朝明宗九年五月丙寅，有星似彗，出北斗魁前，尾长一丈…六月庚午，彗星见于北斗魁前，文昌后，在张星二度许，色白，尾长三、四寸许…辛巳夜，彗星不见。

《朝鲜王朝实录·明宗》卷16

(b) 李朝明宗九年甲寅五月丙寅，彗见于北斗，尾长一尺四五寸，色白，至六月辛巳不见。

《增补文献备考》卷6

公元1555年10月15日—11月13日 ［中国］

明世宗嘉靖三十四年十月，彗星出昴，北行渐至斗。

清顺治 山西《潞安府志》卷15

公元1556年1月31日 ［中国］

明世宗嘉靖三十四年十二月二十日，彗星始见于东南，旋入西北，月余而没，芒长数十丈。

清顺治 甘肃《崇信县志》卷下

公元1556年3月1日 ［中国］

(a) 明世宗嘉靖三十五年正月庚辰，夜，彗星见于进贤星旁，芒长尺，西南指，渐长至三尺余，历扫大微垣次相星，又东北行，入紫微垣，扫天床星。至四月二日始灭。

《明世宗实录》卷431

(b) 明世宗嘉靖三十五年正月庚辰，彗星见进贤旁，长尺许，西南指，渐至三尺余。扫太微垣、次相东北，入紫微垣，犯天床，四月二日灭。

《明史·天文志》卷27

公元1556年3月1日 ［朝鲜］

(a) 李朝明宗十一年正月庚辰夜，有星见于轸星北五六度许，似有芒角，指西…壬午夜…彗星见於轸星东北八九度许，在太微东垣外，尾指西南，长一尺余，色微白…二月庚寅朔，夜一更，彗星见於寅地，自二更至五更，彗星见於招摇星上，在亢宿初度，去北极四十四度，尾指西南，长四尺许，色白…乙未，彗星见紫微宫北传舍星前，气至华盖星，去极二十三度余，尾二尺余，指西南，色白…丁巳夜，彗星见於寅方，在壁星北去北极六十度，尾指西南，长二尺许，色白…三月辛未夜，彗星见於卯地。

《朝鲜王朝实录·明宗》卷20

(b) 李朝明宗十一年丙辰正月庚辰，彗见于轸，历角亢，入紫微。尾长五尺许，色白。至三月庚午抵东壁，形色渐微。

《增补文献备考》卷6

公元1556年3月11日—4月10日　［中国］

明世宗嘉靖三十五年二月，彗星见北方，长丈余，凡三十余日乃灭。

明嘉靖　福建《长泰县志》卷下

公元1557年10月10日　［中国］

(a) 明世宗嘉靖三十六年九月戊辰，彗星见于天市垣内列肆星旁，芒长尺余，东北指，至十月二十三日始灭。

《明世宗实录》卷451

(b) 明世宗嘉靖三十六年九月戊辰，彗星见天市垣列肆旁，东北指，至十月二十三日灭。

《明史·天文志》卷27

公元1557年10月13日　［朝鲜］

(a) 李朝明宗十二年九月辛未，夜二更，彗星见於西方宦者星帝座星车肆星斗星之间，尾指候星，长一尺许，色微白…甲戌，夜一更，彗星见於天市楼星上，尾指宗人星，长二尺许，色微白，二更一点入西方天际…己卯，彗星见於南斗杓第一星西天河星，星光为河所侵，形体熹微…十月癸未，夜一更，彗星见於鳖星上，形体熹微，一更五点入西方天际…甲申夜一更，彗星见於南斗魁第一星，形体稍微，一更五点入西方天际。

《朝鲜王朝实录·明宗》卷23

(b) 李朝明宗十二年丁巳九月辛未，彗见于天市内，尾长一尺许，色白。十月移在南斗，形色熹微。

《增补文献备考》卷6

公元1558年8月8日 [朝鲜]

(a) 李朝明宗十三年七月庚午夜，彗星见於太微垣内，尾长四、五尺许，色白，芒气指天中，没入乾方天际...闰七月庚辰夜，彗星见於太微垣内，尾指郎将星南，长四尺许，色白，入乾方天际...乙未夜，彗星见於天市垣北，尾指贯索星南，长一尺许，色微白，至三更入西方天际...八月乙巳朔夜，彗星见於天市馆内市楼星上，尾指宗人星南，长四、五尺许，色微白，至三更入西方天际。

《朝鲜王朝实录·明宗》卷24

(b) 李朝明宗十三年戊午七月庚午，彗见于太微内，尾长四、五尺，色白。闰月移于天市内，其形熹微。

《增补文献备考》卷6

公元1558年8月9日 [日本]

日本永禄元年闰六月廿五日，天上扫帚星出，直至 七 月 五 、六日。

《细川两家记》

公元1567年1月10日 [朝鲜]

李朝明宗二十一年丙寅十二月丁亥，南方有星状如长星，广三四尺，长五六尺，上广下狭，良久而灭。

《朝鲜王朝实录·明宗》卷33

公元1567年10月27日 [中国]

明世宗嘉靖四十六年九月二十六日，晓，白气见东南，半月而灭。寻有星孛于东方，渐移而北，光长数丈，直亘天中。

清康熙 江苏《常州府志》卷3

公元1567年11月1日—30日 ［中国］
明世宗嘉靖四十六年十月，夜半，有白气数丈，起于东南，月余，彗星抵鬼斗，皆现于卯日，经于铃链二星，乃柘之分野。
清乾隆 河南《柘城县志》卷18

公元1569年11月9日 ［中国］
(a) 明穆宗隆庆三年十月辛丑，彗星见于天市垣，色苍白，芒指东北，长二尺余，至二十日灭。
《明穆宗实录》卷38
(b) 明穆宗隆庆三年十月辛丑，彗星见天市垣，东北指，至庚申灭。
《明史·天文志》卷27

公元1573年9月26日—10月25日 ［中国］
明神宗万历元年九月，夜，有星孛于西南方，氛气散漫，高约十丈余，上锐下阔，经三月方灭。
清康熙 福建《安海志》卷8

公元1574年12月13日—1575年1月11日 ［中国］
明神宗万历二年十二月，火星守奎宿，有尾，光芒长丈余，如彗星，一月乃散。
明天启 江苏《淮安府志》卷24

公元1577年8月10日 ［中国］
明神宗万历五年七月二十七日，有星如白气，长数丈，自西南方现，直指东北，至十一月渐小而没。
清康熙 福建《沙县志》卷11

公元1577年8月13日 ［中国］
明世宗万历五年七月终，彗星见西南方，犯斗度，至十二月没。
清康熙 浙江《缙云县志》卷4

公元1577年9月11日 ［中国］
明神宗万历五年八月三十日，彗星见于西南，光芒直射于车，形如簪，至十二月初旬方散。
清康熙 安徽《直县志》卷5

公元1577年10月14日 ［中国］

(a) 明神宗万历五年九月四日，彗星见西方，形如拖练，长约十余丈，由尾历箕越斗度牛，至十一月十九日止。

清乾隆 江西《龙泉县志》卷末

(b) 明神宗万历五年九月四日，彗星见西方，形如白云，势若拖练，根五丈余，阔三丈余，长约十丈，由尾历箕，越牛渡斗，至十一月二十九日乃止。

清乾隆 江西《大庾县志》卷1

公元1577年11月7日 ［中国］

(a) 明神宗万历五年九月二十八日，西南有一星甚大冲出，紫气弥天。

明天启 浙江《平湖县志》卷18

(b) 明神宗万历五年九月二十八日，酉刻，天西南见一星，从上逆下，光芒尺余，渐长数丈。后数日移逆中天，其影渐灭。

清康熙 浙江《温州府志》卷30

(c) 明神宗万历五年九月二十八日，彗星见西南，扫楚分。十月二日，光芒愈长，如帛百尺，竟西天。

清道光 江苏《璜泾志稿》卷7

公元1577年11月8日 ［中国］

(a) 明神宗万历五年九月二十九日，彗星见西南，光芒三丈余，东指凡半日，至十一月十五日始灭。

清嘉庆 江苏《东台县志》卷7

(b) 明神宗万历五年九月二十九日，彗星见于尾箕，长十余丈，光蔽南斗，芒焰如旗旆然，凡
四十余日而灭。此星色赤而特长，乃拖练星也。

《崔鸣吾纪事》

公元1577年11月8日—9日 ［日本］

(a) 日本天正五年九月廿八日壬午（11月8日），有彗星见于未申，长五丈，至翌年灭。

《续史愚抄》 50

(b) 日本天正五年九月廿九夜（11月9日），至翌年正月中旬，彗星出现百余日，光如明星。十二月起其形如大字，头东尾西，长二十间程。

《异本塔寺长帐》5

公元1577年11月10日—14日　[中国]

(a) 明神宗万历五年十月朔（11月10日），彗星见斗、牛间，尾长数十丈，光芒竟天，状如练，气成白虹。

清康熙　安徽《太平府志》卷3

(b) 明神宗万历五年十月朔（11月10日），彗星见斗、牛间，尾指婺女，长数十丈，光芒竟天，状若练，气成白虹。

《二申野录》卷5

(c) 明神宗万历五年十月朔（11月10日），彗星见于斗、牛间，尾指婺女宿，长数十丈。

清康熙　河北《昌平州志》卷26

(d) 明神宗万历五年十月戊子（11月14日），彗星见西南，光明大如盏，芒苍白色，长数丈， 繇尾、箕、越斗、牛，直逼女宿。

《明神宗实录》卷68

(e) 明神宗万历五年十月戊子（11月14日），彗星见西南，苍白色，长数丈，气成白虹。由尾、箕越斗、牛逼女，经月而灭。

《明史·天文志》卷27

公元1578年秋　[朝鲜]

李朝宣祖十一年秋，长星亘天，状如白练，数月乃灭。

《增补文献备考》卷6

公元1579年9月21日—10月19日　[中国]

明神宗万历七年九月，彗星见，在翼、轸间。

清康熙　浙江《乌青文献》卷3

公元1580年10月9日—11月6日　[中国]

(a) 明神宗万历八年九月，彗星见，在翼、轸间。

明万历　浙江《崇德县志》卷11

(b) 明神宗万历八年九月，彗星出东南，长数丈，至十月终乃没。

明天启　安徽《凤书》卷4

公元1581年4月3日—5月2日　　[中国]

明神宗万历九年三月，彗星见紫微垣，尾光芒射西北。

明天启　江苏《淮安府志》卷24

公元1581年冬　　[中国]

明神宗万历九年冬，脱练星出牛、斗间，长数尺，渐长至数丈，月余乃灭。

清康熙　河南《扶沟县志》卷4

公元1582年5月15日　　[日本]

日本天正十年四月廿三日，彗星出现于乾方，头在戌亥尾在辰巳，长十丈。

《多闻院日记》

公元1582年5月20日　　[中国]

(a) 明神宗万历十年四月丙辰，夜，彗星见于西北，尾指五车，历二十余日始灭。

《明神宗实录》卷123

(b) 明神宗万历十年四月丙辰，彗星见西北，形如匹练，尾指五车，历二十余日灭。

《明史·天文志》卷27

(c) 明神宗万历十年四月丙辰，夜，彗星见西北，尾指五车，浃旬始灭。

《国榷》卷71

公元1582年6月20日　　[中国]

明神宗万历十年六月丁亥，朔三日，彗出五车口，柱星以南。

《二申野录》卷5

公元1585年10月13日 [中国]

(a) 明神宗万历十三年九月戊子，彗星出羽林傍，形如弹丸，尾长尺许，指东北，其色苍白；后每夕移东行，芒体渐缩小…十月癸酉，彗星灭不见。

《明神宗实录》卷165 — 卷167

(b) 明神宗万历十三年九月戊子，彗星出羽林旁，长尺许，每夕东行，渐小，至十月癸酉灭。

《明史·天文志》卷27

(c) 明神宗万历十三年九月戊子，夜，彗出羽林旁，形如弹丸，尾长尺，指东北，其色苍；每夕渐东西缩。十月癸酉夜，彗灭。

《国榷》卷73

公元1585年10月16日 [朝鲜]

(a) 李朝宣祖十八年九月辛卯，夜一更，有星见于羽林星，似有芒气，变异非常…乙未夜一更，彗星见於天仓星西第一星内，在壁宿七度，去北极一百一度半，尾指艮方，长可一尺，色苍白，至五更一点入西方浊气中，不得看候。

《朝鲜王朝实录·宣祖》卷19

(b) 朝宣祖十八年乙酉九月辛卯，彗见于羽林，尾长尺余。

《增补文献备考》卷6

公元1587年10月2日—10月30日 [朝鲜]

李朝宣祖二十年九月，彗见西方，本柄偃曲，长三四十丈，光芒烛地，三阅月乃灭。

《增补文献备考》卷6

公元1589年1月2日 [中国]

明神宗万历十六年十月乙丑，彗星出于氐，长丈余，指东南，渐指西北，入亢度扫北斗、文昌、五车逼紫宫右，十一月甲辰灭。

清同治 湖南《桂阳直隶州志》卷22

注：按干支规律，十月应为十一月；十一月应为十二月。

公元1591年4月13日—23日 ［中国］

(a) 明神宗万历十九年三月丙辰（二十日）（4月13日），西北有星如彗，长尺余。历胃、室、壁，长二尺。闰三月丙寅朔，入娄。

《明史·天文志》卷27

(b) 明神宗万历十九年三月丙辰（二十日）（4月13日），夜，彗星见胃、室间，尾长二尺。闰三月丙寅夜，彗入娄宿。

《国榷》卷75

(c) 明神宗万历十九年三月丙辰（二十日）（4月13日），彗星见西北胃宿度分，尾长尺许，继移东北室、壁宿度分，尾长二尺。

清康熙 河北《昌平州志》卷25

(d) 明神宗万历十九年闰三月丙寅朔（4月23日），彗入娄…癸酉，大学士申时行题；三月二十日一更后，西北方有星如彗，尾长尺余，钦天监云，在胃宿度，近有在东北方室壁宿度，尾长约二尺，不胜惊骇。

《明神宗实录》卷234

公元1591年4月23日—5月22日 ［中国］

明神宗万历十九年闰三月，彗星在西北方室、壁度，尾长二尺。

明崇祯 上海《松江府志》卷47

公元1591年8月19日—9月16日 ［中国］

明神宗万历十九年七月，彗见东井，逆行入紫微。

清乾隆 广西《全州志》卷11

公元1593年7月30日 ［中国］

(a) 明神宗万历二十一年七月乙卯，夜五鼓，彗星见于井度，长三尺余…乙亥，彗星逆行，入紫微垣，犯华盖星。

《明神宗实录》卷262

(b) 明神宗万历二十一年七月乙卯，彗星见东井。乙亥，逆行入紫微垣，犯华盖。

《明史·天文志》卷27

(c) 明神宗万历二十一年七月乙卯，夜，彗见井宿，芒尺余。乙亥夜，彗逆行入紫微垣，犯华盖星。

《国榷》卷76

公元1593年8月3日—4日　[日本]

(a) 日本文禄二年七月七日（8月3日），天晓丑时彗星出……廿日，彗星消失。

《多闻院日记略》

(b) 日本文禄二年七月八日(8月4日)，天晴，夜半过，扫帚星现于丑寅方，长一丈。

《时庆卿记》 5

公元1593年 8月6日　[朝鲜]

(a) 李朝宣祖二十六年七月壬戌，五更有星似彗，出於紫微西垣外近处…色苍白，尾长三四尺许…乙丑，夜五更，彗星渐移东北行，在於八穀星下内阶星前…辛巳，自三更至五更，彗星渐移上行在华盖星外，传舍星内，形体暂微，色苍白，尾长二、三尺许…八月丁亥，彗星移在华盖西第五第六星外，传舍西第八星内，形体微小，色苍白，无尾…丙午，彗星消灭不见。

《朝鲜王朝实录·宣祖》卷41

(b) 李朝宣祖二十六年癸巳七月壬戌，彗见紫微外。辛巳，抵传舍，其形渐微。八月丙午乃灭。

《增补文献备考》卷6

公元1594年1月21日—2月19日　[中国]

明神宗万历二十一年十二月，彗出天津。

清康熙　江苏《淮安府志》卷1

公元1596年5月27日—7月24日　[日本]

(a) 日本庆长元年五月（5月27日—6月25日），彗星见。

《续史愚抄》 51

(b) 日本庆长元年六月廿九日（7月24日），晴，今夜酉戌刻间，戌亥方彗星出现。七月二日，天晴，今夜彗星出，长五尺斗有之…七日，天晴，彗星出。

《孝亮宿弥日次记》1

公元1596年7月19日—25日 ［朝鲜］

(a) 李朝宣祖二十九年六月庚申（7月19日），彗见于井，状如五车大星，至七月甲午乃灭。

《增补文献备考》卷6

(b) 李朝宣祖二十九年七月朔丙寅（7月25日），夜一更，彗星见於中台星东…癸酉，夜一更，彗星见于下台…壬午，夜一更，彗星见於常陈星下…甲午，彗星消灭。

《朝鲜王朝实录·宣祖》卷77

公元1596年7月25日—8月23日 ［中国］

(a) 明神宗万历二十四年七月，彗星见于西北。

清康熙 广西《合浦县志》卷1

清康熙 广西《廉州府志》卷1

(b) 明神宗万历二十四年七月，彗星见北斗。

清乾隆 福建《邵武府志》卷24

公元1596年8月5日 ［中国］

(a) 明神宗万历二十四年七月丁丑，彗星见西北方，如弹丸大，苍白色，芒指东南，入翼宿度，约长尺余，往西北行入。

《明神宗实录》卷299

(b) 明神宗万历二十四年七月丁丑，彗星见西北，如弹丸，入翼，长尺余，西北行。

《明史·天文志》卷27

公元1600年9月2日 ［朝鲜］

李朝宣祖三十三年庚子七月丙寅，彗见于中台。辛卯，移在太微内，光芒渐微。

《增补文献备考》卷6

公元1601年12月20日　［朝鲜］
李朝宣祖三十四年辛丑十一月二十六日，庆尚道熊川县彗见于子丑间，尾长数尺。
《增补文献备考》卷6

公元1602年　［朝鲜］
李朝宣祖三十五年壬寅七月乙巳，彗见于三台北极之间，移于郎位下，尾长七八尺，久而不灭。
《增补文献备考》卷6
注：七月庚申朔，无乙巳。

公元1604年秋　［中国］
明神宗万历三十二年秋，彗星见于天中，其色暗淡不现，长可二三尺，约六日始消。
清乾隆 广东《潮州府志》卷11

公元1607年9月20日　［中国］
(a) 明神宗万历三十五年七月庚申，夜，彗见东井，二尺余。八月丁丑夜，彗历于房。壬午夜，彗自房历于心。
《国榷》卷80
(b) 明神宗万历三十五年七月庚申，夜五鼓，东北方彗星出于井宿，尾指西南，长一尺。
明崇祯 上海《松江府志》卷47

公元1607年9月21日　［中国］
(a) 明神宗万历三十五年八月辛酉朔，彗星见于井度，色苍白，尾指西南，约长二尺，渐往西北…丁丑，彗历于房…壬午，彗历于心。
《明神宗实录》卷437
(b) 明神宗万历三十五年八月辛酉朔，彗星见东井，指西南，渐往西北。壬午，自房历心灭。
《明史·天文志》卷27
(c) 明神宗万历三十五年八月辛酉朔，彗星见井度，长二尺，渐往西北。壬午灭。
清雍正 《陕西通志》卷47

公元1607年9月25日—10月4日 [日本]

(a) 日本庆长十二年八月十四日甲戌（10月4日），东（寅丑角）西彗星出现...九月十一日辛丑，晴。一两夜以来，彗星不见。

《孝亮宿祢日次记》2

(b) 日本庆长十二年八月五日乙丑（9月25日），丑时彗星出现艮，其长八尺白色也，光气指未……十一日辛未，戌时彗星出现兑，其长二五寸，色白，光气指震。九月五日乙未，戌时从艮行坤，其长二丈余，色赤如火也。

《土御门家秘要录》7

公元1607年9月23日 [朝鲜]

(a) 李朝宣祖四十年八月癸亥，夜五更，云破处有星见於上台之南，其色苍白，似有偏指之尾...九月甲午，彗星见於天市垣第一星。

《朝鲜王朝实录·宣祖》卷215—216

(b) 李朝宣祖四十年丁未八月癸亥，彗见三台，尾长七八尺，苍白，移于天市，至九月十八日乃灭。（补）十月，见三台间，尾指文昌及北斗魁，长十余丈，色苍白，逾月乃灭。

《增补文献备考》卷6

公元1607年10月21日—11月18日 [中国]

明神宗万历三十五年九月，彗星见轸。

清乾隆 广西《廉州府志》卷5

公元1614年3月14日 [日本]

(a) 日本庆长十九年二月四日，彗星出于东方，朝野几藉藉，以为弗祥。

《逸史》11

(b) 日本庆长十九年二月口日，或记有彗星，长数丈，芒向西北，至夏云。

《续史愚抄》53 后水尾院上

公元1616年冬 [中国]

明神宗万历四十四年冬，彗见，自东井入紫微垣，南指有光。

清康熙 山西《榆次县续志》卷12

公元1617年6月3日—7月2日 [中国]
明神宗万历四十五年五月，彗入南斗月余，长三丈余。
清康熙 浙江《长兴县志》卷4

公元1618年1月26日—2月24日 [中国]
明神宗万历四十六月正月，白彗出氐、亢。
《二申野录》卷6

公元1618年3月26日—4月24日 [中国]
明神宗万历四十六年三月，彗出于虚。
清乾隆《山东通志》卷33

公元1618年春 [中国]
明神宗万历四十六年春，彗星出东南，冲紫微垣，五十日始没。
清乾隆 山西《解州芮城县志》卷11

公元1618年8月28日—12月14日 [日本]

(a) 日本元和四年七月九日（8月28日）晓，彗星出现于东北，尾三尺…十一月二日晓，彗星出现，由东方移至北方。

《孝亮日次记》

(b) 日本元和四年七月九日乙未（8月28日）晴，此晓，丑寅角彗星出现，三尺…十月廿八日癸未晴，夜半以后彗星自丑寅角出。

《孝亮宿祢日次记》5

(c) 日本元和四年秋八月八日（9月26日），彗星出。

《桑都日记》3

(d) 日本元和四年十月十一日丙寅（11月27日），彗星见东（晨见），日出时长丈余。

《续史愚抄》53

公元1618年9月19日—10月17日 [中国]

(a) 明神宗万历四十六年八月，彗出东南，尾扫西北，八十余日灭。

清顺治 河北《饶阳县志》卷5

(b) 明神宗万历四十六年八月，彗星入娄。

清康熙 广东《潮阳县志》卷3

(c) 明神宗万历四十六年八月，有白气自东亘西，数夕后有巨星长丈余，每夜见于东方，空中悬有白气，形如关刀，蛇枪长方，又如白布，出辰，分角、亢度，其尾冲指奎、娄、壁度，自北铺南竟天，至十月始灭。

清康熙 广西《平乐县志》卷6

公元1618年11月14日 [中国]

(a) 明神宗万历四十六年九月癸丑，广州彗星见，出辰分角、亢度，其尾冲指奎、娄、壁度。先数夜有白气自东亘西，如刀形，与彗星并见，锋芒如帚，两月乃灭。

《二申野录》卷6

(b) 明神宗万历四十六年九月癸丑，彗星见于东方，出辰分角、亢度，其尾冲指奎、娄、壁度，长数丈，壮如箭。

清康熙 广东《三水县志》卷1

清康熙 广东《肇庆府志》卷3

公元1618年11年15日—16日 [中国]

(a) 清太祖天命三年九月甲寅（11月15日），夜，有长星见翼、轸，色白，形如匹布，衡尺余，长可二丈，凡十九日灭。

《国榷》卷83

(b) 明神宗万历四十六年九月乙卯（11月16日），有长星见于东南，其星白气一道，形如匹布，阔尺余，长二丈余，东至轸，西入翼，凡一十九日而灭。

《明神宗实录》卷574

(c) 明神宗万历四十六年九月乙卯（11月16日），东南有白气一道，阔尺余，长二丈余，东至轸，西入翼，十九日灭。

《明史·天文志》卷27

公元1618年11月26日 [中国]

(a) 明神宗万历四十六年十月乙丑，彗星出于氐，如鸡弹大，长丈余，色苍白，尾指东南，后十数日，转指西北，扫犯太阳守星，入亢七度一十分。又三日，渐往西北方行，尾扫北斗、天璇、天玑、文昌、五车等星，逼紫垣，在亢六度，至次月十九日始灭。

《明神宗实录》卷575

(b) 明神宗万历四十六年十月乙丑，彗星出于氐，长丈余，指东南，渐指西北。扫犯太阳守星，入亢度，西北扫北斗、天玑、文昌、五车，逼紫微垣右，至十一月甲辰灭。

《明史·天文志》卷27

(c) 明神宗万历四十六年十月乙丑，夜，彗见氐宿，大如鸡子，长丈余，色苍白，尾指东南。逾数夕，转指西北，犯太阳守星。又三日，渐往西北，尾扫北斗天璇、天玑、文昌、五车等星，逼紫微，至次月十九日灭。

《国榷》卷83

(d) 明神宗万历四十六年十月乙丑，彗星见，出卯分，氐房度，其尾冲指申分觜参度，从下观长有丈余，状如刀，芒焰逾月始灭。

清康熙 广东《三水县志》卷1

清康熙 广东《肇庆府志》卷3

公元1618年11月26日 [朝鲜]

(a) 李朝光海君十年十月乙丑，彗星见於东方氐星度内，长丈余，色黄赤，天明不见...辛巳三更，彗星始见，长可二十丈，广四五尺许，傍犯北斗第二三星，通贯文昌星，形体比前稍减，色也渐淡...十一月丁酉，彗星芒气熹微...癸丑夜，彗星微见形体，芒气几尽消灭。

《朝鲜王朝实录·光海君》卷133—134

(b) 李朝光海君十年戊午十月乙丑，彗见于氐，尾长二十丈，广四五尺，色赤，犯北斗，贯文昌。十一月...癸丑，彗星亦消。

《增补文献备考》卷6

公元1618年11月27日 [中国]

(a) 清太祖天命三年十月十一日（丙寅），五更时，东南更有白气自星出，约宽五尺，直冲明国，至十四日后不见。其出气之星每夜向北斗渐移，至二十九日直越北斗柄，自次以后不见。

《大清满洲实录》

(b) 清太祖天命三年十月十一日（丙寅），彗星见于东方，在先见白气之东，广五尺，其尾直指明国，每夜渐移向北斗，凡十有九日，越北斗尾，方灭。

《清太祖高皇帝实录》卷5

(c) 清太祖天命三年十月十一日（丙寅），彗星见东方，尾长五丈，每夜渐移向北斗，十九日而没。

《清史稿·天文志》卷39

公元1618年12月17日—1619年1月15日 ［中国］

(a) 明神宗万历四十六年十一月，有星孛于轸，星赤如气如刀，长丈余，次年正月始灭。

明崇祯 广东《南海县志》卷2

(b) 明神宗万历四十六年十一月，彗星见东方，下破军星，五尺，光芒拂北斗。

清嘉庆 江苏《扬州府志》卷70

公元1619年2月中 ［中国］

(a) 明神宗万历四十七年正月秒(杪)，彗星见东方，长数百尺，月下射射，末曲而锐，未几而见于东北，又未几见于西。

《明神宗实录》卷578

(b) 明神宗万历四十七年正月杪，彗见东南，长数百尺，光芒下射，末曲而锐，未几见于东北，又未几见于西。

《明史·天文志》卷27

公元1619年春 ［中国］

明神宗万历四十七年春，彗星见，初时午，后射子，入北斗，逾二旬方灭。

清康熙 广东《文昌县志》卷9

公元1622年9月7日 ［日本］

日本元和八年八月二日丙寅，小雨…乾角彗星出现，长一间计有之，子丑刻许出云云。

《孝亮宿祢日次记》7

公元1624年7月27日　[日本]
日本宽永元年七月一日癸丑晴，或人语云，去六月十三（7月27日）、四日（7月28日）满月之夜，南方彗星出现，长五尺许云云。
《孝亮宿祢日次记》8

公元1627年夏　[朝鲜]
李朝仁祖五年夏，星孛于太阴。
《增补文献备考》卷6

公元1628年8月13日　[日本]
日本宽永五年七月十四日癸酉晴，寅刻彗星辰方现。十六日乙亥晴，彗星出现寅刻。
《孝亮宿祢日次记》10

公元1630年2月1日　[日本]
日本宽永六年十二月十九日庚午晴，乾角彗星出现云云。
《孝亮宿祢日次记》10

公元1638年6月12日—7月10日　[朝鲜]
李朝仁宗十六年五月，彗见巽方，光芒甚赤，经月乃灭。
《增补文献备考》卷6

公元1639年8月29日—9月26日　[中国]
明毅宗崇祯十二年八月，彗星入北斗。
清乾隆 山西《介休县志》卷1

公元1639年9月27日—10月25日　[中国]
明毅宗崇祯十二年九月，彗星见参分。
《国榷》卷97

公元1639年秋　[中国]
明毅宗崇祯十二年秋，彗星见参分。
《明史·天文志》卷27

公元1639年11月25日—12月23日 [中国]
明毅宗崇祯十二年冬月，夜，有彗星入北斗。
清康熙 山西《介休县志》卷1

公元1652年11月1日—30日 [中国]
(a) 清世祖顺治九年十月，彗星见，尾长二丈。
清乾隆 广东《顺德县志》卷16
(b) 清世祖顺治九年十月，南有群星团聚，一星大如月，光有芒，盖孛也，由井逆行至毕、昴而没。
清道光 江苏《璜泾志稿》卷7

公元1652年12月16日 [日本]
日本承应元年十一月十六日夜客星出现，星上如属紘，晕指东北五尺许。二十四日已后，晕薄日减去。
《山鹿素行家谱年谱》

公元1652年12月22日 [中国]
清世祖顺治九年十一月庚寅，异星苍白气见于参，西北行入毕。
《清史稿·天文志》卷39

公元1653年9月20日 [朝鲜]
李朝孝宗四年癸巳七月壬戌，彗见紫微西垣外，体小于北河，渐移入觳下，尾指西南，长七八尺，色苍白，至八月二十五日不见。
《增补文献备考》卷6

公元1656年 [朝鲜]
李朝孝宗七年丙申六月庚申，彗见于井，状如五车大星，尾长二三尺，色苍白，至七月二十九日乃灭。
《增补文献备考》卷6
注：六月朔戊寅，无庚申。

公元1659年冬 [中国]
清世祖顺治十六年冬，彗星犯北斗。
清顺治山东《登州府志》卷1

公元1661年2月4日　　[朝鲜]

(a) 李朝显宗二年正月丙辰夜，五更，彗星见於东方牛宿度内河鼓败瓜之间，而星体似河鼓小星，尾长二尺许，指乾方，色白，至十余日乃灭。

《朝鲜王朝实录·显宗》卷4

(b) 李朝显宗二年辛丑正月丙辰，彗见于牛，尾长二尺许，色白。

《增补文献备考》卷6

公元1663年10月31日—11月29日　　[中国]

(a) 清圣祖康熙二年十月，彗星见于翼轸之分，芒如帚，长二丈余，十二月消。

清康熙广东《海阳县志》 灾祥

(b) 清圣祖康熙二年十月，彗星见，起自东南，光烛西北，长丈余，经十昼夜方散，次年三月又见。

清康熙河北《冀州志》卷1

(c) 清圣祖康熙二年十月，彗星见东南，凡五十余日始没。三年二月彗星复见。

清康熙 安徽《霍丘县志》卷10

(d) 清圣祖康熙二年十月，彗星见东南，凡五十余日。三年二月复见。

清康熙江苏《江宁府志》卷3

清康熙 江苏《上元县志》卷13

公元1664年3月28日　　[日本]

日本宽文四年三月朔日起七日止，未申方彗星出现。

《当所年代记》平山日记

公元1664年10月19日—28日　　[中国]

清圣祖康熙三年九月初旬，有星光气如烛，长尺余，随星而上，五鼓见，每夜如帚，至冬末，其光渐短，然犹有四、五寸许，渐。转渐早，遂以晚见，至次年春月乃不见。

清康熙 安徽《巢县志》卷4

公元1664年10月19日—11月17日 [中国]

(a) 清圣祖康熙三年九月，彗星见于翼轸，抵十七宿，仲冬月乃止。

清雍正 湖北《沔阳州志》卷13

(b) 清圣祖康熙三年九月，彗星见翼轸分野，尾向西北，数日后，复向东北，长三丈余，白光黯然，历五十余日而灭。

清嘉庆 湖南《郴州总志》卷41

公元1664年11月18日 [中国]

(a) 清圣祖康熙三年十月己未朔，彗星见轸宿度。丁亥，彗星见翼宿度，指西北方。十一月丁酉，彗星犯张宿。庚子，彗星犯井宿。丙午，彗星犯胃宿，尾指东北方。十二月壬戌，彗星在奎宿度，形渐小。丁卯，刑科给事中杨雍建疏言：本年十月初旬彗星见，及今五十余日，历一十三宿……四年二月己巳，彗星见女宿度。乙酉，彗星在壁宿一度三十分，尾长五尺余。三月己丑，彗星行入奎宿。

《大清圣祖仁皇帝实录》卷13

(b) 清圣祖康熙三年十月己未朔，有星茀于轸，见东方；丁卯，尾长七八寸，苍白，指西南；丁亥，尾长三尺余，指西北，逆行至翼；十一月戊戌，尾长五尺余，指北方，至张；庚子，在井；癸卯，往西北行至昴；乙巳，尾指东北，至胃；庚戌，至娄，尾指东，青色；十二月壬戌，至奎体小，尾长二尺余。四年二月己巳，东南方有异星见于女；甲戌，尾长七寸，指西南，苍白色；丁丑，尾长尺余，往东北顺行至虚；辛巳，至室，体渐大，尾长八尺余；乙酉，至壁，尾长五尺余。

《清史稿·天文志》卷39

(c) 清圣祖康熙三年十月己未朔，彗星见东南方，其体微小，在轸宿右辖星旁，丁卯，见东方，尾迹长七寸余，苍色，指西南。丁亥，逆行西南，其体渐大，尾迹长三尺余，指西北，在翼宿。十一月戊戌，尾迹长五尺余，指北，在张宿。庚子，在井宿。癸卯，逆行西北，在昴宿。乙巳，尾指东，在胃宿。庚戌，在娄宿。十二月壬戌，彗星移奎宿，其体渐小，四年正月癸巳，不复见。二月己巳，复见东南方，其体渐小，在女宿。甲戌，见东北，尾迹长七寸余，苍白色，指西南。丁丑，尾迹长一尺余，在虚宿。辛巳，其体渐大，尾迹长八尺余，在室宿。乙酉，尾迹长五尺余，在壁宿。

《钦定皇朝文献通考·象纬》卷267

公元1664年11月26日 [朝鲜]

(a) 李朝显宗五年十月丁卯夜，有星见於轸宿东侧，有偏指之尾，色白，即彗星…乙酉夜，彗星见於右辖星西，形体渐大…十一月戊子朔夜，彗星见於翼星下…戊戌夜，彗星见於外厨星下弧星东…十二月戊午夜，彗星见娄宿度内天囷星上…乙亥夜，彗星见於奎宿度内…六年正月癸巳，是夜彗灭。

《朝鲜王朝实录·显宗》卷9

(b) 李朝显宗五年甲辰十月丁卯，彗见于轸，似有尾跡，历九宿抵娄，至翌年正月乃灭。

《增补文献备考》卷6

公元1664年11月27日—11665年2月26日 [日本]

(a) 日本宽文四年十月十日（11月27日），彗星出现，自轸左辖指西，其白气二间余，至翌年正月末止。

《山鹿素行家谱年谱》

(b) 日本宽文四年十一月二十九日(12月16日)，半夜就寝一睡之后，寅前出中庭望见，则东南有异星一，光芒青赤如彗，其长数步，指西北，芒本小末大，其分处可余筋，可谓变异也，殆是彗星乎？先是见之者曰，今月上旬此星在参宿之间，顷日退座数舍，自丑刻见其光薄，逐刻增光，寅后甚盛，自寅半其光又薄云云。

《国史馆日录》1

公元1665年3月13日 [日本]

日本宽文五年正月二十七日丙寅，西方有白气，二丈可，如柱，自日初入，每夜见，至戌刻消散，至二月上旬灭。

《续史愚抄》60

公元1665年4月1日 [日本]
日本宽文五年二月十六日癸酉，顷日有彗星见寅方（晨见），至三月上旬灭。
《续史愚抄》60

公元1667年3月11日 [中国]
清圣祖康熙六年二月十七日，彗星见角、亢间。
清雍正 广西《钦州志》卷1

公元1668年3月3日—12日 [中国]
清圣祖康熙七年正月末旬，有彗星如疋布著天，见于轸宿，阅三月余，历十五宿而始没。
清康熙 山东《黄县志》卷7

公元1669年2月1日—1670年1月20日 [中国]
清圣祖康熙八年，彗星芒入斗口。
清康熙 陕西《汉南郡志》卷2

公元1670年1月21日—1671年2月8日 [中国]
(a) 清圣祖康熙九年，彗星吐芒，亘天，入斗口。
清康熙 陕西《洋县志》卷1
(b) 清圣祖康熙九年，彗星见，长丈余，数月方没。
清乾隆 陕西《长治县志》卷21

公元1673年5月9日 [中国]
清圣祖康熙十二年三月癸巳，异星见于娄，大如核桃，色白，尾长尺余，指东方；甲午，仍见。
《清史稿·天文志》卷39
注：原文二月，二月无癸巳，若三月，则为5月9日。

公元1676年11月15日 [日本]
日本延宝四年十月十日，彗星出现轸之左辖，其气渐二三尺。
《山鹿素行家谱年谱》

公元1677年4月28日 ［中国］
清圣祖康熙十六年三月癸卯，东北方有异星见于娄，体色光明润泽，尾长尺余，指西南，占曰含誉。
《清史稿·天文志》卷39

公元1678年1月23日—1679年2月10日 ［中国］
清圣祖康熙十七年，西南一星，上带白气长丈余，形如扫帚，冲入北斗，月余而没。
清顺治 山东《登州府志》卷1

公元1679年9月2日 ［中国］
(a) 清圣祖康熙十八年七月二十八日，有星见于西方，白气竞天，历尾箕，至牛、女，逾月乃没。
清康熙 安徽《桐城县志》卷1
(b) 清圣祖康熙十八年七月二十八日，彗星见。
清道光 河北《保安州志》卷1

公元1679年12月3日—1680年1月1日 ［中国］
清圣祖康熙十八年十一月，西方有一星如帚，有一小星黑浑而白，气自内喷出，扫及奎、娄，经月乃灭。
清康熙 广东《阳春县志》卷1

公元1680年11月14日—1681年1月19日 ［日本］
(a) 日本延宝八年九月廿三日（11月14日）夜，彗星出西南，至十二月初旬不见。
《德川实纪》常宪院殿御实纪二
(b) 日本延宝八年十月九日（11月29日），今晓七时过见彗星、东角星、上相并出，其帚之气及半天，凡此间客星之沙汰有之。当月初出现也，甚迟行，历五三日甚下东，而其吐气亦甚薄。
《山鹿素行家谱年谱》
(c) 日本延宝八年十一月三日（12月23日），长星，十一月三日出西方。初不见星，如引白练，至七日、八日夜星见，光芒半天，次第灭，至月末消。
《分类本朝年代记》1

(d) 日本延宝八年十一月（1680年12月21日—1681年1月19日），长星西现，光芒指东，幅三尺，长数十丈，至十二月没。
《续皇年代略记》

公元1680年11月21日—12月20日 ［中国］
(a) 清圣祖康熙十九年十月，有星放白光，自西北直至东南，长竞天，昏时未高，中夜则移在斗、牛之间，两月乃没。
清康熙 浙江《定海县志》
(b) 清圣祖康熙十九年十月，有星见西方，如练，起女虚入奎四十余日。
清乾隆 安徽《东流县志》卷7
(c) 清圣祖康熙十九年十月，长星见西方，如练，起女虚入奎，凡四十余日。
清嘉庆 江西《九江府志》卷30
(d) 清圣祖康熙十九年十月，长星见于西南，帚长二三丈，东北指天汉，月余始灭。
清康熙《山西通志》卷30

公元1680年11月21日 ［朝鲜］
(a) 李朝肃宗六年十月朔丙戌，夜五更，有星见於太微西垣外，似有尾跡…丁亥夜，昨夜所见之星移在太微右执法星傍翼宿六度，星体如织女大星，色苍白，似有尾跡…癸巳，彗星移见於右角星上，形色比前无异，尾跡几至丈余…十一月朔丙辰昏，白气自西方指天中，状如彗，累日不灭…彗星隐见出没於牛斗参虚危室壁奎娄胃星之上者。
《朝鲜王朝实录·肃宗》卷10
(b) 李朝肃宗六年庚申十月丙戌，彗见于太微西垣外，尾长三四尺，历角亢，在氐不见。十一月丙辰，初昏白气在西，长三四丈，微见星体，而在濛气中不得测侯。
《增补文献备考》卷6

公元1680年11月23日 ［中国］
(a) 清圣祖康熙十九年十月初三（戊子），夜，西南方白气亘天，半月不散，每夜移至正西而息。
清康熙 浙江《临海县志》卷11

(b) 清圣祖康熙十九年十月初三（戊子），彗星见翼宿度。十一月丙辰，彗星夕见于西方。

《大清圣祖仁皇帝实录》卷92 — 卷93

(c) 清圣祖康熙十九年十月初三（戊子），彗星见右执法，色白，尾长尺余，指西方，东行甚速；壬寅，近太阳不见。十一月丙辰朔，尾迹夕见西方；壬戌见星体，色苍白，尾长六丈余，宽二尺余，指东北。

《清史稿·天文志》卷39

公元1680年12月21日 ［中国］

(a) 清圣祖康熙十九年（庚申）十一月丙辰朔，沧州有白气如帚，自西南向东北，浃旬方灭；卢龙有白气如云，长亘向东，越数夕色淡，而高起如帚芒状；绛县夜见白气如虹。

《清史稿·灾异四》卷43

(b) 清圣祖康熙十九年（庚申）十一月丙辰朔，自酉至戌白气如彗，其长竟天，状如大带，自西南起直抵银汉，渐西流至北而化。初四日，根粗梢细，有黄光，中有青气一道。初五，根连一小昏星。初九，色淡白，梢又上星。至十五日更细小，将至天中，星却明亮。其后隐现，不时短长迭变，历五十余日而化。

清康熙 山西《武乡县志》卷4

(c) 清圣祖康熙十九年（庚申）十一月丙辰朔，日初没，有妖星吐白光，自西南起指东北，围大尺余，其长竟天，上下锐，厥状劲直，逾时而灭，至十二月初旬始隐不见。

《三冈识略》卷8

公元1680年12月21日—1681年1月19日 ［中国］

(a) 清圣祖康熙十九年十一月，正西有星，吐白气，直冲东北，抵紫微垣右掖卫，光芒如匹练状，尾开数十丈，经明年春，始渐次没。

清康熙 江西《赣州府志》卷18

(b) 清圣祖康熙十九年十一月，白气自西南冲贯斗、牛，气尾有星。

清康熙 河北《玉田县志》卷8

(c) 清圣祖康熙十九年十一月，彗星见西南，倏变白虹，长数丈，两头皆锐；十三、十四等日又一星吐白芒，横亘天中，久而不灭。

清乾隆 广东《湖州府志》卷11

公元1680年12月22日 [中国]

(a) 清圣祖康熙十九年十一月初二日，镇洋西方见白气亘天，长数丈、移时乃灭；临淄见白气自西而东。

《清史稿·灾异》卷43

(b) 清圣祖康熙十九年十一月初二日，有白气一道，长数十丈，现于西南，如白练一疋，直贯三台，至十二月十日不见。

清康熙 四川《成都府志》卷25

(c) 清圣祖康熙十九年十一月初二日，彗星见女虚，逾月乃灭。

清康熙 湖南《零陵县志》卷14

公元1682年8月24日 [朝鲜]

(a) 李朝肃宗八年七月丁卯，有星似彗，见艮方北河星上...戊辰，彗星暂移於三台上阶星南，是后屡见於角亢摄提星上，至九年始不见。

《朝鲜王朝实录·肃宗》卷13

(b) 李朝肃宗八年壬戌七月丁卯，彗见三台，尾长三四尺，大如织女大星，色白，历右角，抵摄提，至九月乃灭。

《增补文献备考》卷6

公元1682年8月26日 [中国]

(a) 清圣祖康熙二十一年七月己巳，彗星见井宿度，尾长二尺余。

《大清圣祖仁皇帝实录》卷103

(b) 清圣祖康熙二十一年七月己巳，彗星见北河之北，色白，尾长二尺余，指西南，往东北行甚速；壬申，入午宫，尾长六尺余。

《清史稿·天文志》卷39

(c) 清圣祖康熙二十一年七月己巳，彗星见东北方，白色，尾迹长二尺余，指西南，在井宿北河北。壬申，行东北，尾迹长六尺余。

《钦定皇朝文献通考·象纬》卷267

公元1682年9月3日—9月4日 [日本]

(a) 日本天和二年八月二日（9月3日），今晚见彗星，戌方出现，其芒丈余，此间久出现云云。廿日，今夕暮后，西方有彗星出现。

《山鹿素行家谱年谱》

(b) 日本天和二年八月三日（9月4日），今夜彗星见酉戌之间，光辉六尺许。今日予出京，行左大臣殿亭，归路于途中见之。

《尧恕法亲王日记》2

公元1682年9月9日 [中国]

清圣祖康熙二十一年八月初八日，彗星见于角，次于房，经月乃灭。

清康熙 江西《彭泽县志》卷14

注：清乾隆 安徽《东流县志》卷7和清嘉庆 江西《九江府志》卷30有相同记载，但时间是八月。

公元1683年8月2日 [中国]

清圣祖康熙二十二年闰六月庚戌，异星见于五车北八谷东，色白，往西南逆行；戊辰，入五车。

《清史稿·天文志》卷39

公元1684年6月26日 [中国]

清圣祖康熙二十三年五月甲申，异星见太微垣，东属轸，色白明大，往东北顺行；乙酉，行四度余，至右摄提下。

《清史稿·天文志》卷39

公元1686年8月26日 [中国]

清圣祖康熙二十五年七月庚寅，异星见东方，近地平，色白，东行不急；丁酉，凡行十六度，至柳，微有尾迹；壬寅，至星，渐没。

《清史稿·天文志》卷39

公元1686年9月14日 [朝鲜]
李朝肃宗十二年丙寅七月己酉，彗见于张，至八月辛酉乃灭。
《增补文献备考》卷6

公元1689年12月6日 [日本]
(a) 日本元禄二年十月二十五日戊子，有异星见巽。
《续史愚抄》64
(b) 日本元禄二年十月二十五日夜，异星出巽方。
《武江年表》1

公元1695年10月22日后 [中国]
清圣祖康熙三十四年九月十五日后，五更时，有彗星现，自东天起自天河，至天色已明其影方没。至十月初八夜尚未息。
清康熙 上海《历年纪》

公元1695年11月1日 [日本]
日本元禄八年九月二十五日晨，东南有白气，本在轸，末至翼，十月中旬消，此在骏之分野。
《壬癸录》2

公元1698年11月8日 [日本]
日本元禄十一年十月六日丁未 … 即夜，大流星飞入井宿内，垂下数尺，经数刻而消。今夜流星如织，人皆见之。
《壬癸录》2

公元1701年3月28日 [朝鲜]
李朝肃宗二十七年辛巳二月丁丑，彗见于危室，尾长二丈余，状如五车大星，色白，移见于奎，至三月初四日不见。
《增补文献备考》卷6

公元1701年10月28日 [朝鲜]
(a) 李朝肃宗二十七年九月辛亥，（彗）见太微端门外，过翼抵张，尾长二三尺，至十月己巳不见。
《增补文献备考》卷6

(b) 李朝肃宗二十七年九月辛亥夜，有星见於太微垣端门外，似有尾迹…壬子，太微垣端门外所见之星形体如奎大星，尾迹稍长，似是彗星…癸丑夜，彗星见於太微垣端门外翼宿十六度，去北极九十三度，色苍白，长可二三尺，形体与昨无异…十月壬戌，彗星见於张宿内，形体比昨稍微，尾迹则几尽消灭…甲子，彗星见於张宿内，形体比昨尤微。

《朝鲜王朝实录·肃宗》卷35

公元1702年2月25日—3月8日　　［日本］

(a) 日本元禄十五年正月二十九日（2月25日），今夕起彗星出西方。

《新井白石日记》上

(b) 日本元禄十五年正月下弦至二月上弦（2月17日—3月8日）初昏，申酉方有白气，本出于奎十四度天仓内，末至于昴初度天苑内，其野中国西国。

《壬癸录》8

公元1702年2月28日　　［中国］

清圣祖康熙四十一年二月甲寅，酉时，西南白云一道，长二丈余，宽尺余，穿天仓、天苑，入地平，至丁巳，长三丈余。

《清史稿·天文志》卷39

公元1723年10月11日　　［中国］

清世宗雍正元年九月己丑，异星见弧矢下，色白，体微，芒长尺余，指西北，逆行至井。

《清史稿·天文志》卷39

公元1723年10月16日　　［中国］

清世宗雍正元年九月十八日，彗星出牛、女间，至二十四灭。

清雍正　广西《苍梧志》卷4

公元1723年10月19日　　［朝鲜］

(a) 李朝景宗三年癸卯九月丁酉，彗见于女，似有尾跡，移见于牛，至十月十七日不见。

《增补文献备考》卷6

(b) 李朝景宗三年九月丁酉，夜一更，彗星见於女宿度内，色白，有尾跡，十五日而灭。

《朝鲜王朝实录·景宗》卷13

公元1737年2月7日—2月17日 [日本]

(a) 日本元文二年正月初八日（2月7日），是夕彗星见于西方，直长庚之下，连见二十余日。

《紫芝園漫笔》8

(b) 日本元文二年正月十八日（2月17日），孛星见于西天，至下旬。

《一话一言》13

公元1737年2月13日—15日 [朝鲜]

(a) 李朝英祖十三年丁巳正月癸卯（2月13日），彗见于危，尾长三尺许，色白，体大於奎大星，历室壁奎，抵天仓西，至二月壬戌渐微不见。

《增补文献备考》卷6

(b) 李朝英祖十三年正月乙巳（2月15日），彗星见於危宿度，尾长尺许，指巽方，色白...丙午，彗星移见於室宿度，尾长二尺，色白，指巽方...壬子，彗星移见於壁宿度内铁质星上，尾长三尺许，指巽方，色白...丙辰，彗星移见於奎宿度内天仓星之西，尾长二尺许，指巽方，色白...壬戌，彗星见奎宿度内天仓星之西，尾长三四寸，指巽方，色白。

《朝鲜王朝实录·英祖》卷43

公元1737年7月8日 [中国]

清高宗乾隆二年六月丁卯，异星出右更东，色白，属娄，向西南行；丙子，仍见。

《清史稿·天文志》卷39

公元1741年3月17日—4月15日 [中国]

清高宗乾隆六年二月，彗星见西方，长丈余，七月复见，至十二月始灭。

清光绪 山东《登州府志》卷23

公元1741年5月15日—6月12日 [中国]
清高宗乾隆六年四月，西北有星，吐光如彗，九月移近西南，及冬而没。
清乾隆 上海《续淞南志》 灾祥

公元1742年1月24日 [中国]
清高宗乾隆六年十二月十八日，有星如彗芒角，起西北隅，上冲室尾，历次年二月初乃至。
清乾隆 湖南《善化县志》卷17

公元1742年2月26日—3月6日 [日本]
(a) 日本宽保二年正月壬午（2月26日）至今日（二月十二日）（3月18日），异星见于艮，其芒气指坤，长三尺许。
《太史局草》
(b) 日本宽保二年正月（2月5日—3月6日），彗星见于河鼓南及河鼓。
《紫芝園漫笔》8
(c) 日本宽保二年，自正月廿二三日，彗星夜半见于东北，至二月稍薄，中旬犹见之。
《文会杂记》6

公元1742年3月2日 [中国]
清高宗乾隆七年正月丙戌，异星见于斗宿之次，在天弁第二星之上，其色黄白，向西北逆行，四十余日隐伏。
《清朝通志·灾祥》卷122

公元1742年3月3日 [中国]
清高宗乾隆七年正月丁亥，异星见东南方；戊子，出地二十七度余，大如弹丸，色黄，尾长四尺余，指西南，属丑宫，在天市垣徐星外，逆行四旬余不见。
《清史稿·天文志》卷39

公元1742年3月6日 [朝鲜]
(a) 李朝英祖十八年壬戌正月庚寅，彗见于斗，尾长二三尺许，色白，体大如河鼓中星，历牛女，抵紫微，入垣内，至二月壬子乃灭。
《增补文献备考》卷6

(b) 李朝英祖十八年正月庚寅，客星见於斗星度内…二月辛卯，客星移见於牛宿，星体似彗，尾长二尺，色白…壬辰夜，彗星见於东方牛宿度内，尾长二三尺，色白，体如河鼓…癸巳，彗星移见於女宿度内，星体大如牵牛，长二三尺，色白…丁酉夜，彗星见於艮方，三更移于紫微东垣外，色白…戊申夜，彗星移见於紫微垣内，尾长二尺，色白…丁巳，彗星几消灭。

《朝鲜王朝实录·英祖》卷55

公元1742年12月27日—1743年1月25日 [中国]

(a) 清高宗乾隆七年十二月，彗出奎宿，光逾尺，至次年正月乃灭。

清乾隆 山东《惠民县志》卷3

(b) 清高宗乾隆七年十二月，彗出奎宿，光逾尺，至次年正月乃灭。彗长尺许，夕出，指东北；久之则朝出，指西南，前后五十余日。

清光绪 山东《霑化县志》卷4

(c) 清高宗乾隆七年十二月，彗星见，自壁而室、尾及奎、娄，至八年癸亥春正月始没。

清道光 江西《于都县志》卷27

公元1743年11月16日—12月15日 [中国]

(a) 清高宗乾隆八年十月，彗星见于西北方，躔危，室之次，长尺余，光芒斜指东北，两月余乃灭。

清乾隆 广东《潮洲府志》卷11

(b) 清高宗乾隆八年十月，彗星见西方，至甲子正月始没。

清同治 江西《安仁县志》卷34

公元1743年秋 [中国]

(a) 清高宗乾隆八年秋，有星见室、壁，光冲四五丈，自西向东，三月乃止。

清乾隆 江苏《吴县志》卷26

(b) 清高宗乾隆八年秋，彗星见西方，色白，长三丈余，阅月乃没。

清乾隆 山西《浮山县志》卷34

(c) 清高宗乾隆八年十一月壬戌，彗星见女、虚间，历危，上扫室、壁，经五十日没。

清乾隆 山东《掖县志》卷5
注：该月无壬戌。

公元1743年12月20日—1744年1月14日 [日本]

(a) 日本宽保三年十一月五日甲申（12月20日），顷日有彗星见西，至翌年正月中旬灭。

《续史愚抄》73

(b) 日本宽保三年十一月（1743年12月16日—1744年1月14日），彗星见于东壁，光芒斜指奎，二月稍衰。至十二月十八日复甚明，长大倍前，连夕不衰。至明年正月朔夕，又增光明，形益长大，过元宵乃伏而不见。初见时在东壁二星之中，后稍移，而西北向营室下星，最后切近营室下星，而光指东壁上星，凡见者前后五旬而伏。

《紫芝園漫笔》8

公元1744年1月1日—5日 [中国]

(a) 清高宗乾隆八年十一月十七日（1月1日），初昏，彗星见宿（应是奎）、壁之间，至明年正月杪，且见东方。

民国《江苏省通志稿》卷3

(b) 清高宗乾隆八年十一月己亥（1月4日），彗星见奎、壁之间，大如弹丸，色黄白，尾长尺余，向东指，属戌宫，逆行至九年正月辛卯，凡五十三日，行二十九度余。

《清史稿·天文志》卷39

(c) 清高宗乾隆八年十一月己亥（1月4日），彗星见，距奎宿第二星二度，大如弹丸，黄色，尾迹长尺余，每夜逆行指东。

《钦定皇朝文献通考·象纬》卷267

(d) 清高宗乾隆八年十一月己亥（1月4日），彗星见奎、壁之中，距奎宿第二星二度，体如弹丸，其色黄，尾迹长尺余，每夜向西逆行，由戌宫至亥宫。

《清朝通志·灾祥》卷122

(e) 清高宗乾隆八年十一月二十一日（1月5日），夜，彗星现于壁。

清乾隆 广东《揭阳县志》卷7

公元1744年1月6日 [朝鲜]

(a) 李朝英祖十九年癸亥十一月辛丑，彗见于壁，移室抵危，尾长二丈许，色淡赤，体大於太白，至翌年正月辛卯昧爽，反见于东，为曙色所射，不得看候。庚子乃灭。

《增补文献备考》卷6

(b) 李朝英祖十九年十一月辛丑，客星见于壁宿度内，形体似彗...十二月辛亥，彗星移见於壁宿奎宿之间，色淡赤，尾长四五尺，指艮方...癸丑，彗星移见室宿度内...二十年正月庚辰夜，彗星见於室宿度内壁宿之西，大如天狼星，尾长二丈许，色淡赤，指艮方...丙戌，彗星移见於危宿度内室宿西，星体比天狼稍大，尾长二丈许，色淡赤，指艮方...辛卯，昧爽，彗星见东方天际，形色尾跡所在躔度为曙色所射，不得测侯...庚子，彗星微见于日出时若有，星体旋失所在。

《朝鲜王朝实录·英祖》卷58—59

公元1744年2月6日 [中国]

(a) 清高宗乾隆八年十二月二十三日，奎宿度内有星黄赤，本大如小酒杯，末长约六、七寸，大如笔管，二十四至二十八联夜阴雨，暗黑不见。及二十九夜又见之，则进于壁宿度内，末扫东方，其长丈余，大如帚，来年正月天将晓时，犹有见之者。

清乾隆 江西《会昌县志稿》卷34

(b) 清高宗乾隆八年十二月二十三日，夜，有星丽奎宿度内，其光作黄赤色，本大径寸，末小如指，长约六、七寸，二十四至二十八连夜阴雨，暗黑不见，及二十九夜观之，则进于壁室度内，末扫东方，其长丈余，大如帚，次年正月天将晓时，犹有见之者。

清道光 江西《会昌县志》卷27

公元1744年11月17日 [朝鲜]

(a) 李朝英祖二十年十月丁巳夜，客星见於角宿度内...戊午夜，客星见於角宿度内，有尾。

《朝鲜王朝实录·英祖》卷60

(b) 李朝英祖二十年甲子十月丁巳，有星见于角，似有尾跡，二日后不见。

《增补文献备考》卷6

公元1747年2月16日 [中国]

清高宗乾隆十二年正月初八，戌时彗星见，光芒如帚，长丈余，一月之内，见者七次。

清乾隆 四川《江安县志》卷3

公元1747年冬 [中国]

(a) 清高宗乾隆十二年冬，有星见于西方，光芒数十丈，其形如帚。

清乾隆 广西《昭平县志》卷4

(b) 清高宗乾隆十二年冬，彗星见西南。

清乾隆 河南《固始县志》卷11

公元1748年4月26日 [中国]

清高宗乾隆十三年三月癸丑，异星见东方，大如榛子，色黄，尾长二尺余，向西南指，在离宫第三星南，顺行至四月甲寅朔，行三度，尾长尺余，体小光微；壬戌，至螣蛇；乙丑，至王良；丙寅，不见。

《清史稿·天文志》卷39

公元1748年5月2日 [朝鲜]

(a) 李朝英祖二十四年四月己未，客星见于室宿度内…庚申夜，四更，彗星移见於壁宿度内银河之中，星体大如王良星，尾长尺许，色淡赤，指坤方…丁卯夜，彗星移见於奎宿度内王良星之西，而形体大如王良星，尾长一尺，色淡赤。

《朝鲜王朝实录·英祖》卷67

(b) 李朝英祖二十四年戊辰四月己未，彗见于室，大如王良，尾长尺许，色淡赤，过壁在奎。

《增补文献备考》卷6

公元1759年4月1日 [朝鲜]

(a) 李朝英祖三十五年己卯三月乙酉（4月1日），彗见于危，大如河鼓中星，尾长二尺许，色苍白，过虚抵女，至于戊申乃灭。四月癸丑见于翼，大如金星，尾长二尺许，至戊寅乃灭。

《增补文献备考》卷6

(b) 李朝英祖三十五年三月丙戌（4月2日），彗星见于虚星度内，体如河鼓，色苍白...丙申，彗星复见於女宿度内...四月甲寅夜，彗星见翼星下，大如木星，色苍白...丁卯夜，彗星微见於张宿之上...壬申夜，彗星微见於张度四度...己卯，彗星微见。

《朝鲜王朝实录·英祖》卷93

公元1759年4月1日—5月16日 [日本]

(a) 日本宝历九年三月四日（4月1日），彗星出。

《彗星出现年纪》

(b) 日本宝历九年四月一日辛亥（4月27日），有彗星见加贺，长一丈，至中旬不见云。

《续史愚抄》76

(c) 日本宝历九年四月二十日庚午（5月16日）晴，彗星...晨天出虚危之间，光芒五六寸许...三月中旬以来出现云云。

《八槐御记》

公元1759年4月8日 [中国]

清高宗乾隆二十四年三月壬辰，彗星见东南方；甲午，出虚第一星下，大如榛子，色苍白，尾长尺余，指西南，顺行，癸卯，体小光微，尾余三四寸；戊申，全消。四月戊辰，彗星见西南方，在张第二星上；己巳，离张六度，大如榛子，色苍黄，尾光散漫，长二尺余，指东南，顺行；壬申，行迹微小；丁丑，更微；己卯，渐散；五月壬午，全消。

《清史稿·天文志》卷39

公元1759年4月10日 [中国]

(a) 清高宗乾隆二十四年三月甲午，彗星见。

《清史稿·高宗纪》卷12

(b) 清高宗乾隆二十四年三月甲午，彗星见于虚宿之次，色苍白，尾迹长尺余，指西南，每夜顺行，十余日伏不见。四月戊辰复出，在张宿，体势甚微，向东顺行，至五月初隐伏。

《钦定皇朝文献通考·象纬》卷267

(c) 清高宗乾隆二十四年三月甲午，彗星见于虚宿之次，其色苍白，尾迹尺余，指西南，每夜顺行，十余日伏不见。四月戊辰，复出西南方，在张宿第二星之上，体势甚微，向东顺行，至五月初仍隐伏。

《清朝通志·灾祥》卷122

公元1760年1月9日 [中国]

(a) 清高宗乾隆二十四年十一月戊辰，异星见东南方，在井第四星下，大如榛子，色苍黄，向西北行；癸酉，行四度，在胃，微有尾迹；十二月丁丑朔，全消。

《清史稿·天文志》卷39

(b) 清高宗乾隆二十四年十一月戊辰，有客星见东南方，在未宫井宿四星之下，自未宫行至酉宫胃宿之次，寻亦隐伏。

《清朝通志·灾祥》卷122

公元1766年4月7日 [日本]

(a) 日本明和三年二月二十八日昏，西地平彗星出现，大八九寸许，在胃宿之南，左更之东，向西行。三月三日昙，四日不见。

《天变地异书拔》

(b) 日本明和三年二月二十八日昏，彗星见于胃宿，光芒长二度许。其光至渐，三月朔昏昙天，二日昏，彗所见如前，光芒三度许，以远镜窥之，胃宿在彗中。

《三际图说并宽宝以来实测图说》

公元1769年7月17日—10月30日 [日本]

(a) 日本明和六年六月十四日（7月17日），戌刻由西向东如旗状云出现，七月初，光如白 练，幅三尺，长十四五尺，人多见之。

《津轻偏览日记》

(b) 日本明和六年七月廿七日（8月28日）晨，彗星出于东方昴宿大炊北东第一二星之间，过日东行，光芒长大成。八月八日晨，芒长四十八度许。同月十日之晨，芒生岐两股，各长六十五度半许。同月十一日晨，在鬼宿外厨西第一星之南，一股长七十二度半许，一股长三十度许。十六日至廿四日，昙天。二五日晨，清明彗灭不见。

《天变地异书拔》

(c) 日本明和六年己丑，七月彗星经纬度，于武江所测如左（8月24日—10月30日）：

七月廿三日晨	昴七度	去极八十度
八月二日	毕六度	去极八十一度
同 四	毕十三度	去极八十二度
同 六	参二度	八十五度
同 七	参七度	八十六度
同 九	井五度	八十八度
同 十	井十度	八十九度
同 十一日	井十六度半	九十一度半
同 十二日	井廿一度	九十三度
同 十三日	井廿十六度半	九十四度半
同 十五日	井三十六度	去极九十八度

后雨昙天不见，彗星东行渐近日轮，八月下旬，彗星与日轮纬度合。

十月朔日	房三度半
同 二日	心初度

四日五日昙天。

《三际图说并宽宝以来实测图说》

公元1769年8月2日—30日　　［中国］

(a) 清高宗乾隆三十四年七月，夜，彗星出东北，长丈许，西南射。

清乾隆 河北《高邑县志》卷7

(b) 清高宗乾隆三十四年七月，有星自东北光射西南，每于五更时见。

清嘉庆 浙江《嘉善县志》卷20

(c) 清高宗乾隆三十四年七月，长星见于参井之间，光长二丈余，至九月尽方灭。

清光绪 山西《平遥县志》卷12

公元1769年8月28日 ［中国］

(a) 清高宗乾隆三十四年七月二十七日（甲辰），东南方彗星见，星尾指向西南。

《清高宗实录》卷840

(b) 清高宗乾隆三十四年七月二十七日（甲辰），彗星见东南方，在昴下；丁未，大如弹丸，色苍白，尾长三尺，指西南，顺行甚速；八月丁卯，与太阳同宫不见；十月辛亥，见西方，在列肆第二星下，体势微小，尾长一尺；丙子，全消。

《清史稿·天文志》卷39

(c) 清高宗乾隆三十四年七月二十七日（甲辰），彗星见于昴宿之次，体如弹丸，色苍白，尾迹长二尺余，指正西偏南，每夜顺行，八月望后伏不见，十月复移见西方，寻即隐伏。

《清朝通志·灾祥》卷122

《钦定皇朝文献通考·象纬》卷267

公元1769年8月30日 ［朝鲜］

(a) 李朝英祖四十五年七月己酉，有星见于毕宿度内，似彗…八月朔庚戌，彗星见于毕宿度内，体大如毕，色白…甲寅，彗星见于参宿，形体尾长如前…丁巳，彗星移见於井宿度内，形色如前…壬戌，彗星见于鬼宿度内，色白，尾长六五尺…癸亥，彗星见于柳宿，形色如前…九月丁未，有星见于西方房宿之上，似有尾跡…十月辛亥，彗星见于心宿度内，色淡泊…甲寅，彗星见於心宿度内，尾跡若存若无。

《朝鲜王朝实录·英祖》卷113

(b) 李朝英祖四十五年己丑七月己酉，彗见于毕，大如毕大星，尾长三四尺，色白，历参井鬼柳，尾跡稍长。八月丙寅，星体渐微。九月丁未初昏，反见于西方房宿上，尾长尺许。十月甲寅，移在心宿，丁巳乃灭。

《增补文献备考》卷6

公元1769年10月29日—11月27日 [中国]
清高宗乾隆三十四年十月，据钦天监奏：七月内彗星见，光照西南，经日度遂灭。今又见西南，光照东北。
《清高宗实录》卷844

公元1769年11月18日 [日本]
日本明和六年十月二十一日己巳，又有彗星见艮（晨见），长数丈，凡两三日不见云。
《续史愚抄》78

公元1770年6月24日—6月28日 [日本]

(a) 日本明和七年六月二日丁未（6月24日），有孛星，光芒四出一尺所，见牵牛及紫微边，但夜夜异所，凡十日不见。

《续史愚抄》78

(b) 日本明和七年庚寅六月五日（6月27日），有孛星于天弁，实测不正，六日夜以赤道仪测得，孛星斗宿初度，去北极九十度。七日夜，孛星浸入天市垣内，以赤道仪测得，斗宿一度，去北极八十三度。

《续谈海》39

(c) 日本明和七年六月六日（6月28日）夜半前，孛星出现于天市垣东海之东，北行，十日夜入紫微垣，在北辰之北。十一日、十二日昙天，十三日不见。

《天变地异书拔》

公元1770年6月26日 [中国]

(a) 清高宗乾隆三十五年闰五月己酉，异星见东南方，在天弁第一星西，大如弹丸，色苍黄；癸丑，向北行三十二度；乙丑，不见。

《清史稿·天文志》卷39

(b) 清高宗乾隆三十五年闰五月己酉，客星见于斗宿之次，每夜向北行，十余日即隐伏。

《钦定皇朝文献通考·象纬》卷267

(c) 清高宗乾隆三十五年闰五月己酉，客星见于斗宿之次，在天弁第一星之西，其色苍黄；每夜向北行，十余日即隐伏。

《清朝通志·灾祥》卷122

公元1770年12月17日—1771年1月15日 ［中国］
清高宗乾隆三十五年十一月，彗星见于柳宿之次，在第一星之下，色苍白，尾迹长尺余，指南，每夜向北行十余度，七日隐伏。
《钦定皇朝文献通考·象纬》卷267
《清朝通志·灾祥》卷122

公元1771年1月8日 ［中国］
清高宗乾隆三十五年十一月乙丑，彗星见东南方，长尺余；丙寅，在柳第二星下；戊辰，色苍白，尾指东南，每日向西北行十余度；庚午，微暗；辛未，全消。
《清史稿·天文志》卷39

公元1771年1月9—10日 ［日本］
(a) 日本明和七年十一月二十五日丁卯（1月10日），今夜有彗星，见鬼柳间，光芒白夕见，长四五度云。
《续史愚抄》79
(b) 日本明和七年庚寅仲冬，二十四日（1月9日）有星孛于柳，光芒四五度，自西而指东，其光至微。廿五日少旋移于西，在鬼柳之交。廿七日尚左旋浸入井之野，在南河之北，于时日躔斗宿之一十度许度，彗星在井宿末度。初昏太阳没西，彗星现东，相距一百九十有许度……终夜窥之。
《三际图说并宽宝以来实测图说》

公元1773年3月23日—4月21日 ［中国］
清高宗乾隆三十八年寎月，有星大于鸡子，首指东北，尾指西南，光芒散溢，形宛如帚，长数丈，三更起，五更没，至八月始敛。
《明斋小识》卷1
注："寎月"即三月。

公元1776年8月14日—9月12日 ［中国］
清高宗乾隆四十一年七月，长星见，形如匹练，光芒五、六丈，横亘东南，三更始出，五更不见，如是者计十余夜。
清 湖南《洞庭湖志》卷7

公元1784年1月19日—2月20日 [日本]

(a) 日本天明三年旧冬（十二月）二十七日至正月三日（1月19日—1月24日），彗星见于坤方。

《武江年表》1

(b) 日本天明四年甲辰正月（1月22日—2月20日）彗见。六日初昏云间窥之，星体至微，光芒数寸指东，虽属室宿度分，未得详窥其所在。自七日至望夜夜天阴。偶值晴夜，为月明失光芒不得窥测焉。十八日夜月出前窥之，嘒小星在霹雳，有些光芒指东，星体渺微难窥测。

《三际图说并宽宝以来实测图说》

公元1784年1月24日 [朝鲜]

(a) 李朝正祖八年正月己丑，有星如彗，见于室星度内...庚寅，彗星见于室星度内羽林星之东。

《朝鲜王朝实录·正祖》卷17

(b) 李朝正祖八年正月己丑，彗见于羽林星。

《增补文献备考》卷6

Part II. Meteor Showers

约公元前16、17世纪 [中国]

夏癸（桀）十年，夜中，星陨如雨。

《竹书纪年》卷上

公元前687年3月23日 [中国]

(a) 鲁庄公七年四月辛卯，夜，恒星不见。夜中，星陨如雨。

《春秋经传集解》卷3

(b) 鲁庄公七年四月辛卯，昔：恒星不见，夜中星陨如雨。

《春秋谷梁传》卷5

公元前245 [中国]

秦始皇二年三月乙未，有流星大小西行，不可胜数，至晚乃息。

《通志·灾祥略》卷74

注：该月无乙未。若按二月乙未，则按夏历计算为BC245.2.15；按颛顼历计算为2.14。

公元前74年4月7日　　[中国]

(a) 汉昭帝元平元年二月甲申，晨，有流星大如月，众星皆随西行。

《汉书·昭帝纪》卷7

(b) 汉昭帝元平元年二月甲申，晨，有大星如月，有众星皆随而西行。

《汉书·天文志》卷26

公元前15年3月27日　　[中国]

(a) 汉成帝永始二年二月癸未，星陨如雨。

《汉书·成帝纪》卷10

(b) 汉成帝永始二年二月癸未，夜过中，星陨如雨，长一二丈，绎绎未至地灭，至鸡鸣止。

《汉书·五行志》卷27

公元前15年9月23日—10月21日　　[日本]

日本垂仁十五年八月，星下如雨。

《仁寿镜》

公元前12年5月22日　　[中国]

(a) 汉成帝元延元年四月丁酉，无云而雷，声光耀耀，四面下至地，昏止。

《汉书·成帝纪》卷10

(b) 汉成帝元延元年四月丁酉，日晡时，天暒晏，殷殷如雷声，有流星头大如缶，长十余丈，皎然赤白色，从日下东南去。四面或大如盂，或如鸡子，耀耀如雨下，至昏止。

《汉书·天文志》卷26

(c) 汉成帝元延元年四月丁酉，四方众星，白昼流陨。

《汉书·谷永传》卷85

公元34年　　[中国]

汉光武帝建武十年三月癸卯，流星如月，从太微出，入北斗魁第六星，色白。旁有小星射者十余枚，灭则有声如雷，食顷止。

《后汉书·天文志》卷20
注：该月无癸卯。若是二月癸卯，则为3月4日。

公元36年 ［中国］
汉光武帝建武十二年正月己未，小流星百枚以上，或西北，或正北，或东北，二夜止。
《后汉书·天文志》卷20
注：该月无己未，若为乙未，则是3月16日；若为二月己未，则为4月9日。

公元36年7月17日 ［中国］
汉光武帝建武十二年六月戊戌晨，小流星百枚以上，四面行。
《后汉书·天文志》卷20

公元104年2月14日—3月13日 ［朝鲜］
婆娑王二十五年春正月，众星陨如雨，不至地。
《增补文献备考·象纬考》卷7

公元268年7月27日—8月24日 ［中国］
(a) 晋武帝泰始四年七月，众星西流。
《晋书·武帝纪》卷3
(b) 晋武帝泰始四年七月，星陨如雨，皆西流。
《晋书·天文志》卷13
《宋书·天文志》卷23

公元288年9月26日 ［中国］
(a) 晋武帝太康九年八月壬子，星陨如雨。
《晋书·武帝纪》卷3
《晋书·天文志》卷13
(b) 晋武帝太康九年八月壬子，星昼陨如雨。
《通志·灾祥略》卷74

公元308年1月20日 ［中国］
晋怀帝永嘉元年十二月丁亥，星流震散。
《晋书·天文志》卷13
《宋书·天文志》卷24

公元401年4月8日 [中国]

(a) 晋安帝隆安五年三月甲寅，众星西流，历太微。

《晋书·安帝纪》卷10

(b) 晋安帝隆安五年三月甲寅，流星赤色众多，西行经牵牛、虚危、天津、阁道，贯太微、紫宫。

《晋书·天文志》卷13
《宋书·天文志》卷25

公元443年4月9日 [中国]
南朝宋文帝元嘉二十年二月二十四日，有流星大如桃，出天津，入紫宫，须臾有细流星或五或三相续，又有一大流星从紫宫出，入北斗魁，须臾又大流星出贯索中，经天市垣，诸流星并向北行，至晓不可称数。
《宋书·天文志》卷26

公元447年2月1日—3月2日 [中国]
南朝宋文帝元嘉二十四年正月，天星并西流，多细，大不过如鸡子，尾有长短，当有数百，至且日光定乃止，有入北斗紫宫者。
《宋书·天文志》卷26

公元454年2月14日—455年2月2日 [朝鲜]
毗有王二十八年，星陨如雨。星孛于西北，长二丈许。
《增补文献备考·象纬考》卷7

公元458年3月31日—4月28日 [中国]
北魏高宗太安四年三月，流星数万西行。
《魏书·天象志》卷105之3

公元461年3月27日—4月25日 [中国]
南朝宋武帝大明五年三月，有流星数千万，或长或短，或大或小，并西行，至晓而止。
《宋书·天文志》卷26

公元461年4月13日 [中国]
刘宋大明五年辛丑，三月甲戌，夜，众星西流。
清嘉庆 南京《建康志》卷8

公元461年4月20日 ［中国］
北魏高宗和平二年三月辛巳，有长星出天津，色赤，长匹余，灭而复出，大小百数。
《魏书·天象志》卷105之3

公元464年3月24日—4月21日 ［中国］
北魏高宗和平五年三月……月入舆鬼，流星无数西行（不可胜数）。
《古今图书集成·庶征典》卷41

公元465年4月12日—5月10日 ［中国］
北魏高宗和平六年三月，有流星西行，不可胜数，至明乃止。
《魏书·天象志》卷105之3

公元465年7月26日 ［中国］
北魏高宗和平六年六月己卯，又有流星，多西南行。
《魏书·天象志》卷105之3

公元466年4月8日 ［中国］
南朝宋明帝泰始二年三月乙未，有流星大小西行，不可称数，至晓乃息。
《宋书·天文志》卷26

公元466年5月22日 ［中国］
南朝宋明帝泰始二年四月己卯，竟夜有流星百余西南行，一大如瓯，尾长丈余，黑色，从河鼓出。
《宋书·天文志》卷26

公元530年4月9日 ［中国］
北魏庄帝永安三年二月壬申，有大流星相随西北，尾迹不绝，以千计。
《魏书·天象志》卷105之3

公元532年8月28日 ［中国］
南朝梁武中大通四年七月甲辰，星陨如雨。

《梁书·武帝纪》卷3
《南史·梁武帝纪》卷7

公元532年8月28日 ［朝鲜］
圣王十年秋七月甲辰，星陨如雨。
《增补文献备考·象纬考》卷7

公元534年3月1日—29日 ［中国］
北魏孝武帝永熙三年二月，众星北流。
《北史·魏武帝纪》卷5

公元566年6月4日—7月2日 ［朝鲜］
平原王八年五月，星陨如雨。
《增补文献备考·象纬考》卷7

公元581年3月20日 ［朝鲜］
平原王二十三年春二月晦，星陨如雨。
《增补文献备考·象纬考》卷7

公元585年9月23日 ［中国］
隋高祖开皇五年八月戊申，有流星数百，四散而下。
《北史·隋文帝纪》卷11
《隋书·高祖纪》卷1

公元586年5月24日 - 6月21日 ［朝鲜］
(a) 真平王八年夏五月，星陨如雨。
《增补文献备考·象纬考》卷7
(b) 真平王八年夏五月，雷震，星陨如雨。
《三国史记》卷5

公元643年11月1日 ［朝鲜］
宝藏王二年秋九月十五日，夜明不见月，众星西流。
《增补文献备考·象纬考》卷7

公元647年9月5日—10月3日 ［朝鲜］
(a) 真德主元年秋八月，众星北流。

《增补文献备考·象纬考》卷7

(b) 真德主元年秋八月，彗星出于南方，又众星北流。

《三国史记》卷5

公元684年11月13日—12月11日 [朝鲜]

神文王四年冬十月，自昏及曙，流星纵横。

《增补文献备考·象纬考》卷7，《三国史记》卷5

公元685年1月1日 [日本]

日本天武十二年十一月戊辰，昏时七星俱流东北则陨之。

《日本书纪》二十九

公元685年1月3日 [日本]

日本天武十二年十一月庚午，日没时星陨东方，大如瓮，逮于戌，天文悉乱，以星陨如雨。

《日本书纪》二十九

公元706年4月17日—5月16日 [朝鲜]

圣德王五年春三月，众星西流。

《增补文献备考·象纬考》卷7，《三国史记》卷5

公元707年8月2日—31日 [中国]

唐中宗神龙三年秋七月，验近事中宗朝，卫王重俊与李多祚等欲举兵诛锄干纪者武三思等，于时光夜有流星缤纷交乱，不可胜记。向晚，有大流星百数，皆有尾迹，纷驰渐下向西南。

《开元占经》卷71

注：按《唐书·卫王重俊列传》重俊与李多祚起兵攻武三思，事在唐中宗神龙三年秋七月，日期不详。

公元708年7月14日 [中国]

唐中宗景龙二年六月癸未夜，有流星无数，四方奔堕，缤缤纷纷，不可胜记，多出虚、危、河鼓、天津、贯索、织女、羽林、王良、阁道。

《开元占经》卷71

公元710年7月2日—30日 [中国]

(a) 唐中宗景龙四年六月，天星散落如雪。

《日涉编》卷6

(b) 唐中宗景龙四年六月，二鼓，天星散落如雪。

《日知录》

公元713年7月15日 [中国]

唐玄宗先天二年六月庚戌，夜有流星无数，皆向北及西北流，从羽林飞入紫微宫，甚众，亦出河鼓、织女等星。

《开元占经》卷71

公元713年7月21日 [中国]

唐玄宗先天二年六月丙辰，夜，小流星四面交流，极多，不可胜记。

《开元占经》卷71

公元714年7月15日 [中国]

唐玄宗开元二年五月乙卯晦，有星西北流，或如瓮，或如斗，贯北极，小者不可胜数，天星尽摇，至曙乃止。

《新唐书·天文志》卷32

公元718年10月28日—11月26日 [朝鲜]

圣德王十七年冬十月，流星自昴入于奎，众小星随之，天狗陨艮方。

《增补文献备考·象纬考》卷7，《三国史记》

公元735年5月30日 [日本]

日本天平七年五月己未，夜天众星交错乱行，无常所。

《续日本纪》十二

公元764年12月30日 [中国]

(a) 唐代宗广德二年十二月丙寅，自乙夜至曙，星流如雨。

《新唐书·天文志》卷32

(b) 唐代宗广德二年十二月丙寅，众星陨。

《新唐书·代宗纪》卷6

(c) 唐代宗广德二年十二月三日，夜，星流如雨，自亥乃晓。

《旧唐书·天文志》卷36

公元764年12月31日 ［中国］

唐代宗广德二年十二月丁卯，夜，星流如雨。

《旧唐书·代宗纪》卷11

公元765年1月7日 ［朝鲜］

景德王二十三年冬十二月十一日，三月，星孛于东南，流星或大或小，观者不能数。

《增补文献备考·象纬考》卷7

公元767年7月30日—8月28日 ［朝鲜］

恭惠王三年秋七月，三星陨王庭，相击其光如火，迸散。

《增补文献备考·象纬考》卷7

公元773年1月10日 ［日本］

日本宝龟三年十二月己未，星陨如雨。

《续日本纪》三十二

公元784年7月10日 ［中国］

唐德宗兴元元年六月戊午，星或十或伍而陨。

《新唐书·天文志》卷32

公元784年12月24日 ［日本］

日本延历三年十一月八日乙巳日，夜自戌时至于寅时，天星悉动，缤纷而飞迁。

《日本灵异记》下

公元801年9月12日—10月10日 ［朝鲜］

哀庄王二年秋八月，星陨如雨。

《增补文献备考·象纬考》卷7

公元808年2月24日 [中国]
唐宪宗元和三年正月二十五日夜，流星如雨。
《考异》
注：原文误写为“恒星如雨”。

公元814年1月25日—2月23日 [中国]
唐宪宗元和九年正月，有大星如半席，自下而升，有光烛地，群小星随之。
《新唐书·天文志》卷32

公元824年5月3日—31日 [中国]
唐穆宗长庆四年四月，紫微中，星陨者众。
《新唐书·天文志》卷32

公元830年7月22日 [中国]
(a) 唐文宗太和四年六月辛未，夜，自一更至五更，大小流星旁舞，观者不能数。
《旧唐书·文宗纪》卷17
(b) 唐文宗太和四年六月辛未，自昏至戊夜，流星或大或小，观者不能数。
《新唐书·天文志》卷32

公元833年 [中国]
唐文宗太和七年六月戊子，自昏及曙，四方流星，大小纵横百余。
《新唐书·天文志》卷32
注：六月无戊子。若是七月戊子，则为7月23日。

公元835年7月22日 [中国]
(a) 唐文宗太和九年六月丁酉，夜一更至四更，流星纵横旁舞，约二十余处，多近天汉。
《旧唐书·天文志》卷36
(b) 唐文宗太和九年六月丁酉，自昏至丁夜，流星二十余，纵横出没，多近天汉。
《新唐书·天文志》卷32

公元837年 [中国]
唐文宗开成二年九月丁酉，有星如大斗，长五丈，自室壁西北流，入大角下没，行类枉矢，中天有声，小星数百随之。
《新唐书·天文志》卷32
注：该月无丁酉。

公元837年 [日本]
承和四年三月四日丁卯(AD 837.3.22)，彗星见于东南，其光芒东至天涯。壬申(4.17), 彗星犹见，但为月光所夺，其光芒微少耳。
《续日本后纪》

公元839年4月10日 [中国]
唐文宗开成四年二月乙亥（己亥），丁夜至戊夜，四方中天流星大小凡二百余，并西流，有尾迹，长二丈至五丈。
《新唐书·天文志》卷32
原注：原文“己亥”，二月癸丑朔，无己亥，系乙亥误。

公元839年4月13日 [中国]
唐文宗开成四年二月二十六日，自夜四更至五更，四方中央流星大小二百余，并西流，有尾迹，长二丈。
《旧唐书·天文志》卷36

公元839年5月10日 [中国]
唐文宗开成四年三月二十三日，一更到五更上方及四方，有流星大小百余交横出灭。
《唐会要》卷43

公元841年6月23日 [中国]
唐武宗会昌元年六月庚子，夜五更，小流星五十余，傍午流散。
《旧唐书·武帝纪》卷18

公元841年7月21日 [中国]
(a) 唐武宗会昌元年六月二十九日，从一鼓至五鼓，小流星五十余，交横流散。
《旧唐书·天文志》卷36

(b) 唐武宗会昌元年六月戊辰，自昏至戌夜，小星数十，纵横流散。

《新唐书·天文志》卷32

公元848秋 [朝鲜]

(a) 文圣王十年秋，星陨如雨。

《增补文献备考·象纬考》卷7

(b) 文圣王十年冬十月，天有声如雷。

《三国史记》

公元865年8月1日 [中国]

唐懿宗咸通六年七月乙酉，甲夜，有大流星长数丈，光烁如电，群小星随之，自南往北。

《新唐书·天文志》卷32

公元878年6月13日 [日本]

日本元庆二年五月九日甲辰，亥时有大流星，出自氐南入轸翼间，其尾二许丈，色赤有光，众星随行，所过之处，木叶作声。

《三代实录》三十三

公元881年9月10日—18日 [中国]

(a) 唐僖宗中和元年八月己丑，夜，星陨如雨，或如杯碗者，交流如织，庚寅亦如之，至丁酉止。

《新唐书·天文志》卷32

(b) 唐僖宗中和元年八月己丑，众星陨于成都。

《新唐书·僖宗纪》卷9

公元883年12月1日—884年1月1日 [中国]

唐僖宗中和三年十一月，夜，星陨于西北，如雨。

《新唐书·天文志》卷32

公元884年10月28日—29日 [日本]

日本元庆八年八月四日壬辰，自戌至子小星四方流散，殒堕如雨。五日癸巳，自日没至人定，流星或出入紫微宫，犯众星宫官，或出入北斗贯索陵外宿，其数不可胜计。

《三代实录》四十六

公元905年3月9日—4月7日 [朝鲜]
孝恭王九年春二月，星陨如雨。
《增补文献备考·象纬考》卷7

公元905年4月13/日 [中国]
唐哀宗天佑二年三月乙丑，夜中，有大星出中天，如五斗器，流至西北，去地十丈许而止，上有星芒，炎如火，赤而黄，长丈五许，而蛇行，小星皆动而东南，其陨如雨，少顷没，后有苍白气如竹丛，上冲中天，色瞢瞢，占曰："亦枉矢也"。
《新唐书·天文志》卷32

公元924年7月21日 [中国]
五代唐庄宗同光二年六月甲申，众星交流。
《新五代史·司天考》卷59

公元924年7月23日 [中国]
五代唐庄宗同光二年六月丙戌，众星交流。
《新五代史·司天考》卷59

公元925年7月22日 [中国]
(a) 五代唐庄宗同光三年六月庚寅，众星流，自二更尽三更而止。
《新五代史·司天考》卷59
(b) 五代唐庄宗同光三年六月庚寅，夜一鼓后，西南有流星约七十余，皆有尾迹，西南流。
《五代会要》卷10

公元925年7月23日 [中国]
五代唐庄宗同光三年六月辛卯，众小星流于西南。
《新五代史·司天考》卷59

公元926年7月22日 [中国]
五代唐明宗天成元年六月乙未，众小星交流。
《新五代史·司天考》卷59

公元927年4月13日 [中国]
五代唐明宗天成二年三月庚申，众小星流于西北。
《新五代史·司天考》卷59

公元930年9月25日 [中国]
五代唐明宗长兴元年九月辛酉朔，众小星交流而陨。
《新五代史·司天考》卷59

公元931年10月15日 [中国]
(a) 五代唐明宗长兴二年九月丙戌，夜二鼓初，东北方有小流星入北斗魁灭，至五鼓初，西北方次北有流星，状如半升器，初小后大，速流如奎灭。尾迹凝天，屈曲似云而散，光明烛地。又东北有流星如大桃，出下台星，西北速流，至斗柄第三星旁灭。五鼓后至明，中天及四方有小流星百余，流注交横。
《旧五代史·天文志》卷139
(b) 五代唐明宗长兴二年九月丙戌，自五鼓至厥明，中天及四垂有小星百余，流注交横。
《五代会要》卷10
(c) 五代唐明宗长兴二年九月丙戌，众星交流。
《新五代史·司天考》卷59

公元931年10月16日 [中国]
五代唐明宗长兴二年九月丁亥，众星交流而陨。
《新五代史·司天考》卷59

公元933年7月20日 [中国]
五代唐明宗长兴四年六月庚午，众星交流。
《新五代史·司天考》卷59

公元933年7月25日 [中国]
五代唐明宗长兴四年七月乙亥朔，众星交流。
《新五代史·司天考》卷59

公元934年4月13日 [中国]
五代唐闵帝应顺元年二月丁酉，众星流于西北。
《新五代史·司天考》卷59

公元934年10月13日　　[中国]
辽太祖天显九年九月庚子，西南星陨如雨。
《辽史·太祖纪》卷3

公元934年10月14日　　[中国]
(a) 五代唐末帝清泰元年九月辛丑，夜，有星如五斗器，西南流，尾迹长数丈，屈曲如龙形，又众星乱流，不可胜数。
《旧五代史·唐末帝纪》卷46
(b) 五代唐末帝清泰元年九月辛丑，众星交流。
《新五代史·司天考》卷59

公元967年10月14日　　[日本]
(a) 日本康保四年九月九日甲午……同亥时流星如月，自艮亘坤，众星四（原文误写"西"）乱，终夜流散。
《日本纪略》后五
(b) 日本康保四年九月九日，夜始自亥时至于寅时，普天下众星自东流西，走散无间，形如刀剑。
《扶桑略记》二十六

公元970年　　[中国]
宋太祖开宝三年九月庚午，广州民见众星皆北流。
《宋史·天文志》卷57
《宋史·南汉刘氏》卷481
注：九月无庚午。若是十月庚午，则为11月3日。

公元989年7月24日　　[日本]
(a) 日本永祚元年六月十九日戊辰，亥刻鸭御社中门内御殿前大树颠倒，数星自树中出，指南连飞。
《日本纪略》后九
(b) 日本永祚元年六月十九日，鸭社御殿前大树颠倒，数星自树中飞，指南连去。
《百鍊抄》四

公元1002年10月12日　　[中国]
宋真宗咸平五年九月丙申，有星流出东方，西南行，大如斗，有声若牛吼，小星数十随之而陨。

《文献通考·象纬》卷291
注：《宋史·天文志》卷57记为“八月丙申”，而八月无丙申。

公元1002年10月14日 [中国]
(a) 宋真宗咸平五年九月戊戌，有星十数入舆鬼，至中台，凡一大星偕小星数十随之，其间两星如升器大，一至狼星，一至南斗没。
《文献通考·象纬》卷291
(b) 宋真宗咸平五年九月戊戌，大星偕小星数十随至中台，其间两星一至狼星，二至南斗没。
《续通志·灾祥》卷171
注：《宋史·天文志》卷57记为“八月戊戌”，而八月无戊戌。

公元1002年10月14日—15日 [日本]
(a) 日本长保四年九月六日戊戌，今夜……终夜流星。七日己亥，今夜自子时至寅流星。
《日本纪略》后十
(b) 日本长保四年九月七日己亥……今夜有天变，自子时至于寅刻流星。
《本朝世纪》十六
(c) 日本长保四年九月七日，今月六日变异频星，两曜共蚀，众星乱行。
《廷尉故实》

公元1106年7月27日 [朝鲜]
高丽睿宗元年六月乙酉，枉矢飞行，见者皆惊譟 。流星出天津入宗人，大如杯，尾长二丈许；又二流星出虚入九坎，大如鸡子；又自昏至晓众星流四方。
《高丽史》卷47

公元1007年7月20日 [日本]
日本宽弘四年六月四日戊戌，今夜有流星北行，连夜有此变。
《日本纪略》后十一

公元1007年7月24日 [日本]
日本宽弘四年六月八日壬寅……此夜流星数多至晓流云云。

《法成寺摄政记》

公元1007年7月25日 [日本]
日本宽弘四年六月九日癸卯……此夜又有大流星云云。
《法成寺摄政记》

公元1008年3月26日 [中国]
宋真宗大中祥符元年二月戊申，有星十余，急流入浊，色赤黄，有尾迹。
《宋史·天文志》卷57

公元1012年9月11日 [中国]
宋真宗大中祥符五年八月戊午，有星大小二十余，皆有尾迹，北流，又一星光烛地，出紫微垣外，尾丈余，阔三寸许，东北流，至传舍没。
《宋史·天文志》卷57

公元1032年10月14日 [中国]
宋仁宗明道元年九月丙子，星出娄，没于云雨侧，尾迹久方散。食顷，又有星出天大将军，近奎没，尾迹久方散，明烛地。续有星出北辰西北，速行至内阶没。又有星出天苑，没于天园，明烛地。
《宋史·天文志》卷57

公元1035年10月14日 [日本]
日本长元八年九月十一日，夜寅时有流星变。
《扶桑略纪》二十八

公元1037年8月21日 [中国]
(a) 宋仁宗景祐四年七月戊申，有星数百西南流至东壁，大者其光烛地，黑气长丈余，出毕宿下。
《宋史·仁宗纪》卷10
(b) 宋仁宗景祐不四年七月戊申，有星数百皆西南流，其最大者一星，至东壁没，光烛地，久之不散。
《宋史·天文志》卷57

公元1037年10月14日　　[日本]

(a)　日本长历元年九月三日，众星乱落，四方飞散，人莫不惊。

《百錬抄》四

(b)　日本长历元年九月三日，夜子时有流星变。

《扶桑略纪》二十八

(c)　日本长历元年九月四日，去夜有流星天变。

《行亲记》

公元1038年2月7日—3月8日　　[中国]

辽兴宗重熙六年（宋仁宗宝元改元）春正月，有众星西北流。

《契丹国志》卷8

注：《契丹国志》作"重熙六年"，误，公元纪年按宝元元年正月换算。若按重熙六年春正月，则为1037年1月19日—2月17日。

公元1063年8月22日　　[中国]

(a)　宋英宗嘉祐八年七月乙丑，星大小数百西流。

《宋史·英宗纪》卷13

(b)　宋英宗嘉祐八年七月乙丑，星数百，纵横西流。

《宋史·天文志》卷57

公元1042年7月25日　　[朝鲜]

高丽靖宗八年七月丁未，众星流转。

《高丽史》卷47

公元1091年5月19日　　[朝鲜]

高丽宣宗八年四月丁巳，流星或青或赤，尾长十余尺，出房星，入轩辕，有声如雷。

《高丽史》卷47

公元1095年7月25日　　[朝鲜]

高丽献宗元年六月乙酉，流星大如木瓜，色赤，尾长九尺许，出室西，入南斗魁，亦有众小星南流。

《高丽史》卷47

公元1096年7月22日—8月20日　　[日本]

(a)　日本永长元年七月，乾方有星数十，如贯珠，长四五尺。

《皇代历》三
(b) 日本永长元年七（月），乾方星数十，如贯珠。
《皇年代略记》
注：(b)原文为“七”。

公元1102年8月5日 [中国]
宋徽宗崇宁元年闰六月癸酉，星出斗，向西南慢流，至建没，青白，有尾迹，数小星从之。
《宋史·天文志》卷60

公元1103年9月7日 [朝鲜]
高丽肃宗八年八月辛亥，流星，一出北极入天津，大如炬，尾长三丈许；一出内诸侯入上台，大如碗，尾长一丈许；一出王良入文昌，大如木瓜；又小星百余流行。
《高丽史》卷47

公元1106年7月27日 [朝鲜]
高丽睿宗元年六月乙酉，枉矢飞行，见者皆惊譟。流星出天津入宗人，大如杯，尾长二丈许；又二流星出虚入九坎，大如鸡子；又自昏至晓众星流四方。
《高丽史》卷47

公元1111年10月2日 [朝鲜]
高丽睿宗六年八月戊午，众小星分流四方。
《高丽史》卷47

公元1122年6月26日—28日 [中国]
宋徽宗宣和四年五月戊寅、己卯、庚辰，连夕有流星，若盂碗，自紫微、文昌，犯天津、河鼓，武天汉、斗、牛间，皆南向西奔，电光如匹练，十余夕不已。
明嘉靖 河北《雄乘》卷下

公元1134年7月23日—8月21日 [中国]
宋高宗绍兴四年甲寅，秋七月，星陨达旦。
清同治 江西《南城县志》卷10

公元1136年4月3日 [朝鲜]
高丽仁宗十四年三月戊辰朔，众星自东北流西南。
《高丽史》卷47

公元1178年9月17日 [朝鲜]
高丽明宗八年八月乙未，流星一出奎星入土司空，大如木瓜；一出内平入轩辕，尾长一丈许；一出星七星入张，大如木瓜，尾长七尺许；一出阙五星入孤，尾长五尺许；一出相入常陈，大如木瓜，尾长十五尺许；一出东井入五诸侯，大如木瓜，尾长五丈许；一出翼指南入天际，尾长一丈许；又众星流于四方，不可胜数。
《高丽史》卷47

公元1179年4月17日 [朝鲜]
高丽明宗九年三月丁卯，小星百余自东流西。
《高丽史》卷47

公元1179年12月17日 [日本]
日本治承三年十一月十七日辛未……今夜当坤在大小星，星星进寄大星下，入其内同光，又退归，又进寄，又退云云。
《山槐记》

1238年10月18日 [日本]
日本历仁元年九月七日辛巳……又自亥刻迄丑时，流星或七八尺，或三四尺，不知其员，色白赤。
《吾妻镜》三十二

公元1247年4月18日 [日本]
日本宝治元年三月十二日乙丑，戌刻许有光物，其体如大流星云云，乱飞之耀不异燃火，自东入西天之后鸣动，万人惊耳目，古今希类云云。
《百鍊抄》十六

公元1304年9月7日　　[日本]
日本嘉元二年八月七日，夜大雨大风之最中，自春日山光物五着乾方飞去了，其光如日轮。
《兴福寺略年代记》

公元1321年6月17日　　[日本]
日本元亨元年五月二十一日甲午，阴雨降，今夜亥刻许人人云，如流星者飞天云云，其光殆如月云云，或云，圆而色青白，大如桃子者飞入云，又飞连犹有济济云云，今夜有鸣动，声非地震非雷鸣，或云，将军墓鸣动也。
《花园院宸记》

公元1325年12月9日　　[日本]
日本正中二年十一月四日庚戌，晴，今夜有大流星，人人皆云，烧亡云云...六日壬子，泰世内内进天变占文......大流星连夜，如此之变异重叠，不可不恐欤。
《花园院宸记》

公元1339年8月6日—9月3日　　[日本]
日本延元四年七月，有妖种种，流星屡飞，或夜东方有云，其色黑，中有虹，赤青（原文为"赤毒"），其光如昼云云。
《南朝编年纪略》一

公元1363年5月14日—6月11日　　[朝鲜]
恭愍王十二年四月，有星斗于天中。
《增补文献备考·星变》補

公元1368年3月19日—4月17日　　[中国]
元顺帝至正二十八年三月，有星流于东北，众小星随之，其声大震。
《元史·顺帝纪》卷47

公元1410年2月15日　　[朝鲜]
太宗十年正月己卯，火在井东，月在火北，隔一尺。是日夜半，黄州仇畓村，天明方一里，俄而雷起东方。初如蒺藜炮声，次如钟声，后如滩声，众星摇动。
《朝鲜王朝实录·太宗》卷19

公元1414年6月15日　　[朝鲜]
太宗十四年五月庚子，气爽如秋。是昏，流星出天中，东流有声。初如瓶，至终差大，青赤色，火焰四散，明如月光。至夜半，亦如之。
《朝鲜王朝实录·太宗》卷27

公元1436年10月2日　　[中国]
(a) 明英宗正统元年八月乙酉，昏刻至晓，大小流星一百余。
《明英宗实录》卷21
(b) 明英宗正统元年八月乙酉，昏刻至晓，大小流星百余。
《明史·天文志》卷27

公元1436年12月15日　　[日本]
(a) 日本永享八年十一月八日，晴，抑夜亥时，天变飞，其后屋上物之如落悬震动，面面仰天见之。
《看闻御记》
(b) 日本永享八年十一月八日己亥，夜有流星，自北亘南，光中有物如星，乱飞，没而后有声。
《野史》三

公元1437年2月13日　　[日本]
(a) 日本永享九年正月八日未刻，光物镰仓中飞回数度。
《南方纪传》五
(b) 日本永享九年正月八日己亥，此日未刻镰仓中光怪多飞云。
《续史愚抄》三十六

公元1439年10月5日　　[中国]
(a) 明英宗正统四年八月癸卯，自夜达旦，有流星大小二百六十余。
《明英宗实录》卷58

(b) 明英宗正统四年八月癸卯，大小流星数百。
《明史·天文志》卷27

公元1447年7月30日 [中国]
明英宗正统十二年六月戊寅，夜四更，西方有星如碗，流丈余，发光如斗大，色青百，光明烛地，起自天津，东南行至羽林军，尾迹炸散，后有十余小星随之，有声如雷。
《明英宗实录》卷155

公元1451年7月27日 [中国]
(a) 明景帝景泰二年六月丙申，夜大小流星凡八十有五。
《明英宗实录》卷205
(b) 明景帝景泰二年六月丙申，大小流星八十余。
《明史·天文志》卷27

公元1451年9月11日 [中国]
(a) 明景帝景泰二年八月壬午，夜，有流星二：一大如鸡弹；一大如斗。色俱赤，有光烛地。一出紫微西蕃，北行至阴德，三小星随之；一出天津，东南行至南河（原文河南），十余小星随之，尾迹炸散，有声如雷。
《明英宗实录》卷207
(b) 明景帝景泰二年八月壬午，有赤星二：一如桃；一如斗，光烛地。一出紫微西蕃，北行
至阴德，三小星随之；一出天津，东南行至南河（原文河南），十余小星随之，尾迹炸 散，声如雷。
《明史·天文志》卷27
注：“河南”应作“南河”。

公元1457年8月20日—9月18日 [中国]
明英宗天顺丁丑八月，星陨如雨。
清康熙 河北《安肃县志》卷3

公元1457年12月7日 [中国]
明英宗天顺元年十一月辛巳，夜，有流星大如杯，色青白，光明烛地，出华盖，东北行至天沟，十余小星随之，有声隆隆。

《明英宗实录》卷284

公元1462年8月10日 [中国]
明英宗天顺六年七月戊申，夜，有流星大如碗，色青白，有光烛地，出天宛，北行入参宿，十余小星随之。
《明英宗实录》卷342

公元1466年7月28日 [中国]
(a) 明宪宗成化二年六月丙辰，夜，北方有流星如盏大，青白色有光，自正北行东南云中，后十余小星随之。
《明英宗实录》卷31
(b) 明宪宗成化二年六月丙辰，夜，流星自北行东南，数小星随之。
《国榷》卷34

公元1467年6月2日 [中国]
(a) 明宪宗成化三年五月乙丑朔，夜，南方流星如盏大，青白色，光烛地，自河鼓行丈余，发光如碗大，东北行至奎宿，尾迹炸散，后十余小星随之。
《明英宗实录》卷42
(b) 明宪宗成化三年五月乙丑朔，夜，大星自河鼓流至奎宿，数小星随之。
《国榷》卷35

公元1468年4月7日 [日本]
日本应仁二年三月十六庚子……昨日霰时分，自北方光物三，南ヲ指飞了云云。
《经觉私要钞》六十八

公元1477年4月18日 [日本]
(a) 日本文明九年三月五日，京中有光物，东西南北飞光鸣动云云。
《长兴宿祢记》上
(b) 日本文明九年三月五日癸酉，京师有光怪，飞东西南北。

《续史愚抄》四十

(c) 日本文明九年三月五日癸酉，有流星，飞散四方。

《野史》四

公元1498年10月24日 [中国]

(a) 明孝宗弘治十一年十月壬申，晓刻，东方流星大如盏，色赤，起东北，行丈余，发光如斗，光烛地，东南行，小星数十随之。

《明孝宗实录》卷142

(b) 明孝宗弘治十一年十月壬申，晓，东方赤星如碗，行丈余，光烛地，东南行，小星数十随之。

《明史·天文志》卷27

公元1502年2月7日—1503年1月27日 [中国]

明孝宗弘治十五年，星陨如雨。

清康熙 河北《深州志》卷7

公元1504年5月19日 [朝鲜]

燕山十年闰四月丙寅，夜，流星出氐星，入陈车星，出北斗星柄下，入大角星下；出轸星，入角星；出天江星，入傅说星；出土公星，入五纪星；出南斗星下，没于南方天际；出东咸星，入从官星；出更河星下，入折威星，状如梨，尾长七八尺，色赤；出勾陈星下，入天理星，状如大梨，尾长三四尺，色赤。

《朝鲜王朝实录·燕山》卷52

公元1505年6月13日 [日本]

日本永正二年五月十二日，今夜三星合一落，凶事也。

《二水记》

公元1506年3月25日—4月22日 [中国]

(a) 明武宗正德元年三月，星陨如雨。

清康熙 河北《大城县志》卷8

(b) 明武宗正德元年三月，星陨如雨，有大如月者。

《罪惟录》 志1

公元1510年5月11日 [日本]

(a) 日本永正七年四月四日，戌刻山中鸣动，光物数多转变，して四五岁计之秃童数辈侍立神前令怪异之处。

《实隆公记》里文书

(b) 日本永正七年四月四日，夜神前光物数多点变，五六岁秃童数辈在宫前，其后宫中鸣动以外也。

《后慈眼院殿记》

公元1511年6月28日 [中国]

明武宗正德六年六月壬午，江西临江府，夜见火星交流，顷之，火大如车轮，尾长数丈，陨府城西北。吉安府火星如斗，光赤烛天，自西北流东南，坠声如雷。

《明武宗实录》卷76

公元1517年2月2日 [日本]

日本永正十四年正月十二日，今夜三星合一落，凶事也。

《隆康私记》

公元1517年10月11日 [朝鲜]

中宗十二年九月己亥，夜，流星出文昌星，入北斗星，状如拳，尾长五六尺许，色白。流星出天柱星，入天纪星，状如拳，尾长四五尺许， 色赤；流星出天津星，入败瓜星，状如拳，尾长三四尺许，色赤；流星出毕星，入水府星，状如钵，尾长三四尺许，色赤；流星出天樽星，入常陈星，状如拳，尾长五六尺许，色赤。

《朝鲜王朝实录·中宗》卷29

公元1519年7月5日 [朝鲜]

中宗十四年六月辛未，是日夜，庆尚道庆州府有天变，初昏月光甚明，西方微有云气，云间有光，似电非电似火，或如流矢漫空，或如流星突过，或如赤蛇跃，或如火点飞散，或曲如张弓，或歧如钗股，变幻百状。乍见乍隐，交驰急逐，更发迭出，连属不绝。大槩如放砲之状。光甚闪烁，照烛幽室，自西而始，渐向东北，至三更乃灭云。观察使因府尹金安老所报启。

《朝鲜王朝实录·中宗》卷36

公元1520年4月17日 ［朝鲜］
中宗十五年四月一日午戌，夜，流星出八谷星，入五车星，状如钵，尾长四五尺许，色赤；出贯索星，入房星，状如拳，尾长三四尺许，色赤；出太微星东垣，入氐星，状如梨，尾长三四尺许，色青；出女星，入斗星，状如大梨，尾长四五尺许，色白；出织女星，入大角星，状如拳，尾长四五尺，色赤。
《朝鲜王朝实录·中宗》卷39

公元1523年1月17日—1524年2月3日 ［中国］
明世宗嘉靖癸未，夜半，星陨如雨，红光烛天。
清光绪 安徽《泗虹县志》卷19

公元1528年2月11日 ［中国］
明世宗嘉靖七年正月甲午，夜，忽灯火千数，散落海滨，若人持以行，居民大惊为寇至，逾时乃灭，成化十三年尝一见之。
明嘉靖 浙江《海盐县志》卷5

公元1528年10月13日—11月11日 ［中国］
明世宗嘉靖七年十月，河间星陨如雨。
清康熙 河北《河间县志》卷11
《二申野录》卷4
《明会要》卷69

公元1529年11月1日—11月30日 ［中国］
明世宗嘉靖八年十月，星陨如雨。
明万历 山东《沂州志》卷1

公元1531年1月18日—1532年2月5日 ［中国］
(a) 明世宗嘉靖十年，星陨如雨。
明嘉靖 福建《长泰县志》卷下
清康熙 河北《平乡县志》卷3
(b) 明世宗嘉靖十年，大星散落如雨。

清康熙 河北《平山县志》卷1
清康熙 河北《钜鹿县志》卷8

公元1531年11月15日 [中国]
明世宗嘉靖辛卯十月七日丙子，夜，星陨如雨。
清嘉庆 山东《昌乐县志》卷1
注：本年十月七日为丁亥，非丙子。

公元1532年2月6日 [中国]
明世宗嘉靖十一年春正月元旦夜，星陨如斗，声如雷。
《江盈科闻记·纪灾异》

公元1532年10月24日 [朝鲜]
(a) 李朝中宗二十七年壬辰九月辛未夜，星陨如雨。
《增补文献备考·象纬考》卷7
(b) 中宗二十七年九月辛未夜，流星出轩辕星下，入东方天际，长四五尺许，色赤；又出北斗星，入北方天际，长一丈许，色白；又出五车星，入毕星，长三四尺许，色白；又出参星，入九游星，长四五尺许，色赤；又出柳星，入南方天际，长五六尺许，色白；又出毕星，入天仓星，长一丈许，色尺，为色光大而直，渐成屈曲，良久化为云，移时乃灭；又出昴星，入天囷星下，长五六尺许，色赤，尾色光直，良久化为云乃灭。星陨如雨。彗星见于卯地，尾长四五尺许，色白。
《朝鲜王朝实录·中宗》卷73

公元1532年10月25日 [中国]
明世宗嘉靖十一年九月二十七日，夜四鼓，星陨如雨。
清康熙 河北《东安县志》卷1

公元1532年10月28日—11月26日 [中国]
(a) 明世宗嘉靖十一年十月，星陨如雨。
明嘉靖 广东《兴宁县志》卷1

清乾隆 山西《凤台县志》卷12
(b) 明世宗嘉靖十一年十月，夜，星陨如雨。
清顺治 山西《阳城县志》卷7

公元1532年11月3日 [中国]
(a) 明世宗嘉靖十一年十月初七日，夜，星陨如雨，有光，天为赤色，散落四方。
清康熙《山西通志》卷30
清康熙 山西《平阳府志》卷34
(b) 明世宗嘉靖十一年十月初七日，夜，星四乱飞落。
《江盈科闻记·纪灾异》

公元1532年11月27日—12月25日 [中国]
(a) 明世宗嘉靖十一年十一月，夜，星陨如雨。
明万历 山西《汾州府志》卷16
清康熙 山西《灵石县志》 祥异
(b) 明世宗嘉靖十一年十一月，夜，天星散落如雪，其光烛地。
清康熙 山东《济南府志》卷10
清康熙 山东《滨州志》卷8
清康熙 山东《长山县志》卷7

公元1532年12月6日 [中国]
明世宗嘉靖十一年十一月初十，夜，星陨，有赤光散落四方，如掷砖石，至晓乃息。
清康熙 河南《延津县志》卷7

公元1532年冬 [中国]
明世宗嘉靖十一年冬，泽州，阳城、灵石星陨如雨。
清雍正《山西通志》卷163

公元1533年7月22日—8月19日 [中国]
明世宗嘉靖十二年七月，吉安府西北，星陨如雨。
清康熙《江西通志》卷4
清乾隆 江西《庐陵县志》卷1

公元1533年8月20日—9月17日 [中国]

(a) 明世宗嘉靖十二年八月，夜，星陨如雨。

清顺治 浙江《常山县志》卷1

(b) 明世宗嘉靖十二年八月，星陨如雨。

清康熙 浙江《衢州府志》卷30

(c) 明世宗嘉靖十二年八月，星陨井陉如雨。

清嘉靖 河北《真定府志》卷6

(d) 明世宗嘉靖十二年八月，星陨如雨，光若火，天尽赤。

清乾隆 山西《平阳府志》卷34

公元1533年9月18日—10月17日 [中国]

明世宗嘉靖十二年九月，星陨如雨。

清康熙 福建《连城县志》卷1

明万历 安徽《庐州府志》卷1

明万历 安徽《合肥县志》卷上

公元1533年秋 [中国]

明世宗嘉靖十二年秋，西北群星陨如雨。

明万历 江西《永新县志》卷5

清顺治 江西《吉安府志》卷1

清康熙 江西《庐陵县志》卷2

清康熙 江西《安福县志》卷1

清乾隆 江西《龙泉县志》卷末

清乾隆 江西《泰和县志》卷28

公元1533年10月18日—11月15日 [中国]

(a) 明世宗嘉靖十二年十月，星陨如雨。

《明史·郭宗皋传》卷200

(b) 明世宗嘉靖十二年十月，太原、榆次、太谷、祁县、阳曲、盂县、清沅、寿阳、崞县、临汾、洪洞、霍州、长治、襄垣星陨如雨，光若火，天尽赤。

清雍正《山西通志》卷163

(c) 明世宗嘉靖十二年十月，潮州星陨如雨，七县皆同。

《广闻录》卷1

注：山东、河北、河南、甘肃、山西、江西、安徽、广东、福建等数十个地方志对本次记录有相似记载。

公元1533年10月23日 [中国]
明世宗嘉靖十二年十月乙亥，初昏时，恒星不见，抵夜，星陨如雨，经行失次，南北交流。
清顺治山西《高平县志》卷9

公元1533年10月24日 [中国]
(a) 明世宗嘉靖十二年十月丙子（初七），是夜，流星如盏大，赤色，光明照地，起自中台，东北行至近浊，尾迹化为白气而散。四更至五更，大小流星纵横交行，不计其数，至晓乃息。
《明世宗实录》卷155
(b) 明世宗嘉靖十二年十月丙子，流星如盏，光照地，自中台东北行近浊，尾迹化为白气。四更至五更，四方大小流星纵横交行，不计其数，至明乃息。
《明史·天文志》卷27
注：《明史·天文志》作“九月丙子”，查九月无丙子，疑为“十月丙子”之误。
(c) 明世宗嘉靖十二年十月丙子（七日），夜半至晓，星陨如雨。
明万历 山东《武定州志》卷6
明嘉靖 山东《青州府志》卷5
清乾隆 山东《惠民县志》卷3
(d) 明世宗嘉靖十二年十月丙子（七日），夜，星陨有光，天为赤色，自上而下，至晓未已。
明嘉靖 河北《清苑志》卷6
(e) 明世宗嘉靖十二年十月丙子（七日），夜，星陨如雨。
明嘉靖 山西《霍州县志》卷6
明隆庆 江苏《高邮州志》卷2

公元1533年10月24日 [朝鲜]
(a) 李朝中宗二十八年癸巳十月丙子，飞星流星四面如雨。
《增补文献备考·象纬考》卷7

(b) 中宗二十八年十月丙子，三更，青黄白气，出文昌星尾，指王良星，如一匹布许，形如龙，良久而灭；流星出北河星，入东方天际，状如瓶，尾长五六尺许，色白；流星出少微星，入东方天际，状如瓶，尾长六七尺许，色赤；四更，流星出参星，入天苑星，状如瓶，尾长四五尺许，色白；流星出三台星，入艮方天际，状如拳，尾长五六尺许，色赤；五更，流星出星星，入稷星，状如钵，尾长七八尺许，色白；流星出紫微垣，入乾方天际，状如盆，尾长，云蔽而不见，色则亦赤；流星出太微西垣，入於屏星，状如瓶，尾长七八尺许，其光迹良久不灭；流星出三台星，入大微垣，状如瓶，尾长七八尺许，色白；流星出参星下，入九游星，状如盆，尾长四五尺许，色赤，光迹良久不灭，化为云。流星飞星，四面如雨，出入星度，未及测侯，陨亦如雨，电光。

《朝鲜王朝实录·中宗》卷76

公元1533年10月25日 [中国]

(a) 明世宗嘉靖癸巳（十二年）十月八日，夜四鼓，万星纵横流飞，俄陨如雨，至天曙方已。

《吾学编余·灾异》

(b) 明世宗嘉靖十二年十月八日（丁丑），夜半，星陨如雨。

清顺治 河南《械乡县志》卷1

清康熙 陕西《潼关县志》卷2

(c) 明世宗嘉靖十二年十月八日（丁丑），丑时，星陨如雨。

清乾隆 湖北《英山县志》卷26

明万历 安徽《六安州志》卷下

明万历 安徽《舒城县志》卷10

注：清康熙 湖北《英山县志》作"午时"

公元1533年10月25日 [日本]

(a) 日本天文二年十月八日，夜晓，星降如雪云云，南西方。

《公卿补任》

(b) 日本天文二年十月丁丑，星陨如雪（原文误为"如云"）……海陆振动。

《续本朝通鉴》百八十三

(c) 日本天文二年十月八（日）夜，众星流动而陨地。

《续太平记》三十一编年录

(d) 日本天文二年十月八日丁丑，终宵星陨如雪，凡西南行去。

《续史愚抄》四十六

公元1533年10月26日 [中国]

(a) 明世宗嘉靖十二年十月九日，丑时，星陨如雨。

清康熙 山东《济南府志》卷10

清康熙 山东《海丰县志》卷4

清康熙 山东《长山县志》卷7

(b) 明世宗嘉靖十二年十月九日，夜，星陨如雨。

明嘉靖 浙江《温州府志》卷6

公元1533年10月27日 [中国]

(a) 明世宗嘉靖十二年十月初十己卯，夜中，星陨如雨。

清康熙 安徽《巢县志》卷4

(b) 明世宗嘉靖十二年十月初十己卯，夜，星陨如雨，有火光著门户，焰焰如灼。

清光绪 山东《平度县志》卷12

(c) 明世宗嘉靖十二年十月初十己卯，流星起中台，化白气，自四更起，大小星交行，不计其数，至晚没。

《罪惟录》志1

公元1533年10月29日 [中国]

(a) 明世宗嘉靖十二年十月辛巳，京口星陨如雨，琼州、海州亦如之。

《明会要》卷69

(b) 明世宗嘉靖十二年十月辛巳，昼，星陨如雨，京口舟人不敢渡，潮州、琼州星陨亦如之。

《二申野录》卷4

(c) 明世宗嘉靖十二年十月辛巳，星陨如雨。

清顺治 河北《易水志》卷上

清乾隆 江西《新城县志》卷13
清康熙 河南《陈州志》卷4
清乾隆 河南《淮宁县志》卷11
清宣统 河南《项城县志》卷31

公元1533年11月1日 [中国]
明世宗嘉靖十二年十月既望，夜中，星陨如雨。
清乾隆 甘肃《庄浪志略》卷19

公元1533年11月3日 [中国]
(a) 明世宗嘉靖十二年十月十七日，夜，有群星陨于任邱，如掷金石，至晓未已。
清乾隆 河北《河间府志》卷17
(b) 明世宗嘉靖十二年十月十七日，夜，星陨如雨，天色遂赤。
明崇祯《山西通志》卷36
清顺治 山西《潞安府志》卷15
清康熙 山西《长治县志》卷1
(c) 明世宗嘉靖十二年十月十七日，夜，星陨如雨，天色昏迷，云中有变。
清康熙 山西《襄垣县志》卷9

公元1533年11月4日 [中国]
明世宗嘉靖十二年十月丁亥，泛海舟人云，三更后星坠如雨，继而一红火如斗大，有嚇喇声。
《闻见漫录》卷上

公元1533年11月24日 [中国]
明世宗嘉靖十二年冬十一月九日，夜，星陨如雨。
明万历 浙江《温州府志》卷18

公元1533年冬 [中国]
(a) 明世宗嘉靖十二年冬，虹邑星陨如雨。
清乾隆 安徽《泗州志》卷4
(b) 明世宗嘉靖十二年冬，夜半，星陨如雨，红光烛天。
清康熙 安徽《虹县志》卷1

(c) 明世宗嘉靖十二年冬，天星散落如雨。
明隆庆 江苏《仪真县志》卷13
(d) 明世宗嘉靖十二年冬，夜星陨如雨。
清康熙 浙江《永嘉县志》卷14

公元1534年1月14日—1535年2月1日 [中国]
(a) 明世宗嘉靖十三年，星陨如雨。
明嘉靖 河南《光山县志》卷9
明万历 河南《项城县志》卷7
明万历 河南《林县志》卷8
(b) 明世宗嘉靖十三年，星陨如雨，自东而西。
清雍正 北京《密云县志》卷1
(c) 明世宗嘉靖十三年，苏州星陨如雨。
清乾隆 江苏《江南通志》卷197

公元1534年9月8日—10月6日 [中国]
明世宗嘉靖十三年八月，星陨如雨。
明嘉靖 湖南《衡州府志》卷7

公元1534年11月6日—12月4日 [中国]
明世宗嘉靖十三年十月，星陨如雨。
明万历 山西《平阳府志》卷10
清康熙 山西《解州志》卷12
清乾隆 山西《临汾县志》卷9
清康熙 江苏《江南通志》卷5
清光绪 浙江《乐清县志》卷13

公元1534年11月13日 [中国]
明世宗嘉靖十三年十月八日，夜，星陨如雨。
明崇祯 江苏《吴县志》卷11
清康熙 江苏《苏州府志》卷2
清康熙 江苏《长州县志》卷2

公元1534年11月20日 [中国]
明世宗嘉靖十三年十月既望，夜，星陨如雨。
明嘉靖 甘肃《平凉府志》卷8

公元1535年8月28日—9月26日 [中国]
明世宗嘉靖十四年八月，星陨如雨。
清道光 广东《长乐县志》卷7

公元1535年9月27日—10月26日 [中国]
(a) 明世宗嘉靖十四年九月，星陨临城，大如斗，群星数百流，芒入东南，顷刻，天鼓震鸣如雷。
明隆庆 河北《赵州志》卷9
明嘉靖 河北《真定府志》卷9
(b) 明世宗嘉靖十四年九月，星陨如斗，有声如雷。
清康熙 河北《临城县志》卷8
清康熙 河北《真定府志》卷6
(c) 明世宗嘉靖十四年九月，有星大如斗，后列群星数百，自西北流入东南，顷之，天鼓震响如雷。
明嘉靖 河北《顺德志》卷17
明隆庆 河北《任县志》卷7

公元1535年10月10日 [中国]
明世宗嘉靖十四年九月壬申，夜，众星交动。
明隆庆 江苏《高邮州志》卷12

公元1535年10月26日 [中国]
明世宗嘉靖十四年九月戊子，河南开封白昼天鼓鸣，有星如碗大，东南流，众小星从如珠。
《明世宗实录》卷179
《明史·天文志》卷27

公元1536年1月23日—1537年2月9日 [中国]
明世宗嘉靖十五年，星陨深州如雨。
明嘉靖 河北《真定府志》卷6
清康熙 河北《深州志》卷7

公元1536年10月15日—11月13日 [中国]
(a) 明世宗嘉靖十五年十月，夜分，天星乱落如雨。
明嘉靖 河北《威县志》卷1

明嘉靖 河北《广平府志》卷15

(b) 明世宗嘉靖十五年十月，星陨如雨。

清康熙 河北《广平府志》卷19

清康熙 河北《平乡县志》卷3

(c) 明世宗嘉靖十五年十月，夜，星落如雨。

清顺治 河北《鸡泽县志》卷10

(d) 明世宗嘉靖十五年十月，星落如雨。

清乾隆 河北《永年县志》卷17

公元1538年4月29日—5月27日 [中国]

(a) 明世宗嘉靖十七年四月，夜，宁化星陨如雨。

清乾隆 福建《汀州府志》卷45

(b) 明世宗嘉靖十七年四月，汀州、宁化……夜，星陨如雨。

清道光《福建通志》卷271

公元1538年6月26日—7月25日 [中国]

明世宗嘉靖十七年六月，星陨如雨。

清康熙 山东《济南府志》卷10

清康熙 山东《长山县志》卷7

公元1538年10月23日—11月21日 [中国]

明世宗嘉靖十七年十月，长清星陨如雨。

清乾隆 《山东通志》卷33

清康熙 山东《长清县志》卷14

注：《长清县志》无月份。

公元1538年11月7日 [朝鲜]

中宗三十三年戊戌，十月甲辰，流星出四方。

《朝鲜王朝实录·中宗》卷88

公元1539年5月30日 [中国]

明世宗嘉靖十八年五月十三日，夜，星陨如雨。

明崇祯 福建《汀州府志》卷24

公元1539年11月18日　　[中国]
明世宗嘉靖十八年十月初八，夜半，星陨如雨。
清康熙 山西《蒲县新志》卷7

公元1540年5月6日—6月4日　　[中国]
明世宗嘉靖十九年四月，枣强县天鼓鸣，夜，星陨如雨。
《二申野录》卷4

公元1546年1月14日　　[中国]
明世宗嘉靖二十四年十二月十二日，日晡地旋。夜半月无光，星陨如雨。
明万历 陕西《华阴县志》卷7

公元1547年6月12日　　[中国]
(a) 明世宗嘉靖二十六年五月二十五日，星陨如雨，至寅时，天鼓鸣，有火光。
清康熙 山东《济南府志》卷10
清康熙 山东《长山县志》卷7
(b) 明世宗嘉靖二十六年五月二十五日，星陨如雨，天鼓鸣。
清乾隆 山东《武定府志》卷14
清乾隆 山东《惠民县志》卷3

公元1548年8月24日　　[朝鲜]
(a) 李朝明宗三年戊申七月甲午，大小流星纵横四方，不可纪极。
《增补文献备考·象纬考》卷7
(b) 明宗三年戊申七月甲午，夜，流星出河鼓星，入坤方天际，状如钵，尾长二丈许，色白；流星出天市西垣，入西方，状如钵，尾长五六尺许，色白；流星出天津星，入坤方天际，状如拳，尾长七八尺许，色白；流星出奎星下，入巽方天际，状如瓶，尾长二三丈许，色赤；星出室星下，入南方天际，状如盆，尾长五六丈许，色白，光着地，良久乃灭；流星出北斗第一星，入西方天际，状如瓶，尾长二三丈许，色白；流星出败瓜星，入南斗星，状如钵，尾长二丈许，色白；流星出天津星，入箕星下，状如钵，尾长五六尺许，色白；流星出王良星，入壁星，状如瓶，尾长八九尺许，色赤；流星出司空星，入南方天际，状如拳，尾长五六尺许，色白；自一更至二更，大流星大槩如此，而小流星四方纵横者，不可纪极。

《朝鲜王朝实录·明宗》卷8

(c) 明宗三年戊申七月庚辰，日微晕，申时太白见于未地。夜，月微晕。流星出室星，入娄星，状如钵，尾长二丈许，色白；出羽林星，入坤方天际，状如瓶，尾长五六尺许，色白；出奎星下，入危星，状如钵，尾长三四尺许，色赤。出北河星下，入艮方天际，状如钵，尾长五六尺许，色白。

《朝鲜王朝实录·明宗》卷8

公元1553年9月8日—10月6日　［中国］

(a) 明世宗嘉靖三十二年八月，夜，有星大如斗，小星数万，随之自东而西，光焰烛天，直行如矢，数日后复自西至东。

清康熙 河南《柘城县志》卷4

(b) 明世宗嘉靖三十二年八月，夜，有星大如斗，小星无数，随之自东而西，直行如矢，数日后复自西至东。

清康熙 河南《鹿邑县志》卷8

公元1554年10月14日　［朝鲜］

明宗九年九月丙寅，传曰："近者月晕，流星相继而出，今见庆尚监司状启，月晕有异，尤为闷虑。十四日内外，观象监不启何也？候望不谨事，推问可也。"【庆尚监司状启："今月十四日夜，月晕，初起时，未及见之，至亥末子初，始见之。月圆外有晕重匝，而内晕则周四圆，径一丈四五尺许，其晕青红之色相间，光彩烁然如虹。外晕则圆，径二丈许，其色亦似内圆。月旁及内外晕相接处，色稍淡而似相分焉，月圆亦灭小，而其光熹微，精彩全无。至於晕气稍灭之时，如风波荡漾而散乱，或合如前，再三然后稍灭，而晕痕则久而犹存云云。"】

《朝鲜王朝实录·明宗》卷17

公元1556年7月26日 [中国]

(a) 明世宗嘉靖三十五年六月二十日，南方一星，倏吐光焰丈余，夜分，群星三十余南奔，光芒异常。

清顺治 山东《登州府志》卷1

清康熙 山东《福山县志》卷1

(b) 明世宗嘉靖三十五年六月二十日，夜，南方一星忽光吐丈余，群星有三十余南奔。

清康熙 山东《莱阳县志》卷9

清乾隆 山东《海阳县志》卷3

(c) 明世宗嘉靖三十五年六月二十日，南方倏出一星，光可丈余，夜分，群星三十余南奔，光耀烛地。

清康熙 山东《济南府志》卷10

清康熙 山东《长山县志》卷7

清乾隆 山东《惠民县志》卷3

清乾隆 山东《武定府志》卷14

公元1557年8月11日 [中国]

明世宗嘉靖三十六年七月十七日，夜，东南际大小星乱落于海，少顷西北际亦然。

清康熙 浙江《定海县志》卷6

公元1559年2月7日—1560年1月26日 [中国]

明世宗嘉靖三十八年，有星落于东井，如雨。

明万历 河北《磁州志》卷8

公元1559年9月2日—30日 [中国]

明世宗嘉靖三十八年八月，辽阳星陨如雨。

明嘉靖 辽宁《全辽志》卷4

公元1560年6月27日 [中国]

明世宗嘉靖三十九年六月四日，夜，星陨如雨。

清雍正 河北《阜城县志》卷21

公元1560年8月21日—9月19日 [中国]

明世宗嘉靖三十九年八月，星陨如雨。

清乾隆 江西《建昌府志》卷5
清道光 江西《南城县志》卷27

公元1560年8月24日 [中国]
(a) 明世宗嘉靖三十九年八月初四日，星陨如雨。
清康熙 北京《密云县志》卷2
(b) 明世宗嘉靖三十九年八月初四日，夜，星辰散落如雨。
清康熙 山西《定襄县志》卷7

公元1560年8月24日 [朝鲜]
(a) 李朝明宗十五年庚申八月丁酉，飞星流星四方如雨。
《增补文献备考·象纬考》卷7
(b) 明宗十五年八月丁酉，夜，流星出室星，入羽林星，状如瓶，尾长三四尺许；流星出虚星，入天垒星，状如拳，尾长四五尺；流星出壁星，入南方天际，状如拳，尾长四五尺许；流星出室星，入危星，状如拳，尾长三四尺许，其色皆赤，其光良久乃灭；流星出北河星，入东方天际，状如拳，尾长二三尺许。流星出北极星，入北方天际，状如梨，尾长三四尺许。流星出参星，入北河星，状如拳，尾长二三尺许，其色皆赤；飞星出五车星，入八谷星，状如拳，尾长五六尺许，色赤，良久乃灭；四更、五更，流星飞星四方如雨，出入度数，未及测候。
《朝鲜王朝实录·明宗》卷26

公元1560年9月2日 [中国]
明世宗嘉靖三十九年八月十三日，星陨如雨。
明崇祯 江苏《太仓州志》卷15

公元1563年7月3日 [中国]
明世宗嘉靖四十二年六月十四，夜，有流星如雨，自桂林南至西北。
清雍正《广西通志》卷3

公元1565年7月26日 [日本]
日本永禄八年六月二十九日，六晦，南北星飞乱交如萤。
《新撰和汉合图》

公元1566年10月13日—11月11日　[中国]
明世宗嘉靖四十五年十月，西北星陨如雨，其色青。
明万历 江西《永新县志》卷5

公元1566年10月25日—27日　[中国]
明世宗嘉靖四十五年十月十三，夜，星陨如雨，有声，十四、十五皆然。
《新知录》

公元1566年10月26日　[朝鲜]
(a) 李朝明宗二十一年丙寅十月辛未，飞星流星四方如雨。
《增补文献备考·象纬考》卷7
注：原文"明宗十五年庚申八月丁酉，飞星流星四方如雨。二十一年丙寅十月辛未，亦如之。"
(b) 明宗二十一年丙寅十月辛未，夜，流星出轩辕星，入天庙星，状如钵，尾长五六尺许，色赤，良久乃灭；流星出参星，入西方天际，状如钵，尾长七八尺许许，色赤，良久乃灭；小流星，或如梨，或如桃，出入四方，多不可纪。
《朝鲜王朝实录·明宗》卷33

公元1566年11月25日　[中国]
明世宗嘉靖四十五年闰十月辛丑，夜，流星如织，有流星二，大如月。
明隆庆 江苏《高邮州志》卷12

公元1566年12月23日—25日　[中国]
明世宗嘉靖四十五年十一月望前三日，夜，有大星下陨，群星随之，如雨有声，历三晚。
《罪惟录》 志1

公元1566年12月25日　[中国]
(a) 明世宗嘉靖四十五年十一月十五日，四更，有一大星下陨，群星数百，如雨随之。
《二申野录》卷4

《留青日札》卷4

(b) 明世宗嘉靖四十五年十一月十五日，四更有大星夜陨，群星数百随之。 清光绪

浙江《嘉兴府志》卷35

注：清嘉庆《嘉善县志》作“十一月十一日”

公元1578年2月12日 [中国]

(a) 明神宗万历六年正月初六日，夜，众星流至西方，中有一大星。

明万历 浙江《秀水县志》卷10

(b) 明神宗万历六年正月初六日，夜有大星如日西移，众星随之。

清光绪 浙江《嘉兴府志》卷35

(c) 明神宗万历六年正月初六日，夜有一大星如日出自西方，众星环于西。

明万历 福建《建宁府志》卷47

(d) 明神宗万历六年正月初六日，夜，群星流于西，中一星大甚。

清道光 福建《建阳县志》卷19

公元1578年2月22日 [中国]

明神宗万历六年正月戊辰，有大星如日出西方，众星皆西环。

《明史·天文志》卷27

《国榷》卷70

《续通志·灾祥》卷171

公元1581年8月7日 [中国]

明神宗万历九年七月初九，夜，西南星陨如雨。

清康熙 福建《福安县志》卷9

公元1589年8月25日 [中国]

(a) 明神宗万历己丑七月十五（庚申），夜，月中有白小星迸出如珠。

明崇祯 上海《松江府志》卷47

(b) 明神宗万历己丑七月十五（庚申），夜见东山月上，月中有小白星迸出，如珍珠散乱，移刻始息。
《云间杂识》卷2

公元1590年8月6日 [中国]
明神宗万历十八年七月七日，夜，星陨如雨，逾时乃止。
清乾隆 甘肃《狄道州志》卷11

公元1594年11月5日 [中国]
明神宗万历甲午九月廿三日，夜，东北方有星大如鸡子，青白色，西南方有星大如碗，亦青白色，尾迹散光照地，西南行，后有二小星随之，复有流星数千，四面纷纷交错而行。
《二申野录》卷5

公元1601年11月5日 [中国]
(a) 明神宗万历二十九年十月十一日，夜五更，有星变如雨。
明万历 福建《漳州府志》卷32
(b) 明神宗万历二十九年十月十一日，夜五鼓，星陨如雨。
清康熙 福建《海澄县志》卷14
清乾隆 福建《龙溪县志》卷20

公元1602年秋 [中国]
明神宗万历三十年秋，夜，有星如卵，光散照地，后随小星二，复有大小二星飞行梭织。
清乾隆 安徽《铜陵县志》卷13

公元1602年10月15日—11月13日 [中国]
(a) 明神宗万历三十年九月，星陨如雨。
清顺治 广东《潮州府志》卷2
(b) 明神宗万历三十年九月，流星数日，出参宿入天苑。
《罪惟录》志1

公元1602年11月5日 [中国]
明神宗万历三十年九月庚辰，有大小星数百，交错行。

《明神宗实录》卷376
〈国榷〉 卷79

公元1602年11月6日 [中国]

(a) 明神宗万历三十年（壬寅）九月辛巳（二十三），有大小星数百，交错行。

《明史·天文志》卷27
《文献通考·象纬》卷214
《续通志·灾祥》卷171

(b) 明神宗万历三十年（壬寅）九月辛巳（二十三），夜五更，东北有星如雉卵大，青白色，尾有光，起自下台，东北行，至近西南方；有星如碗大，青白色，尾迹炸散，光照地，起自参宿，西南行入天苑星，后二小星随之，又有大小流星数百纷错随行。

《二申野录》卷5

(c) 明神宗万历三十年（壬寅）九月辛巳（二十三），夜，北方有星如鸡蛋大，青白色，尾迹有光，一起自下台星，东北行至近西南方；有星如碗大，尾迹作散，光照耀地，起自参宿，西北方入天苑星，后有二小星随之，后有大小流星数百千枚，四面纷纷交错而行。

清康熙 福建《安海志》卷8

公元1602年11月7日 [中国]

明神宗万历三十年九月二十四日，颍州见流星蟠曲如龙，向西南落，小星万余随之。

清康熙 安徽《凤阳府志》卷3

公元1602年11月12日 [朝鲜]

宣祖三十五年十一月丙寅，庆尚监司李时发驰启曰：“去九月二十八日夜中，恒星动摇，有如移次之状，俄有大星，自西北流于东南，光芒横天，久而乃灭，瞬息之间，满天之星，四面皆陨，有如雨雹之状，大者芒过半空，小者芒长尺许，黎明乃止；道内之人，多有见之者。变异非常事。”

《朝鲜王朝实录·宣祖》卷56

公元1606年2月9日 [中国]
明神宗万历丙午正月朔之三日，夜，见西方星乱坠。
清康熙 湖北《钟祥县志》卷10

公元1610年9月13日 [中国]
明神宗万历三十八年七月二十六日，夜，天北星乱动，移时乃定。
清康熙 安徽《天长县志》星野

公元1621年8月21日 [中国]
明熹宗天启辛酉七月初五，亥时，流星引练，红赤色，后有小星数百随之，起自西北，傍女宿。
《二申野录》卷7
明天启 江苏《淮安府志》卷24

公元1621年秋 [中国]
(a) 明熹宗天启元年（辛酉）秋，历城星陨如雨。
清康熙 山东《济南府志》卷10
(b) 明熹宗天启元年（辛酉）秋，历城星陨如雨，三鼓后，满天星月皆疾行，闪烁不定，如此凡数夜。
清康熙 山东《长山县志》卷7
(c) 明熹宗天启元年（辛酉）秋，星陨如雨，每夜三鼓后，满天星月皆疾行，闪烁不定者久之，如此凡数夜。
清康熙 山东《新城县志》卷10

公元1622年10月5日—11月2 [中国] 日
明熹宗天启二年九月，星陨如雨。
清康熙 甘肃《静宁州志》卷11

公元1622年10月25日 [中国]
明熹宗天启二年九月甲寅，陕西固原州星陨如雨。
《明熹宗实录》卷26

公元1622年12月3日—12月31日 [中国]
(a) 明熹宗天启二年十一月，有大星如斗，尾带数十星，状如连珠，从东北飞至西南而没，时已黄昏，光明照耀，树木屋舍皆可辨。

清康熙 广西《合浦县志》卷1
清雍正 广西《钦州志》卷1
清乾隆 广西《灵山县志》卷3

(b) 明熹宗天启二年十一月，有大星如斗，尾带数十星，状如连珠，从东北飞至西南而没，时已昏夜，其光照耀如昼，房屋窗棂可辨。

清乾隆 广西《廉州府志》卷5

公元1623年10月18日 [中国]
明熹宗天启三年九月壬子（二十五），夜，南方有流星如盏大，赤色，尾迹有光，起自虚宿，东南行入垒壁阵，星后有二十星随之。
《明熹宗实录》卷38

公元1623年10月20日 [中国]
明熹宗天启三年九月甲寅，固原州星陨如雨。
《明史·天文志》卷27
《续文献通考·象纬》卷214
《续通志·灾祥》卷171

公元1625年8月4日 [中国]
明熹宗天启五年七月二日，遂昌有大星自西流东，尾长二十余丈，光芒如月，有声如雷。自是，每夜流星如织。
清光绪 浙江《处州府志》卷25

公元1625年11月6日 [朝鲜]

(a) 仁祖三年十月壬午，夜，流星十余，出於中天，散入四方天际，其大或如拳，或如杯，长各数三丈，光照地；又流星三十余，散入天际，有光。

《朝鲜王朝实录·仁祖》卷10

(b) 仁祖三年十月壬午，夜一更，月犯牵牛南星；二更，月晕。三更，流星十余，出於中天，散入四方天际，其大如拳，或如杯，长各数三丈许，皆赤，光照也。五更、四更，有流星三十余，散入天际，有光。

《承政院日記》

公元1633年2月8日 ［朝鲜］
仁祖十一年癸酉元朝，夜，有星如相斗之状。
《朝鲜王朝实录·仁祖》卷28

公元1636年6月30日 ［中国］
明毅宗崇祯九年五月辛未，州东门外八里河，星陨如雨。
清光绪 江苏《通州直隶州志》卷末

公元1636年11月27日—12月26日 ［中国］
明毅宗崇祯九年十一月，星陨如雨，天鼓鸣。
清道光 山东《临邑县志》卷16

公元1637年1月18日 ［朝鲜］
仁祖十四年丙子，十二月二十三日夜三更，众星聚繞一大星，良久散向西方而去。
《朝鲜王朝实录·仁祖》卷33

公元1640年5月21日—6月18日 ［中国］
(a) 明毅宗崇祯十三年四月，天鼓鸣，自西北流于东南，星陨如雨。
清康熙 山西《灵石县志》 祥异
(b) 明毅宗崇祯十三年四月，灵石天鼓鸣，自西北流于东北，星陨如雨。
四库本《山西通志》卷163

公元1637年1月29日 ［朝鲜］
仁祖十五年正月甲辰，有星斗于东南，移时乃止。
《朝鲜王朝实录·仁祖》卷34

公元1642年5月28日—6月26日 ［中国］
明毅宗崇祯壬午五月，星流如织。
《二申野录》卷8

公元1642年夏 ［中国］
明毅宗崇祯十五年夏，星流如织。

《明史·天文志》卷27
《续文献通考·象纬》卷214
《续通志·灾祥》卷171

公元1642年7月27日—8月24日 [中国]
明毅宗崇祯十五年七月，普天星象悉聚斗、女之间，向东北流去者七次。
清顺治 山东《泗水县志》卷11

公元1642年9月23日 [中国]
明毅宗崇祯十五年八月晦，有大星如斗，小星数十随之，自西北至东南，坠地有声，光芒数十丈，幽暗皆明。
清康熙 浙江《海盐县志》

公元1643年1月18日 [中国]
明毅宗崇祯十五年闰十一月二十八日，夜，星陨如雨。
清乾隆 广东《揭阳县志》卷7

公元1643年7月8日 [朝鲜]
(a) 仁祖二十一年五月乙卯，星辰尽摇。
《朝鲜王朝实录·仁祖》卷44
(b) 仁祖二十一年 五月二十三日乙卯，夜一、二更。天星尽摇；五更，月有小晕，色内赤外青。
《承政院日記》

公元1644年3月9日—4月6日 [中国]
(a) 明毅宗崇祯十七年二月，星陨如雨。
清道光 湖南《永州府志》卷17
(b) 清世祖顺治元年二月，星如雨下。
《小腆纪年附考》卷3

公元1644年6月5日—7月3日 [中国]
明毅宗崇祯十七年五月，星陨如雨，凡二十四日，夜中星斗交飞，或逆或顺，或有声而坠，或无声而隐。
清康熙 浙江《钱塘县志》卷12

公元1645年6月24日—7月22日 [中国]
清世祖顺治二年六月，终夜流星往来如织。
清乾隆 浙江《平湖县志》卷10

公元1645年7月16日 [中国]
清世祖顺治二年六月二十三日，城外见城内天星散落如雨。
清康熙 浙江《嘉兴府志》卷16

公元1645年7月23日—8月20日 [中国]
(a) 清世祖顺治二年闰六月，每夜，星陨如雨，七月渐稀。
清乾隆 江苏《吴县志》卷26
(b) 清世祖顺治二年闰六月，夜半，小星散落如雨。
清道光 江苏《璜泾志稿》卷7

公元1645年7月23日 [中国]
清世祖顺治二年闰六月朔，夜将半，天上小星散落如雪。
清嘉庆 江苏《太仓州志》卷58

公元1645年7月30日 [中国]
清世祖顺治二年闰六月初八日，夜有流星如月，大小相随，光芒甚白。
清康熙 浙江《绍兴府志》卷13

公元1645年8月6日 [中国]
(a) 清世祖顺治二年闰六月十五日，夜，星陨如雨。
清康熙 江苏《通州志》卷1
(b) 清世祖顺治二年闰六月十五日，夜，众星南流。
清乾隆 上海《嘉定县志》卷3
(c) 清世祖顺治二年闰六月十五日，星陨如雨。
清光绪 江苏《通州直隶州志》卷末

公元1645年8月7日 [中国]
(a) 清世祖顺治二年闰六月十六日，众星流移，纵横绎络，各有光芒。
清道光 江苏《璜泾志稿》卷7

(b) 清世祖顺治二年闰六月丙申，月食既，星流竟夕。
《小腆纪年附考》卷10

公元1646年7月13日—8月10日 [中国]
清世祖顺治三年六月，大星陨东南，小星数百随之。
清康熙 浙江《东阳县志》卷4

公元1646年7月23日 [中国]
清世祖顺治三年六月十一日，星陨如雨。
清康熙 浙江《绍兴府志》卷13
清康熙 浙江《江山县志》卷10
清康熙 浙江《山阴县志》卷9
清康熙 浙江《桃源乡志》卷8

公元1646年8月11日—9月8日 [中国]
清世祖顺治三年七月，有星不计其数，自北而流于南。
清康熙 浙江《宁波府志》卷30
清康熙 浙江《鄞县志》卷24
清乾隆 浙江《镇海县志》卷4

公元1647年2月5日—1648年1月24日 [中国]
清世祖顺治四年，有星如锥，众小星随走。
清雍正 云南《建水州志》卷11

公元1648年1月25日—1649年2月10日 [中国]
清世祖顺治五年，星陨东南如雨。
清乾隆 广西《岑溪县志》

公元1648年3月11日 [中国]
清世祖顺治五年二月十八，夜，东北星陨如雨。
清康熙 浙江《长兴县志》卷4
清乾隆 浙江《湖州府志》卷38

公元1648年春 [中国]
清世祖顺治五年春，东南有星乱陨。

清康熙 广西《平乐县志》卷6
清乾隆 广西《昭平县志》卷4
清道光 广西《栖霞寺志》卷上
注：栖霞寺在桂林。

公元1649年6月10日—7月9日 [中国]
清世祖顺治六年五月，陨星逐队飞流，大者先，而小者从之，间有大如斗者。
清乾隆 广东《归善县志》卷2

公元1651年6月19日 [中国]
(a) 清世祖顺治八年五月戊寅，夜，西南有流星如盏，白色，尾迹有光。
《清世祖实录》卷57
(b) 清世祖顺治八年五月戊寅，流星如盏，自亢宿西南行，色白，众小星随之入翼。
《清史稿·天文志》卷39

公元1651年9月15日—10月13日 [中国]
(a) 清世祖顺治八年八月，含山星陨如雨。
清康熙 安徽《和州志》卷4
(b) 清世祖顺治八年八月，有大星自西南至东北，声响如雷，火光四迸，后有小星无数，随之而行。
清康熙 山东《胶州志》卷6

公元1651年10月14日 [中国]
清世祖顺治八年九月初一乙亥，夜半星陨如雨，远近大骇。
民国《浙江续通志稿》卷10

公元1651年10月15日 [中国]
清世祖顺治八年九月初二日丙子，星陨如雨。
民国 《浙江续通志稿》卷10

公元1654年7月21日 [中国]
清世祖顺治十一年六月八日，四更时，有红星光芒丈余，状如帚，众小星相随东南移，至西北陨。
清乾隆 江苏《吴县志》卷26

公元1659年5月21日—6月19日 [中国]
清世祖顺治十六年四月，星陨如雨。
清康熙 河南《内乡县志》卷11

公元1659年6月6日 [中国]
清世祖顺治十六年四月十七日，天东南一星如斗，后有数小星随之，至西南而没，是时赤日当空，四野无云，隐隐有声，如雷在水底鸣。
清道光 江苏《宝应县志》卷9

公元1660年9月5日—10月3日 [中国]
清世祖顺治十七年八月，府境西方，有大星流北方，众星随者无数。
清乾隆 江西《吉安府志》卷1

公元1661年1月30日—1662年2月17日 [中国]
清世祖顺治十八年，有大星陨于西方，小星从者如雨。
清康熙 云南《建水州志》卷17
清乾隆 云南《赵州志》祥异

公元1661年9月23日—10月22日 [中国]
清世祖顺治十八年八月，酉时，西方有大星流北方，小星随者无数。
清康熙 江西《庐陵县志》卷2

公元1661年10月1日 [中国]
(a) F清世祖顺治十八年八月初九，申时，大星见，小星从之百余，自西南飞入北，有声。

清康熙 湖南《零陵县志》卷14

(b) 清世祖顺治十八年八月初九，申刻，有大星，约长十丈，小星从之，白气如鍊，起于西南，向北疾走，轰然有声。

清乾隆 湖南《清泉县志》卷25

清乾隆 湖南《衡州府志》卷29

注："有大星"误写为"有人星"。

公元1661年10月23日—11月21日 ［中国］

清世祖顺治十八年九月，军山飞星竟夜。

清乾隆 江苏《五山全志》卷8

注：五山在南通。

公元1661年10月30日 ［中国］

清世祖顺治十八年九月初八，薄暮，大星起于西南，小星数千随之，白气一带向东北走，有声如雷。

清康熙 湖南《邵阳县志》卷6

清康熙 湖南《宝庆县志》卷22

公元1661年12月11日 ［中国］

清世祖顺治十八年十月二十日，夜半，星陨如雨，从东南起，群陨于西北。

清道光 江苏《璜泾志稿》卷7

公元1665年11月17日 ［中国］

清圣祖康熙四年（乙巳）十月十一日四更，有大星见东南，众小星随之，或上或下，倏左倏右。大星陨，小星随之陨。

《三冈识略》卷5补遗

公元1665年11月18日 ［中国］

清圣祖康熙四年十月十二日，卯时，星陨如雨。

清康熙 浙江《嘉兴府志》卷2

公元1666年7月23日 ［中国］

(a) 清圣祖康熙五年六月辛未（二十二日），夜，天星乱流如织。

清康熙 河北《增补卢龙县志》卷2

(b) 清圣祖康熙五年六月辛未（二十二日），星流如织。
清康熙 河北《永平府志》卷3
(c) 清圣祖康熙五年六月辛未（二十二日），流星如织，自乾趋巽。
清康熙 河北《滦州直隶府志》卷31
(d) 清圣祖康熙五年六月辛未（二十二日），流星如织，州人有见之，后自杭来人语亦同。
清嘉庆 河北《滦州志》卷1
(e) 清圣祖康熙五年六月辛未（二十二日），星陨如雨。
清道光 江苏太仓《璜泾志稿》卷7

公元1666年7月24日 [中国]
清圣祖康熙五年六月二十三日，亥时，有大星随以小星千余，从西北流至东南，坠地。
清康熙 浙江《海盐县志补遗》灾祥

公元1666年秋 [中国]
清圣祖康熙五年秋，西北星陨如雨。
清康熙 湖南《宁乡县志》卷2
清乾隆 湖南《益阳县志》卷1

公元1666年11月7日 [中国]
清圣祖康熙五年十月立冬日，东北方有大星陨地有声，众星散落如雨。
清光绪 浙江《重修嘉善县志》34 卷

公元1667年1月6日 [中国]
清圣祖康熙五年十二月十二日，凌晨，有大星自东南陨至东北，众小星随之而陨。
清道光 江苏《璜泾志稿》卷7

公元1668年4月21日 [中国]
清圣祖康熙七年三月十一日，夜，有流星大如斗，小星围绕无数。
清乾隆 湖北《沔阳州志》卷13

公元1671年5月9日—6月6日 [中国]
清圣祖康熙十年四月，大星陨，星形如车轮，芒角四散，下带数十小星，由西北趋东南，直过桐江，时天已黎明。
清康熙 浙江《钱塘县志》卷12

公元1672年5月11日 [中国]
(a) 清圣祖康熙十一年四月十五日，大星自东北方经天，后有数十小星随之，至西南方而陨，声响如雷。
清康熙 浙江《建德县志》卷9
清雍正 浙江《严州府志》卷22
(b) 清圣祖康熙十一年四月十五日，卯时，都昌有物如蛟，长数十丈，大如桶，赤如火，群星护之，从东北没于西南，其声如雷。
清康熙 江西《都昌县志》卷10

公元1672年5月12日 [中国]
(a) 清圣祖康熙十一年四月十六日，夜，有异星移于西南，状如龙，长丈余，尾带数十星，过处有白迹，逾时乃没。
清康熙 安徽《宿松县志》卷3
(b) 清圣祖康熙十一年四月十六日，五鼓，众星丛流，火光拉杂，自北而南有声。
清康熙 江西《彭泽县志》卷14

公元1675年11月7日 [中国]
清圣祖康熙十四年九月二十日，夜，有大星如斗坠南方，众小星随之，皆陨。
清康熙 福建《漳州府志》卷33
清康熙 福建《龙溪县志》卷20

公元1676年12月14日 [朝鲜]
(a) 肃宗二年，十一月戊子，流星出北斗星上，入紫微东垣，是後连出；太白亦屡见於昼。
《朝鲜王朝实录·肃》卷5

(b) 肃宗二年，十一月十日戊子，巳时，太白见於未地。自一更至三更，月暈；五更，流星出北斗星上，入紫微东垣下，状如瓶，尾长三四尺許，色黄白，光照地。
《承政院日記》卷

公元1679年9月21日 ［中国］
清圣祖康熙十八年八月十七日，申时，有大星如斗，自西北，经天汉，流入东南，群星无数从之，尾拖白练，长数百丈，见者惊异。
清康熙 湖南《永州府志》卷24

公元1680年2月18日 ［朝鲜］
肃宗六年，正月己酉，流星出；是後，连日而出，或间日，至三十余日而止。
《朝鲜王朝实录·肃宗》卷9

公元1682年8月13日—22日 ［中国］
清圣祖康熙二十一年七月中，夜，有大星十余，各曳长尾，其色惨澹，自西南入于箕、尾分野。
清乾隆 《福建通志》卷65
清乾隆 福建《泉州府志》卷73
清乾隆 福建《晋江县志》卷15
清乾隆 福建《同安县志》卷13

公元1682年9月11日 ［中国］
清圣祖康熙二十一年八月初十日，夜二更，大小星数百坠下鸡笼山。
《台湾外纪》卷26

公元1683年10月26日 ［朝鲜］
肃宗九年，九月壬申，流星出紫微西垣内。
肃宗九年，九月甲戌，流星出北極星下，入北方；是后屡见。
《朝鲜王朝实录·肃宗》卷14下

公元1696年10月29日 ［朝鲜］
(a) 肃宗二十二年，十月丁亥，流星见，连夜屡见。
《朝鲜王朝实录·肃宗》卷30

(b) 肃宗二十二年，十月四日丁亥，夜五更，流星出毕星下，入南方天际，状如拳，尾长二三尺许，色赤。

《承政院日記》

公元1698年11月8日 [日本]

日本元禄十一年十月六日丁未即夜，大流星飞入井宿内，垂下数尺，经数刻而消。今夜流星如织，人皆见之。

《壬癸录》

公元1708年9月4日 [中国]

(a) 清圣祖康熙四十七年七月二十日，申时，有星数百甚灿，排列甚齐，如晃旒一般，自东拖到金钟山后。

清康熙 福建《罗源县志》卷10

(b) 清圣祖康熙四十七年七月二十日，福州府申时，有星数百，光芒灿列，自东流入金钟山后。

清乾隆《福建通志》卷65

公元1709年8月20日 [中国]

清圣祖康熙四十八年七月望后，酉时，有星自南来，其行甚缓，尾后拖白光如练，小星无数随之，徐向北去，经时方灭。

清乾隆《湖南通志》卷172

清嘉庆 湖南《宁远县志》卷10

公元1722年11月24日 [中国]

清圣祖康熙六十一年十月十六日，寅刻，有星大如斗，自东堕西北，带落小星约十数个，声如雷，赤光经天。

清雍正 陕西《渭南县志》卷15

公元1729年2月13日 [中国]

清世宗雍正七年正月十六日，夜，见众星摇动，如欲堕状，又或飞或走，群向东行。

《清世宗实录》卷22

公元1731年10月31日 [中国]

清世宗雍正九年十月朔，星陨如雨。

清乾隆 云南《新兴州志》卷1

公元1732年1月27日—1733年2月13日 [中国]
清世宗雍正十年，西南方星陨如雨。
清道光 云南《昆阳州志》卷2

公元1732年12月15日 [中国]
(a) 清世宗雍正十年十月二十八，夜，星陨如雨。
清道光 云南《广西府志》卷23
(b) 清世宗雍正十年十月二十八，夜，星陨如雪。
清道光 《云南通志稿》卷4

公元1733年2月14日—1734年2月3日 [中国]
清世宗雍正十一年，星坠如雨。
清乾隆 云南《开化府志》卷1

公元1733年5月14日—6月11日 [中国]
清世宗雍正十一年四月，星陨如雨。
清道光 《云南通志稿》卷4

公元1733年6月10日 [中国]
(a) 清世宗雍正十一年四月二十八，夜，星斗于寿星之次。
清同治 广西《苍梧县志》卷18
清光绪 广西《迁江县志》卷4
(b) 清世宗雍正十一年四月二十八，夜正初昏时，星陨如雨。
清乾隆 云南《永北府志》卷24

公元1735年1月29日 [中国]
清世宗雍正十三年正月初六日，东南角有星大于盂，从下升上，随有小星五六与作鬥状，大星不动，小星堕而复升，如是数回，忽堕不见，复有众多小星，飞入大星内，亦随之灭不见。
清乾隆 贵州《玉屏县志》卷1

公元1735年4月23日—5月21日 [中国]
清世宗雍正十三年四月，游星遍飞。
清乾隆 云南《永北府志》卷24

公元1748年10月16日 ［中国］
清高宗乾隆十三年八月二十四日，酉刻，东有流星大如月，其光同昼，渐至天中，长丈余，状颈似捣衣杼，渐小星无数随之，筛下火光而没，声若雷鸣。
清嘉庆 安徽《泾县志》卷27

公元1760年5月11日 ［中国］
清高宗乾隆二十五年三月二十六日，戌刻，星陨大如箕，赤色，自西北而东南，尾后数十小星随之，戛然有声，光彩照人。
清乾隆 河北《衡水县志》卷11

公元1768年3月18日 ［中国］
清高宗乾隆三十三年二月己未，夕，星陨如雨。
《清史稿·列传》卷528

公元1769年秋 ［中国］
清高宗乾隆三十四年秋，有星大如车轮，或自北飞，或自南飞北数次。
清道光 《云南通志稿》卷4

公元1769年12月28日—1770年1月26日 ［日本］
日本明和六年十二月，有大星北流，小星数十从之云。
《续史愚抄》七十八

公元1770年3月27日—4月25日 ［中国］
清高宗乾隆三十五年三月，星陨如雨，天鼓鸣。
清道光 山东《临邑县志》卷16
清道光 山东《济南府志》卷20

公元1794年1月31日—3月1日 ［中国］
清高宗乾隆五十九年正月，剑川流星遍野。
清光绪 《云南通志》卷4

公元1794年12月11日 [中国]
清高宗乾隆五十九年十一月十九日，夜，星陨如雨，经亥、子、丑三时方止。
清道光《云南通志》卷4
清光绪 云南《霑益州志》卷4

公元1794年冬 [中国]
清高宗乾隆五十九年冬，雅安见众星西陨，势如飞萤。
民国 四川《雅安县志》卷4

公元1795年12月20日 [日本]
日本宽政七年十一月十日丁巳，冬至前夜，星飞如雨降。
（注：译自日文）
《宝历现来记》六

公元1796年2月3日 [中国]
清高宗乾隆六十年十二月二十五日，戌时，南方有白光一道，旁附大星数十，迁移无定。
清嘉庆 浙江《嘉善县志》卷20

公元1797年2月27日—3月27日 [中国]
清仁宗嘉庆二年二月，天西南角星陨如雨。
清道光 江苏《璜泾志稿》卷7

公元1797年10月20日—11月17日 [中国]
清仁宗嘉庆二年九月，夜，星飞如织。
清嘉庆 福建《福鼎县志》卷7

公元1797年12月7日 [中国]
清仁宗嘉庆二年十月二十日，星陨如雨。
清嘉庆 安徽《绩溪县志》卷12

公元1797年12月16日 [中国]
清仁宗嘉庆二年十月二十九日，星陨如雨。
清道光 山西《太原县志》卷15

清光绪 山西《交城县志》卷1
清光绪 山西《晋祠志》卷41

公元1797年12月17日 ［中国］
清仁宗嘉庆二年十月晦日，夜半，星陨如雨，尽趋西南，火焰蓬勃，久之乃止息。
清光绪 山西《晋祠志》卷41

公元1797年12月18日—1798年1月16日 ［中国］
(a) 清仁宗嘉庆二年冬月某日，夜，北方星陨如雨，复见天北面一半青苍，星影寥寥。
清光绪 浙江《泰顺分疆录》卷10
(b) 清仁宗嘉庆二年十一月，飞星如织，食顷乃定。
清道光 广东《高要县志》卷10

公元1798年2月16日—3月16日 ［中国］
清仁宗嘉庆三年正月，群星纵横乱飞。
清道光 福建《建阳县志》卷19

公元1798年5月16日—6月13日 ［中国］
清仁宗嘉庆三年四月，星陨如雨。
清嘉庆 四川《郫县志》卷41

公元1798年6月14日—7月12日 ［中国］
清仁宗嘉庆三年五月，某夜，十五都星陨如雨，一时始定。
清光绪 湖南《永明县志》卷43

公元1798年10月9日—11月7日 ［中国］
清仁宗嘉庆三年九月，上海青浦县有星移于天，如飞如织，辉光四布，如是者数夜。
《明斋小识》卷6

公元1798年11月5日—6日 ［中国］
清仁宗嘉庆九月二十八日，夜，众星往来如织，唧唧有声，俄陨如雨，凡二夜。
清光绪 浙江《重修嘉善县志》卷34

公元1798年11月7日 [中国]
清仁宗嘉庆三年九月三十日，星陨如雨。
清道光 云南《续修易门县志》卷1

公元1798年11月8日—12月6日 [中国]
(a) 清仁宗嘉庆三年十月，星飞如织。
清嘉庆 广东《龙山乡志》卷14
(b) 清仁宗嘉庆三年十月，星陨如雨。
清嘉庆 云南《景东县志稿》卷1
清道光 云南《晋宁州志》卷11
清道光 云南《澂江府志》卷2
清宣统 云南《续蒙自县志》卷12
(c) 清仁宗嘉庆三年十月，星流如织。
民国 江苏《如皋县志》 第二册

公元1798年11月20日 [中国]
清仁宗嘉庆三年十月十三，夜，星陨如雨，自北而南一夕。
清光绪 上海《江东志》卷1

公元1798年11月26日 [中国]
清仁宗嘉庆三年十月十九日，黄昏，星如雨下，夜半乃止。
清乾隆 陕西《洵阳县志》卷14
清嘉庆 陕西《白河县志》卷14

公元1798年11月27日 [中国]
清仁宗嘉庆三年十月二十日夜，众星交流如织。
民国 浙江《海宁州志稿》卷40

公元1798年11月29日 [中国]
清仁宗嘉庆三年十月二十二日，戌时，东北方众星飞串梭，俄而星陨如雨。
清光绪 云南《江川县志》卷27

公元1798年12月4日 [中国]
(a) 清仁宗嘉庆三年十月二十七日，夜，星移无数。
清道光 河南《鄢陵县志》卷17

清光绪 河南《祥符县志》卷23

(b) 清仁宗嘉庆三年十月二十七日，夜，星移如织。

民国 河南《鄢陵县志》卷29

公元1798年12月5日 [中国]

清仁宗嘉庆三年十月二十八日，夜，众星交流如织。

清嘉庆 浙江《嘉兴府志》卷35

清道光 浙江《石门县志》卷23

清光绪 浙江《海盐县志》卷13

《清仁宗实录》卷42

公元1798年12月6日 [中国]

(a) 清仁宗嘉庆戊午十月二十九日，夜间，众星交流如织。

《清仁宗实录》卷42

《履园丛话》灾祥

(b) 清仁宗嘉庆戊午十月二十九日（晦日），众星东南流如织，逾时乃止。

清嘉庆 浙江《平湖县志》卷3

(c) 清仁宗嘉庆三年十月二十九日（晦日），亥刻，流星如织，经时乃已。

清嘉庆 河北《滦州志》卷1

清道光 河北《蓟州志》卷2

清同治 河北《昌黎县志》卷1

(d) 清仁宗嘉庆三年十月二十九日（晦日），夜，星四散如织。

清嘉庆 山东《长山县志》卷4

(e) 清仁宗嘉庆三年十月二十九日（晦日），初更，星陨如雨。

清宣统 上海《彭浦里志》卷8

(f) 清仁宗嘉庆三年十月二十九日（晦日），夜戌时，东北角流星百十如织，炊许时乃已。

清嘉庆 广东《潮阳县志》卷12

清同治 广东《大埔县志》卷13

(g) 清仁宗嘉庆三年十月二十九日（晦日），星陨西北如雨。

清嘉庆 云南《临安府志》卷17

(h) 清仁宗嘉庆三年十月二十九日（晦日），戌时，星陨如雨，四更方止。
清道光 江苏《昆山新阳两县志》卷39

公元1798年12月6日 [日本]
(a) 日本宽政十年十月廿九日，传闻，今夜流星数多纵横散乱云云。
《实种公记》
(b) 日本宽政十年戊午十月廿九日，天气晴朗，酉刻，予出庭前，见大小众星空中飞奔，东西南北流陨如雨。至丑寅刻亦然，凌晨稍异。
（注：译自日文）
《北窗杂话》
(c) 日本宽政十年十月廿九日，终夜星如雨。
《道川忠治记录》

公元1798年12月7日—1799年1月5日 [中国]
(a) 清仁宗嘉庆三年十一月，群星乱飞，自酉至亥始定。
清道光 福建《罗源县志》卷29
(b) 清仁宗嘉庆三年十一月，乱星纵横如织，移时乃定。
民国 福建《诏安县志》卷5

公元1798年12月9日 [日本]
日本宽政十年十一月三日夜，星飞如前。
（注：译自日文）
《武江年表》七

公元1798年12月12日 [中国]
清仁宗嘉庆三年十一月六日，夜，流星如织，陨如雨。
清 浙江《海昌续载》卷4

公元1798年12月25日 [中国]
清仁宗嘉庆三年十一月十九，夜，星陨如雨。
清道光 云南《赵州志》卷3
清道光 云南《巴州志》卷10

公元1798年12月31日　［中国］
清仁宗嘉庆三年十一月二十五日，星陨如雨。
清道光 福建《顺昌县志》卷9

公元1799年1月6日—2月4日　［中国］
清仁宗嘉庆三年十二月，星陨如雨。
清道光《云南通志稿》卷4

公元1799年2月2日—3日　［中国］
清仁宗嘉庆三年十二月二十八、九日，夜中，众星交流如织。
《清朝续文献通考》卷303

公元1799年6月3日—7月2日　［中国］
清仁宗嘉庆四年五月，十五都，星陨如雨。
清道光 湖南《永明县志》卷13

公元1799年11月22日　［中国］
清仁宗嘉庆四年十月二十五日，夜，星陨如雨。
清道光 云南《永昌府志》卷24

公元1799年11月25日　［中国］
清仁宗嘉庆四年十月二十八，夜间，众星交流如织。
清光绪 山东《宁津县志》卷11

公元1799年11月26日　［中国］
(a) 清仁宗嘉庆四年十月二十九，初昏，西南星陨如织。
清嘉庆 浙江《义乌县志》卷19
(b) 清仁宗嘉庆四年十月二十九，夜间，众星交流如织。
清光绪 山东《宁津县志》卷11

公元1799年11月27日　［中国］
清仁宗嘉庆四年十月三十日，酉时，星交错飞，流如弹。
清嘉庆 河南《续济源县志》卷1
注：十月小，只有二十九天，“十月三十日”应为“十一月一日”。

公元1800年11月17日—12月15日 [中国]
(a) 清仁宗嘉庆五年十月，东方有流星，百十如织，有顷乃已。
清光绪 广东《海阳县志》卷25
(b) 清仁宗嘉庆五年十月，星陨如雨。
民国 云南《宜良县志》卷1

公元1801年5月5日 [中国]
清仁宗嘉庆六年三月二十三日，戌时，有星大如斗，光如月，其后微星不以数计，自北飞南，锵然有声，后闻天鼓大震，同时而见，闻者数省。
清咸丰 贵州《安顺府志》卷12

公元1801年冬 [中国]
清仁宗嘉庆六年冬，铜山陨星如织。
清同治 江苏《徐州府志》卷57

公元1813年2月1日—1814年1月20日 [中国]
清仁宗嘉庆十八年，星陨如雨。
清光绪 山西《蒲县志》卷9

公元1813年12月1日 [日本]
日本文化癸酉（十年）十一月九日……正黎明，微云起北方，忽焉半空有声，举头视之，有物白云中出，其状大者如拍球，小者如拳，数十连络凡二间计，光恰似所谓狼烟之流星。火亦白引条，从东方向西方而飞，其行徐徐如盆上流水银，终入云中失所在。
《天野佐一郎氏所藏文书》

公元1814年8月14日—9月13日 [中国]
清仁宗嘉庆十九年七月，亥时，天忽开朗，划然而启，灿若长虹，横于南北，长丈余，中有一大星，两旁列小星数十，如贯珠然，大星投东而去，小星随隐，众目所睹。
清道光 广东《粤小记》卷4

公元1818年11月8日 [中国]
清仁宗嘉庆二十三年十月乙亥，夜，星南陨如雨。

清道光 甘肃《镇原县志》卷7

公元1822年2月3日 [中国]
清宣宗道光壬午正月十二日，会城有大星一颗，自西南流至东方，下有小星数十随之，光灿亘天。
民国 广东《广州城防志》卷5
注："道光壬午"为"道光二年"

公元1822年9月9日 [中国]
清宣宗道光二年七月二十四日，有大星大如杯，自北向南，数十小星随其后。
清道光 广东《粤小记》卷4

公元1824年12月29日 [中国]
清宣宗道光四年甲申十一月初十，夜，西北方星陨如雨。
《履园丛话》灾祥
注：甲申系宣宗道光四年，原文仁宗甲申为误。

公元1825年3月4日 [中国]
清宣宗道光五年正月十五日，有大星流于东北，后随小星无数，声如雷。
清光绪 山西《忻州直隶州志》卷39

公元1825年11月10日—12月9日 [中国]
清宣宗道光五年十月，星移如织。
清道光 河南《鄢陵县志》卷17
清光绪 河南《祥符县志》卷23

公元1825年12月3日—4日 [中国]
清宣宗道光五年乙酉十月二十四、二十五两夜，夜，星移如织，俱由西北而至东南。
《履园丛话》灾祥
注：乙酉系宣宗道光五年，原文仁宗乙酉系误。

公元1832年4月16日 [中国]

清宣宗道光十二年三月十六日，夜，有赤星大如斗，自东北起，霹雳一声向西坠，后小星无数随之。
清同治 安徽《桐城县志》卷10

公元1837年2月5日—3月6日 [中国]
清宣宗道光十七年正月，西南方星陨如雨。
民国 云南《昆明县志》卷7

公元1837年6月17日 [中国]
清宣宗道光十七年五月望后，群星南流如雨。
清光绪 河北《东光县志》卷11

公元1837年12月27日—1838年1月25日 [中国]
清宣宗道光十七年十二月，星陨无数。
清光绪 云南《云龙州志》卷2

公元1838年1月26日—1839年2月13日 [中国]
清宣宗道光十八年，星陨如雨。
清道光 云南《古越州志》卷1
清咸丰 云南《南宁县志》卷1

公元1838年7月21日—8月19日 [中国]
清宣宗道光十八年六月，星陨如雨，数夜皆然。
民国 福建《屏南县志》卷3

公元1838年8月20日—9月18日 [中国]
清宣宗道光十八年七月，星陨如雨。
民国 云南《嵩明县志》卷4

公元1838年12月6日 [中国]
清宣宗道光十八年十月二十日，府城满天星飞。
清道光 云南《普洱县志》卷2

公元1839年1月15日—2月13日 [中国]
清宣宗道光十八年十二月，星陨如雨。
清光绪 云南《陆凉州志》卷8

民国 云南《罗平县志》卷1

公元1839年8月23日 [中国]
清宣宗道光十九年七月十五日，亥刻，有星三，彼次相过数次，交流如织，数刻乃不见。
清光绪 广东《香山县志》卷22

公元1841年夏 [中国]
(a) 清宣宗道光二十一年夏，每夜，星陨如雨，大小不等。
清宣统 上海《南沙杂识》祥异
(b) 清宣宗道光二十一年夏，每夜，星陨如雨。
民国 上海《南汇县志》卷22

公元1841年10月31日 [中国]
清宣宗道光二十一年九月十七日，子时，星陨如雨。
清光绪 北京《沙河县志》卷上

公元1842年1月11日 [中国]
清宣宗道光二十一年十二月朔，有大星陨于东南，众小星随之。
清光绪 浙江《平湖县志》卷25
清光绪 浙江《嘉善县志》卷34

公元1842年2月9日 [中国]
清宣宗道光二十一年除夕，流星如织。
清光绪 江苏《丹徒县志》卷58

公元1843年3月31日—4月29日 [中国]
清宣宗道光二十三年三月，落一巨星，无数小星从之，光焰烛天。
清光绪 江苏《昆新两县续修合志》卷51

公元1846年8月5日 [中国]
清宣宗道光二十六年六月十四日，子时，星陨如雨。
清光绪 上海《南汇县志》卷22

公元1846年8月6日 [中国]
清宣宗道光二十六年六月望，星陨如雨。
民国 上海《南汇县二区旧五团乡志》卷16

公元1847年7月12日—8月10日 [中国]
清宣宗道光二十七年六月，京师地震，众星陨。
民国《北京市志稿》卷4

公元1847年7月24日 [中国]
清宣宗道光二十七年六月十二日（庚申），众星陨。
清同治《上海县志》卷30
清光绪 上海《川沙县志》卷14
清光绪 上海《娄县续志》卷12
清光绪 上海《松江府志》卷30
清光绪 上海《青浦县志》卷29
注：十二日为7月23日，庚申为7月24日。

公元1849年8月3日 [中国]
清宣宗道光二十九年六月十五日，夜，星陨如雨。
民国 河南《光山县志》卷1

公元1851年8月11日 [中国]
(a) 清文宗咸丰元年七月十五日，夜，星南走如箭，二时许乃止。
清光绪 天津《宁河县志》卷16
(b) 清文宗咸丰元年七月十五日，夜，星陨如雨。
民国 河南《光山县志》卷1

公元1852年5月21日 [中国]
清文宗咸丰二年四月初三，夜见火光无数，自东南来，绕郡城至西乡仙柱峰始散。
清光绪 江西《抚州府志》卷84

公元1853年2月8日—3月9日 [中国]
清文宗咸丰三年正月，通州有火如星如燐，以千百计，自西南趋东北，凡四、五夜始熄。
《清史稿·灾异志》卷41

公元1853年3月3日　［中国］
清文宗咸丰三年正月己巳，既昏，有火如星如燐，以千百计，自西南趋东北，隐隐闻甲马声，凡四、五夜。
清光绪 江苏《通州直隶州志》卷末

公元1853年8月19日　［中国］
清文宗咸丰三年七月十五日，酉时，西南方星陨如雨。
清光绪 山西《寿阳县志》卷13

公元1854年1月29日—1855年2月16日　［中国］
清文宗咸丰四年，流星南飞如雨，数夜不止。
清光绪 山东《平阳县志》卷6

公元1854年7月4日　［中国］
清文宗咸丰四年六月初十，午，天火纷纷堕街衢，至地无迹。
清光绪 广西《贵县志》卷6

公元1855年8月16日　［中国］
清文宗咸丰五年七月初四，夜，流星如钱，旁随小星数十，自西北飞。
清光绪 广东《海阳县志》卷25

公元1856年10月31日　［中国］
清文宗咸丰六年十月朔三日，夜，群星乱坠。
清光绪 云南《腾越厅志》卷1
民国 云南《龙陵县志》卷1

公元1856年12月8日　［中国］
清文宗咸丰六年冬月十一，夜，星陨如雨。
民国《云南通志》卷17

公元1857年3月24日　［中国］
清文宗咸丰七年二月二十九，夜，星陨如雨。
清光绪 福建《光泽县志》卷1

公元1857年9月8日 [中国]
清文宗咸丰七年七月二十日，夜，星陨如雨，由东北向西，星大小不一，纷纷坠落，其速如箭，其多如雨，经两时许。
清光绪 山西《沁源县续志》卷3

公元1858年1月14日 [中国]
清文宗咸丰七年十一月三十日，夜，星陨如雨。
清光绪 云南《永昌府志》卷3

公元1858年8月23日 [中国]
清文宗咸丰八年七月十五，夜，星流如矢，由西北而东南，数以千计，自戌至丑乃息。
清光绪 山西《崞县志》卷8

公元1858年11月1日 [中国]
清文宗咸丰八年九月二十六日，流星昼见，自西而东，自辰至午，两时许，三、五、七、八不等，皆长尺许。
清光绪 江苏《丹徒县志》卷58

公元1859年8月1日 [中国]
清文宗咸丰九年七月三日，初昏时，距地二、三丈，有流星或大或小，倏东倏西，约两时许方歇。
清光绪 上海《崇明县志》卷5

公元1859年8月13日 [中国]
(a) 清文宗咸丰九年七月十五日，夜，流星如织。
民国 河北《晋县志》卷5
(b) 清文宗咸丰九年七月十五日，夜，星陨如箭。
清光绪 山西《平定州志》卷5

公元1860年8月17日 [中国]
清文宗咸丰十年七月初一日，星陨如雨。
清光绪 山西《交城县志》卷1

公元1860年8月23日 [中国]
清文宗咸丰十年七月初七日，夜，西方星陨如雨。
清光绪 陕西《定远厅志》卷24

公元1861年2月10日—1862年1月29日 [中国]
清文宗咸丰十一年，星陨如雨。
清光绪 云南《永北厅志》卷1

公元1861年6月15日 [中国]
清文宗咸丰十一年五月初八，酉刻，天大黑，天心有一星大如斗，色红而暗，数十小星环左右，摇摇欲动，欻若霹雳一声，星从天心散为红光，向西堕去。
民国 四川《武胜县志》卷12

公元1861年6月23日 [中国]
清文宗咸丰十一年五月十六日，夜，星陨如雨，集东北方。
民国 陕西《宝鸡县志》卷16

公元1861年8月6日—9月4日 [中国]
清文宗咸丰十一年七月，流星南飞，密如缕。
清同治 山东《即墨县志》卷11
清文宗咸丰十一年七月，星陨如雨。
民国 山西《临汾县志》卷6
清文宗咸丰十一年七月，星陨如雨，由东北而西南，约有数刻。
民国 山西《辽州志》卷3

公元1861年8月10日 [中国]
清文宗咸丰十一年辛酉，七月五日夜，流星自北而南，络绎不绝。
民国 江苏《太仓州志》卷26

公元1861年8月14日 [中国]
清文宗咸丰十一年七月初九，夜，星陨如雨。
清光绪 浙江《嘉善县志》卷34
清宣统 上海《枫泾小志》卷10

公元1861年8月19日 [中国]
清文宗咸丰十一年七月十四，众星纷纭流于西南，自初昏起，夜分始息。
清光绪 山东《栖霞县志》卷8

公元1861年8月20日 [中国]

(a) 清文宗咸丰十一年七月十五日（望），夜，流星奔西南，黎明方息。

清光绪 河北《南宫县志》卷8

(b) 清文宗咸丰十一年七月十五日（望），星陨如雨。

清光绪 河北《怀安县志》卷3

(c) 清文宗咸丰十一年七月十五日（望），夜，星陨如雨。

清光绪 山西《太平县志》卷14

清光绪 山西《解州全志》卷11

清光绪 山西《直隶绛州志》卷20

(d) 清文宗咸丰十一年七月十五日（望），夜，星陨如雨，坠西南。

民国 山西《虞乡县志》卷10

(e) 清文宗咸丰十一年七月十五日（望），星陨如雨，向南走，隐约有声。

清光绪 山西《清源乡志》卷16

公元1861年9月22日 [中国]

清文宗咸丰十一年八月甲戌，夜，坠一大星，小星陨如雨。

清光绪 上海《青浦县志》卷29

公元1862年1月30日—1863年2月17日 [中国]

清穆宗同治元年，星陨如雨。

民国 云南《景东县志》卷1

公元1862年5月28日—6月26日 [中国]

清穆宗同治元年五月，星陨如雨。

清光绪 陕西《绥德县志》卷3

公元1862年7月11日 [中国]

清穆宗同治元年六月丙寅，夜，众星西南流。

清光绪 江苏《武阳志余》卷5

公元1862年7月27日—8月24日 [中国]

(a) 清穆宗同治元年七月，星陨如雨。

清同治 山西《续猗氏县志》卷4
清光绪 山西《临晋县志》卷下
清光绪 陕西《续千阳县志》卷14
(b) 清穆宗同治元年七月，流星满天，起西南方。
民国 山东《邱县志》卷9

公元1862年8月9日 [中国]
(a) 清穆宗同治元年七月十四日，流星如雨，自东北而西南。
清同治 河北《盐山县志》卷5
清光绪 河北《大城县志》卷10
清光绪 河北《南皮县志》卷5
(b) 清穆宗同治元年七月十四日，午夜，星自东北向西南纷流，陆续不断。
清光绪 山东《惠民县志补遗》
(c) 清穆宗同治元年七月十四日，夕，流星南渡，相连一夜不止。
民国 山东《济阳县续志》卷20
民国 山东《信阳县志》卷2

公元1862年8月10日 [中国]
(a) 清穆宗同治元年中元（七月十五日），亥时，众星纷纷自东北流于西南，不可计数。
清同治《续天津县志》卷1
(b) 清穆宗同治元年中元（七月十五日），夜，星流如织。
清同治 河北《昌黎县志》卷1
(c) 清穆宗同治元年中元（七月十五日），星陨如雨，由西北向西南。
清光绪 山西《平陆县续志》卷下
(d) 清穆宗同治元年中元（七月十五日），黄昏后，流星如雨，向西南，二更止。
清光绪 山西《平遥县志》卷12
(e) 清穆宗同治元年中元（七月十五日），夜，星陨如雨。
清光绪 陕西《米脂县志》卷10

(f) 清穆宗同治元年中元（七月十五日），夜，星陨如雨。
民国 河南《光山县志》卷1

(g) 清穆宗同治元年中元（七月十五日），夜，流星如织。
清光绪 山东《利津县志》卷10

(h) 清穆宗同治元年中元（七月十五日），众星向西南流。
民国《浙江续通志稿》卷1

(i) 清穆宗同治元年中元（七月十五日），夜，有众星自东南流向西北。
民国 广东《始兴县志》卷16

注：该条还有数十个各省地方志有相似记载。

公元1862年8月10日 [日本]

(a) 日本文久二年七月十五日，今夜戌刻星陨如雨。
《嘉永明治年间录》十一

(b) 日本文久壬戌岁七月十五日夜，流星频降，自初更至五更仍不止，其势迅速，皆自东北驰西南，乃随风行耳。夫流星之飞天也，寻常之事，虽不足怪而如此太甚者亦甚稀，盖古所谓星陨如雨是也。
《流星彗星勘文》图书寮藏本

(c) 日本文久二年七月十五日，今月十五日流星最多，其数及百，而其陨也自艮向坤者多焉。
《流星彗星勘文》

公元1862年8月11日 [中国]

清穆宗同治元年七月十六日，初昏，众星交陨，多趋西南，纵横如织，夜分始息。
清光绪 山东《登州府志》卷23
清光绪 山东《蓬莱县续志》卷1
清光绪 山东《福山县志稿》卷8

公元1862年8月12日 [中国]

(a) 清穆宗同治元年七月十七日，众星向西南急行，如矢甚伙。
清光绪 河南《南阳县志》卷12

(b) 清穆宗同治元年七月十七日，满天流星自东北向西南，交射中天，更余始息。

民国 河南《方城县志》卷5

公元1863年2月18日—1864年2月7日 ［中国］

(a) 清穆宗同治二年，城西星陨如雨，隐隐有声。

清光绪 湖北《京山县志》卷1

(b) 清穆宗同治二年，星陨如雨来。

清宣统《湖北通志》卷76

公元1863年8月28日 ［中国］

(a) 清穆宗同治二年七月十五日，夜，向西南流星似箭，直至天明。

清光绪 河北《新城县志》卷10

(b) 清穆宗同治二年七月十五日，流星蔽天如织，向东南流，至天明始熄。

民国 河北《新城县志》卷22

(c) 清穆宗同治二年七月十五日，夜，星陨如雨。

民国 山西《临县志》卷3

(d) 清穆宗同治二年七月十五日，自暮至初更，流星无数如箭，皆向西南行，至二更方止。

民国 北京《平谷县志》卷3

公元1863年11月11日—12月10日 ［中国］

清穆宗同治二年十月，京山城西，星陨如雨。

民国《湖北通志》卷76

公元1864年8月2日 ［中国］

清穆宗同治三年七月朔，夜，星陨如雨。

民国 四川《乐山县志》卷12

公元1864年8月2日—31日 ［中国］

清穆宗同治三年七月，有众星自东南流。

清同治 山西《高平县志》卷4

公元1864年10月1日—10月29日 ［中国］
(a) 清穆宗同治三年九月，天有白气，少间星陨如雨。
清光绪 湖北《京山县志》卷1
注：宣统《湖北通志》作四年九月
(b) 清穆宗同治三年九月，流星交织。
清光绪 云南《霑益州志》卷4

公元1864年10月15日 ［中国］
清穆宗同治三年九月十五日，夜，流星纵横交织。
民国 福建《上杭县志》卷1

公元1865年9月20日—10月19日 ［中国］
清穆宗同治四年八月，星陨如雨。
清同治《上海县志》卷30

公元1866年8月24日 ［中国］
清穆宗同治五年七月十五日，夜，众星南流如飞矢。
清光绪 河南《閿乡县志》卷末

公元1866年11月3日 ［中国］
清穆宗同治五年九月二十六日，夜半，天如昼，金光坠，如雨粟。
清同治 江西《新昌县志》卷4
民国 江西《盐乘》卷11

公元1867年11月12日 ［中国］
清穆宗同治六年十月十七日，夜，西北方有星陨如雨。
清同治 安徽《桐城县志》卷10

公元1867年11月13日 ［日本］
日本庆应三年十月十八日，寅下刻始，流星频见。
（注：译自日文）
《守护国界急务》

公元1869年2月11日—3月12日 ［中国］
清穆宗同治八年正月，星陨如雨，一时无光。
清光绪 云南《呈贡县志》卷5

公元1869年8月22日 ［中国］
清穆宗同治八年七月十五日，流星向西南为雨，有声。
清光绪 河北《开州志》卷1

公元1869年12月15日 ［中国］
清穆宗同治八年十月庚辰，夜，众星自艮方流至坤方，陨如雨。
清光绪 江苏《武阳志余》卷5
注：十月无庚辰，应是十一月。

公元1871年8月10日 ［中国］
清穆宗同治十年六月二十四日，夜，东北星陨如雨。
清宣统 江苏《琼山县志》卷28

公元1872年8月4日—9月2日 ［中国］
清穆宗同治十一年七月，夜，星陨如雨。
清光绪 云南《云南县志》卷1

公元1872年11月1日 ［中国］
清穆宗同治十一年十月一日，夜半，星陨如雨。
清光绪 云南《霑益州志》卷4

公元1872年11月27日 ［中国］
清穆宗同治十一年十月二十七日，夜子刻，星陨如雨。
清光绪 广东《饶平县志》卷13

公元1872年12月27日 ［中国］
清穆宗同治十一年十一月二十七日，夜，群星乱坠。
清光绪 云南《腾越厅志》卷1
民国 云南《龙陵县志》卷1

公元1874年7月31日 ［中国］
清穆宗同治甲戌六月十八日，夜，群星易位，往来如织。

民国 云南《龙陵县志》卷1
民国 湖北《英山县志》卷14

公元1874年8月4日 ［中国］
清穆宗同治十三年六月二十二日，夜，众星纷飞向南方。
清光绪 广东《揭阳县续志》卷4

公元1875年6月9日 ［中国］
清德宗光绪元年五月六日，傍晚，天鼓鸣，流星四溢。
清光绪 四川《井研志》卷32

公元1875年11月17日 ［中国］
清德宗光绪元年十月二十日，夜，星陨如雨。
民国 云南《昆阳县志》卷3

公元1876年11月16日—12月15日 ［中国］
清德宗光绪二年十月，夜，庄浪星陨如雨。
清宣统《甘肃新通志》卷2

公元1877年6月11日—7月10日 ［中国］
清德宗光绪三年五月，星陨如雨。
民国 云南《罗平县志》卷1

公元1877年8月25日—27日 ［中国］
清德宗光绪三年七月十七、十八、十九，三夜，天星过位，往来如箭，满天星斗皆乱。
民国 广西《藤县志》 灾异

公元1879年冬 ［中国］
清德宗光绪五年冬，有流星数十，击射天空，速如激箭，历一分钟始息。
民国 河北《徐水新志》卷10

公元1881年3月30日—4月27日 ［中国］
清德宗光绪七年三月，满天星移，方向无定。
民国 云南《姚安县志》卷66

公元1881年9月23日—10月22日　　［中国］
清德宗光绪七年八月，星陨如雨，有声。
民国 云南《广南县志》卷3

公元1882年9月2日　　［中国］
清德宗光绪八年七月二十日，夜子时，星陨如雨，二三时始止。
民国 云南《宣良县志》卷1
注：民国《禄劝县志》作十二月二十日。

公元1882年9月12日—10月11日　　［中国］
清德宗光绪八年八月，满天大小星皆向南飞。
民国 广西《灵山县志》卷5

公元1882年12月7日　　［中国］
清德宗光绪八年十月庚辰，众星自艮方流至坤方，陨如雨。
清光绪 江苏《武阳志余》卷5

公元1882年12月22日　　［中国］
清德宗光绪八年十一月十三日，夜，流星多如烟火，光满天际。
民国《浙江续通志稿》卷1

公元1883年6月12日　　［中国］
清德宗光绪九年五月八日，夜，流星纷陨如雨。
民国 四川《巴中县志》 第四编

公元1883年11月10日　　［中国］
清德宗光绪九年十月戊午，夜，星陨如雨，四更方止。
民国 云南《续修昆明县志》卷7

公元1883年11月30—12月28日　　［中国］
清德宗光绪九年十一月，夜，流星弥天，四面坠落。
清光绪 河南《南阳县志》卷12

公元1884年6月28日　　［中国］
清德宗光绪十年闰五月初六，夜，星落如雨。
民国 福建《吉田县志》卷3

公元1884年10月11日 [中国]
清德宗光绪十年八月二十三日，群星如雨降，自七时起至九时止。
民国 云南《昭通志稿》卷12

公元1884年11月18日—12月16日 [中国]
(a) 清德宗光绪十年十月，夜，众星流，光芒四射。
民国 山东《博山县志》卷1
(b) 清德宗光绪十年十月，星陨如雨。
民国 云南《昆明县志》 五行

公元1884年12月2日 [中国]
清德宗光绪十年十月十五日，星陨如雨。
民国 甘肃《镇原县志》卷18

公元1884年12月8日 [中国]
(a) 清德宗光绪十年十月二十一日，夜，群星西流，约两时始定。
民国 河南《许昌县志》卷19
民国 河南《临颍县志》卷13
(b) 清德宗光绪十年十月二十一日，夜，星陨如雨。
清光绪 陕西《白河县志》卷13

公元1884年12月9日 [中国]
(a) 清德宗光绪十年十月二十二日，夜，星陨如雨。
清光绪 陕西《米脂县志》卷10
(b) 清德宗光绪十年十月二十二日，星飞如织，光芒如白练，经时方定。
民国 河南《西华县续志》卷1

公元1884年12月9日—10日 [中国]
清德宗光绪十年十月二十二、二十三日，星移如织，有光芒数尺者。
清宣统 河南《项城县志》卷31

公元1884年12月10日 [中国]
清德宗光绪十年十月二十三日，西南星陨如雨。
民国 陕西《紫阳县志》卷5

公元1884年12月11日 [中国]
清德宗光绪十年十月二十四日，夜，满天星陨如雨，前后三夜始止。
民国 云南《嵩明县志》卷4

公元1885年2月15日—1886年2月3日 [中国]
(a) 清德宗光绪十一年，星陨如雨。
民国 云南《姚安县志》卷66
(b) 清德宗光绪十一年，星陨如雨。
民国《云南通志》卷17

公元1885年夏 [中国]
清德宗光绪十一年夏某日，星流如织，陨星碎落如雨。
民国《浙江续通志稿》卷4

公元1885年8月10日—9月8日 [中国]
(a) 清德宗光绪十一年七月，星陨如雨数夕。
民国 陕西《紫阳县志》卷5
(b) 清德宗光绪十一年七月，流星如织，不计其数。
民国 山东《单县志》卷14
(c) 清德宗光绪十一年七月，有流星往来无定，不计其数。
清光绪 山东《肥城县志》卷10

公元1885年8月30日 [中国]
清德宗光绪十一年七月二十一，夜，星坠如雨。
清光绪 广东《澄迈县志》卷12

公元1885年9月9日—10月7日 [中国]
(a) 清德宗光绪十一年八月，满天星乱。
民国 广西《陆川县志》卷2
(b) 清德宗光绪十一年八月，初昏时，平越城中人见天星纷陨，有如雨雪，至三更始止。
清光绪 贵州《平越州志》卷1

公元1885年10月26日　　[中国]
清德宗光绪十一年九月十九日，夜，星陨如雨。
清光绪 江西《南昌县志》卷55

公元1885年10月28日　　[中国]
清德宗光绪十一年九月二十一日，夜，星陨如雨。
清光绪 四川《奉节县志》卷11

公元1885年11月2日　　[中国]
清德宗光绪十一年九月二十六日，星流如织。
清光绪 湖北《沔阳县志》卷1

公元1885年11月3日　　[中国]
清德宗光绪十一年九月二十七日，星陨如雨。
清宣统 江苏《宿迁县志稿》卷7

公元1885年11月7日—12月5日　　[中国]
(a) 清德宗光绪十一年十月，星陨如雨。
清光绪 广西《信宜县志》卷8
清光绪 广西《嵩明州志》卷2
民国 陕西《神木乡土志》卷4
(b) 清德宗光绪十一年十月，星流如织。
清光绪 安徽《寿州志》卷35
(c) 清德宗光绪十一年十月，星落如雨。
清光绪 江西《崇义县志》卷1
(d) 清德宗光绪十一年十月，满天星斗，纵横乱飞。
民国 广西《钦州志》卷首
(e) 清德宗光绪十一年十月，小星数百枚以上，四面行。
民国《云南通志》卷17

公元1885年11月14日　　[中国]
清德宗光绪十一年冬十月初八日，夜半至鸡鸣，华亭县星陨如雨。
清宣统《甘肃新通志》卷2

公元1885年11月15日 [中国]
清德宗光绪十一年十月甲戌，夜，天星散落如雨。
清光绪 贵州《黔江县志》卷5

公元1885年11月19日 [中国]
清德宗光绪十一年十月十三日，晚亥、子之际，星陨如雨，天星暗。移时，咸、长、泾，原等属尤甚。
民国 陕西《周至县志》卷8

公元1885年11月23日—29日 [中国]
清德宗光绪十一年十月十七至二十三日，每夜大小星交流如织，流向西者更多。
清光绪 湖北《德安府志》卷20

公元1885年11月24日—25日 [中国]
清德宗光绪十一年十月十八、九日，夜，星陨如雨。
民国 陕西《华县志》卷1

公元1885年11月26日 [中国]
(a) 清德宗光绪十一年十月二十日，夜，灵台县星陨如雨，顷刻星稀不见。
清宣统《甘肃新通志》卷2
(b) 清德宗光绪十一年十月二十日，夜，星陨如雨。
民国 甘肃《灵台县志》卷3
(c) 清德宗光绪十一年十月二十日，星移如织，连夕不定。
清光绪 河南《祥符县志》卷23
(d) 清德宗光绪十一年十月二十日，夜，群星自西北流于东南。
清光绪 广东《临桂县志》卷18

公元1885年11月27日—29日 [中国]
(a) 清德宗光绪十一年十月二十一、二、三，夜，星陨如雨。
清光绪 湖北《武昌县志》卷10
(b) 清德宗光绪十一年十月二十一、二、三，星陨如雨，半空中作交织状，连三夜皆然。
清光绪 广东《高州府志》卷51
注：光绪《电白县志》作十一月。

(c) 清德宗光绪十一年十月二十一、二、三，上天雨星，至二十三日止。
清光绪 云南《腾越厅志》卷1
民国 云南《龙陵县志》卷1

公元1885年11月27日 [中国]
(a) 清德宗光绪十一年十月二十一（丙戌），星陨如雨，初昏至晓。
清光绪 陕西《富平县志》卷10
清光绪 陕西《蒲城新志》卷13
(b) 清德宗光绪十一年十月二十一（丙戌），星流如织，自申至次日辰刻止。
民国 河南《郑县志》卷1
(c) 清德宗光绪十一年十月二十一（丙戌），夜，流星如织。
清光绪 山东《诸城县续志》 总记
(d) 清德宗光绪十一年十月二十一（丙戌），日落后，流星满天，俱自西南向东北。
民国 江苏《高邮县志》补遗
(e) 清德宗光绪十一年十月二十一（丙戌），星陨起更后，至半夜止，大小以千百计。
清光绪 上海《罗溪镇志》卷8
民国 上海《宝山县续志》卷17
注：《上海县续志》作十一月二十一日。
(f) 清德宗光绪十一年十月二十一（丙戌），夜，星陨如雨。
民国 福建《宁化县志》卷2
(g) 清德宗光绪十一年十月二十一（丙戌），星陨如雨。
清光绪 云南《丽江县志》卷2
民国 云南《陆良县志》卷1
民国 云南《罗平县志》卷1
注：本条记录还有数十种地方志有相似记载。

公元1885年11月28日 [中国]
(a) 清德宗光绪十一年十月二十二日，夜，星陨如雨。

清光绪 陕西《渭南县志》卷11

(b) 清德宗光绪十一年十月二十二日，夜，流星如织。

清光绪 河北《东光县志》卷11

民国 河北《交河县志》卷10

民国 山东《宁津县志》卷11

(c) 清德宗光绪十一年十月二十二日，夜，流星四窜，震动不安，令人视之恐惧。

清光绪 河南《永城县志》卷15

(d) 清德宗光绪十一年十月二十二日，亥时，星陨如雨，光芒映射，错落下地有声，如是者数夜。

清宣统 云南《续蒙自县志》卷12

(e) 清德宗光绪十一年十月二十二日，夜，星陨如雨，有声。

民国 云南《蒙化志稿》卷2

注：民国《建水县志》作十一月二十二日。

公元1885年11月29日 [中国]

(a) 清德宗光绪十一年十月二十三日，夜，流星四窜，震动不安，令人视之恐惧。

清光绪 河南《永城县志》卷15

(b) 清德宗光绪十一年十月二十三日，夜，星陨如雨。

清光绪 广东《吴川县志》卷10

(c) 清德宗光绪十一年十月二十三日，下午八句钟起，满天星陨如雨，至十二句钟，始见星稀。

民国 云南《维西县志》卷4

公元1885年11月30日 [中国]

清德宗光绪十一年十月念四，夜亥时，满天星纵横乱飞。

民国 广西《灵山县志》卷5

公元1885年11月30日—12月1日 [中国]

(a) 清德宗光绪十一年十月二十四、二十五日，两渚诸星飞流，乱落如雨。

清光绪 陕西《靖边志稿》卷4

(b) 清德宗光绪十一年十月二十四、二十五日，夜，安化之诸星乱落甚众。

民国 甘肃《庆阳县志稿》卷4

公元1885年12月6日—12月15日 [中国]

清德宗光绪十一年十一月初旬一夕，九句钟时，城西家信子书塾生徒夜读方休，出现天中有众小星往来如掷梭，阅时乃止，人皆见之。

民国 辽宁《辽阳县志》卷40

公元1885年12月6日—1886年1月4日 [中国]

(a) 清德宗光绪十一年十一月，夜间星流，流火满天，大小不一甚多，无数纵横乱窜，彻夜不息，次日晚间，尚有见者。

民国 河南《信阳县志》卷31

(b) 清德宗光绪十一年十一月，夜，流星如织。

民国 河南《淮阳县志》卷20

民国 河南《中牟县志》卷1

(c) 清德宗光绪十一年十一月，星流如雨。

清光绪 江苏《江浦埤乘》卷39

民国 江苏《锡金续识小录》卷1

清光绪 云南《普洱府志》卷3

公元1885年12月16日 [中国]

清德宗光绪十一年十一月十一日， 下十一句钟，大小星乱陨，顷刻星稀。

民国 陕西《乾县新志》卷8

公元1885年12月26日 [中国]

(a) 清德宗光绪十一年十一月二十一日（乙卯），繁星乱流，自北至东南，终夜有声。

民国 上海《青浦县志》卷23

(b) 清德宗光绪十一年十一月二十一日（乙卯），夜，星陨如雨，自西北至东南。

清光绪 浙江《平湖县志》 祥异

(c) 清德宗光绪十一年十一月二十一日（乙卯），戌刻，星陨如雨，至丑始定。

清光绪 福建《浦城县志》卷42

(d) 清德宗光绪十一年十一月二十一日（乙卯），夜流星满天，自西北至东南，唧唧有声，终夜不息。

清光绪 浙江《嘉善县志》卷34

公元1885年12月28日 [中国]

清德宗光绪十一年冬月二十三日，午后五时许，忽有声自境内东南来，视之则见东南天空星陨乱飞，有大如蛋壳者，有如碗大者，有如籽粒大小者纷纷堕地，若飘雪然，约三、四十分钟，末了，但见忽上忽下，轰然作声遂止。

民国《云南通志》卷17

公元1885年冬 [中国]

(a) 清德宗光绪十一年冬，天中众小星如织，一时许。

民国 辽宁《辽阳县志》卷首

(b) 清德宗光绪十一年冬，星陨如雨。

民国 云南《邱北县志》卷1

(c) 清德宗光绪十一年冬，戌时，东南角流星如雨，如是者三晚。

民国《云南通志》卷17

公元1886年1月2日 [中国]

(a) 清德宗光绪十一年十一月二十八日，星流如织。

清光绪 河南《陕州志》卷10

(b) 清德宗光绪十一年十一月二十八日，星陨如雨，自西北而南。

清光绪 河南《械乡县志》卷末

公元1886年1月15日 [中国]

清德宗光绪十一年十二月十一日，流星多向坤方。

民国 河南《方城县志》卷5

公元1886年2月4日—1887年1月23日 [中国]

清德宗光绪十二年，傍晚，空中流星宛如急雨下降，历两小时始止殆，不可以计数。

民国 《云南通志》卷17

公元1886年7月31日—8月28日 [中国]
清德宗光绪十二年七月某日，夜，星陨如雨。
清光绪 湖北《利川县志》卷9

公元1886年9月4日 [中国]
清德宗光绪十二年八月初七，夜，流星如雨。
江苏《唐市志补遗》

公元1886年11月16日 [中国]
(a) 清德宗光绪十二年十月二十一，夜，星陨如雨。
清光绪 陕西《永寿县志》卷10
(b) 清德宗光绪十二年十月二十一，夜，见众星飞流，皆西而东。
清光绪 广西《贵县志》卷6
(c) 清德宗光绪十二年十月二十一，夜，星流如织，至晨始止。
民国 河南《密县志》卷19

公元1886年11月25日 [中国]
清德宗光绪十二年十月三十日，夜，星陨如雨。
清光绪 四川《巫山县志》卷10

公元1886年11月26日—12月24日 [中国]
清德宗光绪十二年十一月，星交流如织。
民国 山东《临淄县志》卷8

公元1886年12月15日 [中国]
清德宗光绪十二年十一月二十日，夜半，天空众星闪窜缭乱，如陨如织，
移时甫定。
清光绪 安徽《霍山县志》卷15

公元1886年12月16日 [中国]
清德宗光绪十二年十一月二十一日，夜，戌亥之交，星陨如雨，至数小时之久，光芒纷错，四迸如射。
民国 安徽《黟县四志》卷1

公元1887年11月24日 [中国]
清德宗光绪十三年十月癸巳，夜，星乱度纵横冲激，逾刻乃定。
清光绪 浙江《宁海县志》卷23

公元1887年11月29日 [中国]
清德宗光绪十三年十月十五，夜，西北方星陨如雨。
民国 福建《建宁县志》卷27

公元1888年3月16日 [中国]
清德宗光绪十四年二月丙戌，星陨如雨，三日始复。
清光绪 山东《峄县志》卷15

公元1888年4月11日—5月10日 [中国]
(a) 清德宗光绪十四年三月，夜，流星无数，南北相斗，如箭激射。
清光绪 广西《玉林州志》卷4
民国 广西《陆川县志》卷2
(b) 清德宗光绪十四年三月，夜，流星无数，南北相冲，如箭互射。
民国 广西《北流县志》 编12

公元1888年8月22日 [中国]
清德宗光绪十四年七月十五日，星陨如雨，自北而南。
清光绪 河南《阌乡县志》卷末

公元1888年11月4日—12月2日 [中国]
清德宗光绪十四年十月，众星流东南，数夜乃止。
清宣统 山东《蒙阴县志》卷下

公元1888年冬 [中国]
清德宗光绪十四年冬，星陨如雨。
清光绪 云南《浪穹县志略》卷1

公元1890年1月21日—1891年2月8日 [中国]
清德宗光绪十六年，夜，有群星自西南趋东北，光芒数丈。
民国 山东《博山县志》卷1

公元1890年8月16日—9月13日 [中国]
清德宗光绪十六年七月，某晚，星陨如雨，空际星稀，大地闇然无光。
民国 陕西《乾县新志》卷8

公元1891年4月9日—5月7日 [中国]
清德宗光绪十七年三月，满天星陨，至空际忽隐。
民国 云南《姚安县志》卷66

公元1891年4月16日 [中国]
清德宗光绪十七年三月八日，亥时，星陨如雨，霎然有声，天乍昏黑，经数刻复开朗如故。
民国 陕西《蓝田县志》卷3

公元1891年11月16日 [中国]
清德宗光绪十七年十月既望，庄浪星陨如雨。
民国 《甘肃通志稿》 变异

公元1891年12月1日—30日 [中国]
清德宗光绪十七年冬月，有夜星如雨。
清光绪 河北《南宫县志》卷8
注：冬月为十一月。

公元1891年12月27日 [中国]
清德宗光绪十七年十一月二十七日，夜，星陨如雨，数小时始定。
民国 河南《正阳县志》卷3

公元1892年3月4日 [中国]
清德宗光绪十八年二月初六日，夜，满天星飞如箭射。
民国 山东《定陶县志》卷9

公元1892年8月22日—9月20日 [中国]
清德宗光绪十八年七月，夜，静宁东北，灵州西南，星陨如雨。
清宣统 《甘肃新通志》卷2

公元1892年12月19日—1893年1月17日 [中国]
清德宗光绪十八年冬月，夜，有星如雨。
清光绪 河北《南宫县志》卷8

公元1893年2月17日—1894年2月5日 [中国]
清德宗光绪十九年，星陨如雨。
民国 陕西《镇安县志》卷10

公元1893年10月10日—11月7日 [中国]
(a) 清德宗光绪十九年九月，星陨如雨，红光满天。
民国 陕西《澄城续志》卷11
(b) 清德宗光绪十九年九月，南天空，星陨如雨。
民国 云南《嵩明县志》卷4

公元1893年11月10日 [中国]
清德宗光绪十九年初三，夜，山丹县星陨如雨，光芒烛地，越一时始灭。
清宣统 《甘肃新通志》卷2

公元1894年4月6日—5月4日 [中国]
清德宗光绪二十年三月，星陨如雨。
清宣统 江苏《宿迁县志》卷7

公元1894年10月29日—11月26日 [中国]
清德宗光绪二十年十月，庄浪星陨如雨。
清宣统 《甘肃新通志》卷2

公元1897年3月3日— 4月1日 [中国]
清德宗光绪二十三年二月，群星昼见，皋兰南方尤多。
清宣统 《甘肃新通志》卷2

公元1897年11月14日 [中国]
清德宗光绪二十三年十月二十日，夜，众星西流如雨。
民国 安徽《太湖县志》卷40

公元1898年9月1日 ［中国］
清德宗光绪二十四年七月十六日，夜，枉矢星如梭织。
民国 安徽《蒙城县志》卷12

公元1899年8月12日 ［中国］
清德宗光绪二十五年七月七日，夜，有大星如斗，自西北飞至东南，其光烛天，旁有无数小星随之。
清光绪 浙江《寿昌县志》卷11

公元1899年11月14日 ［中国］
(a) 清德宗光绪二十五年十月十二日（丙戌），夜丑刻，星陨如雨。
清光绪 上海《蒸里志略》卷12
清宣统 上海《枫泾小志》卷10
(b) 清德宗光绪二十五年十月十二日（丙戌），星陨如雨。
民国 上海《青浦县续志》卷23

1900年1月31日—1901年2月18日 ［中国］
清德宗光绪二十六年，夜间，西方群星堕落如雨。
民国《云南通志》卷17

公元1901年4月19日—5月17日 ［中国］
清德宗光绪二十七年三月，星陨如雨。
民国 甘肃《临泽县志》卷4

公元1902年7月23日 ［中国］
清德宗光绪二十八年六月十九日，夜，有星向东散落如雨。
民国 四川《万源县志》卷10

公元1902年9月2日—10月1日 ［中国］
清德宗光绪二十八年八月，群星昼见。
民国 辽宁《凤城县志》卷15

公元1902年10月23日 ［中国］
清德宗光绪二十八年九月二十二日，夜，星陨如雨。
民国 陕西《紫阳县志》卷5

公元1903年2月27日—3月28日 [中国]
(a) 清德宗光绪二十九年二月，星陨如雨。
民国《云南通志》卷17
(b) 清德宗光绪二十九年二月，众星南流，如雨而陨。
清宣统《甘肃新通志》卷2

公元1904年秋 [中国]
清德宗光绪三十年秋，有星大如斗，自北流，数十小星随之，至西而没。
民国 广西《灵山县志》卷5

公元1906年8月13日 [中国]
清德宗光绪三十二年六月二十三日，夜，众星飞走如雨，自北飞南。
清光绪 广东《澄迈县志》卷12
清光绪 广西《合浦县志》卷5

公元1906年10月18日—11月15日 [中国]
清德宗光绪三十二年九月，星陨如雨连数夜，内中有一大者，径有三、四尺许。
民国《云南通志》卷17

公元1906年10月24日 [中国]
清德宗光绪三十二年九月初七日，夜明，声自西南来，空中流星乱落。
民国 山东《临沂县志》卷1

公元1909年1月31日 [中国]
清末帝宣统元年正月初十日，满天星移，方向无定。
民国 云南《姚安县志》卷66

公元1910年2月10日—1911年1月29日 [中国]
清末帝宣统二年，西北方陨星如雨，时余如故。
民国 陕西《乾县新志》卷8

公元1910年5月9日—6月6日 [中国]
清末帝宣统二年四月，夜，流星如织。
民国 江苏《丹徒县续志》卷41

INDEX